Visit classzone.com and get connected

Online resources provide instruction, practice, and learning support correlated to your text.

- **Misconceptions database** provides solutions for common student misconceptions about science and their world.
- **Professional development links**, including SciLinks, offer additional teaching resources.
- **Animations and visualizations** help improve comprehension.
- **Math Tutorial** helps strengthen students' math skills.
- **Flashcards** help students review vocabulary.
- **State test practice** prepares students for assessments.

You have immediate access to *ClassZone's* teacher resources.

MCDTCOWDMSSZ

Use this code to create your own username and password.

Also visit *ClassZone* to learn more about these innovative and updated online resources.

- eEdition Plus Online
- eTest Plus Online
- EasyPlanner Plus Online
- Content Review Online

Now it all clicks!™

McDougal Littell

TEACHER'S EDITION

McDougal Littell Science

Motion and Forces

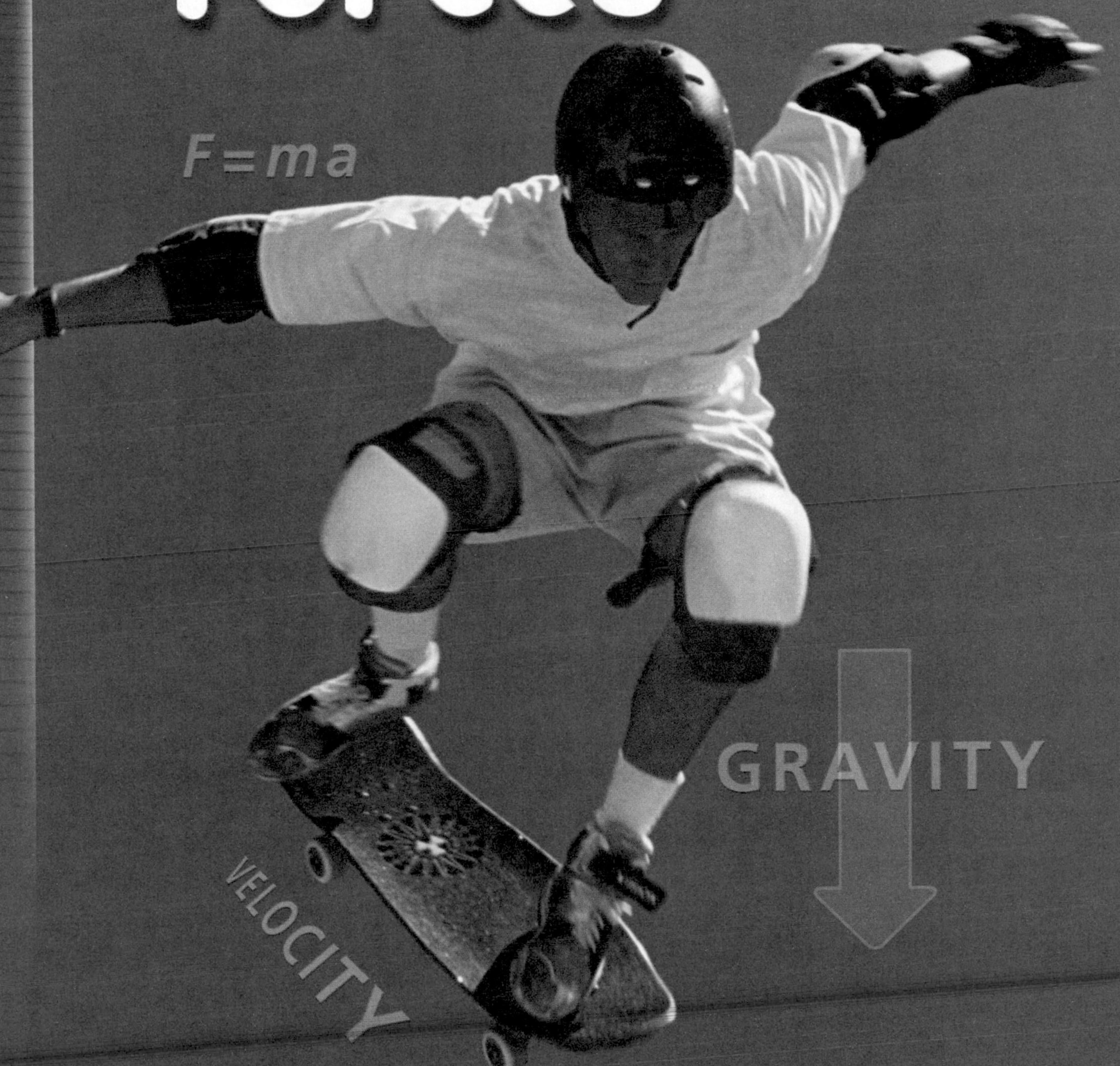

Credits
37C, 73B, 73C, 111B Photographs by Sharon Hoogstraten; **141C** Illustration by Dan Stuckenschneider.

Acknowledgements
Excerpts and adaptations from National Science Education Standards by the National Academy of Sciences. Copyright © 1996 by the National Academy of Sciences. Reprinted with permission from the National Academies Press, Washington, D.C.

ISBN: 0-618-33443-2 1 2 3 4 5 6 7 8 VJM 08 07 06 05 04

Internet Web Site: http://www.mcdougallittell.com

McDougal Littell Science

Effective Science Instruction Tailored for Middle School Learners

Motion and Forces Teacher's Edition Contents

Program Consultants and Reviewers	T4
Research Based Solutions for Your Classroom	T6
Content Organized Around Big Ideas	T8
Many Ways to Learn	T10
Differentiated Instruction	T12
Effective Assessment	T14
Program Overview	T16
Teaching Resources	T18
Correlation to National Science Education Standards	T20
Correlation to Project 2061 Benchmarks	T22
Planning the Unit	T24
Planning the Chapter	T26
Planning the Lesson	T28
Lab Materials List	T30

Consultants and Reviewers

Science Consultants

Chief Science Consultant

James Trefil, Ph.D. is the Clarence J. Robinson Professor of Physics at George Mason University. He is the author or co-author of more than 25 books, including *Science Matters* and *The Nature of Science*. Dr. Trefil is a member of the American Association for the Advancement of Science's Committee on the Public Understanding of Science and Technology. He is also a fellow of the World Economic Forum and a frequent contributor to *Smithsonian* magazine.

Rita Ann Calvo, Ph.D. is Senior Lecturer in Molecular Biology and Genetics at Cornell University, where for 12 years she also directed the Cornell Institute for Biology Teachers. Dr. Calvo is the 1999 recipient of the College and University Teaching Award from the National Association of Biology Teachers.

Kenneth Cutler, M.S. is the Education Coordinator for the Julius L. Chambers Biomedical Biotechnology Research Institute at North Carolina Central University. A former middle school and high school science teacher, he received a 1999 Presidential Award for Excellence in Science Teaching.

Instructional Design Consultants

Douglas Carnine, Ph.D. is Professor of Education and Director of the National Center for Improving the Tools of Educators at the University of Oregon. He is the author of seven books and over 100 other scholarly publications, primarily in the areas of instructional design and effective instructional strategies and tools for diverse learners. Dr. Carnine also serves as a member of the National Institute for Literacy Advisory Board.

Linda Carnine, Ph.D. consults with school districts on curriculum development and effective instruction for students struggling academically. A former teacher and school administrator, Dr. Carnine also co-authored a popular remedial reading program.

Sam Miller, Ph.D. is a middle school science teacher and the Teacher Development Liaison for the Eugene, Oregon, Public Schools. He is the author of curricula for teaching science, mathematics, computer skills, and language arts.

Donald Steely, Ph.D. serves as principal investigator at the Oregon Center for Applied Science (ORCAS) on federal grants for science and language arts programs. His background also includes teaching and authoring of print and multimedia programs in science, mathematics, history, and spelling.

Vicky Vachon, Ph.D. consults with school districts throughout the United States and Canada on improving overall academic achievement with a focus on literacy. She is also co-author of a widely used program for remedial readers.

Content Reviewers

John Beaver, Ph.D.
Ecology
Professor, Director of Science Education Center
College of Education and Human Services
Western Illinois University
Macomb, IL

Donald J. DeCoste, Ph.D.
Matter and Energy, Chemical Interactions
Chemistry Instructor
University of Illinois
Urbana-Champaign, IL

Dorothy Ann Fallows, Ph.D., MSc
Diversity of Living Things, Microbiology
Partners in Health
Boston, MA

Michael Foote, Ph.D.
The Changing Earth, Life Over Time
Associate Professor
Department of the Geophysical Sciences
The University of Chicago
Chicago, IL

Lucy Fortson, Ph.D.
Space Science
Director of Astronomy
Adler Planetarium and Astronomy Museum
Chicago, IL

Elizabeth Godrick, Ph.D.
Human Biology
Professor, CAS Biology
Boston University
Boston, MA

Isabelle Sacramento Grilo, M.S.
The Changing Earth
Lecturer, Department of the Geological Sciences
Montana State University
Bozeman, MT

David Harbster, MSc
Diversity of Living Things
Professor of Biology
Paradise Valley Community College
Phoenix, AZ

Richard D. Norris, Ph.D.
Earth's Waters
Professor of Paleobiology
Scripps Institution of Oceanography
University of California, San Diego
La Jolla, CA

Donald B. Peck, M.S.
Motion and Forces; Waves, Sound, and Light; Electricity and Magnetism
Director of the Center for Science Education (retired)
Fairleigh Dickinson University
Madison, NJ

Javier Penalosa, Ph.D.
Diversity of Living Things, Plants
Associate Professor, Biology Department
Buffalo State College
Buffalo, NY

Raymond T. Pierrehumbert, Ph.D.
Earth's Atmosphere
Professor in Geophysical Sciences (Atmospheric Science)
The University of Chicago
Chicago, IL

Brian J. Skinner, Ph.D.
Earth's Surface
Eugene Higgins Professor of Geology and Geophysics
Yale University
New Haven, CT

Nancy E. Spaulding, M.S.
Earth's Surface, The Changing Earth, Earth's Waters
Earth Science Teacher (retired)
Elmira Free Academy
Elmira, NY

Steven S. Zumdahl, Ph.D.
Matter and Energy, Chemical Interactions
Professor Emeritus of Chemistry
University of Illinois
Urbana-Champaign, IL

Susan L. Zumdahl, M.S.
Matter and Energy, Chemical Interactions
Chemistry Education Specialist
University of Illinois
Urbana-Champaign, IL

Safety Consultant

Juliana Texley, Ph.D.
Former K–12 Science Teacher and School Superintendent
Boca Raton, FL

English Language Advisor

Judy Lewis, M.A.
Director, State and Federal Programs for reading proficiency and high risk populations
Rancho Cordova, CA

Research-Based Solutions for Your Classroom

The distinguished program consultant team and a thorough, research-based planning and development process assure that *McDougal Littell Science* supports all students in learning science concepts, acquiring inquiry skills, and thinking scientifically.

Standards-Based Instruction

Concepts and skills were selected based on careful analysis of national and state standards.

- National Science Education Standards
- Project 2061 Benchmarks for Science Literacy
- Comprehensive database of state science standards

Standards and Benchmarks

Each chapter in **Motion and Forces** covers some of the learning goals that are described in the *National Science Education Standards* (NSES) and the Project 2061 *Benchmarks for Science Literacy*. Selected content and skill standards are shown below in shortened form. The following National Science Education Standards are covered on pages xii–xxvii, in Frontiers in Science, and in Timelines in Science, as well as in chapter features and laboratory investigations: Understandings About Scientific Inquiry (A.9), Understandings About Science and Technology (E.6), Science and Technology in Society (F.5), Science as a Human Endeavor (G.1), Nature of Science (G.2), and History of Science (G.3).

Content Standards

1 Motion

National Science Education Standards

B.2.a An object's motion
- is described by its position, direction of motion, and speed
- can be measured and shown on a graph

Project 2061 Benchmarks

10.A.1 An object's motion is relative to a certain object or point in space.

2 Forces

National Science Education Standards

B.2.b An object will move in a straight line at a constant speed unless a force acts on it.

B.2.c If more than one force acts on an object,
- the forces can act together
- the forces can act against each other depending on the size and direction of the forces

Project 2061 Benchmarks

4.F.3 An unbalanced force
- will change an object's speed and/or direction
- can cause an object to orbit the center of the force

10.B.1 Newton's Laws describe motion everywhere in the universe.

11.C.2 A system may stay the same because
- no forces are acting on the system
- forces are acting on the system, but they all cancel each other out

3 Gravity, Friction, and Pressure

National Science Education Standards

B.1.a Substances have certain properties, including density.

D.3.c Gravity is the force that
- keeps planets in orbit around the Sun
- governs motion within the solar system
- holds us to Earth's surface
- produces tides

x Unit: Motion and Forces

CHAPTER 1

Motion

the BIG idea

The motion of an object can be described and predicted.

Where will these people be in a few seconds? How do you know?

Key Concepts

SECTION 1.1 **An object in motion changes position.** Learn about measuring position from reference points, and about relative motion.

SECTION 1.2 **Speed measures how fast position changes.** Learn to calculate speed and how velocity depends on speed and direction.

SECTION 1.3 **Acceleration measures how fast velocity changes.** Learn about acceleration and how to calculate it.

Internet Preview

CLASSZONE.COM

Chapter 1 online resources: Content Review, Visualization, Simulation, two Resource Centers, Math Tutorial, Test Practice

D 6 Unit: Motion and Forces

Internet Activity: Relative Motion

Go to **ClassZone.com** to examine motion from different points of view. Learn how your motion makes a difference in what you observe.

Observe and Think How does the way you see motion depend on your point of view?

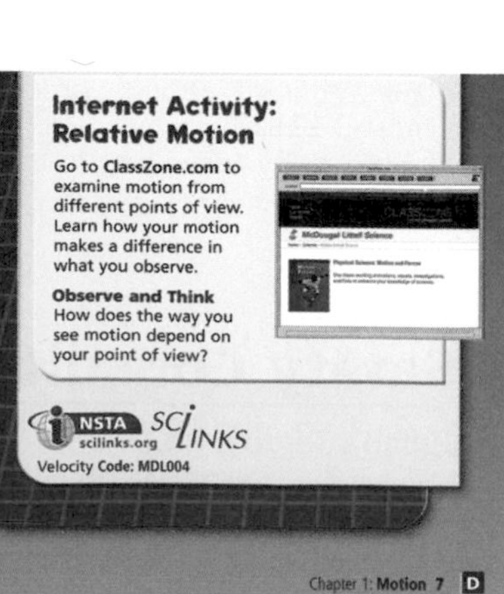

NSTA scilinks.org SCILINKS

Velocity Code: MDL004

Chapter 1: Motion 7 D

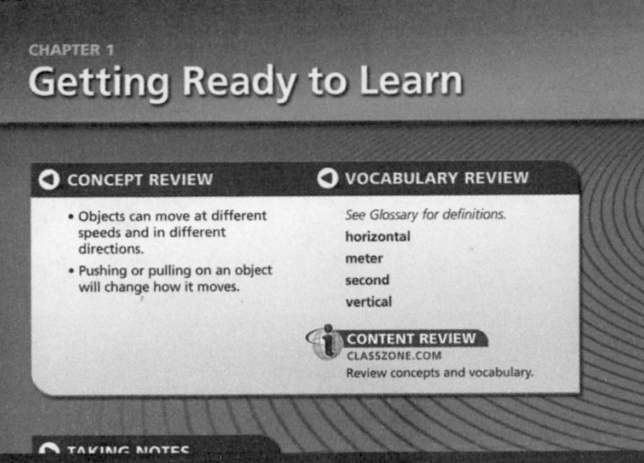
CHAPTER 1

Getting Ready to Learn

CONCEPT REVIEW

- Objects can move at different speeds and in different directions.
- Pushing or pulling on an object will change how it moves.

VOCABULARY REVIEW

See Glossary for definitions.

horizontal
meter
second
vertical

CONTENT REVIEW
CLASSZONE.COM
Review concepts and vocabulary.

TAKING NOTES

Effective Instructional Strategies

McDougal Littell Science incorporates strategies that research shows are effective in improving student achievement. These strategies include

- Notetaking and nonlinguistic representations (Marzano, Pickering, and Pollock)
- A focus on big ideas (Kameenui and Carnine)
- Background knowledge and active involvement (Project CRISS)

Robert J. Marzano, Debra J. Pickering, and Jane E. Pollock, *Classroom Instruction that Works; Research-Based Strategies for Increasing Student Achievement* (ASCD, 2001)

Edward J. Kameenui and Douglas Carnine, *Effective Teaching Strategies that Accommodate Diverse Learners* (Pearson, 2002)

Project CRISS (Creating Independence through Student Owned Strategies)

WHAT DO YOU THINK?
What kinds of information must you give another person when you are trying to describe a location?

Position describes the location of an object.

Have you ever gotten lost while looking for a specific place? If so, you probably know that accurately describing where a place is can be very important. The **position** of a place or an object is the location of that place or object. Often you describe where something is by comparing its position with where you currently are. You might say, for example, that a classmate sitting next to you is about a meter to your right, or that a mailbox is two blocks south of where you live. Each time you identify the position of an object, you are comparing the location of the object with the location of another object or place.

VOCABULARY
Make a description wheel in your notebook for *position*.

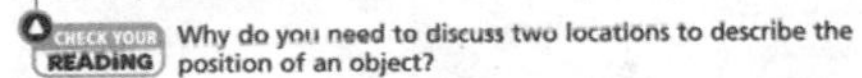
CHECK YOUR READING Why do you need to discuss two locations to describe the position of an object?

Chapter 1: **Motion** 9

Gravity, Friction, and Pressure *continued*

	Project 2061 Benchmarks
4.B.3	Everything on or near Earth is pulled toward Earth's center by the force of gravity.
4.G.1	Every object exerts the force of gravity, but the force • depends on the mass and distance of objects • is difficult to detect unless an object contains a lot of mass
4.G.2	In the solar system • the Sun's gravity keeps the planets in orbit around the Sun • the planets' gravity keeps their moons in orbit around the planets

4 Work and Energy

	National Science Education Standards
B.3.a	Energy is transferred in many ways. Energy is often associated with heat, sound, and mechanical motion.
	Project 2061 Benchmarks
4.E.1	Energy cannot be created or destroyed, but it can be changed from one form to another.
4.E.4	Energy has many different forms, including • heat—the disorderly motion of particles • chemical—the arrangement of atoms in matter • mechanical—the kinetic plus the potential energy of an object • gravitational—the separation of objects that attract each other

5 Machines

	National Science Education Standards
E.6.c	• Science and technology often work together. • Science helps drive technology. • Technology is used to improve scientific investigations.
E.6.d	Perfectly designed solutions do not exist. All technology has risks and trade-offs, such as cost, safety, and efficiency.
E.6.e	All designs have limits, including those having to do with material properties and availability, safety, and environmental protection.
	Project 2061 Benchmarks
8.B.4	The use of robots has changed the nature of work in many fields, including manufacturing.

Process and Skill Standards

	National Science Education Standards
A.2	Design and conduct a scientific investigation.
A.3	Use appropriate tools and techniques to gather and interpret data.
A.4	Use evidence to describe, predict, explain, and model.
A.5	Use critical thinking to find relationships between results and interpretations.
A.7	Communicate procedures, results, and conclusions.
A.8	Use mathematics in scientific investigations.
E.1	Identify a problem to be solved.
E.2	Design a solution or product.
E.3	Implement the proposed solution.
E.4	Evaluate the solution or design.

	Project 2061 Benchmarks
11.C.4	Use equations to summarize observed changes.
12.C.3	Using appropriate units, use and read instruments that measure length, volume, weight, time, rate, and temperature.
12.D.1	Use tables and graphs to organize information and identify relationships.
12.D.2	Read, interpret, and describe tables and graphs.
12.D.4	Understand information that includes different types of charts and graphs, including circle charts, bar graphs, line graphs, data tables, diagrams, and symbols.

Standards and Benchmarks xi

Comprehensive Research, Review, and Field Testing

An ongoing program of research and review guided the development of *McDougal Littell Science.*

- Program plans based on extensive data from classroom visits, research surveys, teacher panels, and focus groups
- All pupil edition activities and labs classroom-tested by middle school teachers and students
- All chapters reviewed for clarity and scientific accuracy by the Content Reviewers listed on page T5
- Selected chapters field-tested in the classroom to assess student learning, ease of use, and student interest

Content Organized Around Big Ideas

Each chapter develops a big idea of science, helping students to place key concepts in context.

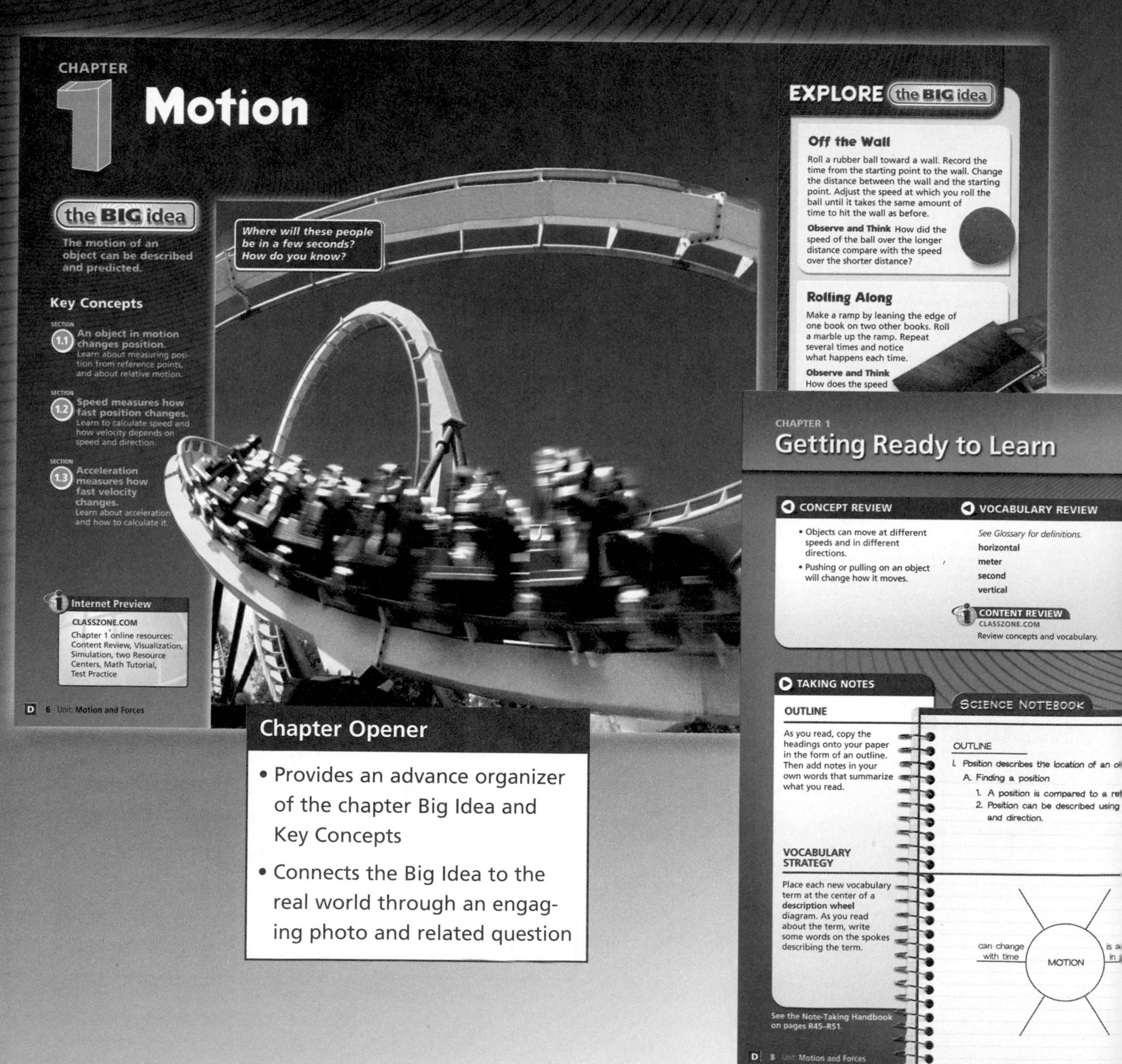
CHAPTER 1
Motion

the BIG idea
The motion of an object can be described and predicted.

Key Concepts

SECTION 1.1 An object in motion changes position. Learn about measuring position from reference points, and about relative motion.

SECTION 1.2 Speed measures how fast position changes. Learn to calculate speed and how velocity depends on speed and direction.

SECTION 1.3 Acceleration measures how fast velocity changes. Learn about acceleration and how to calculate it.

Internet Preview
CLASSZONE.COM
Chapter 1 online resources: Content Review, Visualization, Simulation, two Resource Centers, Math Tutorial, Test Practice

Where will these people be in a few seconds? How do you know?

D 6 Unit: Motion and Forces

EXPLORE the BIG idea

Off the Wall
Roll a rubber ball toward a wall. Record the time from the starting point to the wall. Change the distance between the wall and the starting point. Adjust the speed at which you roll the ball until it takes the same amount of time to hit the wall as before.
Observe and Think How did the speed of the ball over the longer distance compare with the speed over the shorter distance?

Rolling Along
Make a ramp by leaning the edge of one book on two other books. Roll a marble up the ramp. Repeat several times and notice what happens each time.
Observe and Think How does the speed

CHAPTER 1
Getting Ready to Learn

CONCEPT REVIEW
- Objects can move at different speeds and in different directions.
- Pushing or pulling on an object will change how it moves.

VOCABULARY REVIEW
See Glossary for definitions.
horizontal
meter
second
vertical

CONTENT REVIEW
CLASSZONE.COM
Review concepts and vocabulary.

TAKING NOTES

OUTLINE
As you read, copy the headings onto your paper in the form of an outline. Then add notes in your own words that summarize what you read.

VOCABULARY STRATEGY
Place each new vocabulary term at the center of a description wheel diagram. As you read about the term, write some words on the spokes describing the term.

See the Note-Taking Handbook on pages R45–R51.

D 8 Unit: Motion and Forces

SCIENCE NOTEBOOK
OUTLINE
I. Position describes the location of an o
A. Finding a position
1. A position is compared to a re
2. Position can be described using and direction.
can change with time
MOTION

Chapter Opener

- Provides an advance organizer of the chapter Big Idea and Key Concepts
- Connects the Big Idea to the real world through an engaging photo and related question

Visual Summary

- Summarizes Key Concepts using both text and visuals
- Reinforces the connection of Key Concepts to the Big Idea

Section Opener

- Highlights the Key Concept
- Connects new learning to prior knowledge
- Previews important vocabulary

The Big Idea Questions

- Help students connect their new learning back to the Big Idea
- Prompt students to synthesize and apply the Big Idea and Key Concepts

1 Chapter Review

the BIG idea

The motion of an object can be described and predicted.

CONTENT REVIEW CLASSZONE.COM

KEY CONCEPTS SUMMARY

1.1 An object in motion changes position.

VOCABULARY
position p. 9
reference point p. 10
motion p. 11

1.2 Speed measures how fast position changes.

- Speed is how fast positions change with time.
- Velocity is speed in a specific direction.

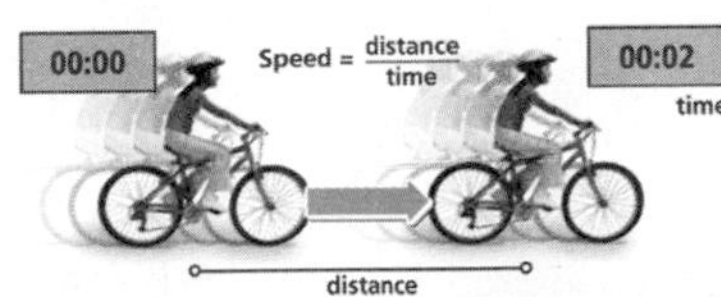

VOCABULARY
speed p. 16
velocity p. 22
vector p. 22

1.3 Acceleration measures how fast velocity changes.

acceleration = final velocity – i... / tim...

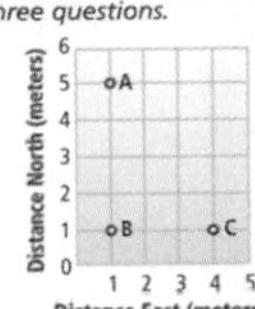

VOCABULARY
acceleration p. 25

Reviewing Vocabulary

Copy and complete the chart below. If the left column is blank, give the correct term. If the right column is blank, give a brief description.

Term	Description
1.	speed in a specific direction
2.	a change of position over time
3. speed	
4.	an object's location
5. reference point	
6.	the rate at which velocity changes over time
7.	a quantity that has both size and direction

Reviewing Key Concepts

Multiple Choice *Choose the letter of the best answer.*

8. A position describes an object's location compared to
   a. its motion
   b. a reference point
   c. its speed
   d. a vector

...What ...alk?

Thinking Critically

Use the following graph to answer the next three questions.

19. **OBSERVE** Describe the location of point A. Explain what you used as a reference point for your location.
20. **COMPARE** Copy the graph into your notebook. Draw two different paths an object could take when moving from point B to point C. How do the lengths of these two paths compare?
21. **ANALYZE** An object moves from point A to point C in the same amount of time that another object moves from point B to point C. If both objects traveled in a straight line, which one had the greater speed?

Read the following paragraph and use the information to answer the next three questions.

In Aesop's fable of the tortoise and the hare, a slow-moving tortoise races a fast-moving hare. The hare, certain it can win, stops to take a long nap. Meanwhile, the tortoise continues to move toward the finish line at a slow but steady speed. When the hare wakes up, it runs as fast as it can. Just as the hare is about to catch up to the tortoise, however, the tortoise wins the race.

22. **ANALYZE** How does the race between the tortoise and the hare show the difference between average speed and instantaneous speed?
23. **MODEL** Assume the racetrack was 100 meters long and the race took 40 minutes. Create a possible distance-time graph for both the tortoise and the hare.
24. **COMPARE** If the racetrack were circular, how would the tortoise's speed be different from its velocity?
25. **APPLY** How might a person use a floating stick to measure the speed at which a river flows?
26. **CONNECT** Describe a frame of reference other than the ground that you might use to measure motion. When would you use it?

Using Math Skills in Science

27. José skated 50 m in 10 s. What was his speed?
28. Use the information in the photograph below to calculate the speed of the ant as it moves down the branch.

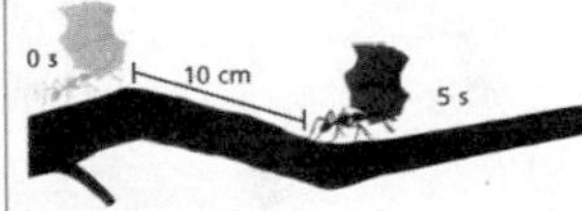

29. While riding her bicycle, Jamie accelerated from 7 m/s to 2 m/s in 5 s. What was her acceleration?

the BIG idea

30. **PREDICT** Look back at the picture at the beginning of the chapter on pages 6–7. Predict how the velocity of the roller coaster will change in the next moment.
31. **WRITE** A car is traveling east at 40 km/h. Use this information to predict where the car will be in one hour. Discuss the assumptions you made to reach your conclusion and the factors that might affect it.

UNIT PROJECTS

If you are doing a unit project, make a folder for your project. Include in your folder a list of the resources you will need, the date on which the project is due, and a schedule to keep track of your progress. Begin gathering data.

KEY CONCEPT

1.1 An object in motion changes position.

BEFORE, you learned
- Objects can move in different ways
- An object's position can change

NOW, you will learn
- How to describe an object's position
- How to describe an object's motion

VOCABULARY
position p. 9
reference point p. 10
motion p. 11

EXPLORE Location

How do you describe the location of an object?

PROCEDURE

1. Choose an object in the classroom that is easy to see.
2. Without pointing to, describing, or naming the object, give directions to a classmate for finding it.
3. Ask your classmate to identify the object using your directions. If your classmate does not correctly identify the object, try giving directions in a different way. Continue until your classmate has located the object.

WHAT DO YOU THINK?
What kinds of information must you give another person when you are trying to describe a location?

Position describes the location of an object.

Have you ever gotten lost while looking for a specific place? If so, you probably know that accurately describing where a place is can be very important. The **position** of a place or an object is the location of that place or object. Often you describe where something is by comparing its position with where you currently are. You might say, for example, that a classmate sitting next to you is about a meter to your right, or that a mailbox is two blocks south of where you live. Each time you identify the position of an object, you are comparing the location of the object with the location of another object or place.

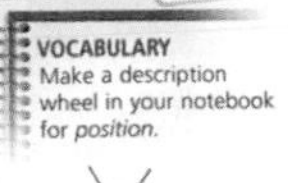

VOCABULARY
Make a description wheel in your notebook for *position*.

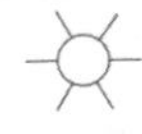

CHECK YOUR READING Why do you need to discuss two locations to describe the position of an object?

Many Ways to Learn

Because students learn in so many ways, *McDougal Littell Science* gives them a variety of experiences with important concepts and skills. Text, visuals, activities, and technology all focus on Big Ideas and Key Concepts.

Considerate Text

- Clear structure of meaningful headings
- Information clearly connected to main ideas
- Student-friendly writing style

You will find many examples of Newton's first law around you. For instance, if you throw a stick for a dog to catch, you are changing the motion of the stick. The dog changes the motion of the stick by catching it and by dropping it at your feet. You change the motion of a volleyball when you spike it, a tennis racket when you swing it, a paintbrush when you make a brush stroke, and an oboe when you pick it up to play or set it down after playing. In each of these examples, you apply a force that changes the motion of the object.

VOCABULARY
Make a magnet word diagram for *inertia* in your notebook.

Inertia

Inertia (ih-NUR-shuh) is the resistance of an object to a change in the speed or the direction of its motion. Newton's first law, which describes the tendency of objects to resist changes in motion, is also called the law of inertia. Inertia is closely related to mass. When you measure the mass of an object, you are also measuring its inertia. You know from experience that it is easier to push or pull an empty box than it is to push or pull the same box when it is full of books. Likewise, it is easier to stop or to turn an empty wagon than to stop or turn a wagon full of sand. In both of these cases, it is harder to change the motion of the object that has more mass.

Hands-on Learning

- Activities that reinforce Key Concepts
- Skill Focus for important inquiry and process skills
- Multiple activities in every chapter, from quick Explores to full-period Chapter Investigations

INVESTIGATE Inertia

Which ball has more inertia?

DESIGN — YOUR OWN — EXPERIMENT

Two balls have different masses and therefore different amounts of inertia. Use what you know about force and inertia to design an experiment that shows which ball has more inertia. Your procedure cannot include lifting the balls, weighing the balls, or touching the balls with your hands.

SKILL FOCUS
Designing experiments

MATERIALS
- 2 balls of unknown masses
- string
- block
- meter stick

TIME
30 minutes

PROCEDURE

1. Figure out how to use the meter stick or other materials to compare the inertia of the two balls.
2. Write up your procedure.
3. Test your procedure.

WHAT DO YOU THINK?

- What were the results of your experiment? Did it work? Why or why not?
- What was the variable? What were the constants?
- How does your experiment demonstrate the property of inertia?

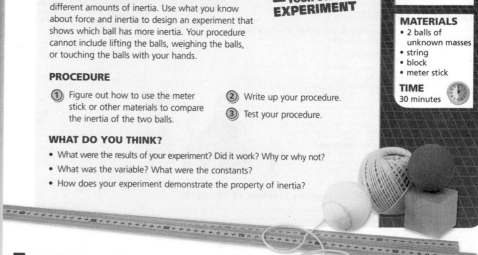

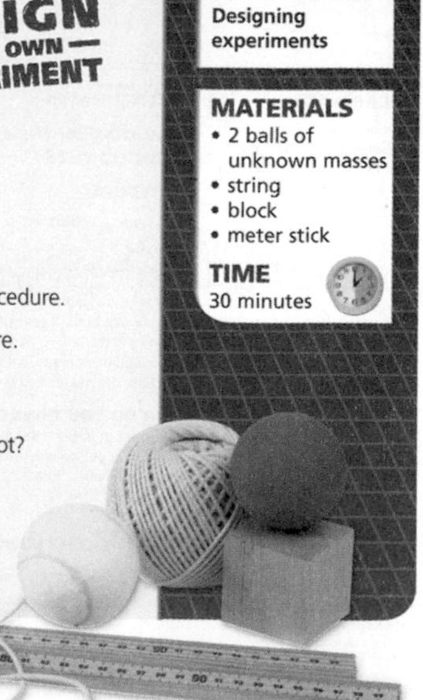

Inertia is the reason that people in cars need to wear seat belts. A moving car has inertia, and so do the riders inside it. When the driver applies the brakes, an unbalanced force is applied to the car. Normally, the bottom of the seat applies an unbalanced force—friction—which slows the riders down as the car slows. If the driver stops the car suddenly, however, this force is not exerted over enough time to stop the motion of the riders. Instead, the riders continue moving forward with most of their original speed because of their inertia.

Find out more about inertia.

1 As a car moves forward, the driver—shown here as a crash-test dummy—moves forward with the same velocity as the car.

2 When the driver hits the brakes, the car stops. If the stop is sudden and the driver is not wearing a seat belt, the driver keeps moving forward.

3 Finally, the windshield applies an unbalanced force that stops the driver's forward motion.

If the driver is wearing a seat belt, the seat belt rather than the windshield applies the unbalanced force that stops the driver's forward motion. The force from the seat belt is applied over a longer time, so the force causes less damage. In a collision, seat belts alone are sometimes not enough to stop the motion of drivers or passengers. Air bags further cushion people from the effects of inertia in an accident.

If a car makes a sudden stop, what happens to a passenger riding in the back seat who is not wearing a seat belt?

2.1 Review

KEY CONCEPTS

1. Explain the difference between balanced and unbalanced forces.
2. What is the relationship between force and motion described by Newton's first law?
3. What is inertia? How is the inertia of an object related to its mass?

CRITICAL THINKING

4. **Infer** Once a baseball has been hit into the air, what forces are acting upon it? How can you tell that any forces are acting upon the ball?
5. **Predict** A ball is at rest on the floor of a car moving at a constant velocity. What will happen to the ball if the car swerves suddenly to the left?

CHALLENGE

6. **Synthesize** What can the changes in an object's position tell you about the forces acting on that object? Describe an example from everyday life that shows how forces affect the position of an object.

Integrated Technology

- Interaction with Key Concepts through Simulations and Visualizations
- Easy access to relevant Web resources through Resource Centers and SciLinks
- Opportunities for review through Content Review and Math Tutorials

Visuals that Teach

- Information-rich visuals directly connected to the text
- Thoughtful pairing of diagrams and real-world photos
- Reading Visuals questions to support student learning

Differentiated Instruction

A full spectrum of resources for differentiating instruction supports you in reaching the wide range of learners in your classroom.

1.1 INSTRUCT

Teach from Visuals

To point out different ways to describe a location, use the "Describing Position" visual and ask:

- What does the map on the left show? *using Brasília as a reference point for finding Santiago*
- What does the map on the right show? *using latitude and longitude, a grid system, to locate Santiago*

Teacher Demo

Make a tube out of graph paper and tape it so it will stay closed. Draw a point on the tube. Ask students how they could use the graph paper squares to locate the point. List the different suggestions on the board, and discuss the similarities and differences in the methods. Point out how two reference lines, one vertical and one horizontal, are needed to describe the location. Discuss how longitude and latitude are like the lines on the tube.

Ongoing Assessment

Describe an object's position.

Ask: If you describe positions on a map by comparing them with the position of landmark Q, what is Q called? *a reference point*

READING VISUALS *Answer: The left-hand map uses a distance and a direction from Brasília to describe Santiago's location. The right-hand map uses the number of degrees from the equator and the prime meridian to describe Santiago's location.*

D 10 Unit: Motion and Forces

RESOURCE CENTER CLASSZONE.COM
Learn more about how people find and describe position.

Describing a Position

You might describe the position of a city based on the location of another city. A location to which you compare other locations is called a **reference point.** You can describe where Santiago, Chile, is from the reference point of the city Brasília, Brazil, by saying that Santiago is about 3000 kilometers (1860 mi) southwest of Brasília. (A)

You can also describe a position using a method that is similar to describing where a point on a graph is located. For example, in the longitude and latitude system, locations are given by two numbers—longitude and latitude. Longitude describes how many degrees east or west a location is from the prime meridian, an imaginary line running north-south through Greenwich, England. Latitude describes how many degrees north or south a location is from the equator, the imaginary circle that divides the northern and southern hemispheres. Having a standard way of describing location, such as longitude and latitude, makes it easier for people to compare locations.

Describing Position

There are several different ways to describe a position. The way you choose may depend on your reference point.

1. Reference Point: Brasília

To describe where Santiago is, using Brasília as a reference point, you would need to know how far Santiago is from Brasília and in what direction it is.

2. Reference Point: 0° longitude, 0° latitude

In the longitude and latitude system, a location is described by how many degrees north or south it is from the equator and how many degrees east or west it is from the prime meridian.

READING VISUALS Compare and contrast the two ways of describing the location of Santiago as shown here.

Measuring Distance

If you were to travel from Brasília to Santiago, you would end up about 3000 kilometers from where you started. The actual distance you traveled, however, would depend on the exact path you took. If you took a route that had many curves, the distance you traveled would be greater than 3000 kilometers.

(B) The way you measure distance depends on the information you want. Sometimes you want to know the straight-line distance between two positions. Sometimes, however, you might need to know the total length of a certain path between those positions. During a hike, you are probably more interested in how far you have walked than in how far you are from your starting point.

When measuring either the straight-line distance between two points or the length of a path between those points, scientists use a standard unit of measurement. The standard unit of length is the meter (m), which is 3.3 feet. Longer distances can be measured in kilometers (km), and shorter distances in centimeters (cm).

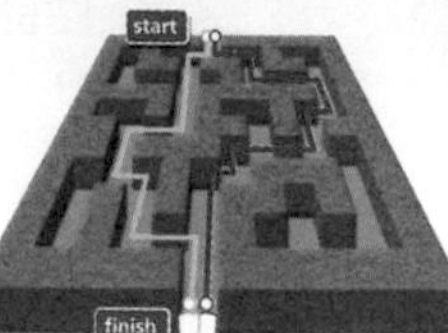

COMPARE How does the distance each person has walked compare with the distance each is from the start of the maze?

Motion is a change in position.

The illustration below shows an athlete at several positions during a long jump. If you were to watch her jump, you would see that she is in motion. **Motion** is the change of position over time. As she jumps, both her horizontal and vertical positions change. If you missed the motion of the jump, you would still know that motion occurred because of the distance between her starting and ending positions. A change in position is evidence that motion happened.

REMINDER *Horizontal* and *vertical* describe directions, as shown.
vertical
horizontal

DIFFERENTIATE INSTRUCTION

More Reading Support

A What would you call a location with which other locations can be compared? *a reference point*

English Learners English learners may need help understanding sentences that imply *if/then* construction, such as *If you were to travel from Brasília to Santiago, you would end up about 3000 kilometers from where you started* (p. 11). This sentence begins with *If* but does not contain *then,* and therefore requires a reader to infer the cause-and-effect relationship. Point out other examples of this construction and make sure English learners recognize that *then* is implied.

DIFFERENTIATE INSTRUCTION

More Reading Support

B List two ways to measure distance. *measure a straight-line distance between two positions; measure the total length of a path between two positions*

Advanced Have students refer to the map on page 10 that shows a connecting line between Santiago and Brasília. Ask: How could you estimate the distance of this diagonal line using grid units—that is, the distance south and the distance west? *You could measure the vertical distance and the horizontal distance; then use the Pythagorean theorem to determine the length of the hypotenuse.*

R Challenge and Extension, p. 19

Teacher's Edition

- More Reading Support for below-level readers
- Strategies for below-level and advanced learners, English learners, and inclusion students

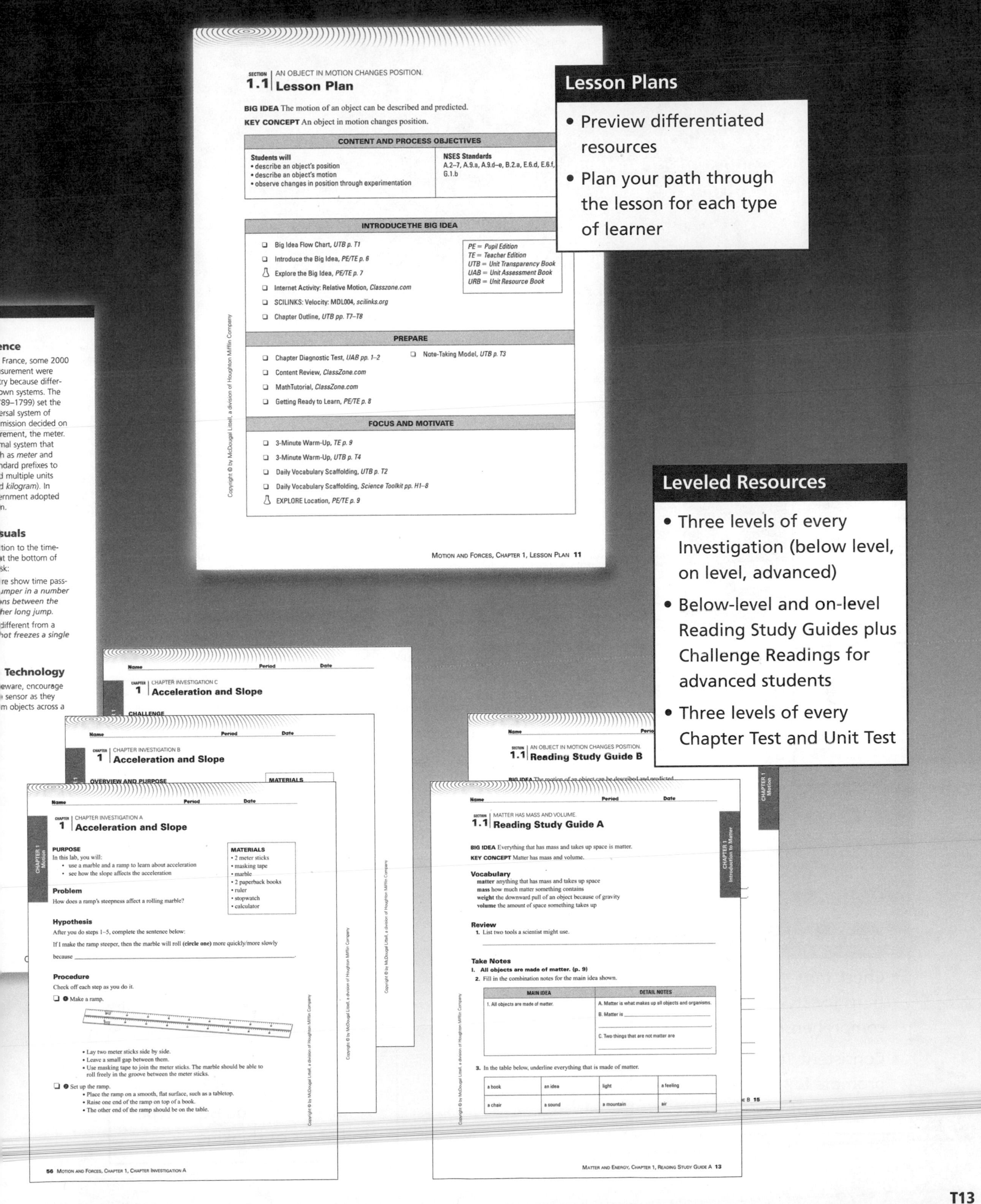

SECTION 1.1 | AN OBJECT IN MOTION CHANGES POSITION.
Lesson Plan

BIG IDEA The motion of an object can be described and predicted.
KEY CONCEPT An object in motion changes position.

CONTENT AND PROCESS OBJECTIVES	
Students will • describe an object's position • describe an object's motion • observe changes in position through experimentation	**NSES Standards** A.2–7, A.9.a, A.9.d–e, B.2.a, E.6.d, E.6.f, G.1.b

INTRODUCE THE BIG IDEA
- ❑ Big Idea Flow Chart, *UTB p. T1*
- ❑ Introduce the Big Idea, *PE/TE p. 6*
- Explore the Big Idea, *PE/TE p. 7*
- ❑ Internet Activity: Relative Motion, *Classzone.com*
- ❑ SCILINKS: Velocity: MDL004, *scilinks.org*
- ❑ Chapter Outline, *UTB pp. T7–T8*

PE = Pupil Edition
TE = Teacher Edition
UTB = Unit Transparency Book
UAB = Unit Assessment Book
URB = Unit Resource Book

PREPARE
- ❑ Chapter Diagnostic Test, *UAB pp. 1–2*
- ❑ Content Review, *ClassZone.com*
- ❑ MathTutorial, *ClassZone.com*
- ❑ Getting Ready to Learn, *PE/TE p. 8*
- ❑ Note-Taking Model, *UTB p. T3*

FOCUS AND MOTIVATE
- ❑ 3-Minute Warm-Up, *TE p. 9*
- ❑ 3-Minute Warm-Up, *UTB p. T4*
- ❑ Daily Vocabulary Scaffolding, *UTB p. T2*
- ❑ Daily Vocabulary Scaffolding, *Science Toolkit pp. H1–8*
- EXPLORE Location, *PE/TE p. 9*

MOTION AND FORCES, CHAPTER 1, LESSON PLAN 11

Lesson Plans

- Preview differentiated resources
- Plan your path through the lesson for each type of learner

Leveled Resources

- Three levels of every Investigation (below level, on level, advanced)
- Below-level and on-level Reading Study Guides plus Challenge Readings for advanced students
- Three levels of every Chapter Test and Unit Test

Name Period Date

CHAPTER 1 | CHAPTER INVESTIGATION A
Acceleration and Slope

PURPOSE
In this lab, you will:
- use a marble and a ramp to learn about acceleration
- see how the slope affects the acceleration

MATERIALS
- 2 meter sticks
- masking tape
- marble
- 2 paperback books
- ruler
- stopwatch
- calculator

Problem
How does a ramp's steepness affect a rolling marble?

Hypothesis
After you do steps 1–5, complete the sentence below:
If I make the ramp steeper, then the marble will roll **(circle one)** more quickly/more slowly because ______.

Procedure
Check off each step as you do it.

❑ ❶ Make a ramp.
- Lay two meter sticks side by side.
- Leave a small gap between them.
- Use masking tape to join the meter sticks. The marble should be able to roll freely in the groove between the meter sticks.

❑ ❷ Set up the ramp.
- Place the ramp on a smooth, flat surface, such as a tabletop.
- Raise one end of the ramp on top of a book.
- The other end of the ramp should be on the table.

56 MOTION AND FORCES, CHAPTER 1, CHAPTER INVESTIGATION A

Name Period Date

SECTION 1.1 | MATTER HAS MASS AND VOLUME.
Reading Study Guide A

BIG IDEA Everything that has mass and takes up space is matter.
KEY CONCEPT Matter has mass and volume.

Vocabulary
matter anything that has mass and takes up space
mass how much matter something contains
weight the downward pull of an object because of gravity
volume the amount of space something takes up

Review
1. List two tools a scientist might use.

Take Notes
I. **All objects are made of matter. (p. 9)**
2. Fill in the combination notes for the main idea shown.

MAIN IDEA	DETAIL NOTES
1. All objects are made of matter.	A. Matter is what makes up all objects and organisms. B. Matter is ______ C. Two things that are not matter are ______

3. In the table below, underline everything that is made of matter.

a book	an idea	light	a feeling
a chair	a sound	a mountain	air

MATTER AND ENERGY, CHAPTER 1, READING STUDY GUIDE A 13

Effective Assessment

McDougal Littell Science incorporates a comprehensive set of resources for assessing student knowledge and performance before, during, and after instruction.

Diagnostic Tests

- Assessment of students' prior knowledge
- Readiness check for concepts and skills in the upcoming chapter

Ongoing Assessment

CHECK YOUR READING *Sample answer: You observe motion relative to your own position. If you are on a train, for example, you see the ground outside the train moving past you.*

EXPLORE the BIG idea

Revisit "Internet Activity: Relative Motion" on p. 7. Have students explain the reasons for their results.

Reinforce the BIG idea

Have students relate the section to the Big Idea.

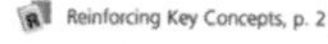

Reinforcing Key Concepts, p. 21

1.1 ASSESS & RETEACH

Assess

Section 1.1 Quiz, p. 3

Reteach

Put two marbles randomly on a checkerboard. Have students describe the positions, and then roll the two marbles at different speeds. Emphasize that a good description of motion includes speed. Finally, have students view the rolling of a marble from very different positions. Solicit students' observations. Discuss relative motion.

Technology Resources

Have students visit **ClassZone.com** for reteaching of Key Concepts.

CONTENT REVIEW

CONTENT REVIEW CD-ROM

When you ride in a train, a bus, or an airplane, you think of yourself as moving and the ground as standing still. That is, you usually consider the ground as the frame of reference for your motion. If you traveled between two cities, you would say that you had moved, not that the ground had moved under you in the opposite direction.

If you cannot see the ground or objects on it, it is sometimes difficult to tell if a train you are riding in is moving. If the ride is very smooth and you do not look out the window at the scenery, you might never realize you are moving at all.

Suppose you are in a train, and you cannot tell if you are stopped or moving. Outside the window, another train is slowly moving forward. Could you tell which of the following situations is happening?

- Your train is stopped, and the other train is moving slowly forward.
- The other train is stopped, and your train is moving slowly backward.
- Both trains are moving forward, with the other train moving a little faster.
- Your train is moving very slowly backward, and the other train is moving very slowly forward.

Actually, all four of these possibilities would look exactly the same to you. Unless you compared the motion to the motion of something outside the train, such as the ground, you could not tell the difference between these situations.

APPLY In the top picture, the train is moving compared with the camera and the ground. Describe the relative motion of the train, camera, and ground in the bottom picture.

CHECK YOUR READING *How does your observation of motion depend on your own motion?*

1.1 Review

KEY CONCEPTS

1. What information do you need to describe an object's location?
2. Describe how your position changes as you jump over an object.
3. Give an example of how the apparent motion of an object depends on the observer's motion.

CRITICAL THINKING

4. **Infer** Kyle walks 3 blocks south from his home to school, and Jana walks 2 blocks north from her home to Kyle's home. How far and in what direction is the school from Jana's home?
5. **Predict** If you sit on a moving bus and toss a coin straight up into the air, where will it land?

CHALLENGE

6. **Infer** Jamal is in a car going north. He looks out his window and thinks that the northbound traffic is moving very slowly. Ellen is in a car going south. She thinks the northbound traffic is moving quickly. Explain why Jamal and Ellen have different ideas about the motion of the traffic.

D 14 Unit: **Motion and Forces**

ANSWERS

1. a reference point

2. As a person jumps over an object, both his horizontal and vertical positions change.

3. Sample answer: A person in an airplane will think of her or his seat as not moving, but an observer on the ground will think the seats are moving with the airplane.

4. one block south

5. It should land back in your hand (or directly beneath where you threw it).

6. Jamal and Ellen have different ideas about motion because both compare the northbound traffic with their own motion. From Jamal's point of view, traffic is going in the same direction as he is; therefore, traffic seems to be moving slowly. Ellen sees oncoming cars and thinks they are moving quickly.

D 14 Unit: **Motion and Forces**

Ongoing Assessment

- Check Your Reading questions for student self-check of comprehension
- Consistent Teacher Edition prompts for assessing understanding of Key Concepts

Reviewing Vocabulary

Copy and complete the chart below. If the left column is blank, give the correct term. If the right column is blank, give a brief description.

Term	Description
1.	speed in a specific direction
2.	a change of position over time
3. speed	
4.	an object's location
5. reference point	
6.	the rate at which velocity changes over time
7.	a quantity that has both size and direction

Reviewing Key Concepts

Multiple Choice *Choose the letter of the best answer.*

8. A position describes an object's location compared to
   a. its motion
   b. a reference point
   c. its speed
   d. a vector
9. Maria walked 2 km in half an hour. What was her average speed during her walk?
   a. 1 km/h
   b. 2 km/h
   c. 4 km/h
   d. 6 km/h
10. A vector is a quantity that has
    a. speed
    b. acceleration
    c. size and direction
    d. position and distance
11. Mary and Keisha run with the same constant speed but in opposite directions. The girls have
    a. the same position
    b. different accelerations
    c. different speeds
    d. different velocities
12. A swimmer increases her speed as she approaches the end of the pool. Her acceleration is
    a. in the same direction as her motion
    b. in the opposite direction of her motion
    c. at right angles to her motion
    d. zero
13. A cheetah can go from 0 m/s to 20 m/s in 2 s. What is the cheetah's acceleration?
    a. 5 m/s^2
    b. 10 m/s^2
    c. 20 m/s^2
    d. 40 m/s^2
14. Jon walks for a few minutes, then runs for a few minutes. During this time, his average speed is
    a. the same as his final speed
    b. greater than his final speed
    c. less than his final speed
    d. zero
15. A car traveling at 40 m/s slows down to 20 m/s. During this time, the car has
    a. no acceleration
    b. positive acceleration
    c. negative acceleration
    d. constant velocity

Short Answer *Write a short answer to each question.*

16. Suppose you are biking with a friend. How would your friend describe your relative motion as he passes you?
17. Describe a situation where an object has a changing velocity but constant speed.
18. Give two examples of an accelerating object.

Chapter 1: **Motion** 35

Section and Chapter Reviews

- Focus on Key Concepts and critical thinking skills
- A full range of question types and levels of thinking

Leveled Chapter and Unit Tests

- Three levels of test for every chapter and unit
- Same Big Ideas, Key Concepts, and essential skills assessed on all levels

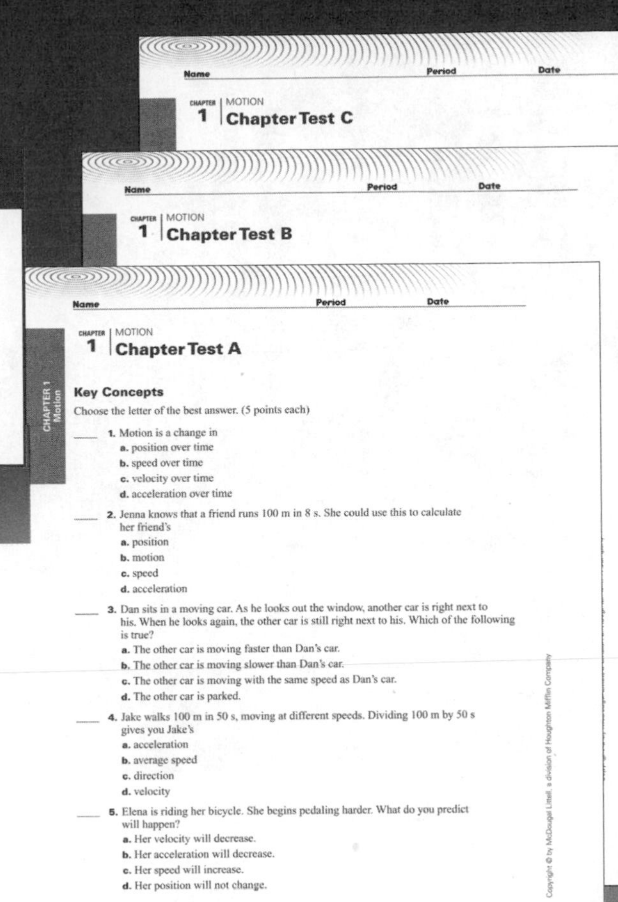

Name Period Date

CHAPTER 1 | MOTION
Chapter Test A

Key Concepts

Choose the letter of the best answer. (5 points each)

___ 1. Motion is a change in
a. position over time
b. speed over time
c. velocity over time
d. acceleration over time

___ 2. Jenna knows that a friend runs 100 m in 8 s. She could use this to calculate her friend's
a. position
b. motion
c. speed
d. acceleration

___ 3. Dan sits in a moving car. As he looks out the window, another car is right next to his. When he looks again, the other car is still right next to his. Which of the following is true?
a. The other car is moving faster than Dan's car.
b. The other car is moving slower than Dan's car.
c. The other car is moving with the same speed as Dan's car.
d. The other car is parked.

___ 4. Jake walks 100 m in 50 s, moving at different speeds. Dividing 100 m by 50 s gives you Jake's
a. acceleration
b. average speed
c. direction
d. velocity

___ 5. Elena is riding her bicycle. She begins pedaling harder. What do you predict will happen?
a. Her velocity will decrease.
b. Her acceleration will decrease.
c. Her speed will increase.
d. Her position will not change.

6 Motion and Forces, Chapter 1, Chapter Test A

Thinking Critically

Use the following graph to answer the next three questions.

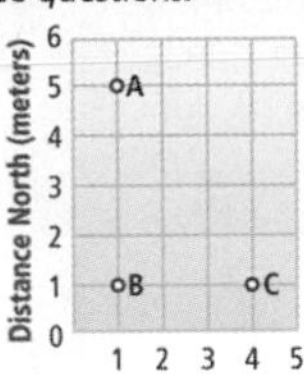

19. **OBSERVE** Describe the location of point A. Explain what you used as a reference point for your location.
20. **COMPARE** Copy the graph into your notebook. Draw two different paths an object could take when moving from point B to point C. How do the lengths of these two paths compare?
21. **ANALYZE** An object moves from point A to point C in the same amount of time that another object moves from point B to point C. If both objects traveled in a straight line, which one had the greater speed?

Read the following paragraph and use the information to answer the next three questions.

In Aesop's fable of the tortoise and the hare, a slow-moving tortoise races a fast-moving hare. The hare, certain it can win, stops to take a long nap. Meanwhile, the tortoise continues to move toward the finish line at a slow but steady speed. When the hare wakes up, it runs as fast as it can. Just as the hare is about to catch up to the tortoise, however, the tortoise wins the race.

22. **ANALYZE** How does the race between the tortoise and the hare show the difference between average speed and instantaneous speed?
23. **MODEL** Assume the racetrack was 100 meters long and the race took 40 minutes. Create a possible distance-time graph for both the tortoise and the hare.
24. **COMPARE** If the racetrack were circular, how would the tortoise's speed be different from its velocity?
25. **APPLY** How might a person use a floating stick to measure the speed at which a river flows?
26. **CONNECT** Describe a frame of reference other than the ground that you might use to measure motion. When would you use it?

Using Math Skills in Science

27. José skated 50 m in 10 s. What was his speed?
28. Use the information in the photograph below to calculate the speed of the ant as it moves down the branch.

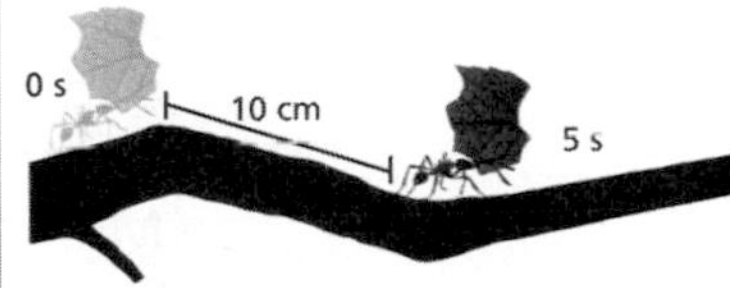

29. While riding her bicycle, Jamie accelerated from 7 m/s to 2 m/s in 5 s. What was her acceleration?

the BIG idea

30. **PREDICT** Look back at the picture at the beginning of the chapter on pages 6–7. Predict how the velocity of the roller coaster will change in the next moment.
31. **WRITE** A car is traveling east at 40 km/h. Use this information to predict where the car will be in one hour. Discuss the assumptions you made to reach your conclusion and the factors that might affect it.

UNIT PROJECTS

D 36 Unit: Motion and Forces

Rubrics

- Rubrics in Teacher Edition for all extended response questions
- Rubrics for all Unit Projects
- Alternative Assessment with rubric for each chapter
- A wide range of additional rubrics in the Science Toolkit

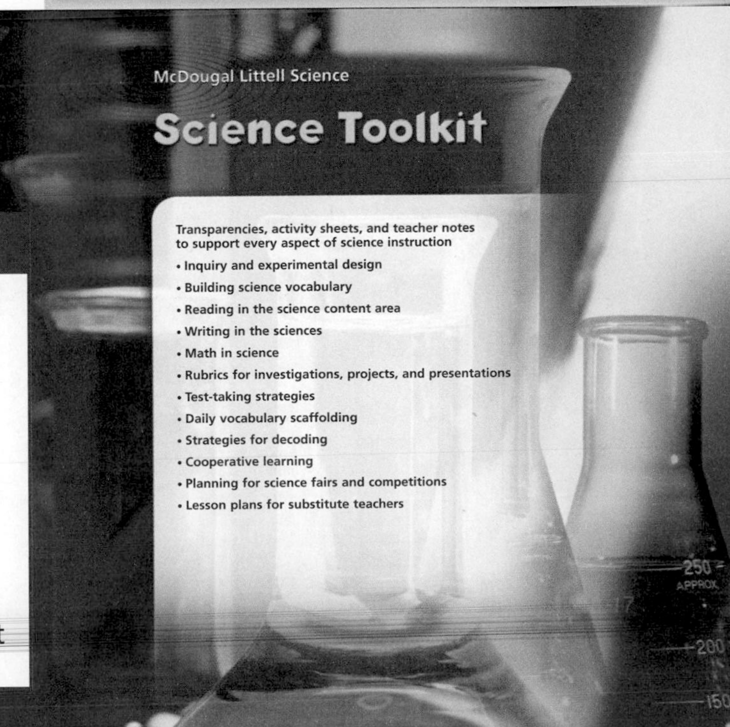

McDougal Littell Science
Science Toolkit
Transparencies, activity sheets, and teacher notes to support every aspect of science instruction
• Inquiry and experimental design
• Building science vocabulary
• Reading in the science content area
• Writing in the sciences
• Math in science
• Rubrics for investigations, projects, and presentations
• Test-taking strategies
• Daily vocabulary scaffolding
• Strategies for decoding
• Cooperative learning
• Planning for science fairs and competitions
• Lesson plans for substitute teachers

McDougal Littell Science Modular Series

McDougal Littell Science lets you choose the titles that match your curriculum. Each module in this flexible 15-book series takes an in-depth look at a specific area of life, earth, or physical science.

- Flexibility to match your curriculum
- Convenience of smaller books
- Complete Student Resource Handbooks in every module

Life Science Titles

A ▶ Cells and Heredity
1. The Cell
2. How Cells Function
3. Cell Division
4. Patterns of Heredity
5. DNA and Modern Genetics

B ▶ Life Over Time
1. The History of Life on Earth
2. Classification of Living Things
3. Population Dynamics

C ▶ Diversity of Living Things
1. Single-Celled Organisms and Viruses
2. Introduction to Multicellular Organisms
3. Plants
4. Invertebrate Animals
5. Vertebrate Animals

D ▶ Ecology
1. Ecosystems and Biomes
2. Interactions Within Ecosystems
3. Human Impact on Ecosystems

E ▶ Human Biology
1. Systems, Support, and Movement
2. Absorption, Digestion, and Exchange
3. Transport and Protection
4. Control and Reproduction
5. Growth, Development, and Health

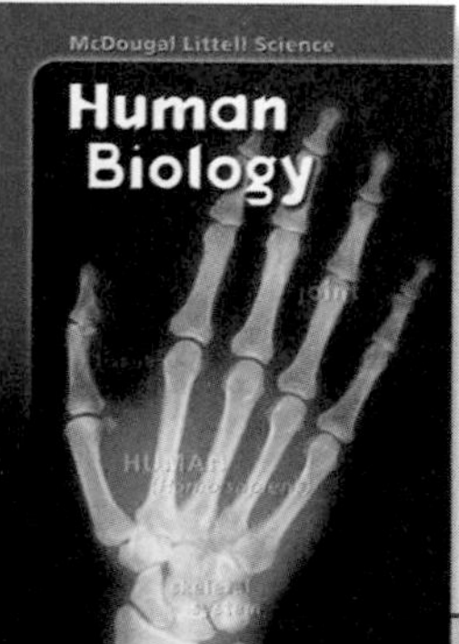

Earth Science Titles

A ▶ Earth's Surface
1. Views of Earth Today
2. Minerals
3. Rocks
4. Weathering and Soil Formation
5. Erosion and Deposition

B ▶ The Changing Earth
1. Plate Tectonics
2. Earthquakes
3. Mountains and Volcanoes
4. Views of Earth's Past
5. Natural Resources

C ▶ Earth's Waters
1. The Water Planet
2. Freshwater Resources
3. Ocean Systems
4. Ocean Environments

D ▶ Earth's Atmosphere
1. Earth's Changing Atmosphere
2. Weather Patterns
3. Weather Fronts and Storms
4. Climate and Climate Change

E ▶ Space Science
1. Exploring Space
2. Earth, Moon, and Sun
3. Our Solar System
4. Stars, Galaxies, and the Universe

Physical Science Titles

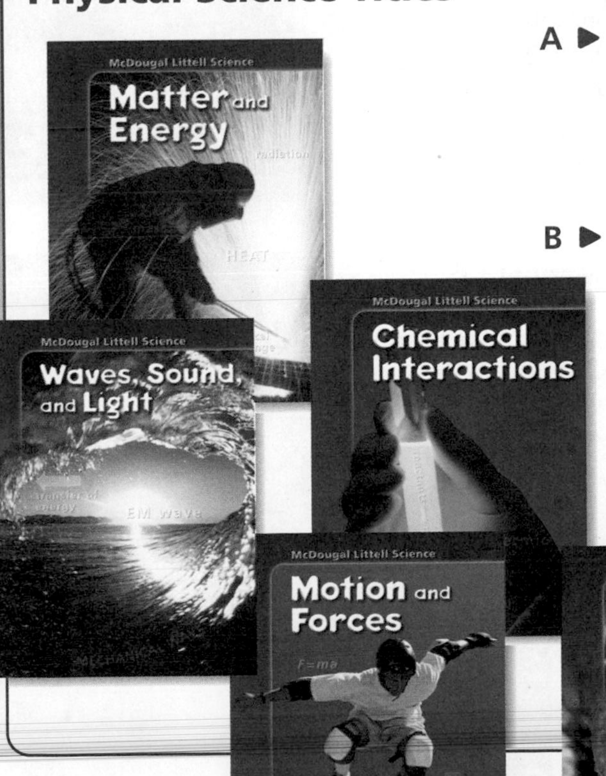

A ▶ Matter and Energy
1. Introduction to Matter
2. Properties of Matter
3. Energy
4. Temperature and Heat

B ▶ Chemical Interactions
1. Atomic Structure and the Periodic Table
2. Chemical Bonds and Compounds
3. Chemical Reactions
4. Solutions
5. Carbon in Life and Materials

C ▶ Motion and Forces
1. Motion
2. Forces
3. Gravity, Friction, and Pressure
4. Work and Energy
5. Machines

D ▶ Waves, Sound, and Light
1. Waves
2. Sound
3. Electromagnetic Waves
4. Light and Optics

E ▶ Electricity and Magnetism
1. Electricity
2. Circuits and Electronics
3. Magnetism

Teaching Resources

A wealth of print and technology resources help you adapt the program to your teaching style and to the specific needs of your students.

Book-Specific Print Resources

Unit Resource Book provides all of the teaching resources for the unit organized by chapter and section.

- Family Letters
- *Scientific American Frontiers* Video Guide
- Unit Projects
- Lesson Plans
- Reading Study Guides (Levels A and B)
- Spanish Reading Study Guides
- Challenge Readings
- Challenge and Extension Activities
- Reinforcing Key Concepts
- Vocabulary Practice
- Math Support and Practice
- Investigation Datasheets
- Chapter Investigations (Levels A, B, and C)
- Additional Investigations (Levels A, B, and C)
- Summarizing the Chapter

McDougal Littell Science

Unit
Resource Book

Motion and Forces

- Family Letters (English and Spanish)
- *Scientific American Frontiers* Video Guides
- Unit Projects (with Rubrics)
- Lesson Plans
- Reading Study Guides (Levels A and B and Spanish)
- Challenge Activities and Readings
- Reinforcing Key Concepts
- Vocabulary Practice and Decoding Support
- Math Support and Practice
- Investigation Datasheets
- Chapter Investigations (Levels A, B, and C)
- Additional Investigations (Levels A, B, and C)

Unit Assessment Book contains complete resources for assessing student knowledge and performance.

- Chapter Diagnostic Tests
- Section Quizzes
- Chapter Tests (Levels A, B, and C)
- Alternative Assessments
- Unit Tests (Levels A, B, and C)

McDougal Littell Science

Unit
Assessment Book

Motion and Forces

Unit Transparency Book includes instructional visuals for each chapter.

- Three-Minute Warm-Ups
- Note-Taking Models
- Daily Vocabulary Scaffolding
- Chapter Outlines
- Big Idea Flow Charts
- Chapter Teaching Visuals

Unit Lab Manual

Unit Note-Taking/Reading Study Guide

Program-Wide Print Resources

Process and Lab Skills

Problem Solving and Critical Thinking

Standardized Test Practice

Science Toolkit

City Science

Visual Glossary

Multi-Language Glossary

English Learners Package

***Scientific American Frontiers* Video Guide**

How Stuff Works Express
This quarterly magazine offers opportunities to explore current science topics.

Technology Resources

***Scientific American Frontiers* Video Program**
Each specially-tailored segment from this award-winning PBS series correlates to a unit; available on VHS and DVD

Audio CDs Complete chapter texts read in both English and Spanish

Lab Generator CD-ROM
A searchable database of all activities from the program plus additional labs for each unit; edit and print your own version of labs

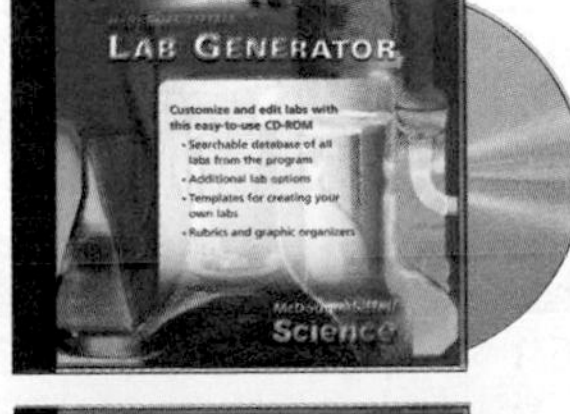

Test Generator CD-ROM

eEdition CD-ROM

EasyPlanner CD-ROM

Content Review CD-ROM

Power Presentations CD-ROM

Online Resources

- ClassZone.com
- Content Review Online
- eEdition Plus Online
- EasyPlanner Plus Online
- eTest Plus Online

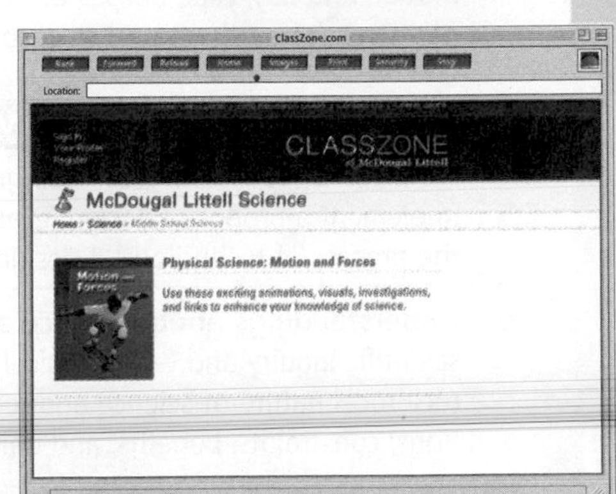

Correlation to National Science Education Standards

This chart provides an overview of how the five Physical Science modules of *McDougal Littell Science* address the National Science Education Standards.

A Matter and Energy
B Chemical Interactions
C Motion and Forces
D Waves, Sound, and Light
E Electricity and Magnetism

A. Science as Inquiry	Book, Chapter, and Section
A.1– A.8 **Abilities necessary to do scientific inquiry** Identify questions for investigation; design and conduct investigations; use evidence; think critically and logically; analyze alternative explanations; communicate; use mathematics.	All books (pp. R2–R44), All Chapter Investigations, All Think Science features
A.9 **Understandings about scientific inquiry** Different kinds of investigations for different questions; investigations guided by current scientific knowledge; importance of mathematics and technology for data gathering and analysis; importance of evidence, logical argument, principles, models, and theories; role of legitimate skepticism; scientific investigations lead to new investigations.	All books (pp. xxii–xxv) A3.1, B2.2, C4.2, D3.2, E3.1

B. Physical Science	Book, Chapter, and Section
B.1 **Properties and changes of properties in matter** Physical properties; substances, elements, and compounds; chemical reactions.	A1.1, A1.2, A1.3, A1.4, A2.1, A2.2, B1, B3.2, B4.1, B4.2, B4.3, B5.1, B5.3, C3.4
B.2 **Motions and forces** Position, speed, direction of motion; balanced and unbalanced forces.	C1.1, C1.2, C1.3, C2.1, C2.3, C3.1, C3.2, C3.3, C3.4, C4.1, E5.1
B.3 **Transfer of energy** Energy transfer; forms of energy; heat and light; electrical circuits; sun as source of Earth's energy.	A3.1, A3.2, A3.3, A4.1, A4.2, A4.3, B3.3, C4.2, D1.1, D3.3, D3.4, D4.1, D4.2, D4.3, E1.1, E1.2, E1.3, E2.1, E2.2

C. Life Science	Book, Chapter, and Section
C.1 **Structure and function in living systems** Systems; structure and function; levels of organization; cells and cell activities; specialization; human body systems; disease.	B1.1 (Connecting Sciences), B5.2, C5.2 (Connecting Sciences)

D. Earth and Space Science	Book, Chapter, and Section
D.1 **Earth's changing atmosphere**	B4.2 (Connecting Sciences)
D.3 **Earth in the solar system** Sun, planets, asteroids, comets; regular and predictable motion and day, year, phases of the moon, and eclipses; gravity and orbits; sun as source of energy for earth; cause of seasons.	A3.2, A3.3, C3.1, D3.3

E. Science and Technology	Book, Chapter, and Section
E.1– E.5 **Abilities of technological design** Identify problems; design a solution or product; implement a proposed design; evaluate completed designs or products; communicate the process of technological design.	A2.3, A3.1, A4.3, B (p. 5), B4.2, C1.2, C2.1, C3.2, C5.3, D2.4, D3.1, D3.3, D4.4, E2.3
E.6 **Understandings about science and technology** Similarities and differences between scientific inquiry and technological design; contributions of people in different cultures; reciprocal nature of science and technology; nonexistence of perfectly designed solutions; constraints, benefits, and unintended consequences of technological designs.	All books (pp. xxvi–xxvii) All books (Frontiers in Science, Timelines in Science) A.1.2, A3.3, B3.4, B3.1, B3.3, C5.3, D4.4, E1.2, E1.3

F. Science in Personal and Social Perspectives		Book, Chapter, and Section
F.1	**Personal health** Exercise; fitness; hazards and safety; tobacco, alcohol, and other drugs; nutrition; STDs; environmental health	B4.2, B5.2
F.2	**Populations, resources, and environments** Overpopulation and resource depletion; environmental degradation.	A3.1
F.3	**Natural hazards** Earthquakes, landslides, wildfires, volcanic eruptions, floods, storms; hazards from human activity; personal and societal challenges.	B3.2, D3.2, E1.2
F.4	**Risks and benefits** Risk analysis; natural, chemical, biological, social, and personal hazards; decisions based on risks and benefits.	B4.3
F.5	**Science and technology in society** Science's influence on knowledge and world view; societal challenges and scientific research; technological influences on society; contributions from people of different cultures and times; work of scientists and engineers; ethical codes; limitations of science and technology.	All books (Timelines in Science) A1.2, A3.2, A3.3, B3.4, B4.4, C5.3, D2, D3.2, D3.3, D4.4, E1.2, E1.3

G. History and Nature of Science		Book, Chapter, and Section
G.1	**Science as a human endeavor** Diversity of people w.orking in science, technology, and related fields; abilities required by science	All books (pp. xxii–xxv; Frontiers in Science)
G.2	**Nature of science** Observations, experiments, and models; tentative nature of scientific ideas; differences in interpretation of evidence; evaluation of results of investigations, experiments, observations, theoretical models, and explanations; importance of questioning, response to criticism, and communication.	B1.2, B2.1, B2.3, E3.2
G.3	**History of science** Historical examples of inquiry and relationships between science and society; scientists and engineers as valued contributors to culture; challenges of breaking through accepted ideas.	All books (Frontiers in Science; Timelines in Science) B1.2, B3.2, C2.1, D2.4

Correlations to Benchmarks

This chart provides an overview of how the five Physical Science modules of *McDougal Littell Science* address the National Science Education Standards.

A Matter and Energy
B Chemical Interactions
C Motion and Forces
D Waves, Sound, and Light
E Electricity and Magnetism

1. The Nature of Science	Book, Chapter, and Section
	The Nature of Science (pp. xxii–xxv); E2.3; Think Science Features: A3.1, B2.2, C2.1, C4.2, D3.2, E3.1; Scientific Thinking Handbook (pp. R2–R9); Lab Handbook (pp. R10–R35)

3. The Nature of Technology	Book, Chapter, and Section
	The Nature of Technology (pp. xxvi–xxvii); A3.3, B4.4, D4.4, E1, E2.3, E3.2, E3.3, E3.4; Timelines in Science Features

4. The Physical Setting	Book, Chapter, and Section
4.B THE EARTH	A3.1, A4.3, C3.1
4.D STRUCTURE OF MATTER	
4.D.1 All matter is made of atoms; atoms of any element are alike but different from atoms of other elements; different arrangements of atoms into groups compose all substances.	A1.2, A1.3, B1.1, B2.1, B2.2
4.D.2 Equal volumes of different substances usually have different weights.	A2.1, A2.3
4.D.3 Atoms and molecules are perpetually in motion; increased temperature means greater average energy of motion; states of matter: solids, liquids, gases.	A1.2, A4.1
4.D.4 Temperature and acidity of a solution influence reaction rates. Many substances dissolve in water, which may facilitate reactions between them.	B3.1, B4.2, B4.3
4.D.5 Greek philosopheres' scientific ideas about elements; most elements tend to combine with others, so few elements are found in their pure form.	B1.1
4.D.6 Groups of elements have similar properties; oxidation; some elements, like carbon and hydrogen, don't fit into any category and are essential elements of living matter.	B1.2, B1.3, B3.1, B5.1, B5.2
4.D.7 Conservation of matter: the total weight of a closed system remains the same because the total number of atoms stays the same regardless of how they interact with one another.	B3.2
4.E ENERGY TRANSFORMATIONS	
4.E.1 Energy cannot be created or destroyed, but only changed from one form into another.	A3.2, C4.2
4.E.2 Most of what goes on in the universe involves energy transformations.	A3, A4.2, A4.3
4.E.3 Heat can be transferred through materials by the collisions of atoms or across space by radiation; convection currents transfer heat in fluid materials.	A4.2, A4.3
4.E.4 Energy appears in many different forms, including heat energy, chemical energy, mechanical energy, and gravitational energy.	A3.1, A4.2, A4.3, C4.2
4.F MOTION	
4.F.1 Light from the Sun is made up of many different colors of light; objects that give off or reflect light have a different mix of colors.	D3.3, D3.4
4.F.2 Something can be "seen" when light waves emitted or reflected by it enter the eye.	D4.1, D4.3

4.F.3 An unbalanced force acting on an object changes its speed or direction of motion, or both. If the force acts toward a single center, the object's path may curve into an orbit around the center.	C2.1, C2.2, C3.1
4.F.4 Vibrations in materials set up wavelike disturbances (such as sound) that spread away from the source; waves move at different speeds in different materials.	D1, D2.1, D2.2, D3.1, D3.4
4.F.5 Human eyes respond to only a narrow range of wavelengths of electromagnetic radiation—visible light. Differences of wavelengths within that range are perceived as differences in color.	D3.2, D3.4, D4.3
4.G FORCES OF NATURE	
4.G.1 Objects exerts gravitational forces on one another, but these forces depend on the mass and distance of objects, and may be too small to detect.	C3.1
4.G.2 The Sun's gravitational pull holds Earth and other planets in their orbits; planets' gravitational pull keeps their moons in orbit around them.	C3.1
4.G.3 Electric currents and magnets can exert a force on each other.	E3.1, E3.2, E3.3
5. The Living Environment	**Book, Chapter, and Section**
5.E Flow of Matter and Energy	B5.2
8. The Designed World	**A2.1, A2.3, A3.3, B3.4, B4.4, B5.3, C5.3, E2.3, E3.2**
9. The Mathematical World	**All Math in Science Features, E2.3**
10. Historical Perspectives	**B1, B2, B3.2, C1.1, C2, D4.4**
12. Habits of Mind	**Book, Chapter, and Section**
12.A VALUES AND ATTITUDES	Think Science Features: A3.1, B2.2, C2.1, C4.2, D3.2, E3.1
12.B Computation and Estimation	All Math in Science Features, Lab Handbook (pp. R10–R35)
12.C Manipulation and Observation	All Investigates and Chapter Investigations
12.D Communication Skills	All Chapter Investigations, Lab Handbook (pp. R10–R35)
12.E Critical-Response Skills	Think Science Features: A3.1, B2.2, C2.1, C4.2, D3.2, E3.1; Scientific Thinking Handbook (pp. R2–R9)

Planning the Unit

The Pacing Guide provides suggested pacing for all chapters in the unit as well as the two unit features shown below.

Frontiers in Science

- Features cutting-edge research as an engaging point of entry into the unit
- Connects to an accompanying *Scientific American Frontiers* video and viewing guide
- Introduces three options for unit projects.

VIDEO SUMMARY

SCIENTIFIC AMERICAN FRONTIERS

TEETERING TO VICTORY This 10-minute video documents an exciting annual contest held at MIT—Massachusetts Institute of Technology—in which students build machines using principles of force and motion. Their devices compete against an opponent's machine to tilt their end of a teeter-totter beam—and they have 45 seconds to defeat that opponent's machine. Students design their own technology with very specific constraints: they are given identical kits of parts, each machine can weigh no more than 10 pounds, and it must fit back into the box the parts came in. The top contenders are carpet grabbers, bulldozers, and mobile jacks. As different competitors eliminate each other, viewers get caught up in the contest while learning about technological design.

National Science Education Standards

A.9.a–d Understandings About Scientific Inquiry
E.6.a–f Understandings About Science and Technology
F.5.a–e Science and Technology in Society
G.1.a–b Science as a Human Endeavor
G.2.a Nature of Science

FRONTIERS in Science

ROBOTS on Mars

If you could design a robot to explore Mars, what would you want it to be able to do?

SCIENTIFIC AMERICAN FRONTIERS

View the video "Teetering to Victory" to learn about a competition that challenges students to use their knowledge of motion and forces to design a machine.

NASA's Mars Exploration Rover (MER) shown in a computer-simulated Martian landscape

The Design Challenge

It's still not possible to send scientists to M... of water, but in 1999 a team of scientists a... design two robots for NASA's 2004 mission... worked, they relied on their scientific und... forces, and machines to create and test a s...

To identify their goals, the team starte... what scientists would want to do if they c... they would want to look around the land... areas to study. Then they would need to t... and collect rock samples for analysis. Final... variety of tools to analyze the rocks, inter... communicate their findings back to Earth... basic plan for the Mars Exploration Rovers (MERs).

As you can see in the photograph, the MER team designed a rover with cameras for viewing the surface, wheels for moving around the landscape, and an extendable arm in front equipped with tools for collecting and studying rock samples. The rover also has a computer to process information, a radio antenna to communicate with Earth, and batteries and solar panels to provide energy for everything.

As in any technology project, the MER team had to work within specific constraints, or limits. The most basic constraints were time and money. They had to design rovers that could be built within NASA's budget and that would be ready in time for launch in 2003. But the team also faced some more challenging

Motion and Forces — Frontiers in Science 3

develop robots. *Did life ever exist on Mars? Did Mars ever have water?*

Teach from Visuals

Point out parts of the MER robot in the picture. The camera holder on top can rotate 360° and is mounted on a mast that can extend 5 feet high. The jointed arm in front maneuvers the RAT, the Rock Abrasion Tool that drills into rock samples. Two antennae are visible: a long one parallel to the mast, and the disc-shaped one. Point out the jointed legs and the size of the wheels (10 inches in diameter).

Technology Design

Focus students' attention on "The Design Challenge." Elicit from them, and write on the board, a list of the tasks the technology needs to accomplish. Next to that, ask what type of technology was designed to meet that need. *Examples: take pictures—camera; energy source—batteries, solar panels*

ADDITIONAL RESOURCES

Technology Resources

Scientific American Frontiers Video: *Teetering to Victory:* 10-minute video segment that introduces the unit

ClassZone.com
CAREER LINK: physicists, mechanical engineer

Guide student viewing and comprehension of the video:

Frontiers in Science Teaching Guide, pp. 1–2; Viewing Guide, p. 3, Video Wrap-Up, p. 4

Scientific American Frontiers Video Guide, pp. 9–12

Unit projects procedures and rubrics:

Unit Projects, pp. 5–10

DIFFERENTIATE INSTRUCTION

More Reading Support

A What is the purpose of the robots? *to explore Mars*

B What does this paragraph describe? *the MER robot that will go to Mars*

Advanced Have students research the MER project on the NASA Web site or in periodicals and draw a diagram of a MER's parts. Add labels to include additional information about the robot. As the MER begins to explore Mars, have students keep the class informed of its progress as reported on NASA's Web site and in the news.

D 2 Unit: Motion and Forces — Frontiers in Science 3 D

Timelines in Science

- Traces the history of key scientific discoveries
- Highlights interactions between science and technology.

TIMELINES in Science

FOCUS

Set Learning Goals

Students will

- Compare ancient and modern ideas of force and motion.
- Observe how progress in science aided progress in technology and vice versa.
- Examine the development of technological design and its limits.

National Science Education Standards

A.9.a–g Understandings About Scientific Inquiry
E.6.a–c Understandings About Science and Technology
F.5.a–e, F.5.g Science and Technology in Society
G.1.a–b Science as a Human Endeavor
G.2.a Nature of Science
G.3.a–c History of Science

INSTRUCT

The timeline shows major developments ... civilizations, are still used today.

TIMELINES in Science

UNDERSTANDING FORCES

In ancient times, people thought that an object would not move unless it was pushed. Scientists came up with ingenious ways to explain how objects like arrows stayed in motion. Over time, they came to understand that all motion could be described by three basic laws. Modern achievements such as suspension bridges and space exploration are possible because of the experiments with motion and forces performed by scientists and philosophers over hundreds of years.

This timeline shows just a few of the many steps on the path toward understanding forces. Notice how scientists used the observations and ideas of previous thinkers as a springboard for developing new theories. The boxes below the timeline show how technology has led to new insights and to applications of those ideas.

350 B.C. *Aristotle Discusses Motion*
The Greek philosopher Aristotle states that the natural condition of an object is to be at rest. A force is necessary to keep the object in motion. The greater the force, the faster the object moves.

250 B.C. *Levers and Buoyancy Explained*
The Greek inventor Archimedes uses a mathematical equation to explain how a small weight can balance a much larger weight near a lever's fulcrum. He also explains buoyancy, which provides a way of measuring volume.

A.D. 1121 *Force Acting on Objects Described*
Persian astronomer al-Khazini asserts that a force acts on all objects to pull them toward the center of Earth. This force varies, he says, depending on whether the object moves through air, water, or another medium. His careful notes and drawings illustrate these principles.

1150 *Perpetual-Motion Machine Described*
Indian mathematician and physicist Bhaskara describes a wheel that uses closed containers of liquid to turn forever without stopping. If it worked, his idea would promise an unending source of power that does not rely on an external source.

EVENTS

400 B.C. 350 B.C. 300 B.C. 250 B.C. A.D. 1100 1150 1200

APPLICATIONS AND TECHNOLOGY

TECHNOLOGY
Catapulting into History
As early as 400 B.C., armies were using objects in motion to do work. Catapults, or machines for hurling stones and spears, were used as military weapons. Five hundred years later, the Roman army used catapults mounted on wheels. In the Middle Ages, young trees were sometimes bent back, loaded with an object, and then released like a large slingshot. Today catapult technology is used to launch airplanes from aircraft carriers. A piston powered by steam propels the plane along the deck of the aircraft carrier until it reaches takeoff speed.

APPLICATION
The First Steam-Powered Engine
In the first century A.D., Hero of Alexandria, a Greek inventor, created the first known steam engine, called the aeolipile. It was a hollow ball with two cylinders jutting out in opposite directions. The ball was suspended above a kettle that was filled with water and placed over a fire. As the water boiled, steam caused the ball to spin. The Greeks never used this device for work. In 1690, Sir Isaac Newton formulated the principle of the aeolipile in scientific terms in his third law of motion. A steam engine designed for work was built in 1698. The aeolipile is the earliest version of steam-powered pumps, steam locomotives, jet engines, and rockets.

108 — Timelines in Science 109

DIFFERENTIATE INSTRUCTION

Below Level To give students a better idea of the amount of time that passed between discoveries, draw a long line across the blackboard. Label the left end 350 B.C., the middle A.D. 750, and the right end 2000. Mark off other year divisions. As you discuss the timeline with students, point to the section of the blackboard timeline in which that event took place.

Advanced Encourage students to trace the development of ideas from Aristotle to Leonardo to Newton to Einstein. Students might create a visual that represents each new idea as "building on" or "knocking down" the previous idea.

Scientific Process

Stress to students that we should not think of the science of the past as being wrong or of the people of the past as being less intelligent. The ideas they formed were based on the observations available to them at that time. Throughout history, people have tried to explain how the natural world worked; their explanations were simply different from the ones we have today. As important as any discovery was the development of the scientific methods of thinking.

Social Studies Connection

A.D. 800 TO 1300 Present-day Iran was once the heart of the Persian Empire. Have students find Iran on a world map. Point out that al-Khazini was an astronomer and physician who lived in the Persian Empire. Emphasize that from about the 800s to the 1300s, some of the world's most innovative science was being done in this region. Ask how geographical features might have helped this civilization advance. *(Ideas from Greece, Egypt, China, and India spread throughout the Persian Empire via sea, river, and overland trade routes.)*

Application

AEOLIPILE Help students make a connection between the simple device shown on page 109 and Watt's steam engine on page 110—and with modern-day steam locomotives and rockets. The aeolipile (EE-uh-lih-PYL) showed the effects produced by steam under pressure. A steam engine converts thermal energy from steam into mechanical energy by allowing the steam to expand and cool. A rocket engine burns fuel, which comes out of the rocket at high speed, creating the thrust needed to propel the rocket.

D 108 Unit: Motion and Forces — Timelines in Science 109 D

Motion and Forces Pacing Guide

The following pacing guide shows how the chapters in ***Motion and Forces*** can be adapted to fit your specific course needs.

	TRADITIONAL SCHEDULE (DAYS)	BLOCK SCHEDULE (DAYS)
Frontiers in Science: Robots on Mars	1	0.5
Chapter 1 Motion		
1.1 An object in motion changes position.	2	1
1.2 Speed measures how fast positions change.	2	1
1.3 Acceleration measures how fast velocity changes.	3	1.5
Chapter Investigation	1	0.5
Chapter 2 Forces		
2.1 Forces change motion.	2	1
2.2 Force and mass determine acceleration.	2	1
2.3 Forces act in pairs.	2	1
2.4 Forces transfer momentum.	3	1.5
Chapter Investigation	1	0.5
Chapter 3 Gravity, Friction, and Pressure		
3.1 Gravity is a force exerted by masses.	2	1
3.2 Friction is a force that opposes motion.	2	1
3.3 Pressure depends on force and area.	2	1
3.4 Fluids can exert a force on objects.	3	1.5
Chapter Investigation	1	0.5
Timelines in Science: Understanding Forces	1	0.5
Chapter 4 Work and Energy		
4.1 Work is the use of force to move an object.	2	1
4.2 Energy is transferred when work is done.	2	1
4.3 Power is the rate at which work is done.	3	1.5
Chapter Investigation	1	0.5
Chapter 5 Machines		
5.1 Machines help people do work.	2	1
5.2 Six simple machines have many uses.	2	1
5.3 Modern technology uses compound machines.	3	1.5
Chapter Investigation	1	0.5
Total Days for Module	**46**	**23**

Planning the Chapter

Complete planning support precedes each chapter.

Previewing Content

- Section-by-section science background notes
- Common Misconceptions notes

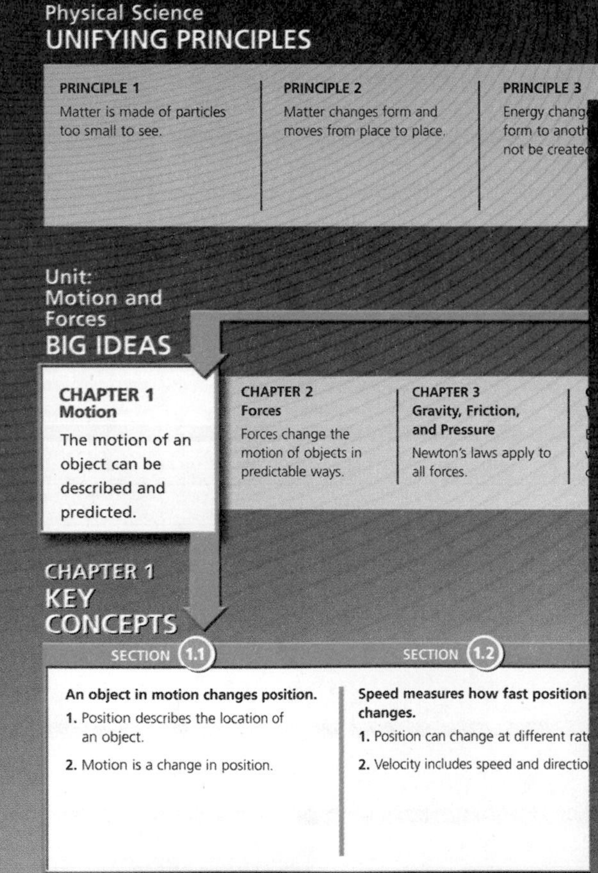
CHAPTER 1 **Motion**

Physical Science
UNIFYING PRINCIPLES

PRINCIPLE 1	PRINCIPLE 2	PRINCIPLE 3	PRINCIPLE 4
Matter is made of particles too small to see.	Matter changes form and moves from place to place.	Energy chang… form to anoth… not be create…	

Unit: Motion and Forces
BIG IDEAS

CHAPTER 1 Motion
The motion of an object can be described and predicted.

CHAPTER 2 Forces
Forces change the motion of objects in predictable ways.

CHAPTER 3 Gravity, Friction, and Pressure
Newton's laws apply to all forces.

CHAPTER 1
KEY CONCEPTS

SECTION 1.1
An object in motion changes position.
1. Position describes the location of an object.
2. Motion is a change in position.

SECTION 1.2
Speed measures how fast position changes.
1. Position can change at different rat…
2. Velocity includes speed and directio…

The Big Idea Flow Chart is available on p. T1 in the **UNIT TRANSPARENCY BOO…**

D 5A Unit: **Motion and Forces**

Previewing Content

SECTION

An object in motion changes position. pp. 9–15

1. Position describes the location of an object.
The **position,** or location, of an object is described relative to a **reference point.** The choice of reference point affects how the position is described. For example, a city can be located by measuring its direction and distance from another city, or by using a grid system such as the longitude-latitude system…

SECTION 1.2
1. Positi…
Speed…
partic…
Speed…

Previewing Content

SECTION

Acceleration measures how fast velocity changes. pp. 25–33

1. Speed and direction can change with time.
Acceleration is the rate at which velocity changes with time. Contrary to a popular misconception, acceleration is not limited to increases in velocity but includes *any* change in velocity. The following are examples of acceleration:

- speed increases
- speed decreases
- direction changes (regardless of speed)

2. Acceleration can be calculated from velocity and time.
You determine acceleration from the change in velocity and how long the change took. The formula for calculating acceleration is shown below.

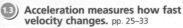

$$a = \frac{v_{final} - v_{initial}}{t}$$

Negative acceleration is a decrease in velocity during a specific period of time. The acceleration formula yields a negative result when the final velocity is less than the initial velocity.
A velocity-time graph shows how both velocity and acceleration change with time. The graph on the right below is a velocity-time graph; compare it with the distance-time graph for the same set of data, which is shown below. The graphs show a boy (1) starting and speeding up on his scooter, (2) coasting (velocity is constant), and (3) slowing to a stop.

Distance-Time Graph

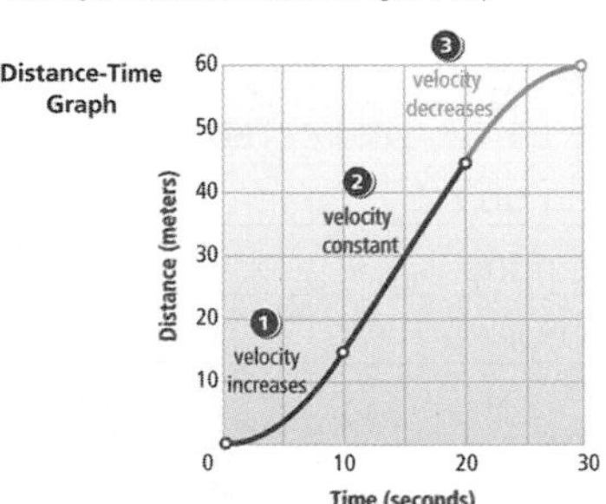

Velocity-Time Graph

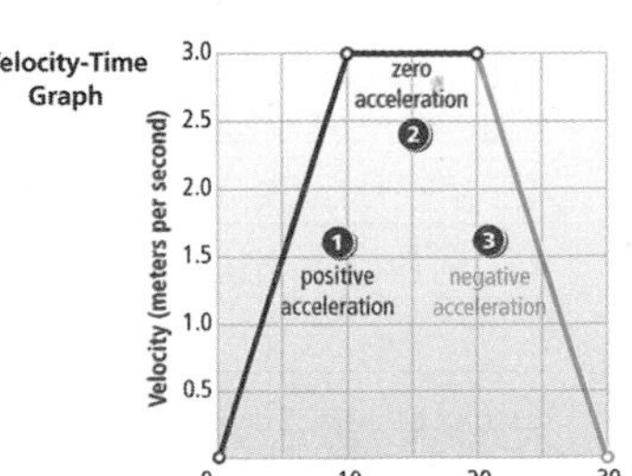

Common Misconceptions

MISCONCEPTION DATABASE
CLASSZONE.COM Background on student misconceptions

ACCELERATION EQUATED WITH SPEEDING UP Acceleration describes *any* change in velocity. However, students may hold the misconception that acceleration means only an increase in velocity.

TE This misconception is addressed on p. 26.

NEGATIVE ACCELERATION The change in velocity that occurs during a slowing down is described by the term *negative acceleration.* Students may confuse negative acceleration with negative direction, which involves reversing direction.

TE This misconception is addressed on p. 29.

D 5C Unit: **Motion and Forces**

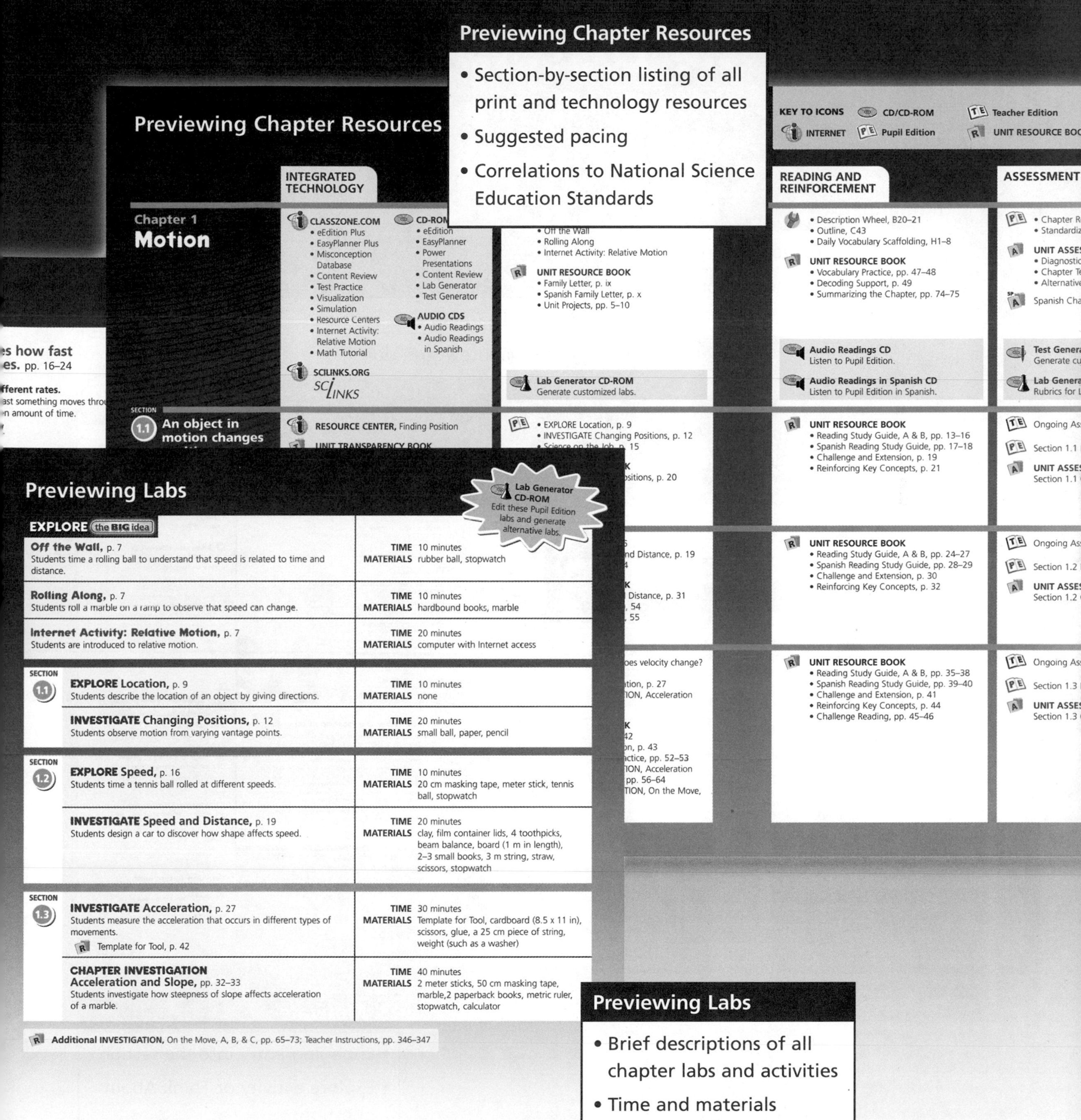

Previewing Chapter Resources

- Section-by-section listing of all print and technology resources
- Suggested pacing
- Correlations to National Science Education Standards

Previewing Chapter Resources

KEY TO ICONS: CD/CD-ROM · Teacher Edition · INTERNET · Pupil Edition · UNIT RESOURCE BOO

Chapter 1
Motion

INTEGRATED TECHNOLOGY		READING AND REINFORCEMENT	ASSESSMENT
CLASSZONE.COM • eEdition Plus • EasyPlanner Plus • Misconception Database • Content Review • Test Practice • Visualization • Simulation • Resource Centers • Internet Activity: Relative Motion • Math Tutorial CD-ROM • eEdition • EasyPlanner • Power Presentations • Content Review • Lab Generator • Test Generator AUDIO CDS • Audio Readings • Audio Readings in Spanish SCILINKS.ORG	• Off the Wall • Rolling Along • Internet Activity: Relative Motion UNIT RESOURCE BOOK • Family Letter, p. ix • Spanish Family Letter, p. x • Unit Projects, pp. 5–10 Lab Generator CD-ROM Generate customized labs.	• Description Wheel, B20–21 • Outline, C43 • Daily Vocabulary Scaffolding, H1–8 UNIT RESOURCE BOOK • Vocabulary Practice, pp. 47–48 • Decoding Support, p. 49 • Summarizing the Chapter, pp. 74–75 Audio Readings CD Listen to Pupil Edition. Audio Readings in Spanish CD Listen to Pupil Edition in Spanish.	• Chapter Re • Standardiz UNIT ASSES • Diagnostic • Chapter Te • Alternative Spanish Cha Test Genera Generate cus Lab Genera Rubrics for L
SECTION 1.1 An object in motion changes RESOURCE CENTER, Finding Position UNIT TRANSPARENCY BOOK	• EXPLORE Location, p. 9 • INVESTIGATE Changing Positions, p. 12 • Science on the Job, p. 15 … K … sitions, p. 20	UNIT RESOURCE BOOK • Reading Study Guide, A & B, pp. 13–16 • Spanish Reading Study Guide, pp. 17–18 • Challenge and Extension, p. 19 • Reinforcing Key Concepts, p. 21	Ongoing Ass Section 1.1 R UNIT ASSES Section 1.1 C
	…nd Distance, p. 19 … K … Distance, p. 31 … 54 … 55	UNIT RESOURCE BOOK • Reading Study Guide, A & B, pp. 24–27 • Spanish Reading Study Guide, pp. 28–29 • Challenge and Extension, p. 30 • Reinforcing Key Concepts, p. 32	Ongoing Ass Section 1.2 R UNIT ASSES Section 1.2 C
	…oes velocity change? …tion, p. 27 …ION, Acceleration … K … 42 …n, p. 43 …ctice, pp. 52–53 …ION, Acceleration … pp. 56–64 …TION, On the Move,	UNIT RESOURCE BOOK • Reading Study Guide, A & B, pp. 35–38 • Spanish Reading Study Guide, pp. 39–40 • Challenge and Extension, p. 41 • Reinforcing Key Concepts, p. 44 • Challenge Reading, pp. 45–46	Ongoing Ass Section 1.3 R UNIT ASSES Section 1.3 C

…s how fast …es. pp. 16–24
…fferent rates.
…ast something moves thro… …n amount of time.

Previewing Labs

Lab Generator CD-ROM
Edit these Pupil Edition labs and generate alternative labs.

EXPLORE the BIG idea

Activity	Time and Materials
Off the Wall, p. 7 Students time a rolling ball to understand that speed is related to time and distance.	**TIME** 10 minutes **MATERIALS** rubber ball, stopwatch
Rolling Along, p. 7 Students roll a marble on a ramp to observe that speed can change.	**TIME** 10 minutes **MATERIALS** hardbound books, marble
Internet Activity: Relative Motion, p. 7 Students are introduced to relative motion.	**TIME** 20 minutes **MATERIALS** computer with Internet access
SECTION 1.1 **EXPLORE Location,** p. 9 Students describe the location of an object by giving directions.	**TIME** 10 minutes **MATERIALS** none
INVESTIGATE Changing Positions, p. 12 Students observe motion from varying vantage points.	**TIME** 20 minutes **MATERIALS** small ball, paper, pencil
SECTION 1.2 **EXPLORE Speed,** p. 16 Students time a tennis ball rolled at different speeds.	**TIME** 10 minutes **MATERIALS** 20 cm masking tape, meter stick, tennis ball, stopwatch
INVESTIGATE Speed and Distance, p. 19 Students design a car to discover how shape affects speed.	**TIME** 20 minutes **MATERIALS** clay, film container lids, 4 toothpicks, beam balance, board (1 m in length), 2–3 small books, 3 m string, straw, scissors, stopwatch
SECTION 1.3 **INVESTIGATE Acceleration,** p. 27 Students measure the acceleration that occurs in different types of movements. Template for Tool, p. 42	**TIME** 30 minutes **MATERIALS** Template for Tool, cardboard (8.5 x 11 in), scissors, glue, a 25 cm piece of string, weight (such as a washer)
CHAPTER INVESTIGATION Acceleration and Slope, pp. 32–33 Students investigate how steepness of slope affects acceleration of a marble.	**TIME** 40 minutes **MATERIALS** 2 meter sticks, 50 cm masking tape, marble,2 paperback books, metric ruler, stopwatch, calculator

Additional INVESTIGATION, On the Move, A, B, & C, pp. 65–73; Teacher Instructions, pp. 346–347

Chapter 1: Motion 5D D

Previewing Labs

- Brief descriptions of all chapter labs and activities
- Time and materials required for each activity

Planning the Lesson

Point-of-use support for each lesson provides a wealth of teaching options.

1. Prepare

- Concept and vocabulary review
- Note-taking and vocabulary strategies

2. Focus

- Set Learning Goals
- 3-Minute Warm-up

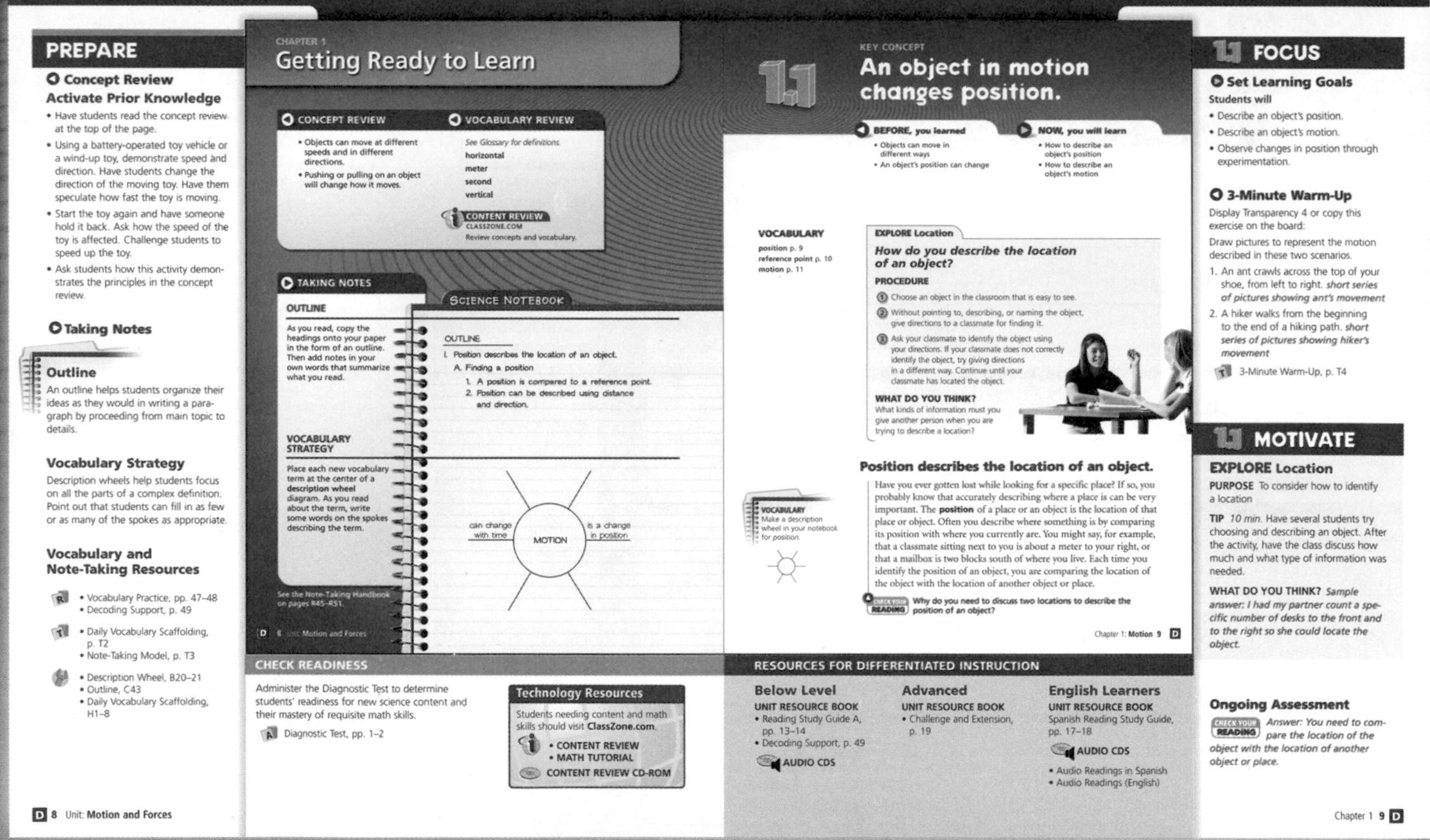

3. Motivate

- Engaging entry into the section
- Explore activity or Think About question

1.1 INSTRUCT

Teach from Visuals
To point out different ways to describe a location, use the "Describing Position" visual and ask:
- What does the map on the left show? *using Brasilia as a reference point for finding Santiago*
- What does the map on the right show? *using latitude and longitude, a grid system, to locate Santiago*

Teacher Demo
Make a tube out of graph paper and tape it so it will stay closed. Draw a point on the tube. Ask students how they could use the graph paper squares to locate the point. List the different suggestions on the board, and discuss the similarities and differences in the methods. Point out how two reference lines, one vertical and one horizontal, are needed to describe the location. Discuss how longitude and latitude are like the lines on the tube.

Ongoing Assessment
Describe an object's position.
Ask: If you describe positions on a map by comparing them with the position of landmark Q, what is Q called? *a reference point*

READING VISUALS *Answer: The left-hand map uses a distance and a direction from Brasilia to describe Santiago's location. The right-hand map uses the number of degrees from the equator and the prime meridian to describe Santiago's location.*

Describing a Position

RESOURCE CENTER CLASSZONE.COM Learn more about how people find and describe position.

You might describe the position of a city based on the location of another city. A location to which you compare other locations is called a **reference point.** You can describe where Santiago, Chile, is from the reference point of the city Brasilia, Brazil, by saying that Santiago is about 3000 kilometers (1860 mi) southwest of Brasilia.

You can also describe a position using a method that is similar to describing where a point on a graph is located. For example, in the longitude and latitude system, locations are given by two numbers—longitude and latitude. Longitude describes how many degrees east or west a location is from the prime meridian, an imaginary line running north-south through Greenwich, England. Latitude describes how many degrees north or south a location is from the equator, the imaginary circle that divides the northern and southern hemispheres. Having a standard way of describing location, such as longitude and latitude, makes it easier for people to compare locations.

Describing Position
There are several different ways to describe a position. The way you choose may depend on your reference point.

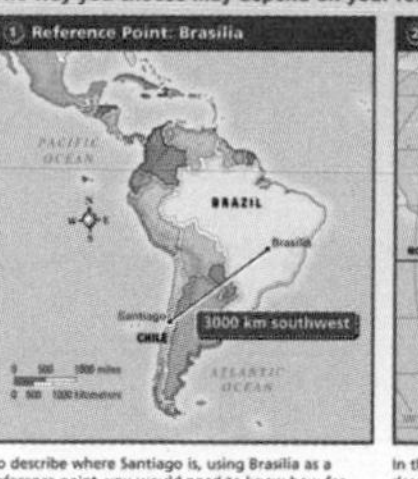

To describe where Santiago is, using Brasilia as a reference point, you would need to know how far Santiago is from Brasilia and in what direction it is.

In the longitude and latitude system, a location is described by how many degrees north or south it is from the equator and how many degrees east or west it is from the prime meridian.

READING VISUALS Compare and contrast the two ways of describing the location of Santiago as shown here.

Measuring Distance

If you were to travel from Brasilia to Santiago, you would end up about 3000 kilometers from where you started. The actual distance you traveled, however, would depend on the exact path you took. If you took a route that had many curves, the distance you traveled would be greater than 3000 kilometers.

The way you measure distance depends on the information you want. Sometimes you want to know the straight-line distance between two positions. Sometimes, however, you might need to know the total length of a certain path between those positions. During a hike, you are probably more interested in how far you have walked than in how far you are from your starting point.

COMPARE How does the distance each person has walked compare with the distance each is from the start of the maze?

When measuring either the straight-line distance between two points or the length of a path between those points, scientists use a standard unit of measurement. The standard unit of length is the meter (m), which is 3.3 feet. Longer distances can be measured in kilometers (km), and shorter distances in centimeters (cm).

Motion is a change in position.

The illustration below shows an athlete at several positions during a long jump. If you were to watch her jump, you would see that she is in motion. **Motion** is the change of position over time. As she jumps, both her horizontal and vertical positions change. If you missed the motion of the jump, you would still know that motion occurred because of the athlete between her starting and ending positions. A change in position is evidence that motion happened.

REMINDER Horizontal and vertical describe directions, as shown.
vertical
horizontal

starting position
ending position

History of Science
In eighteenth-century France, some 2000 different units of measurement were used across the country because different locales had their own systems. The French Revolution (1789–1799) set the stage for a new, universal system of measurement. A commission decided on a basic unit of measurement, the meter. They designed a decimal system that uses base names (such as *meter* and *gram*) and affixes standard prefixes to indicate fractional and multiple units (such as *milligram* and *kilogram*). In 1795 the French government adopted the new metric system.

Teach from Visuals
Direct students' attention to the time-exposure illustration at the bottom of the page, and then ask:
- How does the picture show time passing? *It shows the jumper in a number of different positions between the start and finish of her long jump.*
- How is the picture different from a snapshot? *A snapshot freezes a single moment in time.*

Teaching with Technology
If students have probeware, encourage them to use a motion sensor as they push various classroom objects across a flat, smooth surface.

DIFFERENTIATE INSTRUCTION

More Reading Support
A What would you call a location with which other locations can be com-

English Learners English learners may need help understanding sentences that imply *if/then* construction, such as *If you were to travel from Brasilia to Santiago, you would end up about 3000 kilometers from where you started* (p. 11). This sentence begins with *if* but does not contain *then*, and therefore ... reader to infer the cause-and-effect relationship. Point ... examples of this construction and make sure English ... recognize that *then* is implied.

DIFFERENTIATE INSTRUCTION

More Reading Support
B List two ways to measure distance. *measure a straight-line distance between two positions; measure the*

Advanced Have students refer to the map on page 10 that shows a connecting line between Santiago and Brasilia. Ask: How could you estimate the distance of this diagonal line using grid units—that is, the distance south and the distance west? *You could measure the vertical distance and the horizontal distance; then use the Pythagorean theorem to determine the length of the hypotenuse.*

4. Instruct
- Teaching strategies
- Reading support
- Ongoing assessment
- Addressing misconceptions
- Differentiated instruction activities and tips

Ongoing Assessment
CHECK YOUR READING *Sample answer: You observe motion relative to your own position. If you are on a train, for example, you see the ground outside the train moving past you.*

EXPLORE (the BIG idea)
Revisit "Internet Activity: Relative Motion" on p. 7. Have students explain the reasons for their results.

Reinforce (the BIG idea)
Have students relate the section to the Big Idea.
Reinforcing Key Concepts, p. 21

1.1 ASSESS & RETEACH

Assess
Section 1.1 Quiz, p. 3

Reteach
Put two marbles randomly on a checkerboard. Have students describe the positions, and then roll the two marbles at different speeds. Emphasize that a good description of motion includes speed. Finally, have students view the rolling of a marble from very different positions. Solicit students' observations. Discuss relative motion.

Technology Resources
Have students visit ClassZone.com for reteaching of Key Concepts.
CONTENT REVIEW
CONTENT REVIEW CD-ROM

When you ride in a train, a bus, or an airplane, you think of yourself as moving and the ground as standing still. That is, you usually consider the ground as the frame of reference for your motion. If you traveled between two cities, you would say that you had moved, not that the ground had moved under you in the opposite direction.

If you cannot see the ground or objects on it, it is sometimes difficult to tell if a train you are riding in is moving. If the ride is very smooth and you do not look out the window at the scenery, you might never realize you are moving at all.

Suppose you are in a train, and you cannot tell if you are stopped or moving. Outside the window, another train is slowly moving forward. Could you tell which of the following situations is happening?
- Your train is stopped, and the other train is moving slowly forward.
- The other train is stopped, and your train is moving slowly backward.
- Both trains are moving forward, with the other train moving a little faster.
- Your train is moving very slowly backward, and the other train is moving very slowly forward.

Actually, all four of these possibilities would look exactly the same to you. Unless you compared the motion to the motion of something outside the train, such as the ground, you could not tell the difference between these situations.

APPLY In the top picture, the train is moving compared with the camera and the ground. Describe the relative motion of the train, camera, and ground in the bottom picture.

CHECK YOUR READING How does your observation of motion depend on your own motion?

1.1 Review

KEY CONCEPTS
1. What information do you need to describe an object's location?
2. Describe how your position changes as you jump over an object.
3. Give an example of how the apparent motion of an object depends on the observer's motion.

CRITICAL THINKING
4. **Infer** Kyle walks 3 blocks south from his home to school, and Jana walks 2 blocks north from her home to Kyle's home. How far and in what direction is the school from Jana's home?
5. **Predict** If you sit on a moving bus and toss a coin straight up into the air, where will it land?

CHALLENGE
6. **Infer** Jamal is in a car going north. He looks out his window and thinks that the northbound traffic is moving very slowly. Ellen is in a car going south. She thinks the northbound traffic is moving quickly. Explain why Jamal and Ellen have different ideas about the motion of the traffic.

ANSWERS
1. *a reference point*
2. *As a person jumps over an object, both his horizontal and vertical positions change.*
3. *Sample answer: A person in an airplane will think of her or his seat as not moving, but an observer on the ground will think the seats are moving with the airplane.*
4. *one block south*
5. *It should land back in your hand (or directly beneath where you threw it).*
6. *Jamal and Ellen have different ideas about motion because both compare the northbound traffic with their own motion. From Jamal's point of view, traffic is going in the same direction as he is; therefore, traffic seems to be moving slowly. Ellen sees oncoming cars and thinks they are moving quickly.*

5. Assess & Reteach
- Answers to Section Review
- Reteaching activity
- Resources for review and assessment

Lab Materials List

The following charts list the consumables, nonconsumables, and equipment needed for all activities. Quantities are per group of four students. Lab aprons, goggles, water, books, paper, pens, pencils, and calculators are assumed to be available for all activities.

Materials kits are available. For more information, please call McDougal Littell at 1-800-323-5435.

Consumables

Description	Quantity per Group	Explore *page*	Investigate *page*	Chapter Investigation *page*
bottle, 1 liter plastic	2			62, 96
bottle, 1/2 liter plastic	2			62, 96
cardboard, 8.5" x 11"	1		27	
clay, modeling	3 sticks		19	62
cup, clear plastic	5	41, 98, 130	100	
cup, paper, 6 oz	1		82	
feather	1	41		
food coloring	1 bottle		100	
glue stick	1		27	
marker, permanent black	1		100	
newspaper, sheet	1		54	
paper, construction, 8.5" x 11"	1			62
paper clip	20	49, 98		
plate, stiff paper	1		54	
poster paint	1 oz		54	
sandpaper	1		151	
straw, clear drinking	4		19, 100	62
straw, jumbo drinking	1			62
string	6 meters	49, 98	19, 27, 46, 133	136
tape, masking	1 roll	16	125, 133	32
tape, transparent	1 roll			62
toothpick	4		19	

Nonconsumables

Description	Quantity per Group	Explore *page*	Investigate *page*	Chapter Investigation *page*
balance, triple beam	1		19, 125	
ball, golf	1	77		
ball, baseball	1	64		
ball, basketball	1	64		
ball, racquetball	1		46, 125	
ball, table tennis	1	77		
ball, tennis	1	16, 41	12, 46	
chair	1			136
coffee can with lid, 1 lb	1			96
coin, quarter	1	41		
dishpan, plastic	1		82	96
dowel rod, 1/4" diameter	24"			170
lid, film canister	4		19	
machine, small (stapler, screw driver, can opener, etc.)	1	145		
marble, metal	10	130	54, 66	32
meter stick	2	16	46, 118, 133, 151	32, 62, 96, 136, 170
nail, small	1			96
paint brush	1		54	
pulley cord	4 ft		157	170
pulley, large	2			170
pulley, medium	2		157	170
pulley, small	2			170
ring stand with ring	1		157	
ruler, metric	2		66, 100, 125	32, 96
scissors	1		19, 27, 54	62
spiral notebook	1		118	
spring scale	2		58, 118, 133, 151, 157	136, 170
stopwatch	1	16, 130	19, 125, 133	32, 136
Styrofoam board, 2 cm x 20 cm x 20 cm	1	91		

Description	Quantity per Group	Explore *page*	Investigate *page*	Chapter Investigation *page*
toy car, large	1			136
washer, metal 1"	1		27	
weight, hooked, 100 gram	1		133, 157	170
weight, hooked, 500 gram	1			170
wood block, 2"	1		46	
wood block, 2" with eye hook	1		151	
wood board, 1 m x 15 cm	1		19, 151	136, 170
wood board, 30 cm x 30 cm	1		125	

Unit Resource Book Datasheets

Description		Explore *page*	Investigate *page*	Chapter Investigation *page*
Tool Template			27	

McDougal Littell Science

Motion and Forces

PHYSICAL SCIENCE

A ▶ Matter and Energy
B ▶ Chemical Interactions
C ▶ Motion and Forces
D ▶ Waves, Sound, and Light
E ▶ Electricity and Magnetism

LIFE SCIENCE

A ▶ Cells and Heredity
B ▶ Life Over Time
C ▶ Diversity of Living Things
D ▶ Ecology
E ▶ Human Biology

EARTH SCIENCE

A ▶ Earth's Surface
B ▶ The Changing Earth
C ▶ Earth's Waters
D ▶ Earth's Atmosphere
E ▶ Space Science

ISBN: 0-618-33442-4 2 3 4 5 6 7 8 VJM 08 07 06 05 04

Internet Web Site: http://www.mcdougallittell.com

ii

Science Consultants

Chief Science Consultant

James Trefil, Ph.D. is the Clarence J. Robinson Professor of Physics at George Mason University. He is the author or co-author of more than 25 books, including *Science Matters* and *The Nature of Science.* Dr. Trefil is a member of the American Association for the Advancement of Science's Committee on the Public Understanding of Science and Technology. He is also a fellow of the World Economic Forum and a frequent contributor to *Smithsonian* magazine.

Rita Ann Calvo, Ph.D. is Senior Lecturer in Molecular Biology and Genetics at Cornell University, where for 12 years she also directed the Cornell Institute for Biology Teachers. Dr. Calvo is the 1999 recipient of the College and University Teaching Award from the National Association of Biology Teachers.

Kenneth Cutler, M.S. is the Education Coordinator for the Julius L. Chambers Biomedical Biotechnology Research Institute at North Carolina Central University. A former middle school and high school science teacher, he received a 1999 Presidential Award for Excellence in Science Teaching.

Instructional Design Consultants

Douglas Carnine, Ph.D. is Professor of Education and Director of the National Center for Improving the Tools of Educators at the University of Oregon. He is the author of seven books and over 100 other scholarly publications, primarily in the areas of instructional design and effective instructional strategies and tools for diverse learners. Dr. Carnine also serves as a member of the National Institute for Literacy Advisory Board.

Linda Carnine, Ph.D. consults with school districts on curriculum development and effective instruction for students struggling academically. A former teacher and school administrator, Dr. Carnine also co-authored a popular remedial reading program.

Donald Steely, Ph.D. serves as principal investigator at the Oregon Center for Applied Science (ORCAS) on federal grants for science and language arts programs. His background also includes teaching and authoring of print and multimedia programs in science, mathematics, history, and spelling.

Sam Miller, Ph.D. is a middle school science teacher and the Teacher Development Liaison for the Eugene, Oregon, Public Schools. He is the author of curricula for teaching science, mathematics, computer skills, and language arts.

Vicky Vachon, Ph.D. consults with school districts throughout the United States and Canada on improving overall academic achievement with a focus on literacy. She is also co-author of a widely used program for remedial readers.

iii

Content Reviewers

John Beaver, Ph.D.
Ecology
Professor, Director of Science Education Center
College of Education and Human Services
Western Illinois University
Macomb, IL

Donald J. DeCoste, Ph.D.
Matter and Energy, Chemical Interactions
Chemistry Instructor
University of Illinois
Urbana-Champaign, IL

Dorothy Ann Fallows, Ph.D., MSc
Diversity of Living Things, Microbiology
Partners in Health
Boston, MA

Michael Foote, Ph.D.
The Changing Earth, Life Over Time
Associate Professor
Department of the Geophysical Sciences
The University of Chicago
Chicago, IL

Lucy Fortson, Ph.D.
Space Science
Director of Astronomy
Adler Planetarium and Astronomy Museum
Chicago, IL

Elizabeth Godrick, Ph.D.
Human Biology
Professor, CAS Biology
Boston University
Boston, MA

Isabelle Sacramento Grilo, M.S.
The Changing Earth
Lecturer, Department of the Geological Sciences
Montana State University
Bozeman, MT

David Harbster, MSc
Diversity of Living Things
Professor of Biology
Paradise Valley Community College
Phoenix, AZ

Richard D. Norris, Ph.D.
Earth's Waters
Professor of Paleobiology
Scripps Institution of Oceanography
University of California, San Diego
La Jolla, CA

Donald B. Peck, M.S.
Motion and Forces; Waves, Sound, and Light; Electricity and Magnetism
Director of the Center for Science Education (retired)
Fairleigh Dickinson University
Madison, NJ

Javier Penalosa, Ph.D.
Diversity of Living Things, Plants
Associate Professor, Biology Department
Buffalo State College
Buffalo, NY

Raymond T. Pierrehumbert, Ph.D.
Earth's Atmosphere
Professor in Geophysical Sciences (Atmospheric Science)
The University of Chicago
Chicago, IL

Brian J. Skinner, Ph.D.
Earth's Surface
Eugene Higgins Professor of Geology and Geophysics
Yale University
New Haven, CT

Nancy E. Spaulding, M.S.
Earth's Surface, The Changing Earth, Earth's Waters
Earth Science Teacher (retired)
Elmira Free Academy
Elmira, NY

Steven S. Zumdahl, Ph.D.
Matter and Energy, Chemical Interactions
Professor Emeritus of Chemistry
University of Illinois
Urbana-Champaign, IL

Susan L. Zumdahl, M.S.
Matter and Energy, Chemical Interactions
Chemistry Education Specialist
University of Illinois
Urbana-Champaign, IL

Safety Consultant

Juliana Texley, Ph.D.
Former K–12 Science Teacher and School Superintendent
Boca Raton, FL

English Language Advisor

Judy Lewis, M.A.
Director, State and Federal Programs for reading proficiency and high risk populations
Rancho Cordova, CA

iv

Teacher Panel Members

Carol Arbour
Tallmadge Middle School, Tallmadge, OH

Patty Belcher
Goodrich Middle School, Akron, OH

Gwen Broestl
Luis Munoz Marin Middle School, Cleveland, OH

Al Brofman
Tehipite Middle School, Fresno, CA

John Cockrell
Clinton Middle School, Columbus, OH

Jenifer Cox
Sylvan Middle School, Citrus Heights, CA

Linda Culpepper
Martin Middle School, Charlotte, NC

Kathleen Ann DeMatteo
Margate Middle School, Margate, FL

Melvin Figueroa
New River Middle School, Ft. Lauderdale, FL

Doretha Grier
Kannapolis Middle School, Kannapolis, NC

Robert Hood
Alexander Hamilton Middle School, Cleveland, OH

Scott Hudson
Coverdale Elementary School, Cincinnati, OH

Loretta Langdon
Princeton Middle School, Princeton, NC

Carlyn Little
Glades Middle School, Miami, FL

Ann Marie Lynn
Amelia Earhart Middle School, Riverside, CA

James Minogue
Lowe's Grove Middle School, Durham, NC

Joann Myers
Buchanan Middle School, Tampa, FL

Barbara Newell
Charles Evans Hughes Middle School, Long Beach, CA

Anita Parker
Kannapolis Middle School, Kannapolis, NC

Greg Pirolo
Golden Valley Middle School, San Bernardino, CA

Laura Pottmyer
Apex Middle School, Apex, NC

Lynn Prichard
Booker T. Washington Middle Magnet School, Tampa, FL

Jacque Quick
Walter Williams High School, Burlington, NC

Robert Glenn Reynolds
Hillman Middle School, Youngstown, OH

Theresa Short
Abbott Middle School, Fayetteville, NC

Rita Slivka
Alexander Hamilton Middle School, Cleveland, OH

Marie Sofsak
B F Stanton Middle School, Alliance, OH

Nancy Stubbs
Sweetwater Union Unified School District, Chula Vista, CA

Sharon Stull
Quail Hollow Middle School, Charlotte, NC

Donna Taylor
Okeeheelee Middle School, West Palm Beach, FL

Sandi Thompson
Harding Middle School, Lakewood, OH

Lori Walker
Audubon Middle School & Magnet Center, Los Angeles, CA

Teacher Lab Evaluators

Jill Brimm-Byrne
Albany Park Academy, Chicago, IL

Gwen Broestl
Luis Munoz Marin Middle School, Cleveland, OH

Al Brofman
Tehipite Middle School, Fresno, CA

Michael A. Burstein
The Rashi School, Newton, MA

Trudi Coutts
Madison Middle School, Naperville, IL

Jenifer Cox
Sylvan Middle School, Citrus Heights, CA

Larry Cwik
Madison Middle School, Naperville, IL

Jennifer Donatelli
Kennedy Junior High School, Lisle, IL

Paige Fullhart
Highland Middle School, Libertyville, IL

Sue Hood
Glen Crest Middle School, Glen Ellyn, IL

Ann Min
Beardsley Middle School, Crystal Lake, IL

Aileen Mueller
Kennedy Junior High School, Lisle, IL

Nancy Nega
Churchville Middle School, Elmhurst, IL

Oscar Newman
Sumner Math and Science Academy, Chicago, IL

Marina Penalver
Moore Middle School, Portland, ME

Lynn Prichard
Booker T. Washington Middle Magnet School, Tampa, FL

Jacque Quick
Walter Williams High School, Burlington, NC

Seth Robey
Gwendolyn Brooks Middle School, Oak Park, IL

Kevin Steele
Grissom Middle School, Tinley Park, IL

v

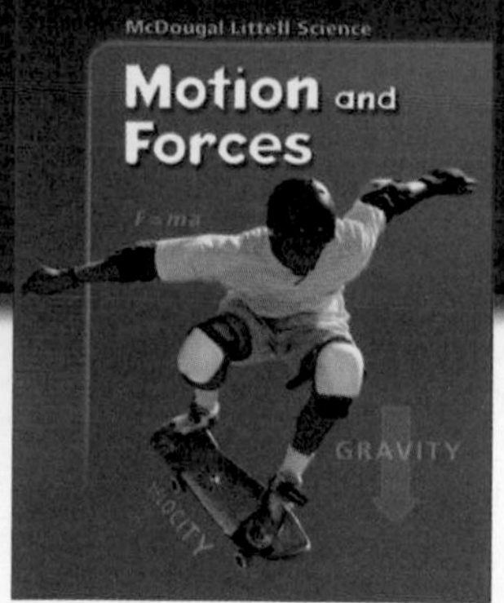

Motion and Forces

Standards and Benchmarks x
Introducing Physical Science xii
Unifying Principles of Physical Science xiii
The Nature of Science xxii
The Nature of Technology xxvi
Using McDougal Littell Science xxviii

Unit Features

SCIENTIFIC AMERICAN

FRONTIERS IN SCIENCE *Robots on Mars* 2

TIMELINES IN SCIENCE *Understanding Forces* 108

1 Motion 6

the BIG idea
The motion of an object can be described and predicted.

1.1 An object in motion changes position. 9
SCIENCE ON THE JOB *Physics for Rescuers* 15

1.2 Speed measures how fast position changes. 16
MATH IN SCIENCE *Working with Units* 24

1.3 Acceleration measures how fast velocity changes. 25
CHAPTER INVESTIGATION *Acceleration and Slope* 32

2 Forces 38

the BIG idea
Forces change the motion of objects in predictable ways.

2.1 Forces change motion. 41
THINK SCIENCE *Why Do These Rocks Slide?* 48

2.2 Force and mass determine acceleration. 49
MATH IN SCIENCE *Using Significant Figures* 56

2.3 Forces act in pairs. 57
CHAPTER INVESTIGATION *Newton's Laws of Motion* 62

2.4 Forces transfer momentum. 64

What must happen for a team to win this tug of war? page 38

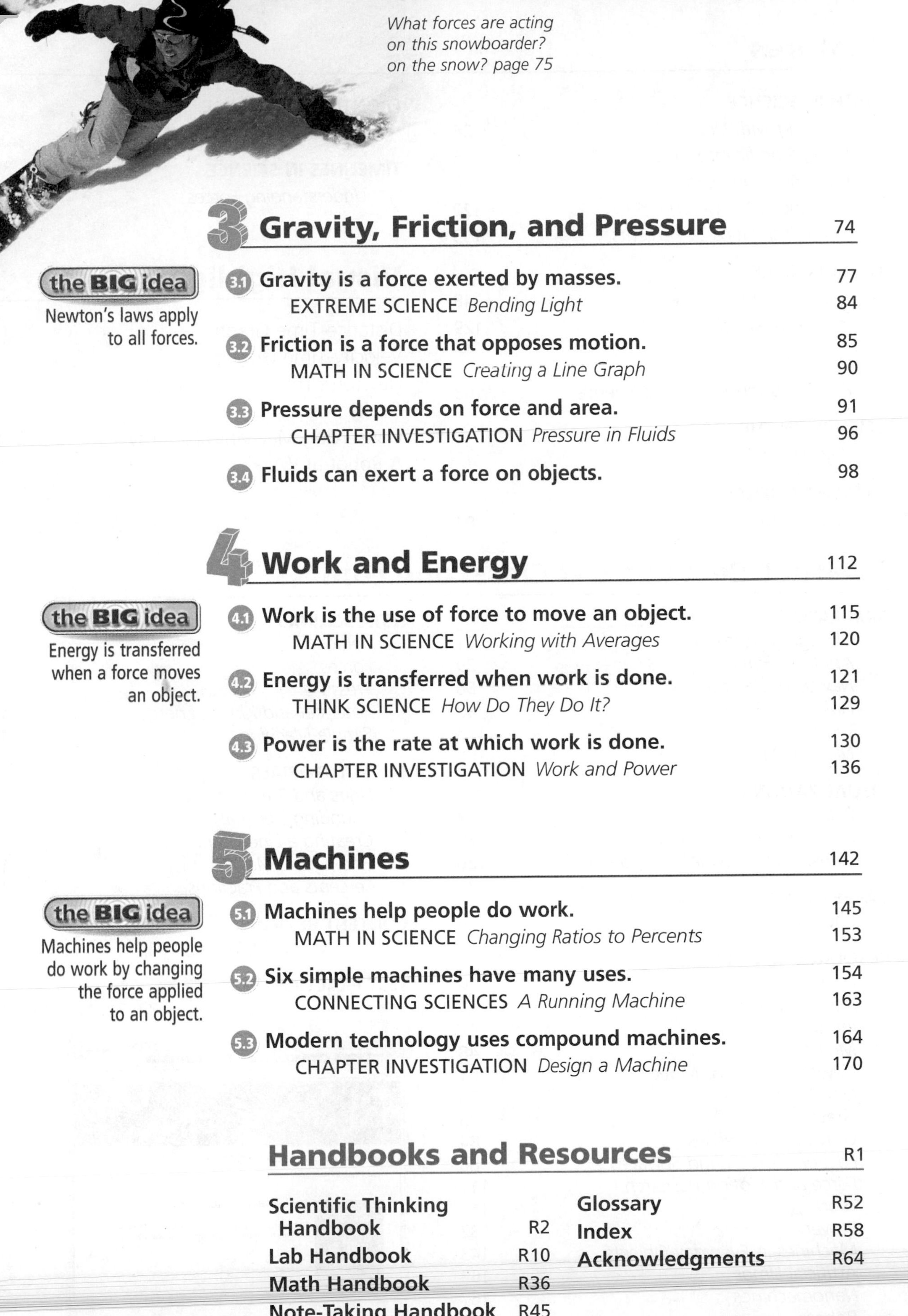

What forces are acting on this snowboarder? on the snow? page 75

3 Gravity, Friction, and Pressure 74

the BIG idea
Newton's laws apply to all forces.

3.1 Gravity is a force exerted by masses. 77
EXTREME SCIENCE *Bending Light* 84

3.2 Friction is a force that opposes motion. 85
MATH IN SCIENCE *Creating a Line Graph* 90

3.3 Pressure depends on force and area. 91
CHAPTER INVESTIGATION *Pressure in Fluids* 96

3.4 Fluids can exert a force on objects. 98

4 Work and Energy 112

the BIG idea
Energy is transferred when a force moves an object.

4.1 Work is the use of force to move an object. 115
MATH IN SCIENCE *Working with Averages* 120

4.2 Energy is transferred when work is done. 121
THINK SCIENCE *How Do They Do It?* 129

4.3 Power is the rate at which work is done. 130
CHAPTER INVESTIGATION *Work and Power* 136

5 Machines 142

the BIG idea
Machines help people do work by changing the force applied to an object.

5.1 Machines help people do work. 145
MATH IN SCIENCE *Changing Ratios to Percents* 153

5.2 Six simple machines have many uses. 154
CONNECTING SCIENCES *A Running Machine* 163

5.3 Modern technology uses compound machines. 164
CHAPTER INVESTIGATION *Design a Machine* 170

Handbooks and Resources R1

Scientific Thinking Handbook R2
Lab Handbook R10
Math Handbook R36
Note-Taking Handbook R45
Glossary R52
Index R58
Acknowledgments R64

Features

MATH IN SCIENCE
- *Working with Units* 24
- *Using Significant Figures* 56
- *Creating a Line Graph* 90
- *Working with Averages* 120
- *Changing Ratios to Percents* 153

THINK SCIENCE
- *Evaluating Hypotheses* 48
- *Isolating Variables* 129

CONNECTING SCIENCES
- *Physical Science and Life Science* 163

SCIENCE ON THE JOB
- *Physics for Rescuers* 15

EXTREME SCIENCE
- *Bending Light* 84

FRONTIERS IN SCIENCE
- *Robots on Mars* 2

TIMELINES IN SCIENCE
- *Understanding Forces* 108

Visual Highlights

- Distance-Time Graph 21
- Velocity-Time Graphs 30
- Newton's Three Laws of Motion 60
- Orbits 81
- Conserving Mechanical Energy 127
- A Robot at Work 168

Internet Resources @ ClassZone.com

SIMULATIONS
- *Changing Acceleration* 31
- *Applying Force* 39
- *Newton's Second Law* 50
- *Fluids and Pressure* 93
- *Work* 113
- *Mechanical Advantage* 161

VISUALIZATIONS
- *Relative Motion* 7
- *Effect of Gravity in a Vacuum* 79
- *Transfer of Potential and Kinetic Energy* 126

CAREER CENTER
- *Physics and Engineering* 5

RESOURCE CENTERS
- *Finding Position* 10
- *Acceleration* 29
- *Inertia* 47
- *Moving Rocks* 48
- *Newton's Laws of Motion* 61
- *Momentum* 65
- *Gravity* 75
- *Gravitational Lenses* 84
- *Friction, Forces, and Surfaces* 86
- *Force and Motion Research* 111
- *Work* 116
- *Power* 132
- *Machines in Everyday Objects* 143
- *Artificial Limbs* 163
- *Nanomachines* 167
- *Robots* 169

NSTA SCILINKS
- *Velocity* 7
- *Forces* 39
- *Pressure* 75
- *Potential and Kinetic Energy* 113
- *Simple Machines* 143

MATH TUTORIALS
- *Units and Rates* 24
- *Rounding Decimals* 56
- *Creating a Line Graph* 90
- *Finding the Mean* 120
- *Percents and Fractions* 153

CONTENT REVIEW
8, 34, 40, 70, 76, 104, 114, 138, 144, 172

TEST PRACTICE
37, 73, 107, 141, 175

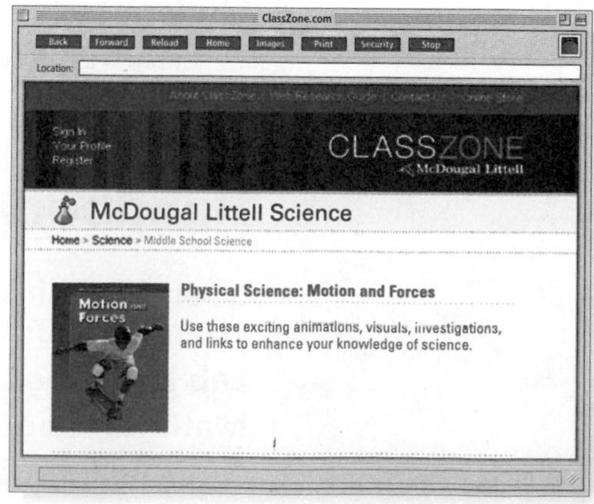

INVESTIGATIONS AND ACTIVITIES

EXPLORE THE BIG IDEA

Chapter Opening Inquiry

1. Off the Wall, Rolling Along, Internet Activity: Relative Motion 7
2. Popping Ping-Pong Balls, Take Off!, Internet Activity: Forces 39
3. Let It Slide, Under Pressure, Internet Activity: Gravity 75
4. Bouncing Ball, Power Climbing, Internet Activity: Work 113
5. Changing Direction, Shut the Door!, Internet Activity: Machines 143

CHAPTER INVESTIGATION

Full-Period Labs

1. Acceleration and Slope 32
2. Newton's Laws of Motion 62
3. Pressure in Fluids 96
4. Work and Power 136
5. Design a Machine ***Design Your Own*** 170

EXPLORE

Introductory Inquiry Activities

Location 9
Speed 16
Changing Motion 41
Acceleration 49
Collisions 64
Downward Acceleration 77
Pressure 91
Forces in Liquid 98
Work 115
Power 130
Machines 145
Changing Forces 154

INVESTIGATE

Skill Labs

Changing Position *Observing* 12
Speed and Distance ***Design Your Own*** 19
Acceleration *Measuring* 27
Inertia ***Design Your Own*** 46
Motion and Force *Hypothesizing* 54
Newton's Third Law *Observing* 58
Momentum *Observing* 66
Gravity *Predicting* 82
Friction in Air ***Design Your Own*** 88
Bernoulli's Principle *Observing* 100
Work *Measuring* 118
Mechanical Energy *Analyzing Data* 125
Power *Inferring* 133
Efficiency *Analyzing Data* 151
Pulleys *Inferring* 157

Standards and Benchmarks

Each chapter in **Motion and Forces** covers some of the learning goals that are described in the *National Science Education Standards* (NSES) and the Project 2061 *Benchmarks for Science Literacy*. Selected content and skill standards are shown below in shortened form. The following National Science Education Standards are covered on pages xii–xxvii, in Frontiers in Science, and in Timelines in Science, as well as in chapter features and laboratory investigations: Understandings About Scientific Inquiry (A.9), Understandings About Science and Technology (E.6), Science and Technology in Society (F.5), Science as a Human Endeavor (G.1), Nature of Science (G.2), and History of Science (G.3).

Content Standards

1 Motion

National Science Education Standards

B.2.a An object's motion
- is described by its position, direction of motion, and speed
- can be measured and shown on a graph

Project 2061 Benchmarks

10.A.1 An object's motion is relative to a certain object or point in space.

2 Forces

National Science Education Standards

B.2.b An object will move in a straight line at a constant speed unless a force acts on it.

B.2.c If more than one force acts on an object,
- the forces can act together
- the forces can act against each other depending on the size and direction of the forces

Project 2061 Benchmarks

4.F.3 An unbalanced force
- will change an object's speed and/or direction
- can cause an object to orbit the center of the force

10.B.1 Newton's Laws describe motion everywhere in the universe.

11.C.2 A system may stay the same because
- no forces are acting on the system
- forces are acting on the system, but they all cancel each other out

3 Gravity, Friction, and Pressure

National Science Education Standards

B.1.a Substances have certain properties, including density.

D.3.c Gravity is the force that
- keeps planets in orbit around the Sun
- governs motion within the solar system
- holds us to Earth's surface
- produces tides

Project 2061 Benchmarks

4.B.3 Everything on or near Earth is pulled toward Earth's center by the force of gravity.

4.G.1 Every object exerts the force of gravity, but the force
- depends on the mass and distance of objects
- is difficult to detect unless an object contains a lot of mass

4.G.2 In the solar system
- the Sun's gravity keeps the planets in orbit around the Sun
- the planets' gravity keeps their moons in orbit around the planets

4 Work and Energy

National Science Education Standards

B.3.a Energy is transferred in many ways. Energy is often associated with heat, sound, and mechanical motion.

Project 2061 Benchmarks

4.E.1 Energy cannot be created or destroyed, but it can be changed from one form to another.

4.E.4 Energy has many different forms, including
- heat—the disorderly motion of particles
- chemical—the arrangement of atoms in matter
- mechanical—the kinetic plus the potential energy of an object
- gravitational—the separation of objects that attract each other

5 Machines

National Science Education Standards

E.6.c
- Science and technology often work together.
- Science helps drive technology.
- Technology is used to improve scientific investigations.

E.6.d Perfectly designed solutions do not exist. All technology has risks and trade-offs, such as cost, safety, and efficiency.

E.6.e All designs have limits, including those having to do with material properties and availability, safety, and environmental protection.

Project 2061 Benchmarks

8.B.4 The use of robots has changed the nature of work in many fields, including manufacturing.

Process and Skill Standards

National Science Education Standards

A.2 Design and conduct a scientific investigation.

A.3 Use appropriate tools and techniques to gather and interpret data.

A.4 Use evidence to describe, predict, explain, and model.

A.5 Use critical thinking to find relationships between results and interpretations.

A.7 Communicate procedures, results, and conclusions.

A.8 Use mathematics in scientific investigations.

E.1 Identify a problem to be solved.

E.2 Design a solution or product.

E.3 Implement the proposed solution.

E.4 Evaluate the solution or design.

Project 2061 Benchmarks

11.C.4 Use equations to summarize observed changes.

12.C.3 Using appropriate units, use and read instruments that measure length, volume, weight, time, rate, and temperature.

12.D.1 Use tables and graphs to organize information and identify relationships.

12.D.2 Read, interpret, and describe tables and graphs.

12.D.4 Understand information that includes different types of charts and graphs, including circle charts, bar graphs, line graphs, data tables, diagrams, and symbols.

Introducing Physical Science

Scientists are curious. Since ancient times, they have been asking and answering questions about the world around them. Scientists are also very suspicious of the answers they get. They carefully collect evidence and test their answers many times before accepting an idea as correct.

In this book you will see how scientific knowledge keeps growing and changing as scientists ask new questions and rethink what was known before. The following sections will help get you started.

Unifying Principles of Physical Science xiii

What do scientists know about matter and energy? These pages introduce four unifying principles that will give you a big picture of physical science.

The Nature of Science xxii

How do scientists learn? This section provides an overview of scientific thinking and the processes that scientists use to ask questions and to find answers.

The Nature of Technology xxvi

How do we use what scientists learn? These pages introduce you to how people develop and use technologies to design solutions to real-world problems.

Using McDougal Littell Science xxviii

How can you learn more about science? This section provides helpful tips on how to learn and use science from the key parts of this program—the text, the visuals, the activities, and the Internet resources.

UNIFYING PRINCIPLES of Physical Science

What Is Physical Science?

In the simplest terms, physical science is the study of what things are made of and how they change. It combines the studies of both physics and chemistry. Physics is the science of matter, energy, and forces. It includes the study of topics such as motion, light, and electricity and magnetism. Chemistry is the study of the structure and properties of matter, and it especially focuses on how substances change into different substances.

The text and pictures in this book will help you learn key concepts and important facts about physical science. A variety of activities will help you investigate these concepts. As you learn, it helps to have a big picture of physical science as a framework for this new information. The four unifying principles listed below will give you this big picture. Read the next few pages to get an overview of each of these principles and a sense of why they are so important.

- **Matter is made of particles too small to see.**
- **Matter changes form and moves from place to place.**
- **Energy changes from one form to another, but it cannot be created or destroyed.**
- **Physical forces affect the movement of all matter on Earth and throughout the universe.**

the BIG idea

Each chapter begins with a big idea. Keep in mind that each big idea relates to one or more of the unifying principles.

Matter is made of particles too small to see.

This simple statement is the basis for explaining an amazing variety of things about the world. For example, it explains why substances can exist as solids, liquids, and gases, and why wood burns but iron does not. Like the tiles that make up this mosaic picture, the particles that make up all substances combine to make patterns and structures that can be seen. Unlike these tiles, the individual particles themselves are far too small to see.

What It Means

To understand this principle better, let's take a closer look at the two key words: *matter* and *particles.*

Matter

Objects you can see and touch are all around you. The materials that these objects are made of are called **matter.** All living things—even you—are also matter. Even though you can't see it, the air around you is matter too. Scientists often say that matter is anything that has mass and takes up space. **Mass** is a measure of the amount of matter in an object. We use the word **volume** to refer to the amount of space an object or a substance takes up.

Particles

The tiny particles that make up all matter are called **atoms.** Just how tiny are atoms? They are far too small to see, even through a powerful microscope. In fact, an atom is more than a million times smaller than the period at the end of this sentence.

There are more than 100 basic kinds of matter called **elements.** For example, iron, gold, and oxygen are three common elements. Each element has its own unique kind of atom. The atoms of any element are all alike but different from the atoms of any other element.

Many familiar materials are made of particles called molecules. In a **molecule,** two or more atoms stick together to form a larger particle. For example, a water molecule is made of two atoms of hydrogen and one atom of oxygen.

Why It's Important

Understanding atoms and molecules makes it possible to explain and predict the behavior of matter. Among other things, this knowledge allows scientists to

- explain why different materials have different characteristics
- predict how a material will change when heated or cooled
- figure out how to combine atoms and molecules to make new and useful materials

xv

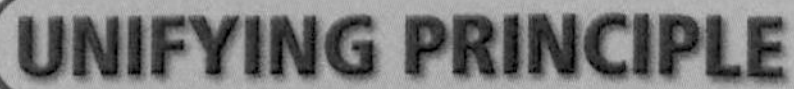

Matter changes form and moves from place to place.

You see matter change form every day. You see the ice in your glass of juice disappear without a trace. You see a black metal gate slowly develop a flaky, orange coating. Matter is constantly changing and moving.

What It Means

Remember that matter is made of tiny particles called atoms. Atoms are constantly moving and combining with one another. All changes in matter are the result of atoms moving and combining in different ways.

Matter Changes and Moves

You can look at water to see how matter changes and moves. A block of ice is hard like a rock. Leave the ice out in sunlight, however, and it changes into a puddle of water. That puddle of water can eventually change into water vapor and disappear into the air. The water vapor in the air can become raindrops, which may fall on rocks, causing them to weather and wear away. The water that flows in rivers and streams picks up tiny bits of rock and carries them from one shore to another. Understanding how the world works requires an understanding of how matter changes and moves.

Matter Is Conserved

No matter was lost in any of the changes described above. The ice turned to water because its molecules began to move more quickly as they got warmer. The bits of rock carried away by the flowing river were not gone forever. They simply ended up farther down the river. The puddles of rainwater didn't really disappear; their molecules slowly mixed with molecules in the air.

Under ordinary conditions, when matter changes form, no matter is created or destroyed. The water created by melting ice has the same mass as the ice did. If you could measure the water vapor that mixes with the air, you would find it had the same mass as the water in the puddle did.

Why It's Important

Understanding how mass is conserved when matter changes form has helped scientists to

- describe changes they see in the world
- predict what will happen when two substances are mixed
- explain where matter goes when it seems to disappear

xvii

Energy changes from one form to another, but it cannot be created or destroyed.

When you use energy to warm your food or to turn on a flashlight, you may think that you "use up" the energy. Even though the camp-stove fuel is gone and the flashlight battery no longer functions, the energy they provided has not disappeared. It has been changed into a form you can no longer use. Understanding how energy changes forms is the basis for understanding how heat, light, and motion are produced.

What It Means

Changes that you see around you depend on energy. **Energy,** in fact, means the ability to cause change. The electrical energy from an outlet changes into light and heat in a light bulb. Plants change the light energy from the Sun into chemical energy, which animals use to power their muscles.

Energy Changes Forms

Using energy means changing energy. You probably have seen electric energy changing into light, heat, sound, and mechanical energy in household appliances. Fuels like wood, coal, and oil contain chemical energy that produces heat when burned. Electric power plants make electrical energy from a variety of energy sources, including falling water, nuclear energy, and fossil fuels.

Energy Is Conserved

Energy can be converted into forms that can be used for specific purposes. During the conversion, some of the original energy is converted into unwanted forms. For instance, when a power plant converts the energy of falling water into electrical energy, some of the energy is lost to friction and sound.

Similarly, when electrical energy is used to run an appliance, some of the energy is converted into forms that are not useful. Only a small percentage of the energy used in a light bulb, for instance, produces light; most of the energy becomes heat. Nonetheless, the total amount of energy remains the same through all these conversions.

The fact that energy does not disappear is a law of physical science. The **law of conservation of energy** states that energy cannot be created or destroyed. It can only change form.

Why It's Important

Understanding that energy changes form but does not disappear has helped scientists to

- predict how energy will change form
- manage energy conversions in useful ways
- build and improve machines

Physical forces affect the movement of all matter on Earth and throughout the universe.

What makes the world go around? The answer is simple: forces. Forces allow you to walk across the room, and forces keep the stars together in galaxies. Consider the forces acting on the rafts below. The rushing water is pushing the rafts forward. The force from the people paddling helps to steer the rafts.

What It Means

A **force** is a push or a pull. Every time you push or pull an object, you're applying a force to that object, whether or not the object moves. There are several forces—several pushes and pulls—acting on you right now. All these forces are necessary for you to do the things you do, even sitting and reading.

- You are already familiar with the force of gravity. **Gravity** is the force of attraction between two objects. Right now gravity is at work pulling you to Earth and Earth to you. The Moon stays in orbit around Earth because gravity holds it close.
- A contact force occurs when one object pushes or pulls another object by touching it. If you kick a soccer ball, for instance, you apply a contact force to the ball. You apply a contact force to a shopping cart that you push down a grocery aisle or a sled that you pull up a hill.
- **Friction** is the force that resists motion between two surfaces pressed together. If you've ever tried to walk on an icy sidewalk, you know how important friction can be. If you lightly rub your finger across a smooth page in a book and then across a piece of sandpaper, you can feel how the different surfaces produce different frictional forces. Which is easier to do?
- There are other forces at work in the world too. For example, a compass needle responds to the magnetic force exerted by Earth's magnetic field, and objects made of certain metals are attracted by magnets. In addition to magnetic forces, there are electrical forces operating between particles and between objects. For example, you can demonstrate electrical forces by rubbing an inflated balloon on your hair. The balloon will then stick to your head or to a wall without additional means of support.

Why It's Important

Although some of these forces are more obvious than others, physical forces at work in the world are necessary for you to do the things you do. Understanding forces allows scientists to

- predict how objects will move
- design machines that perform complex tasks
- predict where planets and stars will be in the sky from one night to the next

xxi

The Nature of Science

You may think of science as a body of knowledge or a collection of facts. More important, however, science is an active process that involves certain ways of looking at the world.

Scientific Habits of Mind

Scientists are curious. They are always asking questions. Scientists have asked questions such as, "What is the smallest form of matter?" and "How do the smallest particles behave?" These and other important questions are being investigated by scientists around the world.

Scientists are observant. They are always looking closely at the world around them. Scientists once thought the smallest parts of atoms were protons, neutrons, and electrons. Later, protons and neutrons were found to be made of even smaller particles called quarks.

Scientists are creative. They draw on what they know to form possible explanations for a pattern, an event, or an interesting phenomenon that they have observed. Then scientists create a plan for testing their ideas.

Scientists are skeptical. Scientists don't accept an explanation or answer unless it is based on evidence and logical reasoning. They continually question their own conclusions and the conclusions suggested by other scientists. Scientists trust only evidence that is confirmed by other people or methods.

Scientists cannot always make observations with their own eyes. They have developed technology, such as this particle detector, to help them gather information about the smallest particles of matter.

Scientists ask questions about the physical world and seek answers through carefully controlled procedures. Here a researcher works with supercooled magnets.

Science Processes at Work

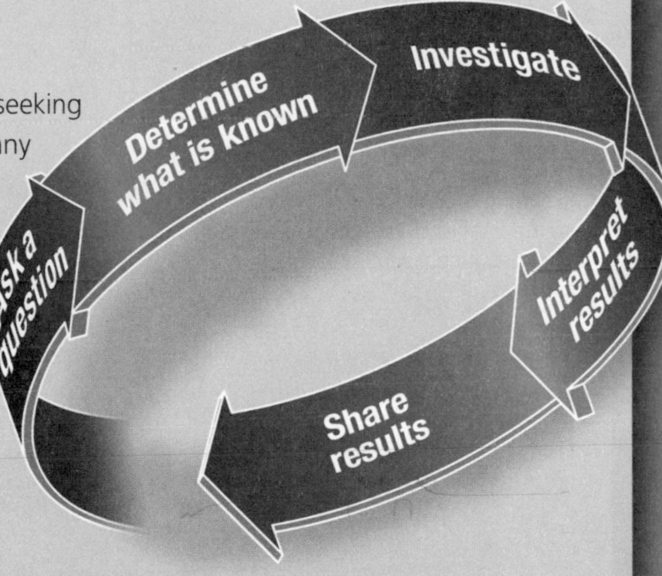

You can think of science as a continuous cycle of asking and seeking answers to questions about the world. Although there are many processes that scientists use, scientists typically do each of the following:

- Ask a question
- Determine what is known
- Investigate
- Interpret results
- Share results

Ask a Question

It may surprise you that asking questions is an important skill. A scientific process may start when a scientist asks a question. Perhaps scientists observe an event or a process that they don't understand, or perhaps answering one question leads to another.

Determine What Is Known

When beginning an inquiry, scientists find out what is already known about a question. They study results from other scientific investigations, read journals, and talk with other scientists. A scientist working on subatomic particles is most likely a member of a large team using sophisticated equipment. Before beginning original research, the team analyzes results from previous studies.

Investigate

Investigating is the process of collecting evidence. Two important ways of investigating are observing and experimenting.

Observing is the act of noting and recording an event, a characteristic, or anything else detected with an instrument or with the senses. A researcher may study the properties of a substance by handling it, finding its mass, warming or cooling it, stretching it, and so on. For information about the behavior of subatomic particles, however, a researcher may rely on technology such as scanning tunneling microscopes, which produce images of structures that cannot be seen with the eye.

An **experiment** is an organized procedure to study something under controlled conditions. In order to study the effect of wing shape on the motion of a glider, for instance, a researcher would need to conduct controlled studies in which gliders made of the same materials and with the same masses differed only in the shape of their wings.

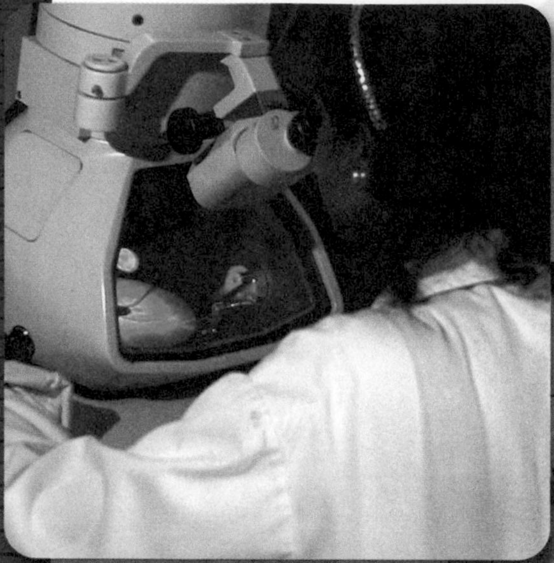

Scanning tunneling microscopes create images that allow scientists to observe molecular structure.

Physical chemists have found a way to observe chemical reactions at the atomic level. Using lasers, they can watch bonds breaking and new bonds forming.

Forming hypotheses and making predictions are two of the skills involved in scientific investigations. A **hypothesis** is a tentative explanation for an observation, a phenomenon, or a scientific problem that can be tested by further investigation. For example, in the mid-1800s astronomers noticed that the planet Uranus departed slightly from its expected orbit. One astronomer hypothesized that the irregularities in the planet's orbit were due to the gravitational effect of another planet—one that had not yet been detected. A **prediction** is an expectation of what will be observed or what will happen. A prediction can be used to test a hypothesis. The astronomers predicted that they would discover a new planet in the position calculated, and their prediction was confirmed with the discovery of the planet Neptune.

Interpret Results

As scientists investigate, they analyze their evidence, or data, and begin to draw conclusions. **Analyzing data** involves looking at the evidence gathered through observations or experiments and trying to identify any patterns that might exist in the data. Scientists often need to make additional observations or perform more experiments before they are sure of their conclusions. Many times scientists make new predictions or revise their hypotheses.

Often scientists use computers to help them analyze data. Computers reveal patterns that might otherwise be missed.

Scientists use computers to create models of objects or processes they are studying. This model shows carbon atoms forming a sphere.

Share Results

An important part of scientific investigation is sharing results of experiments. Scientists read and publish in journals and attend conferences to communicate with other scientists around the world. Sharing data and procedures gives them a way to test one another's results. They also share results with the public through newspapers, television, and other media.

xxv

The Nature of Technology

When you think of technology, you may think of cars, computers, and cell phones, as well as refrigerators, radios, and bicycles. Technology is not only the machines and devices that make modern lives easier, however. It is also a process in which new methods and devices are created. Technology makes use of scientific knowledge to design solutions to real-world problems.

Science and Technology

Science and technology go hand in hand. Each depends upon the other. Even designing a device as simple as a toaster requires knowledge of how heat flows and which materials are the best conductors of heat. Just as technology based on scientific knowledge makes our lives easier, some technology is used to advance scientific inquiry itself. For example, researchers use a number of specialized instruments to help them collect data. Microscopes, telescopes, spectrographs, and computers are just a few of the tools that help scientists learn more about the world. The more information these tools provide, the more devices can be developed to aid scientific research and to improve modern lives.

The Process of Technological Design

The process of technology involves many choices. For example, how does an automobile engineer design a better car? Is a better car faster? safer? cheaper? Before designing any new machine, the engineer must decide exactly what he or she wants the machine to do as well as what may be given up for the machine to do it. A faster car may get people to their destinations more quickly, but it may cost more and be less safe. As you study the technological process, think about all the choices that were made to build the technologies you use.

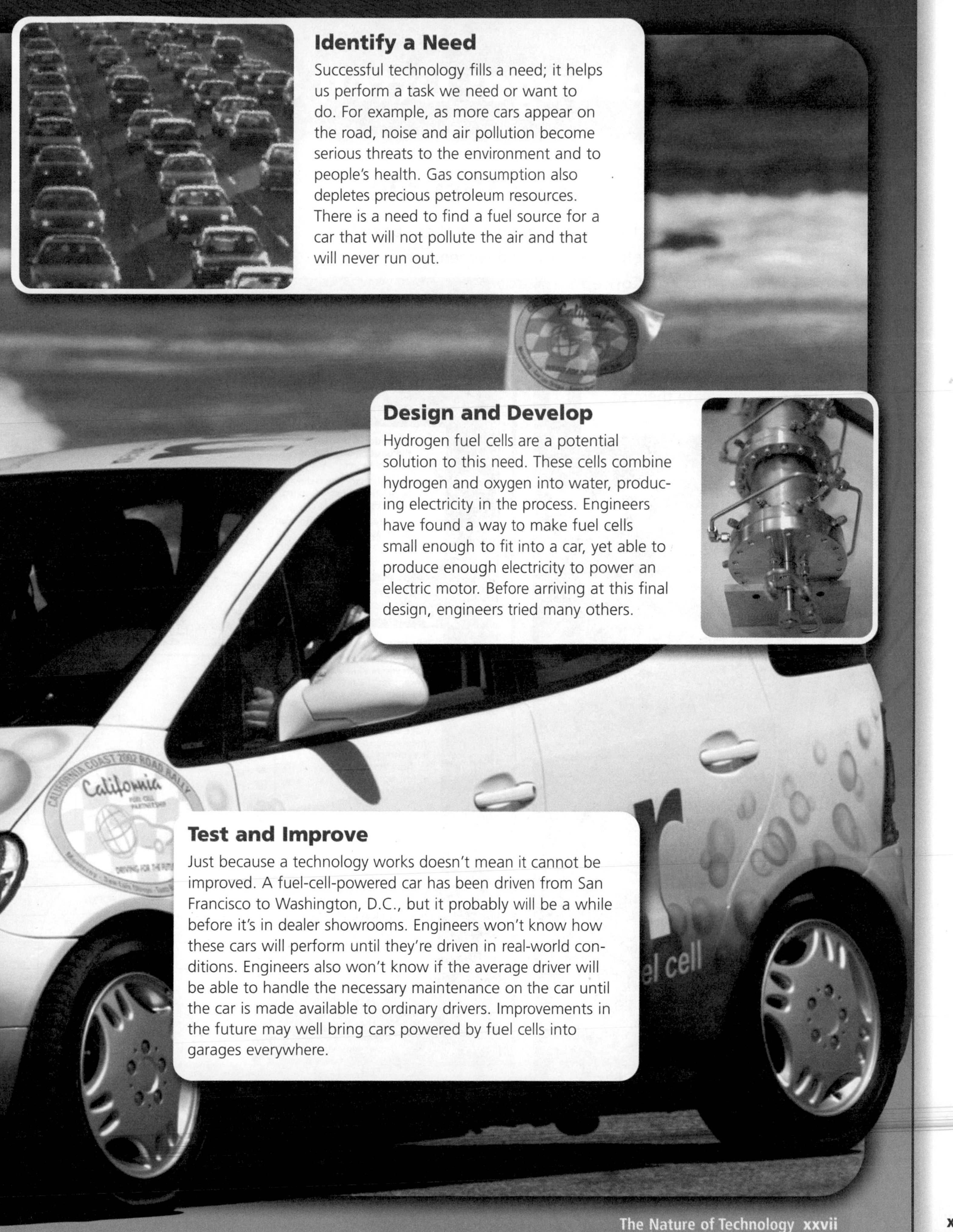

Identify a Need

Successful technology fills a need; it helps us perform a task we need or want to do. For example, as more cars appear on the road, noise and air pollution become serious threats to the environment and to people's health. Gas consumption also depletes precious petroleum resources. There is a need to find a fuel source for a car that will not pollute the air and that will never run out.

Design and Develop

Hydrogen fuel cells are a potential solution to this need. These cells combine hydrogen and oxygen into water, producing electricity in the process. Engineers have found a way to make fuel cells small enough to fit into a car, yet able to produce enough electricity to power an electric motor. Before arriving at this final design, engineers tried many others.

Test and Improve

Just because a technology works doesn't mean it cannot be improved. A fuel-cell-powered car has been driven from San Francisco to Washington, D.C., but it probably will be a while before it's in dealer showrooms. Engineers won't know how these cars will perform until they're driven in real-world conditions. Engineers also won't know if the average driver will be able to handle the necessary maintenance on the car until the car is made available to ordinary drivers. Improvements in the future may well bring cars powered by fuel cells into garages everywhere.

Using McDougal Littell Science

Reading Text and Visuals

This book is organized to help you learn. Use these boxed pointers as a path to help you learn and remember the **Big Ideas** and **Key Concepts.**

Take notes.

Use the strategies on the **Getting Ready to Learn** page.

Read the Big Idea.

As you read **Key Concepts** for the chapter, relate them to **the Big Idea.**

CHAPTER 2
Forc

the BIG idea
Forces change the motion of objects in predictable ways.

Key Concepts

SECTION 2.1 Forces change motion.
Learn about inertia and Newton's first law of motion.

SECTION 2.2 Force and mass determine acceleration.
Learn to calculate force through Newton's second law of motion.

SECTION 2.3 Forces act in pairs.
Learn about action forces and reaction forces through Newton's third law of motio

SECTION 2.4 Forces transfer momentum.
Learn about momentum an how it is affected in collisio

Internet Preview
CLASSZONE.COM
Chapter 2 online resources: Content Review, two Simulations, four Resource Centers, Math Tutorial, Test Practice

C 38 Unit: Motion and Forces

CHAPTER 2
Getting Ready to Learn

CONCEPT REVIEW
- All motion is relative to the position and motion of an observer.
- An object's motion is described by position, direction, speed, and acceleration.
- Velocity and acceleration can be measured.

VOCABULARY REVIEW
velocity p. 22
vector p. 22
acceleration p. 25
mass See Glossary.

CONTENT REVIEW
CLASSZONE.COM
Review concepts and vocabulary.

TAKING NOTES

COMBINATION NOTES
When you read about a concept for the first time, take notes in two ways. First, make an outline of the information. Then make a sketch to help you understand and remember the concept. Use arrows to show the direction of forces.

VOCABULARY STRATEGY
Think about a vocabulary term as a **magnet word** diagram. Write the other terms or ideas related to that term around it.

See the Note-Taking Handbook on pages R45–R51.

SCIENCE NOTEBOOK

NOTES
Types of forces
- contact force
- gravity
- friction

forces on a box being pushed

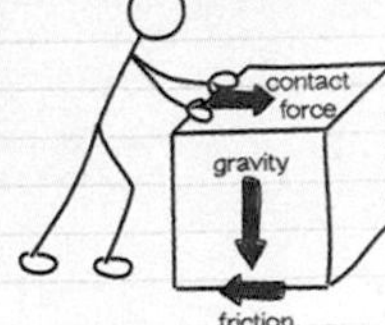

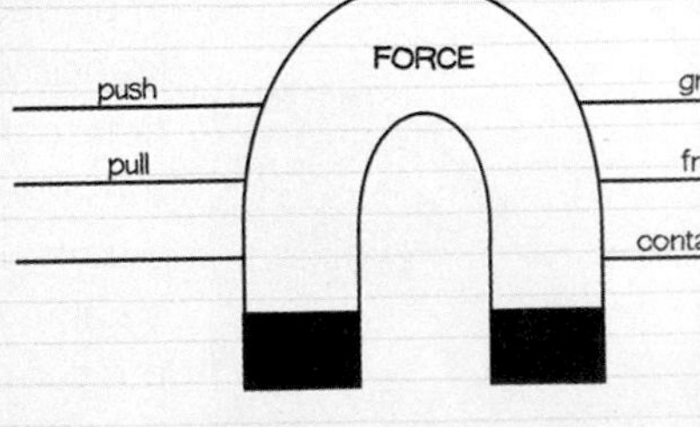

C 40 Unit: Motion and Forces

Read each heading.

See how it fits into the outline of the chapter.

KEY CONCEPT

2.1 Forces change motion.

Remember what you know.

Think about concepts you learned earlier and preview what you'll learn now.

BEFORE, you learned
- The velocity of an object is its change in position over time
- The acceleration of an object is its change in velocity over time

NOW, you will learn
- What a force is
- How unbalanced forces change an object's motion
- How Newton's first law allows you to predict motion

Try the activities.

They will introduce you to science concepts.

VOCABULARY

force p. 41
net force p. 43
Newton's first law p. 45
inertia p. 46

EXPLORE Changing Motion

How can you change an object's motion?

PROCEDURE

1. Choose an object from the materials list and change its motion in several ways, from
   - not moving to moving
   - moving to not moving
   - moving to moving faster
   - moving to moving in a different direction
2. Describe the actions used to change the motion.
3. Experiment again with another object. First, decide what you will do; then predict how the motion of the object will change.

MATERIALS
- quarter
- book
- tennis ball
- cup
- feather

WHAT DO YOU THINK?
In step 3, how were you able to predict the motion of the object?

A force is a push or a pull.

Think about what happens during an exciting moment at the ballpark. The pitcher throws the ball across the plate, and the batter hits it high up into the stands. A fan in the stands catches the home-run ball. In this example, the pitcher sets the ball in motion, the batter changes the direction of the ball's motion, and the fan stops the ball's motion. To do so, each must use a **force,** or a push or a pull.

REMINDER

Motion is a change in position over time.

Learn the vocabulary.

Take notes on each term.

You use forces all day long to change the motion of objects in your world. You use a force to pick up your backpack, to open or close a car door, and even to move a pencil across your desktop. Any time you change the motion of an object, you use a force.

Chapter 2: **Forces** **41** C

Reading Text and Visuals

Objects at rest and objects in motion both resist changes in motion. That is, objects at rest tend to stay at rest, and objects that are moving tend to continue moving unless a force acts on them. Galileo reasoned there was no real difference between an object that is moving at a constant velocity and an object that is standing still. An object at rest is simply an object with zero velocity.

How were Galileo's ideas about objects in motion different from the ideas of the ancient Greeks?

Newton's First Law

Newton restated Galileo's conclusions as his first law of motion. **Newton's first law** states that objects at rest remain at rest, and objects in motion remain in motion with the same velocity, unless acted upon by an unbalanced force. You can easily observe the effects of unbalanced forces, both on the ball at rest and the ball in motion, in the pictures below.

Newton's First Law

Objects at rest remain at rest, and objects in motion remain in motion with the same velocity, unless acted upon by an unbalanced force.

An Object at Rest

An object at rest (the ball) remains at rest unless acted upon by an unbalanced force (from the foot).

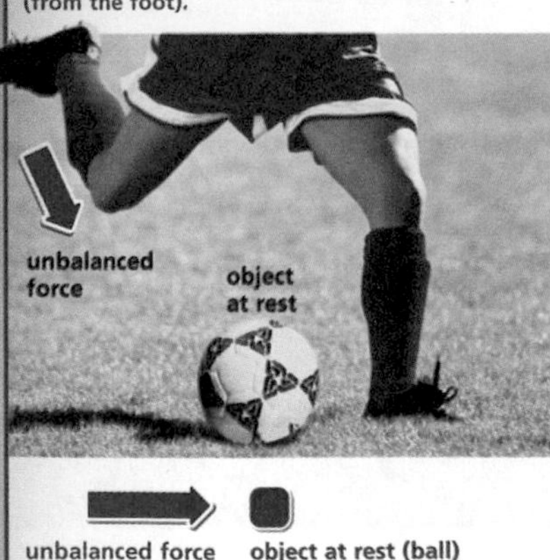

unbalanced force (from the foot) — object at rest (ball)

An Object in Motion

An object in motion (the ball) remains in motion with the same velocity, unless acted upon by an unbalanced force (from the hand).

object in motion (ball) — unbalanced force (from the hand)

READING VISUALS What will happen to the ball's motion in both pictures? Why?

Chapter 2: **Forces** **45** C

Read one paragraph at a time.

Look for a topic sentence that explains the main idea of the paragraph. Figure out how the details relate to that idea. One paragraph might have several important ideas; you may have to reread to understand.

Answer the questions.

Check Your Reading questions will help you remember what you read.

Study the visuals.

- Read the title.
- Read all labels and captions.
- Figure out what the picture is showing. Notice colors, arrows, and lines.
- Answer the question. **Reading Visuals** questions will help you understand the picture.

Doing Labs

To understand science, you have to see it in action. Doing labs helps you understand how things really work.

① Read the entire lab first.

② Form a hypothesis.

③ Follow the procedure.

④ Record the data.

CHAPTER INVESTIGATION

Newton's Laws of Motion

OVERVIEW AND PURPOSE As you know, rocket engineers consider Newton's laws when designing rockets and planning rocket flights. In this investigation you will use what you have learned about Newton's laws to
- build a straw rocket
- improve the rocket's performance by modifying one design element

Problem Write It Up

What aspects of a model rocket affect the distance it flies?

Hypothesize Write It Up

After step 8 in the procedure, write a hypothesis to explain what you predict will happen during the second set of trials. Your hypothesis should take the form of an "If . . . , then . . . , because . . ." statement.

MATERIALS
- 2 straws with different diameters
- several plastic bottles, in different sizes
- modeling clay
- scissors
- construction paper
- meter stick
- tape

Procedure

1. Make a data table like the one shown on the sample notebook page.
2. Insert the straw with the smaller diameter into one of the bottles. Seal the mouth of the bottle tightly with modeling clay so that air can escape only through the straw. This is the rocket launcher.
3. Cut two thin strips of paper, one about 8 cm long and the other about 12 cm long. Connect the ends of the strips to make loops.
4. To create the rocket, place the straw with the larger diameter through the smaller loop and tape the loop to the straw at one end. Attach the other loop to the other end of the straw in the same way. Both loops should be attached to the same side of the straw to stabilize your rocket in flight.

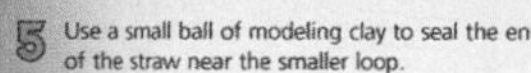

5. Use a small ball of modeling clay to seal the end of the straw near the smaller loop.
6. Slide the open end of the rocket over the straw on the launcher. Place the bottle on the edge of a table so that the rocket is pointing away from the table.
7. Test launch your rocket by holding the bottle with two hands and squeezing it quickly. Measure the distance the rocket lands from the edge of the table. Practice the launch several times. Remember to squeeze with equal force each time.
8. Launch the rocket four times. Keep the amount of force you use constant. Measure the distance the rocket travels each time, and record the results in your data table.
9. List all the variables that may affect the distance your rocket flies. Change the rocket or launcher to alter one variable. Launch the rocket and measure the distance it flies. Repeat three more times, and record the results in your data table.

Observe and Analyze Write It Up

1. **RECORD OBSERVATIONS** Draw a diagram of both of your bottle rockets. Make sure your data table is complete.
2. **IDENTIFY VARIABLES** What variables did you identify, and what variable did you modify?

Conclude Write It Up

1. **COMPARE** How did the flight distances of the original rocket compare with those of the modified rocket?
2. **ANALYZE** Compare your results with your hypothesis. Do the results support your hypothesis?
3. **IDENTIFY LIMITS** What possible limitations or errors did you experience or could you have experienced?
4. **APPLY** Use Newton's laws to explain why the rocket flies.
5. **APPLY** What other real life example can you think of that demonstrates Newton's laws?

INVESTIGATE Further

CHALLENGE Why does the rocket have paper loops taped to it? Determine how the flight of the rocket is affected if one or both loops are completely removed. Hypothesize about the function of the paper loops and design an experiment to test your hypothesis.

Newton's Laws of Motion

Problem What aspects of a model rocket affect the distance it flies?

Hypothesize

Observe and Analyze

Table 1. Flight Distances of Original and Modified Rocket

Trial Number	Original Rocket: Distance Rocket Flew (cm)	Modified Rocket: Distance Rocket Flew (cm)
1		
2		
3		
4		

Conclude

C 62 Unit: Motion and Forces

Chapter 2: Forces 63 C

⑤ Analyze your results.

⑥ Write your lab report.

Using Technology

The Internet is a great source of information about up-to-date science. The ClassZone Web site and SciLinks have exciting sites for you to explore. Video clips and simulations can make science come alive.

Look for red banners.

Go to **ClassZone.com** to see simulations, visualizations, resource centers and content review.

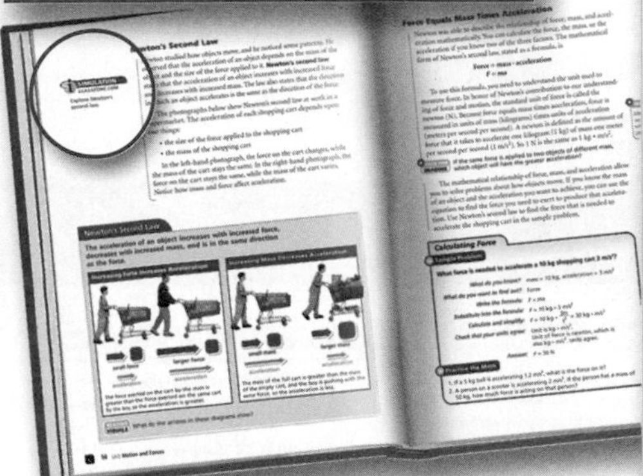

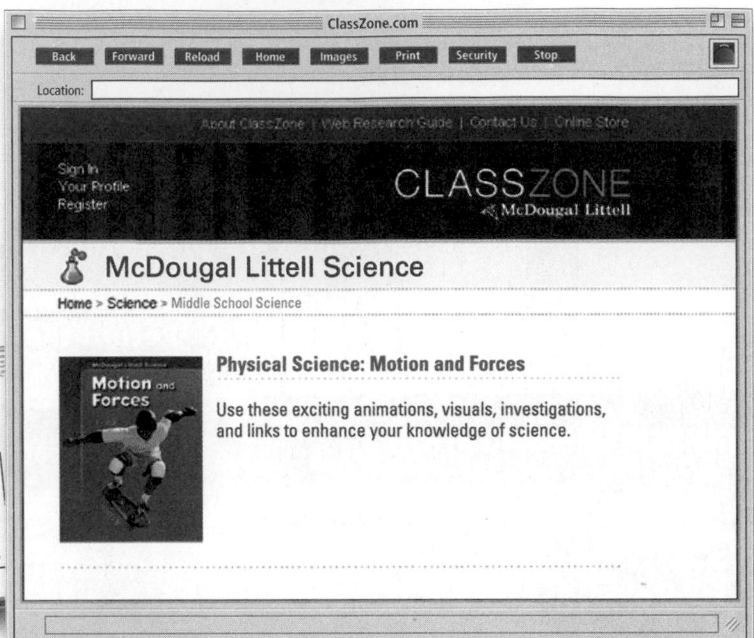

Watch the videos.

Look for the **Scientific American Frontiers video** summary in the unit opener.

Look up SciLinks.

Go to **scilinks.org** to explore the topic.

Forces **Code: MDL005**

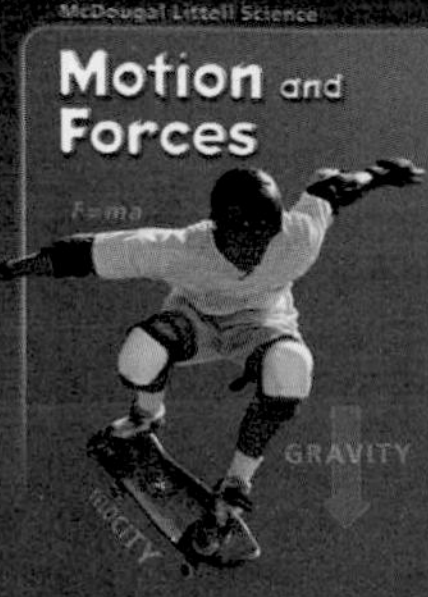

Motion and Forces Contents Overview

Unit Features

FRONTIERS IN SCIENCE Robots on Mars 2
TIMELINES IN SCIENCE Understanding Forces 108

1 Motion 6

the BIG idea
The motion of an object can be described and predicted.

2 Forces 38

the BIG idea
Forces change the motion of objects in predictable ways.

3 Gravity, Friction, and Pressure 74

the BIG idea
Newton's laws apply to all forces.

4 Work and Energy 112

the BIG idea
Energy is transferred when a force moves an object.

5 Machines 142

the BIG idea
Machines help people do work by changing the force applied to an object.

FRONTIERS in Science

VIDEO SUMMARY

SCIENTIFIC AMERICAN FRONTIERS

TEETERING TO VICTORY This 12-minute video documents an exciting annual contest held at MIT—Massachusetts Institute of Technology—in which students build machines using principles of force and motion. Their devices compete against an opponent's machine to pull down their end of a teeter-totter beam—and they have 45 seconds to defeat that opponent's machine. Students design their own technology with very specific constraints: they are given identical kits of parts, each machine can weigh no more than 10 pounds, and it must fit back into the box the parts came in. The top contenders are carpet grabbers, bulldozers, and mobile jacks. As different competitors are eliminated, viewers get caught up in the contest while learning about technological design.

National Science Education Standards

A.9.a–d Understandings About Scientific Inquiry

E.6.a–f Understandings About Science and Technology

F.5.a–e Science and Technology in Society

G.1.a–b Science as a Human Endeavor

G.2.a Nature of Science

FRONTIERS in Science

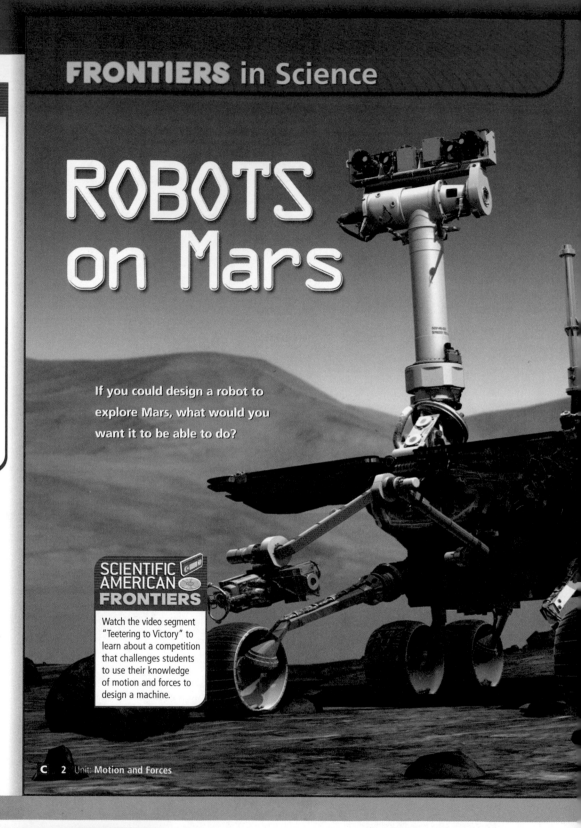

ROBOTS on Mars

If you could design a robot to explore Mars, what would you want it to be able to do?

SCIENTIFIC AMERICAN FRONTIERS

Watch the video segment "Teetering to Victory" to learn about a competition that challenges students to use their knowledge of motion and forces to design a machine.

C 2 Unit: Motion and Forces

ADDITIONAL RESOURCES

Technology Resources

Scientific American Frontiers Video: *Teetering to Victory:* 12-minute video segment that introduces the unit

ClassZone.com
CAREER LINK: physicists, mechanical engineer

Guide student viewing and comprehension of the video:

Frontiers in Science Teaching Guide, pp. 1–2; Viewing Guide, p. 3, Video Wrap-Up, p. 4

Scientific American Frontiers Video Guide, pp. 47–50

Unit projects procedures and rubrics:

Unit Projects, pp. 5–10

The surface of Mars looks rocky and barren today, but scientists have long wondered if life might have existed on Mars long ago. That would have been possible only if Mars once had water, which is necessary for all forms of life.

The Design Challenge

It's still not possible to send scientists to Mars to search for signs of water, but in 1999 a team of scientists and engineers began to design two robots for NASA's 2004 mission to Mars. As the team worked, they relied on their scientific understanding of motion, forces, and machines to create and test a successful design. A

To identify their goals, the team started by thinking about what scientists would want to do if they could go to Mars. First they would want to look around the landscape to find good areas to study. Then they would need to travel to those areas and analyze rock samples. Finally they would use a variety of tools to analyze the rocks, interpret their data, and communicate their findings back to Earth. Those goals set the basic plan for the Mars Exploration Rovers (MERs).

NASA's Mars Exploration Rover (MER) shown in a computer-simulated Martian landscape

As you can see in the photograph, the MER team designed a rover with cameras for viewing the surface, wheels for moving around the landscape, and an extendable arm in front equipped with tools for drilling into, observing, and identifying rocks. The rover also has a computer to process information, an antenna for radio communication with Earth, and batteries and solar panels to provide energy for everything. B

As in any technology project, the MER team had to work within specific constraints, or limits. The most basic constraints were time and money. They had to design rovers that could be built within NASA's budget and that would be ready in time for launch in 2003. But the team also faced some more challenging

DIFFERENTIATE INSTRUCTION

More Reading Support

A What is the purpose of the robots? *to explore Mars*

B What does this paragraph describe? *the MER robot that went to Mars*

Advanced Have students research the MER project on the NASA Web site or in periodicals and draw a diagram of a MER's parts. Add labels to include additional information about the robot. Have interested students prepare a presentation to the class featuring images and other data sent by the MERs *Spirit* and *Opportunity*. Have them answer the question: What tentative conclusions have scientists reached in answer to their questions?

FOCUS

Set Learning Goals

Students will

- Learn how scientists explore the surface of Mars using robots.
- Observe how scientists design technology to answer questions.
- Explore the purposes and constraints of technology design.

Introduce

The MERs *Spirit* and *Opportunity* successfully landed on Mars in January 2004 and immediately began sending images of their surroundings to scientists on Earth. Every bit of data collected from the rovers is analyzed for the new information it yields.

INSTRUCT

Scientific Process

Ask students what question scientists were asking that caused them to develop the Mars robots. *Did life ever exist on Mars? Did Mars ever have liquid water?*

Teach from Visuals

Point out parts of the MER robot in the picture. The camera holder on top can rotate 360° and is mounted on a mast that stands almost 5 feet high. The jointed arm in front maneuvers the RAT, the Rock Abrasion Tool that drills into rock samples. Two antennae are visible: a long one parallel to the mast, and the disc-shaped one. Point out the jointed legs and the size of the wheels (10 inches in diameter).

Technology Design

Focus students' attention on "The Design Challenge." Elicit from them, and write on the board, a list of the tasks the technology needed to accomplish. For each requirement, ask what type of technology was designed to meet it. *Examples: take pictures—camera; energy source—batteries, solar panels*

Technology Design

Sharing results and learning from the past help scientists design better technology. Ask students what design change to *Sojourner* allows a MER to make more useful visual observations. *The Sojourner was very close to the ground. A MER is larger and has cameras on top of a mast that give it a view similar to what a person would see standing on the surface of Mars.*

Scientific Process

The MER team built and used models such as FIDO to conduct testing. Ask students what the purpose of their testing models was. *to make sure all the parts worked* Then ask students to brainstorm why scientists used models instead of a completed MER. *Sample answers: cost, might want to make sure an individual part worked, no need to use whole robot to test one part*

Members of the project team stand with a MER and a replica of the much smaller *Sojourner.*

constraints. The rover must survive a rocket launch from Earth as well as a landing on the surface of Mars. This means it must be both lightweight and compact. Engineers designed the MER to fold up into a pyramid-shaped protective compartment, which drops down onto Mars by parachute. Air bags surrounding the compartment absorb the impact, and then the compartment opens and the MER moves down the compartment panels to the planet's surface.

C Scientists built on some valuable lessons learned from an earlier robot, *Sojourner,* which explored the surface of Mars for 12 weeks in 1997. At the left you see one of the MERs next to a replica of *Sojourner,* which was only about 28 centimeters (about 11 in) tall. MER's mast rises up to 1.4 meters (almost 5 ft), giving the cameras, which can be angled up or down, a view similar to what a person would see when standing on the surface of Mars.

Testing the Model

Every part of the MER had to be tested to be sure it would work properly in the harsh conditions on Mars. For example, consider the Rock Abrasion Tool (RAT) at the end of the rover's extendable arm. The RAT is designed to grind off the weathered surface of rock, exposing a fresh surface for examination. Tests with the RAT showed that it worked fine on hard rocks, but its diamond-tipped grinding wheel became clogged with pieces of soft rock. The solution: Add brushes to clean the RAT automatically after each use. ? D

Scientists were also concerned that the RAT's diamond grinding wheel might wear out if it had to grind a lot of hard rocks. An entry from the design team's status report explains why that turned out not to be a problem:

View the "Teetering to Victory" segment of your Scientific American Frontiers video to learn how some students solved a much simpler design challenge.

IN THIS SCENE FROM THE VIDEO **MIT students prepare to test their machines.**

BATTLE OF MACHINES Each year more than 100 engineering students at the Massachusetts Institute of Technology (MIT) compete in a contest to see who can design and build the best machine. The challenge this time is to build a machine that starts out sitting on a teeter-totter beam and within 45 seconds manages to tilt its end down against an opponent trying to do the same thing.

Just as the Mars rover designers had to consider the constraints of space travel and Mars' harsh environment, the students had constraints on their designs. They all started with the same kit of materials, and their finished machines had to weigh less than 10 pounds as well as fit inside the box the materials came in. Within these constraints, the student designers came up with an amazing variety of solutions.

DIFFERENTIATE INSTRUCTION

More Reading Support

C What earlier robot explored Mars? *Sojourner*

D What parts of MER had to be tested to make sure they worked? *all parts*

Below Level Students may have trouble with the word *constraint*. Make sure they understand that a constraint is a limit. Have students list constraints the MER team faced. *time, money, difficulties of a rocket launch from Earth and a landing on Mars, space travel, harsh environment of Mars* Then have students list the constraints the MIT students had. *same materials, build machine less than 10 pounds that fits in box*

The big question, of course, was how things would work under the very cold, dry, low-pressure atmospheric conditions on Mars. We put a RAT into a test chamber recently, took it to real Martian conditions for the first time, and got a very pleasant surprise. The rate at which our diamond studded teeth wear away slowed way down! We're still figuring out why, but it turns out that when you put this Martian RAT into its natural environment, its teeth don't wear down nearly as fast.

Engineers also needed to test the system by which scientists on Earth would communicate with and control the rovers on Mars. For this purpose, they built a smaller version of the real robot, nicknamed FIDO. In tests FIDO successfully traveled to several locations, dug trenches, and observed and measured rock samples.

Goals of the Mission

Technology like the Mars Exploration Rovers extends the power of scientists to gather data and answer questions about our solar system. One main goal of the MER missions is to study different kinds of rock and soils that might indicate whether water was ever present on Mars. From the data gathered by the MERs, scientists hope to find out what factors shaped the Martian landscape. They also hope to check out areas that have been studied only from far away so that the scientists can confirm their hypotheses about Mars.

UNANSWERED Questions

As scientists learn more and more about Mars, new questions always arise.

- What role, if any, did water, wind, or volcanoes play in shaping the landscape of Mars?
- Were the conditions necessary to support life ever present on Mars?
- Could there be bacteria-like life forms surviving below the surface of Mars today?

UNIT PROJECTS

As you study this unit, work alone or with a group on one of these projects.

Build a Mechanical Arm

Design and build a mechanical arm to perform a simple task.

- Plan and sketch an arm that could lift a pencil from the floor at a distance of one meter.
- Collect materials and assemble your arm.
- Conduct trials and improve your design.

Multimedia Presentation

Create an informative program on the forces involved in remote exploration.

- Collect information about the Galileo mission to Jupiter or a similar expedition.
- Learn how engineers use air resistance, gravity, and rocket thrusters to maneuver the orbiter close to the planet and its moons.
- Give a presentation describing what you learned using mixed media, such as a computer slide show and a model.

Design an Experiment

Design an experiment to determine the pressure needed to crush a small object.

- Select a small object, such as a vitamin C tablet, to use in your experiment.
- Collect other materials of your choosing.
- Plan and conduct a procedure to test the pressure required to crush the object. Vary the procedure until you can crush the object using the least amount of force.

CAREER CENTER CLASSZONE.COM

Learn more about careers in physics and engineering.

UNANSWERED Questions

Have students read the questions and think of some of their own. Remind them that scientists always end up with more questions—that inquiry is the driving force of science.

- With the class, generate on the board a list of new questions.
- Students can add to the list after they watch the Scientific American Frontiers Video.
- Students can use the list as a springboard for choosing their Unit Projects.

UNIT PROJECTS

Encourage students to pick the project that most appeals to them. Point out that each is long-term and will take several weeks to complete. You might group or pair students to work on projects, and in some cases guide student choice. Some of the projects have student choice built into them. Each project has two worksheet pages, including a rubric. Use the pages to guide students through criteria, process, and schedule.

 Unit Projects, pp. 5–10

Technology Resources

Visit **ClassZone.com** for project procedures and for science career direction.

 **RESOURCE CENTER,** Unit Projects

REVISIT concepts introduced in this article:

Chapter 2, Forces change motion, pp. 41–47

Chapter 3, Gravity is a force exerted by masses, pp. 77–83

Chapter 4, Work is the use of force to move an object, pp. 115–119

Chapter 5, Machines help people do work, pp. 145–152; Six simple machines have many uses, pp. 154–162; Modern technology uses compound machines, pp. 164–169

DIFFERENTIATE INSTRUCTION

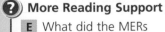

More Reading Support

E What did the MERs study that might indicate whether water was ever present? *rocks and soil*

Differentiate Unit Projects Projects are appropriate for varying abilities. Allow students to choose the ones that interest them most. Encourage them to vary the products they produce throughout the year. Below-level students might try "Multimedia Presentation." Challenge advanced students to try "Design an Experiment."

Motion

Physical Science UNIFYING PRINCIPLES

PRINCIPLE 1
Matter is made of particles too small to see.

PRINCIPLE 2
Matter changes form and moves from place to place.

PRINCIPLE 3
Energy changes from one form to another, but it cannot be created or destroyed.

PRINCIPLE 4
Physical forces affect the movement of all matter on Earth and throughout the universe.

Unit: Motion and Forces BIG IDEAS

CHAPTER 1 Motion
The motion of an object can be described and predicted.

CHAPTER 2 Forces
Forces change the motion of objects in predictable ways.

CHAPTER 3 Gravity, Friction, and Pressure
Newton's laws apply to all forces.

CHAPTER 4 Work and Energy
Energy is transferred when a force moves an object.

CHAPTER 5 Machines
Machines help people do work by changing the force applied to an object.

CHAPTER 1 KEY CONCEPTS

SECTION 1.1

An object in motion changes position.

1. Position describes the location of an object.
2. Motion is a change in position.

SECTION 1.2

Speed measures how fast position changes.

1. Position can change at different rates.
2. Velocity includes speed and direction.

SECTION 1.3

Acceleration measures how fast velocity changes.

1. Speed and direction can change with time.
2. Acceleration can be calculated from velocity and time.

The Big Idea Flow Chart is available on p. T1 in the **UNIT TRANSPARENCY BOOK.**

Previewing Content

SECTION

An object in motion changes position. pp. 9–15

1. Position describes the location of an object.
The **position,** or location, of an object is described relative to a **reference point.** The choice of reference point affects how the position is described. For example, a city can be located by measuring its direction and distance from another city, or by using a grid system such as the longitude-latitude system.

There are two ways to measure the distance an object has traveled. One is to measure the length of the path the object followed. Another is to measure the straight-line distance of an object from its starting point. This straight-line distance is called the displacement of the object.

2. Motion is a change in position.
Motion is a change in position over time. How quickly or slowly the position changes depends on the object's speed.

How motion is observed depends upon the observer's point of view. The observed motion of an object is measured by comparing the object's motion relative to the observer's frame of reference. Suppose a person throws a ball forward on a moving train. The motion of the ball is measured differently by observers on the train and by observers on the ground outside. An observer on the ground would measure the motion of the ball as being faster than the motion of the train, as shown in the diagram below.

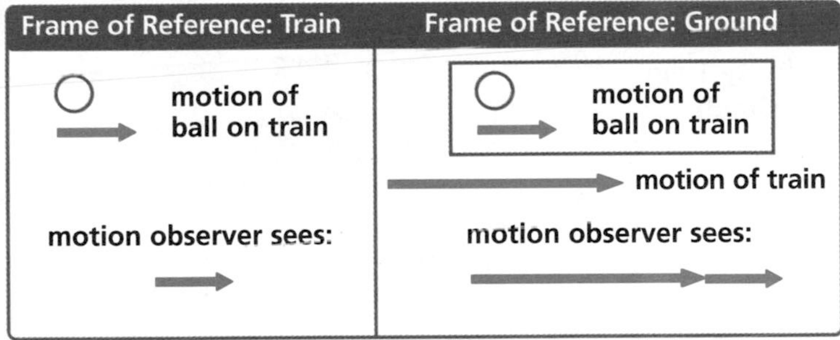

If the ball is thrown backward on the train, an observer on the ground outside would measure the motion of the ball as being slower than the motion of the train.

SECTION

Speed measures how fast position changes. pp. 16–24

1. Position can change at different rates.
Speed is a measure of how fast something moves through a particular distance over a given amount of time.

$$\text{Speed} = \frac{\text{distance}}{\text{time}}, \text{ or } S = \frac{d}{t}$$

Average speed is the average of several instantaneous speeds whose measurements are taken over a specific period of time. A distance-time graph shows how both distance and speed change with time. You can use these graphs to determine the speed of an object by calculating the slope of the line. A positive slope means the object is moving away from its starting point. A negative slope means an object is moving back toward its starting point.

$$\text{slope} = \frac{\text{change in distance}}{\text{change in time}} = \text{speed}$$

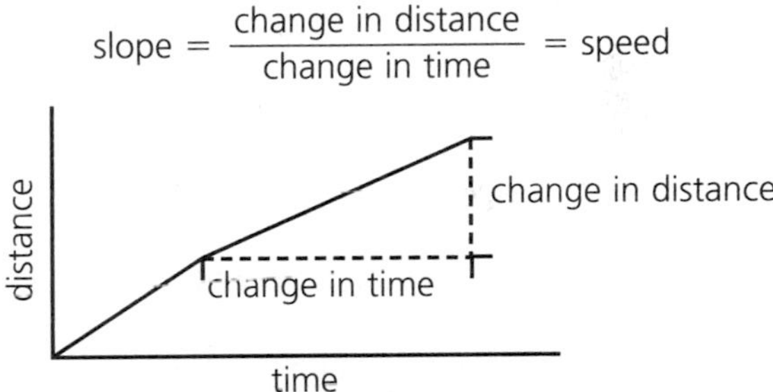

2. Velocity includes speed and direction.
Velocity is speed in a specific direction. Velocity is an example of a vector. A **vector** is a quantity that has both size and direction. Vectors are shown by arrows. The longer the arrow, the faster the speed. The direction of the arrow indicates the direction of motion.

Speed and velocity are not the same. If two runners run at the same speed in opposite directions, they will have identical speed but different velocities.

Previewing Content

SECTION

1.3 Acceleration measures how fast velocity changes. pp. 25–33

1. Speed and direction can change with time.
Acceleration is the rate at which velocity changes with time. Contrary to a popular misconception, acceleration is not limited to increases in velocity but includes *any* change in velocity. The following are examples of acceleration:

- speed increases
- speed decreases
- direction changes (regardless of speed)

2. Acceleration can be calculated from velocity and time.
You determine acceleration from the change in velocity and how long the change took. The formula for calculating acceleration is shown below.

$$a = \frac{v_{final} - v_{initial}}{t}$$

Negative acceleration is a decrease in velocity during a specific period of time. The acceleration formula yields a negative result when the final velocity is less than the initial velocity.
A velocity-time graph shows how both velocity and acceleration change with time. The graph on the right below is a velocity-time graph; compare it with the distance-time graph for the same set of data, which is shown below. The graphs show a boy (1) starting and speeding up on his scooter, (2) coasting (velocity is constant), and (3) slowing to a stop.

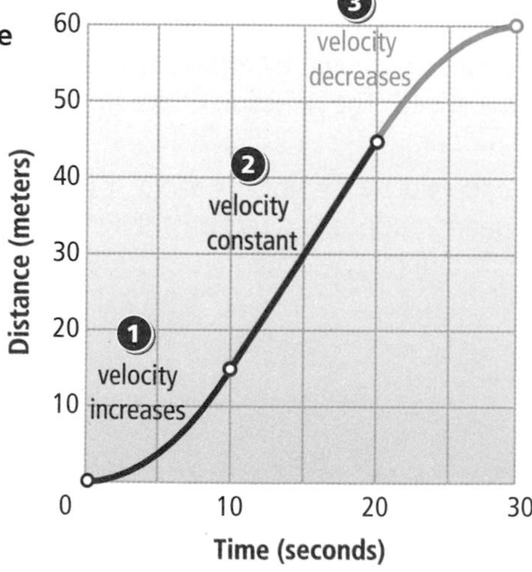

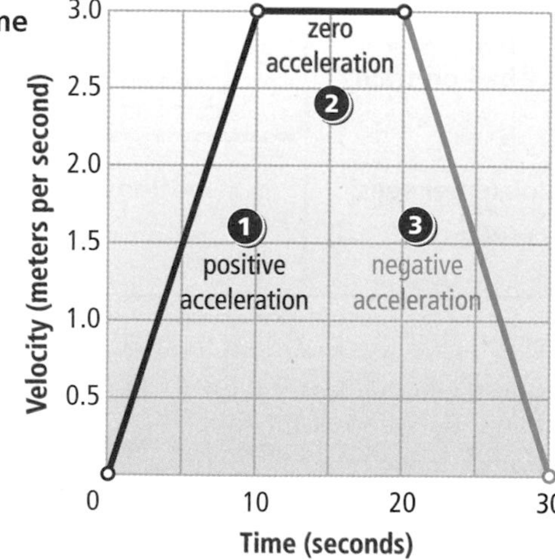

Common Misconceptions

MISCONCEPTION DATABASE
CLASSZONE.COM Background on student misconceptions

ACCELERATION EQUATED WITH SPEEDING UP Acceleration describes *any* change in velocity. However, students may hold the misconception that acceleration means only an increase in velocity.

This misconception is addressed on p. 26.

NEGATIVE ACCELERATION The change in velocity that occurs during a slowing down is described by the term *negative acceleration.* Students may confuse negative acceleration with negative direction, which involves reversing direction.

This misconception is addressed on p. 29.

Previewing Labs

Lab Generator CD-ROM
Edit these Pupil Edition labs and generate alternative labs.

EXPLORE the BIG idea

Off the Wall, p. 7 Students time a rolling ball to understand that speed is related to time and distance.	**TIME** 10 minutes **MATERIALS** rubber ball, stopwatch
Rolling Along, p. 7 Students roll a marble on a ramp to observe that speed can change.	**TIME** 10 minutes **MATERIALS** hardbound books, marble
Internet Activity: Relative Motion, p. 7 Students are introduced to relative motion.	**TIME** 20 minutes **MATERIALS** computer with Internet access

SECTION 1.1

EXPLORE Location, p. 9 Students describe the location of an object by giving directions.	**TIME** 10 minutes **MATERIALS** none
INVESTIGATE Changing Positions, p. 12 Students observe motion from varying vantage points.	**TIME** 20 minutes **MATERIALS** small ball, paper, pencil

SECTION 1.2

EXPLORE Speed, p. 16 Students time a tennis ball rolled at different speeds.	**TIME** 10 minutes **MATERIALS** 20 cm masking tape, meter stick, tennis ball, stopwatch
INVESTIGATE Speed and Distance, p. 19 Students design a car to discover how shape affects speed.	**TIME** 20 minutes **MATERIALS** clay, film container lids, 4 toothpicks, beam balance, board (1 m in length), 2–3 small books, 3 m string, straw, scissors, stopwatch

SECTION 1.3

INVESTIGATE Acceleration, p. 27 Students measure the acceleration that occurs in different types of movements. R Template for Tool, p. 42	**TIME** 30 minutes **MATERIALS** Template for Tool, cardboard (8.5 x 11 in), scissors, glue, a 25 cm piece of string, weight (such as a washer)
CHAPTER INVESTIGATION **Acceleration and Slope,** pp. 32–33 Students investigate how steepness of slope affects acceleration of a marble.	**TIME** 40 minutes **MATERIALS** 2 meter sticks, 50 cm masking tape, marble,2 paperback books, metric ruler, stopwatch, calculator

R **Additional INVESTIGATION,** On the Move, A, B, & C, pp. 65–73; Teacher Instructions, pp. 346–347

Previewing Chapter Resources

Chapter 1 Motion

INTEGRATED TECHNOLOGY

CLASSZONE.COM
- eEdition Plus
- EasyPlanner Plus
- Misconception Database
- Content Review
- Test Practice
- Visualization
- Simulation
- Resource Centers
- Internet Activity: Relative Motion
- Math Tutorial

SCILINKS.ORG

SCILINKS

CD-ROMS
- eEdition
- EasyPlanner
- Power Presentations
- Content Review
- Lab Generator
- Test Generator

AUDIO CDS
- Audio Readings
- Audio Readings in Spanish

LABS AND ACTIVITIES

EXPLORE the Big Idea, p. 7
- Off the Wall
- Rolling Along
- Internet Activity: Relative Motion

UNIT RESOURCE BOOK
- Family Letter, p. ix
- Spanish Family Letter, p. x
- Unit Projects, pp. 5–10

Lab Generator CD-ROM
Generate customized labs.

SECTION

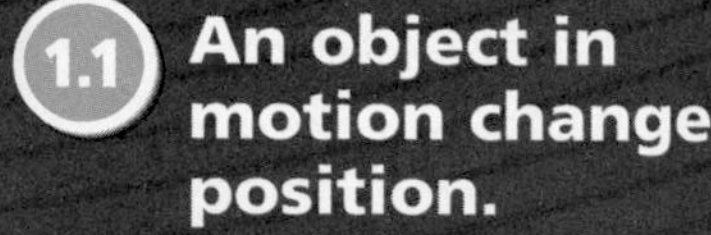

1.1 An object in motion changes position.
pp. 9–15

Time: 2 periods (1 block)

 Lesson Plan, pp. 11–12

Integrated Technology

RESOURCE CENTER, Finding Position

UNIT TRANSPARENCY BOOK
- Big Idea Flow Chart, p. T1
- Daily Vocabulary Scaffolding, p. T2
- Note-Taking Model, p. T3
- 3-Minute Warm-Up, p. T4

Labs and Activities

- EXPLORE Location, p. 9
- INVESTIGATE Changing Positions, p. 12
- Science on the Job, p. 15

UNIT RESOURCE BOOK
Datasheet, Changing Positions, p. 20

SECTION

1.2 Speed measures how fast position changes.
pp. 16–24

Time: 2 periods (1 block)

 Lesson Plan, pp. 22–23

Integrated Technology

MATH TUTORIAL

UNIT TRANSPARENCY BOOK
- Daily Vocabulary Scaffolding, p. T2
- 3-Minute Warm-Up, p. T4
- "Distance Time Graph," Visual, p. T6

Labs and Activities

- EXPLORE Speed, p. 16
- INVESTIGATE Speed and Distance, p. 19
- Math in Science, p. 24

UNIT RESOURCE BOOK
- Datasheet, Speed and Distance, p. 31
- Math Support, pp. 50, 54
- Math Practice, pp. 51, 55

SECTION

1.3 Acceleration measures how fast velocity changes.
pp. 25–33

Time: 4 periods (2 blocks)

 Lesson Plan, pp. 33–34

Integrated Technology

- **RESOURCE CENTER,** Acceleration
- **SIMULATION,** Changing Acceleration

UNIT TRANSPARENCY BOOK
- Big Idea Flow Chart, p. T1
- Daily Vocabulary Scaffolding, p. T2
- 3-Minute Warm-Up, p. T5
- Chapter Outline, pp. T7–T8

Labs and Activities

- THINK ABOUT How does velocity change? p. 25
- INVESTIGATE Acceleration, p. 27
- CHAPTER INVESTIGATION, Acceleration and Slope, pp. 32–33

UNIT RESOURCE BOOK
- Template for Tool, p. 42
- Datasheet, Acceleration, p. 43
- Math Support and Practice, pp. 52–53
- CHAPTER INVESTIGATION, Acceleration and Slope, A, B, & C, pp. 56–64
- Additional INVESTIGATION, On the Move, A, B, & C, pp. 65–73

 KEY TO ICONS  CD/CD-ROM 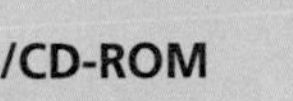 Teacher Edition  UNIT TRANSPARENCY BOOK 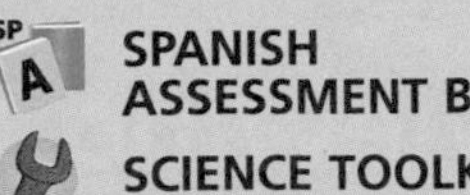 SPANISH ASSESSMENT BOOK

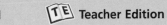

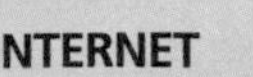 INTERNET Pupil Edition UNIT RESOURCE BOOK UNIT ASSESSMENT BOOK SCIENCE TOOLKIT

READING AND REINFORCEMENT	ASSESSMENT	STANDARDS
 • Description Wheel, B20–21 • Outline, C43 • Daily Vocabulary Scaffolding, H1–8  **UNIT RESOURCE BOOK** • Vocabulary Practice, pp. 47–48 • Decoding Support, p. 49 • Summarizing the Chapter, pp. 74–75  **Audio Readings CD** Listen to Pupil Edition.  **Audio Readings in Spanish CD** Listen to Pupil Edition in Spanish.	 • Chapter Review, pp. 35–36 • Standardized Test Practice, p. 37  **UNIT ASSESSMENT BOOK** • Diagnostic Test, pp. 1–2 • Chapter Test, A, B, & C, pp. 6–17 • Alternative Assessment, pp. 18–19  Spanish Chapter Test, pp. 257–260  **Test Generator CD-ROM** Generate customized tests.  **Lab Generator CD-ROM** Rubrics for Labs	 National Standards A.2–8, A.9.a, A.9.c–e, B.2.a, E.2–5, E.6.d–f, G.1.b See p. 6 for the standards.
 **UNIT RESOURCE BOOK** • Reading Study Guide, A & B, pp. 13–16 • Spanish Reading Study Guide, pp. 17–18 • Challenge and Extension, p. 19 • Reinforcing Key Concepts, p. 21	 Ongoing Assessment, pp. 9, 10, 13, 14  Section 1.1 Review, p. 14  **UNIT ASSESSMENT BOOK** Section 1.1 Quiz, p. 3	National Standards A.2–7, A.9.a–b, A.9.d–f, B.2.a, E.6.d, E.6.f, G.1.b
 **UNIT RESOURCE BOOK** • Reading Study Guide, A & B, pp. 24–27 • Spanish Reading Study Guide, pp. 28–29 • Challenge and Extension, p. 30 • Reinforcing Key Concepts, p. 32	 Ongoing Assessment, pp. 16–18, 20–23  Section 1.2 Review, p. 23  **UNIT ASSESSMENT BOOK** Section 1.2 Quiz, p. 4	National Standards A.2–8, A.9.a–f, B.2.a, E.2–5, G.1.b
 **UNIT RESOURCE BOOK** • Reading Study Guide, A & B, pp. 35–38 • Spanish Reading Study Guide, pp. 39–40 • Challenge and Extension, p. 41 • Reinforcing Key Concepts, p. 44 • Challenge Reading, pp. 45–46	 Ongoing Assessment, pp. 25, 26, 28–31  Section 1.3 Review, p. 31  **UNIT ASSESSMENT BOOK** Section 1.3 Quiz, p. 5	National Standards A.2–8, A.9.a–f, B.2.a, E.6.e, G.1.b

Previewing Resources for Differentiated Instruction

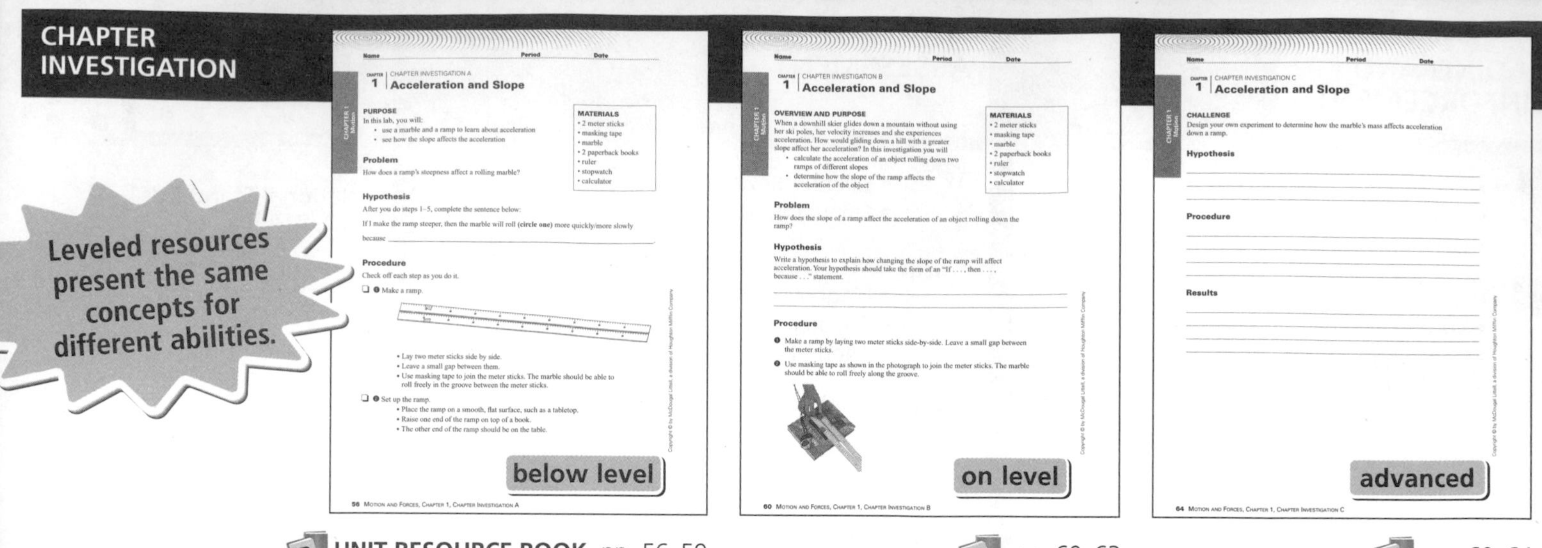

R **UNIT RESOURCE BOOK,** pp. 56–59

R pp. 60–63

R pp. 60–64

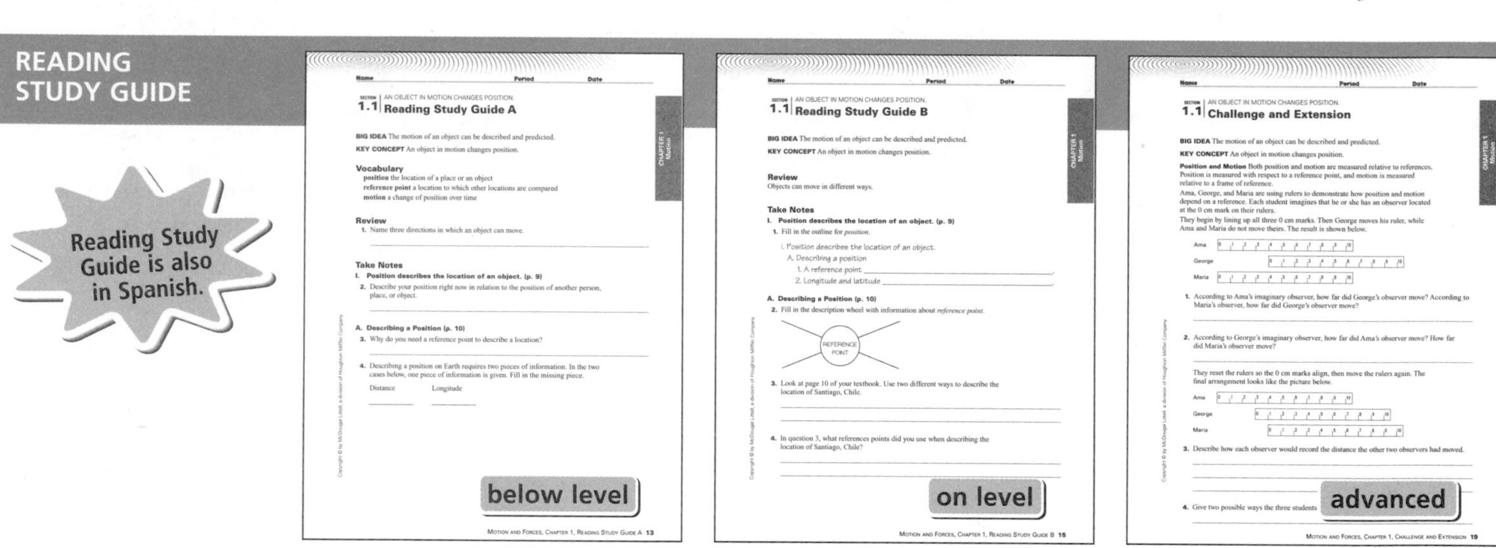

R **UNIT RESOURCE BOOK,** pp. 13–14

R pp. 15–16

R p. 19

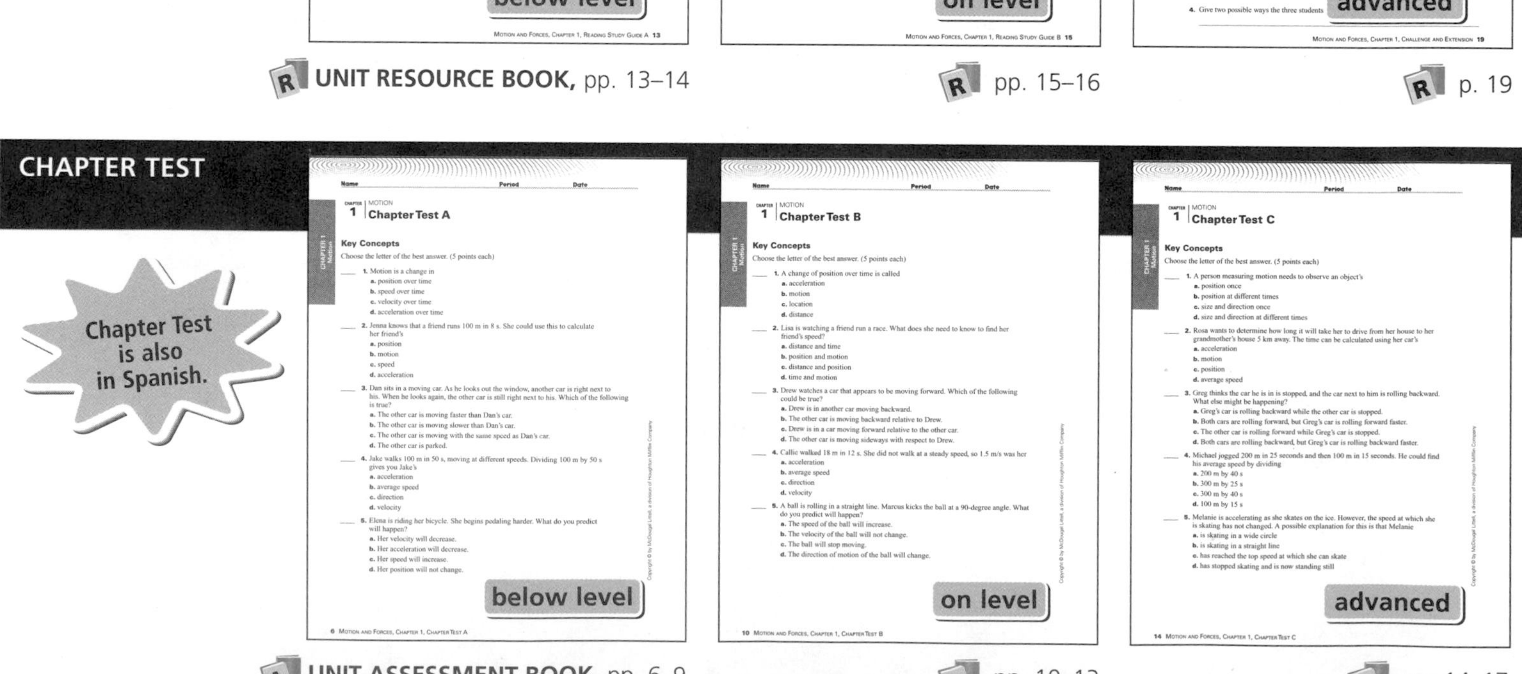

A **UNIT ASSESSMENT BOOK,** pp. 6–9

A pp. 10–13

A pp. 14–17

TECHNOLOGY

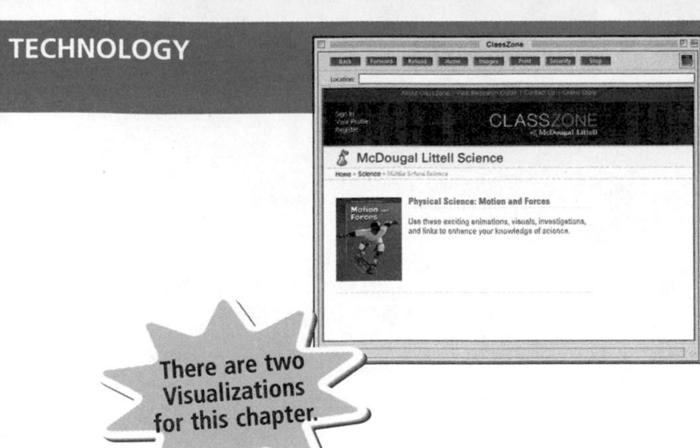

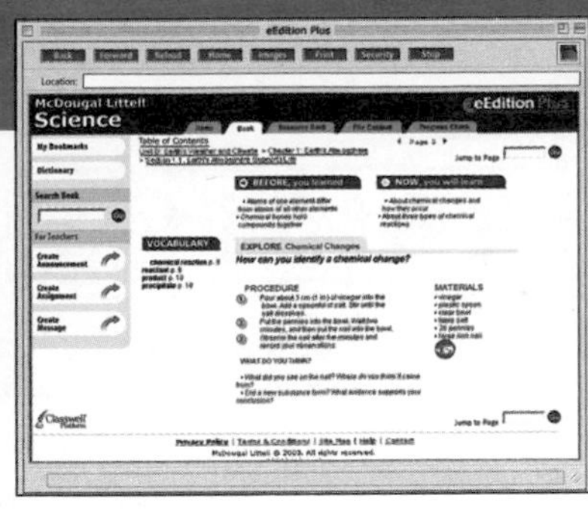

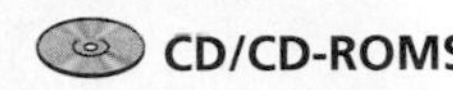

VISUAL CONTENT

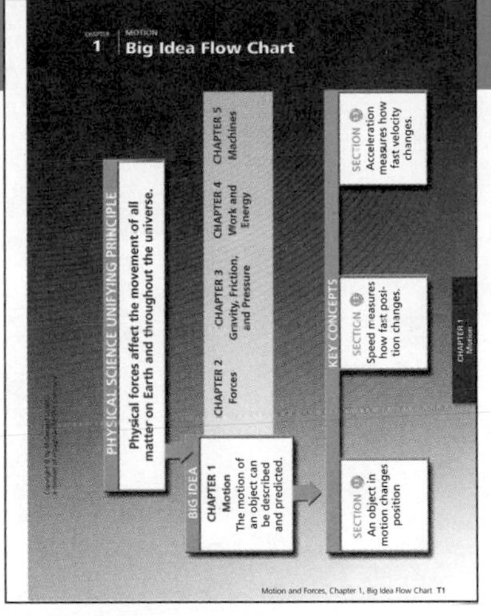

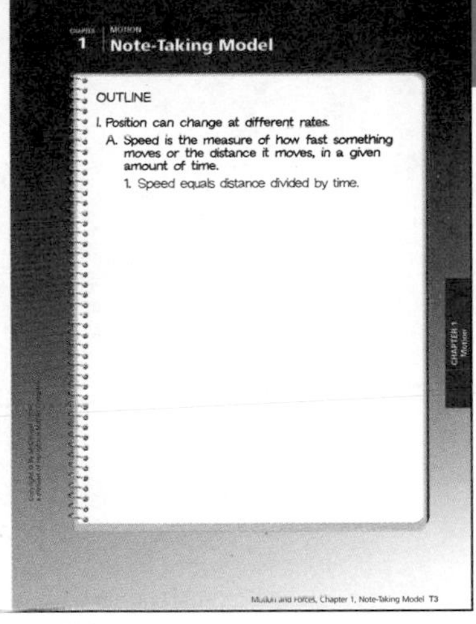

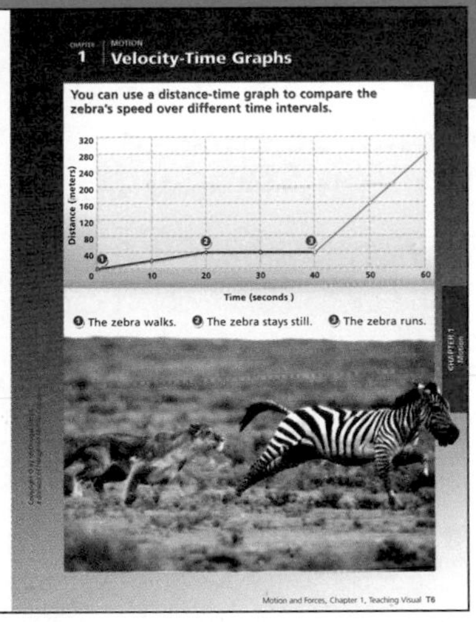

UNIT TRANSPARENCY BOOK, p. T1

p. T3

p. T6

MORE SUPPORT

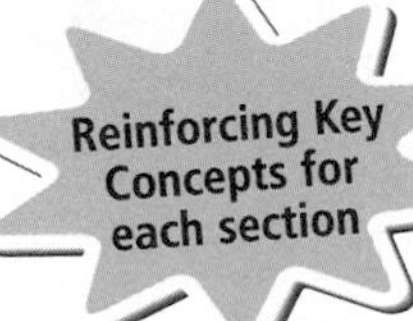

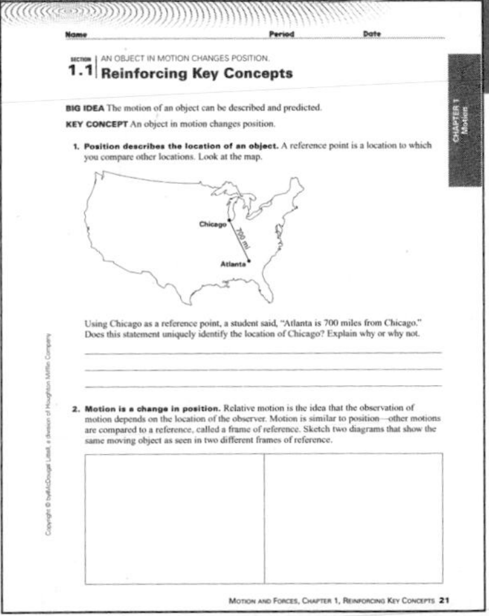

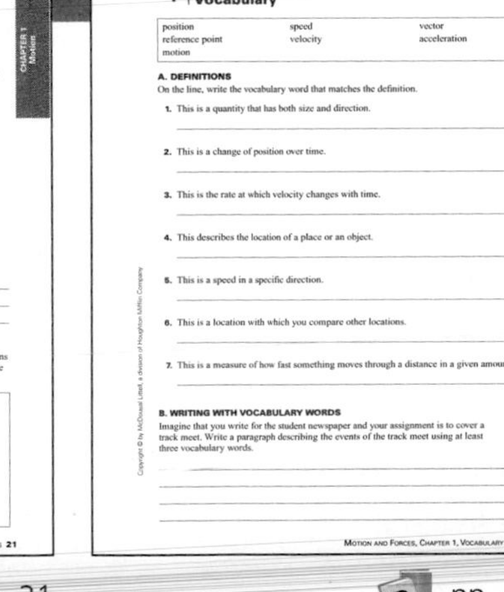

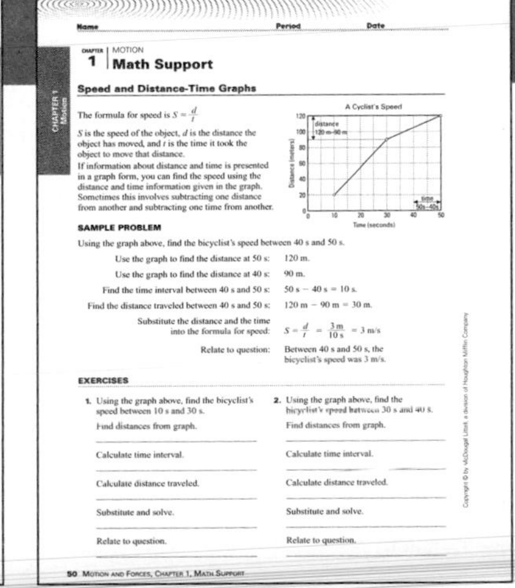

UNIT RESOURCE BOOK, p. 21

pp. 47–48

p. 50

INTRODUCE

the BIG idea

Have students look at the photograph of the roller coaster and discuss how the question in the box links to the Big Idea:

- Do you expect that the speed of the car on the roller coaster will change from time to time? Describe how it might change.
- Will the direction of the track change?
- How does the motion depicted here differ from other familiar motions, such as walking or riding in a car?

National Science Education Standards

Content

B.2.a The motion of an object can be described by its position, direction of motion, and speed. That motion can be measured and represented on a graph.

Process

A.2–8 Design and conduct an investigation; use tools to gather and interpret data; use evidence to describe, predict, explain, model; think critically to make relationships between evidence and explanation; recognize different explanations and predictions; communicate scientific procedures and explanations; use mathematics.

A.9.a–f Understand scientific inquiry by using different investigations, methods, mathematics, technology, and explanations based on logic, evidence, and skepticism.

E.2–5 Design, implement, and evaluate a solution or product; communicate technological design.

E.6.d–f Understandings about science and technology

G.1.b Science as human endeavor

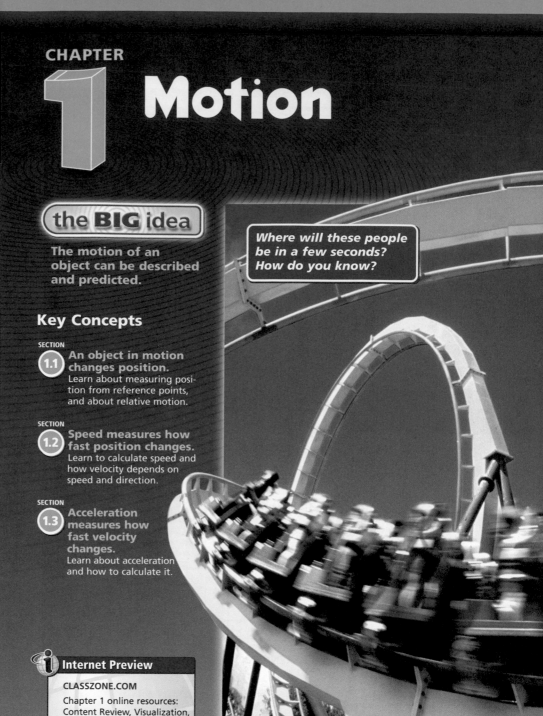

CHAPTER 1 Motion

the BIG idea

The motion of an object can be described and predicted.

Where will these people be in a few seconds? How do you know?

Key Concepts

SECTION 1.1 **An object in motion changes position.** Learn about measuring position from reference points, and about relative motion.

SECTION 1.2 **Speed measures how fast position changes.** Learn to calculate speed and how velocity depends on speed and direction.

SECTION 1.3 **Acceleration measures how fast velocity changes.** Learn about acceleration and how to calculate it.

Internet Preview

CLASSZONE.COM

Chapter 1 online resources: Content Review, Visualization, Simulation, two Resource Centers, Math Tutorial, Test Practice

INTERNET PREVIEW

CLASSZONE.COM For student use with the following pages:

Review and Practice

- Content Review, pp. 8, 34
- Math Tutorial: Units and Rates, p. 24
- Test Practice, p. 37

Activities and Resources

- Internet Activity: Relative Motion, p. 7
- Resource Centers: Finding Position, p. 10; Acceleration, p. 29
- Simulation: Changing Acceleration, p. 31

Velocity **Code: MDL004**

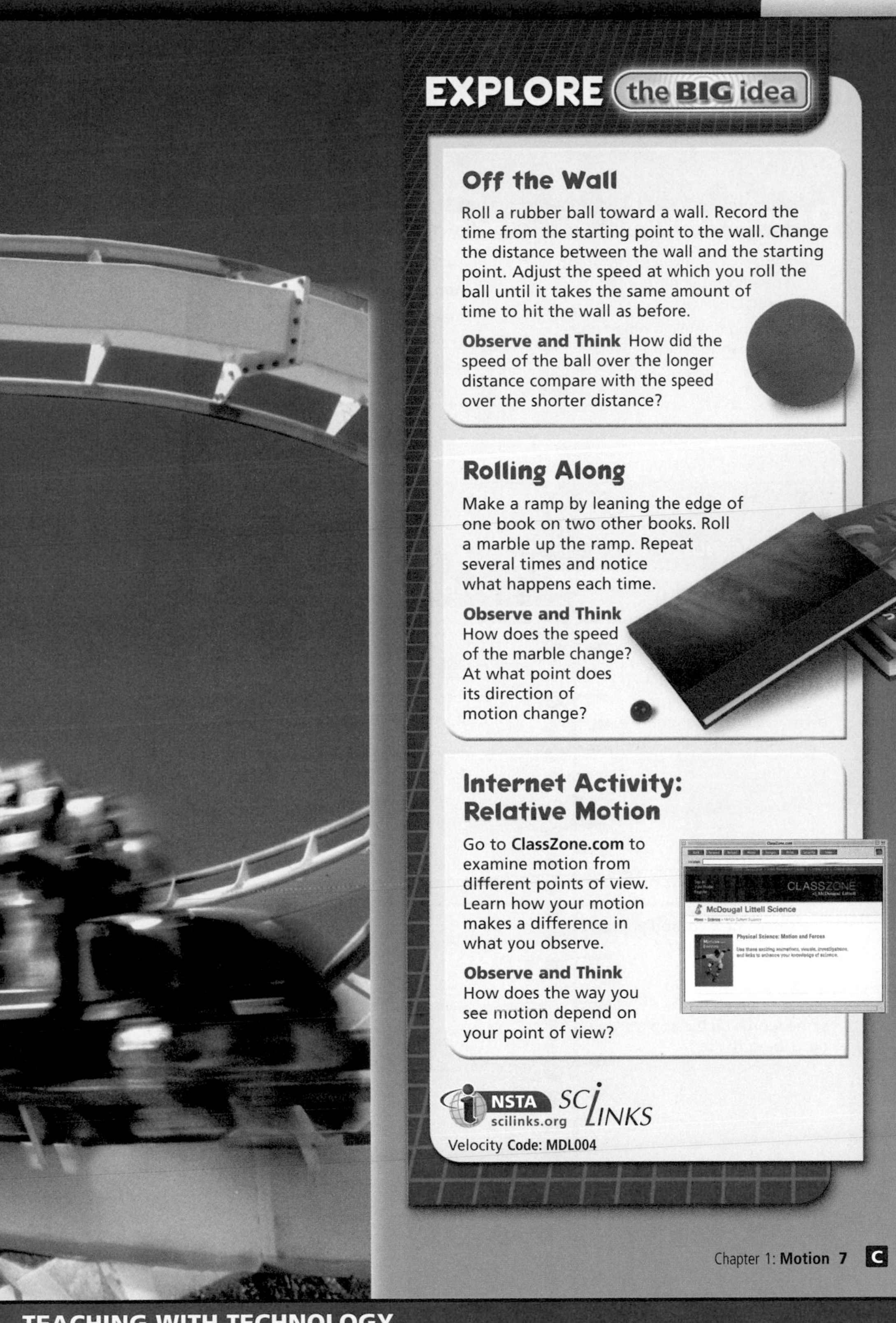

EXPLORE the BIG idea
Off the Wall
Roll a rubber ball toward a wall. Record the time from the starting point to the wall. Change the distance between the wall and the starting point. Adjust the speed at which you roll the ball until it takes the same amount of time to hit the wall as before.
Observe and Think How did the speed of the ball over the longer distance compare with the speed over the shorter distance?
Rolling Along
Make a ramp by leaning the edge of one book on two other books. Roll a marble up the ramp. Repeat several times and notice what happens each time.
Observe and Think How does the speed of the marble change? At what point does its direction of motion change?
Internet Activity: Relative Motion
Go to ClassZone.com to examine motion from different points of view. Learn how your motion makes a difference in what you observe.
Observe and Think How does the way you see motion depend on your point of view?
NSTA scilinks.org SCILINKS
Velocity Code: MDL004
Chapter 1: Motion 7 C

TEACHING WITH TECHNOLOGY

CBL and Probeware If students have probeware, have them use a motion sensor as they push various objects. Motion is introduced on p. 11.

Video Camera You might want to film short clips of students performing variations of the investigation on p. 12. Use them to reinforce the concept that change in position depends on the perspective of the observer.

EXPLORE the BIG idea

These inquiry-based activities are appropriate for use at home or as a supplement to classroom instruction.

Off the Wall

PURPOSE To introduce students to the idea that speed is related to time and distance.

TIP *10 min.* Remind students to study their data for patterns as they contemplate the questions.

Answer: The speed over the longer distance was higher than the speed over the shorter distance.

REVISIT after p. 17.

Rolling Along

PURPOSE To introduce students to the idea that speed can change.

TIP *10 min.* Use a hardbound book so there is no bow or dip in the ramp.

Answer: The marble loses speed while it is rolling up. It changes direction after it stops on its way up the ramp. After it stops, it rolls back down.

REVISIT after p. 26.

Internet Activity: Relative Motion

PURPOSE To introduce students to the idea that how motion is perceived depends on the observer's reference point.

TIP *20 min.* Be sure students understand the difference between a reference point and a frame of reference.

Answer: The way you perceive motion depends on your location and whether you are in motion.

REVISIT after p. 14.

PREPARE

Concept Review

Activate Prior Knowledge

- Have students read the concept review at the top of the page.
- Using a battery-operated toy vehicle or a wind-up toy, demonstrate speed and direction. Have students change the direction of the moving toy. Have them speculate how fast the toy is moving.
- Start the toy again and have someone hold it back. Ask how the speed of the toy is affected. Challenge students to speed up the toy.
- Ask students how this activity demonstrates the principles in the concept review.

Taking Notes

Outline

An outline helps students organize their ideas as they would in writing a paragraph by proceeding from main topic to details.

Vocabulary Strategy

Description wheels help students focus on all the parts of a complex definition. Point out that students can fill in as few or as many of the spokes as appropriate.

Vocabulary and Note-Taking Resources

- Vocabulary Practice, pp. 47–48
- Decoding Support, p. 49

- Daily Vocabulary Scaffolding, p. T2
- Note-Taking Model, p. T3

- Description Wheel, B20–21
- Outline, C43
- Daily Vocabulary Scaffolding, H1–8

CHAPTER 1

Getting Ready to Learn

CONCEPT REVIEW

- Objects can move at different speeds and in different directions.
- Pushing or pulling on an object will change how it moves.

VOCABULARY REVIEW

See Glossary for definitions.

horizontal
meter
second
vertical

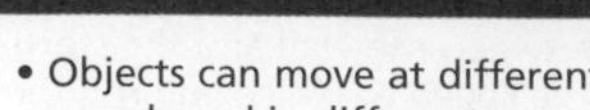

CONTENT REVIEW
CLASSZONE.COM
Review concepts and vocabulary.

TAKING NOTES

OUTLINE

As you read, copy the headings onto your paper in the form of an outline. Then add notes in your own words that summarize what you read.

VOCABULARY STRATEGY

Place each new vocabulary term at the center of a **description wheel** diagram. As you read about the term, write some words on the spokes describing the term.

See the Note-Taking Handbook on pages R45–R51.

SCIENCE NOTEBOOK

OUTLINE

I. Position describes the location of an object.
  A. Finding a position
    1. A position is compared to a reference point.
    2. Position can be described using distance and direction.

CHECK READINESS

Administer the Diagnostic Test to determine students' readiness for new science content and their mastery of requisite math skills.

 Diagnostic Test, pp. 1–2

Technology Resources

Students needing content and math skills should visit **ClassZone.com**.

- CONTENT REVIEW
- MATH TUTORIAL

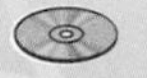 CONTENT REVIEW CD-ROM

KEY CONCEPT

An object in motion changes position.

BEFORE, you learned

- Objects can move in different ways
- An object's position can change

NOW, you will learn

- How to describe an object's position
- How to describe an object's motion

VOCABULARY

position p. 9
reference point p. 10
motion p. 11

EXPLORE Location

How do you describe the location of an object?

PROCEDURE

1. Choose an object in the classroom that is easy to see.
2. Without pointing to, describing, or naming the object, give directions to a classmate for finding it.
3. Ask your classmate to identify the object using your directions. If your classmate does not correctly identify the object, try giving directions in a different way. Continue until your classmate has located the object.

WHAT DO YOU THINK?
What kinds of information must you give another person when you are trying to describe a location?

Position describes the location of an object.

Have you ever gotten lost while looking for a specific place? If so, you probably know that accurately describing where a place is can be very important. The **position** of a place or an object is the location of that place or object. Often you describe where something is by comparing its position with where you currently are. You might say, for example, that a classmate sitting next to you is about a meter to your right, or that a mailbox is two blocks south of where you live. Each time you identify the position of an object, you are comparing the location of the object with the location of another object or place.

VOCABULARY
Make a description wheel in your notebook for *position*.

CHECK YOUR READING Why do you need to discuss two locations to describe the position of an object?

RESOURCES FOR DIFFERENTIATED INSTRUCTION

Below Level

UNIT RESOURCE BOOK
- Reading Study Guide A, pp. 13–14
- Decoding Support, p. 49

AUDIO CDS

Advanced

UNIT RESOURCE BOOK
- Challenge and Extension, p. 19

English Learners

UNIT RESOURCE BOOK
Spanish Reading Study Guide, pp. 17–18

AUDIO CDS
- Audio Readings in Spanish
- Audio Readings (English)

1.1 FOCUS

Set Learning Goals

Students will
- Describe an object's position.
- Describe an object's motion.
- Observe changes in position through experimentation.

3-Minute Warm-Up

Display Transparency 4 or copy this exercise on the board:

Draw pictures to represent the motion described in these two scenarios.

1. An ant crawls across the top of your shoe, from left to right. *short series of pictures showing ant's movement*
2. A hiker walks from the beginning to the end of a hiking path. *short series of pictures showing hiker's movement*

3-Minute Warm-Up, p. T4

1.1 MOTIVATE

EXPLORE Location

PURPOSE To consider how to identify a location

TIP *10 min.* Have several students try choosing and describing an object. After the activity, have the class discuss how much and what type of information was needed.

WHAT DO YOU THINK? *Sample answer: I had my partner count a specific number of desks to the front and to the right so she could locate the object.*

Ongoing Assessment

CHECK YOUR READING *Answer: You need to compare the location of the object with the location of another object or place.*

1.1 INSTRUCT

Teach from Visuals

To point out different ways to describe a location, use the "Describing Position" visual and ask:

- What does the map on the left show? *using Brasília as a reference point for finding Santiago*
- What does the map on the right show? *using latitude and longitude, a grid system, to locate Santiago*

Teacher Demo

Make a tube out of graph paper and tape it so it will stay closed. Draw a point on the tube. Ask students how they could use the graph paper squares to locate the point. List the different suggestions on the board, and discuss the similarities and differences in the methods. Point out how two reference lines, one vertical and one horizontal, are needed to describe the location. Discuss how longitude and latitude are like the lines on the tube.

Ongoing Assessment

Describe an object's position.

Ask: If you describe positions on a map by comparing them with the position of landmark Q, what is Q called? *a reference point*

READING VISUALS *Answer: The left-hand map uses a distance and a direction from Brasília to describe Santiago's location. The right-hand map uses the number of degrees from the equator and the prime meridian to describe Santiago's location.*

Learn more about how people find and describe position.

Describing a Position

You might describe the position of a city based on the location of another city. A location to which you compare other locations is called a **reference point.** You can describe where Santiago, Chile, is from the reference point of the city Brasília, Brazil, by saying that Santiago is about 3000 kilometers (1860 mi) southwest of Brasília.

You can also describe a position using a method that is similar to describing where a point on a graph is located. For example, in the longitude and latitude system, locations are given by two numbers—longitude and latitude. Longitude describes how many degrees east or west a location is from the prime meridian, an imaginary line running north-south through Greenwich, England. Latitude describes how many degrees north or south a location is from the equator, the imaginary circle that divides the northern and southern hemispheres. Having a standard way of describing location, such as longitude and latitude, makes it easier for people to compare locations.

Describing Position

There are several different ways to describe a position. The way you choose may depend on your reference point.

① Reference Point: Brasília

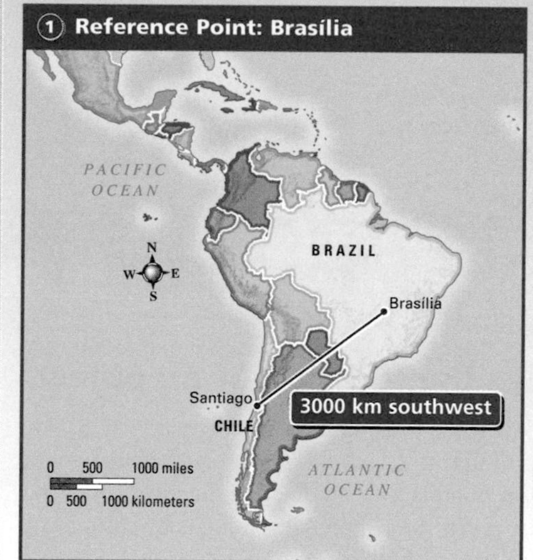

To describe where Santiago is, using Brasília as a reference point, you would need to know how far Santiago is from Brasília and in what direction it is.

② Reference Point: 0° longitude, 0° latitude

In the longitude and latitude system, a location is described by how many degrees north or south it is from the equator and how many degrees east or west it is from the prime meridian.

READING VISUALS Compare and contrast the two ways of describing the location of Santiago as shown here.

DIFFERENTIATE INSTRUCTION

More Reading Support

A What would you call a location with which other locations can be compared? *a reference point*

English Learners English learners may need help understanding sentences that imply *if/then* construction, such as *If you were to travel from Brasília to Santiago, you would end up about 3000 kilometers from where you started* (p. 11). This sentence begins with *If* but does not contain *then,* and therefore requires a reader to infer the cause-and-effect relationship. Point out other examples of this construction and make sure English learners recognize that *then* is implied.

Measuring Distance

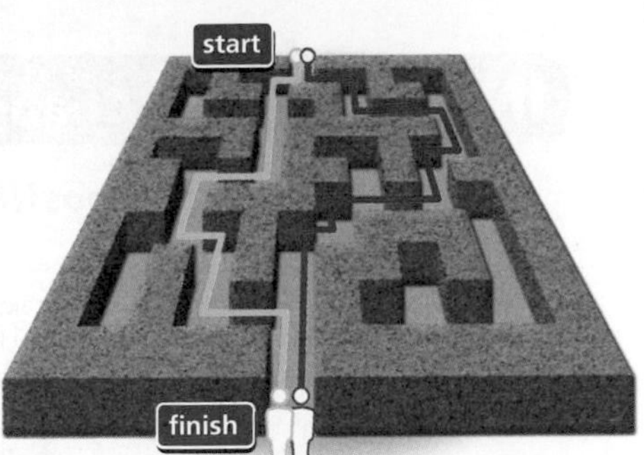

COMPARE How does the distance each person has walked compare with the distance each is from the start of the maze?

If you were to travel from Brasília to Santiago, you would end up about 3000 kilometers from where you started. The actual distance you traveled, however, would depend on the exact path you took. If you took a route that had many curves, the distance you traveled would be greater than 3000 kilometers.

? B The way you measure distance depends on the information you want. Sometimes you want to know the straight-line distance between two positions. Sometimes, however, you might need to know the total length of a certain path between those positions. During a hike, you are probably more interested in how far you have walked than in how far you are from your starting point.

When measuring either the straight-line distance between two points or the length of a path between those points, scientists use a standard unit of measurement. The standard unit of length is the meter (m), which is 3.3 feet. Longer distances can be measured in kilometers (km), and shorter distances in centimeters (cm).

Motion is a change in position.

The illustration below shows an athlete at several positions during a long jump. If you were to watch her jump, you would see that she is in motion. **Motion** is the change of position over time. As she jumps, both her horizontal and vertical positions change. If you missed the motion of the jump, you would still know that motion occurred because of the distance between her starting and ending positions. A change in position is evidence that motion happened.

REMINDER

Horizontal and *vertical* describe directions, as shown.

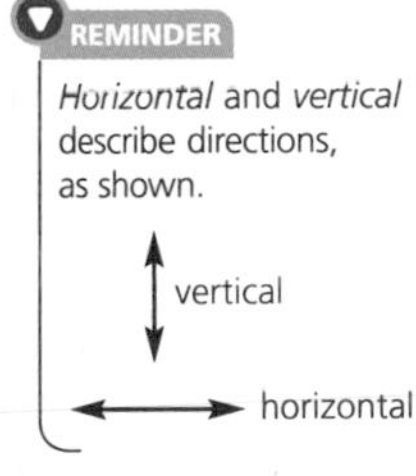

History of Science

In eighteenth-century France, some 2000 different units of measurement were used across the country because different locales had their own systems. The French Revolution (1789–1799) set the stage for a new, universal system of measurement. A commission decided on a basic unit of measurement, the meter. They designed a decimal system that uses base names (such as *meter* and *gram*) and affixes standard prefixes to indicate fractional and multiple units (such as *milligram* and *kilogram*). In 1795 the French government adopted the new metric system.

Teach from Visuals

Direct students' attention to the time-exposure illustration at the bottom of the page, and then ask:

- How does the picture show time passing? *It shows the jumper in a number of different positions between the start and finish of her long jump.*
- How is the picture different from a snapshot? *A snapshot freezes a single moment in time.*

Teaching with Technology

If students have probeware, encourage them to use a motion sensor as they push various classroom objects across a flat, smooth surface.

DIFFERENTIATE INSTRUCTION

More Reading Support

B List two ways to measure distance. *measure a straight-line distance between two positions; measure the total length of a path between two positions*

Advanced Have students refer to the map on page 10 that shows a connecting line between Santiago and Brasília. Ask: How could you estimate the distance of this diagonal line using grid units—that is, the distance south and the distance west? *You could measure the vertical distance and the horizontal distance; then use the Pythagorean theorem to determine the length of the hypotenuse.*

R Challenge and Extension, p. 19

INVESTIGATE
Changing Positions

PURPOSE To explore how change in position depends on the perspective of the observer

TIPS *20 min.*

- Use tennis balls or rubber balls of about the same diameter.
- Model the walking/ball-tossing activity. Then encourage volunteers to try it.
- Check that students, when sketching their own ball tosses, aren't adding their motion to what they see but are recording only what they observe.

WHAT DO YOU THINK? *The path of the ball looked straight up and down when I threw it; when I watched another student throwing the ball, it looked curving. The same motion can look different to two people who are in different locations relative to the motion.*

CHALLENGE *as though it were moving straight up and down*

 Datasheet, Changing Positions, p. 20

Technology Resources

Customize this student lab as needed or look for an alternative. Print rubrics to assess student lab reports.

 **Lab Generator CD-ROM**

INVESTIGATE Changing Positions

How are changes in position observed?

PROCEDURE

1. Begin walking while tossing a ball straight up and catching it as it falls back down toward your hand. Observe the changes in the position of the ball as you toss it while walking a distance of about 4 m.
2. Make a sketch showing how the position of the ball changed as you walked. Use your own position as a reference point for the ball's position.
3. Watch while a classmate walks and tosses the ball. Observe the changes in the position of the ball using your own position as a reference point. Make a sketch showing how the ball moved based on your new point of view.

WHAT DO YOU THINK?

- Compare your two sketches. How was the change in position of the ball you tossed different from the change in position of the ball that your partner tossed?
- How did your change in viewpoint affect what you observed? Explain.

CHALLENGE How would the change in position of the ball appear to a person standing 4 m directly in front of you?

SKILL FOCUS
Observing

MATERIALS
- small ball
- paper
- pencil

TIME
20 minutes

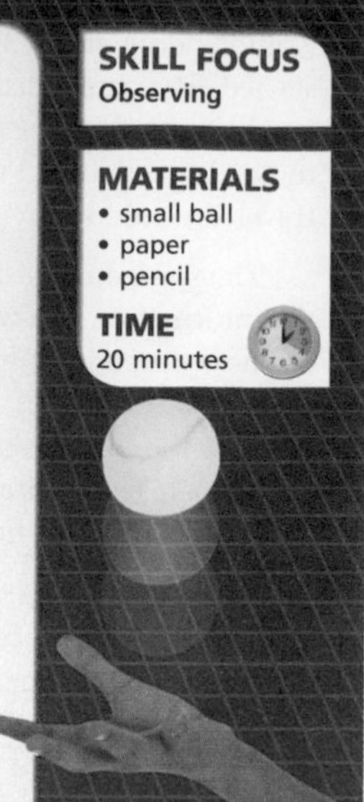

Describing Motion

A change in an object's position tells you that motion took place, but it does not tell you how quickly the object changed position. The speed of a moving object is a measure of how quickly or slowly the object changes position. A faster object moves farther than a slower moving object would in the same amount of time.

The way in which an object moves can change. As a raft moves along a river, its speed changes as the speed of the river changes. When the raft reaches a calm area of the river, it slows down. When the raft reaches rapids, it speeds up. The rafters can also change the motion of the raft by using paddles. You will learn more about speed and changing speed in the following sections.

APPLY Describe the different directions in which the raft is moving.

Metacognitive Strategy

Ask students if they would have accepted the conclusions of the investigation without actually participating in it. Have students write a paragraph explaining the importance of observations.

Teaching with Technology

As students perform the ball-tossing activity, videotape them from various perspectives, such as coming toward, going away from, moving parallel to, and so forth. Have participants label each clip. Have groups view the videos and discuss the change in position of the ball in regard to motion and perspective.

DIFFERENTIATE INSTRUCTION

More Reading Support

C What determines how quickly or slowly a moving object changes position? *the speed of the object*

Alternative Assessment Have students write a paragraph explaining the differences between the perspectives of two people viewing a moving bicycle rider, where one observer is in a car keeping pace with the cyclist, and the other observer is standing at the side of a bicycle path as the cyclist whizzes by.

Relative Motion

If you sit still in a chair, you are not moving. Or are you? The answer depends on the position and motion of the person observing you. You do not notice your position changing compared with the room and the objects in it. But if an observer could leave Earth and look at you from outer space, he could see that you are moving along with Earth as it travels around the Sun. How an observer sees your motion depends on how it compares with his own motion. Just as position is described by using a reference point, motion is described by using a frame of reference. You can think of a frame of reference as the location of an observer, who may be in motion.

OUTLINE
Add relative motion to your outline, along with supporting details.

I. Main idea
   A. Supporting idea
      1. Detail
      2. Detail
   B. Supporting idea

Consider a student sitting behind the driver of a moving bus. The bus passes another student waiting at a street sign to cross the street.

1. To the observer on the bus, the driver is not changing his position compared with the inside of the bus. The street sign, however, moves past the observer's window. From this observer's point of view, the driver is not moving, but the street sign is.
2. To the observer on the sidewalk, the driver is changing position along with the bus. The street sign, on the other hand, is not changing position. From this observer's point of view, the street sign is not moving, but the driver is.

Relative Motion

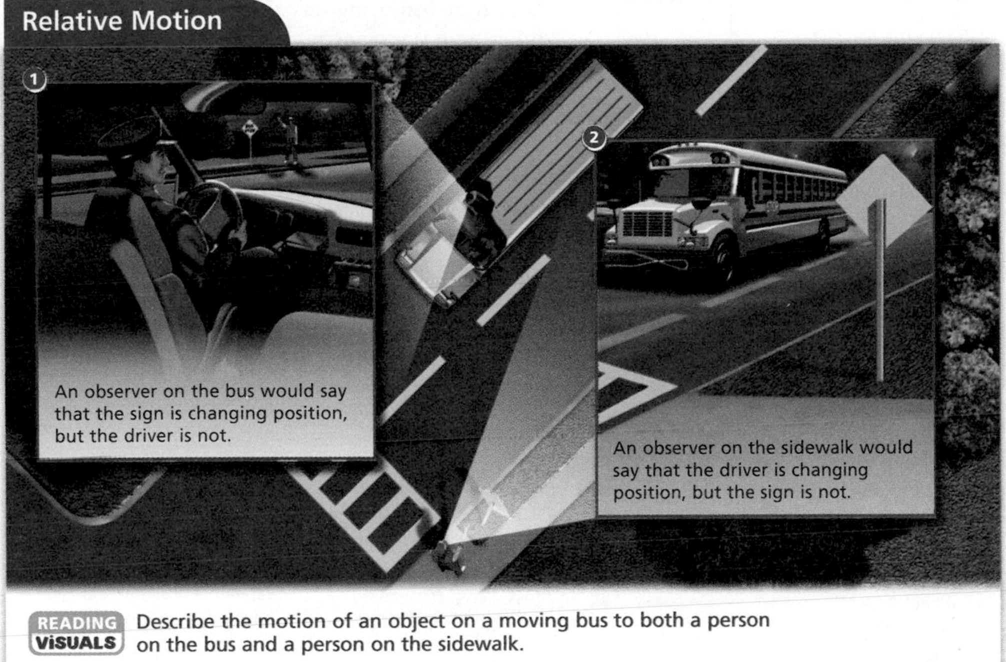

1
An observer on the bus would say that the sign is changing position, but the driver is not.
2
An observer on the sidewalk would say that the driver is changing position, but the sign is not.

READING VISUALS Describe the motion of an object on a moving bus to both a person on the bus and a person on the sidewalk.

Teach Difficult Concepts

Students may have a hard time understanding that all motion is measured relative to an observer's frame of reference. Explain that an object that is not moving in one frame of reference will be moving in another.

Have students think about sitting on an airplane that is in the air. Discuss whether or not the seat is moving. Have students consider the airplane and Earth as frames of reference. Encourage students to think about other frames of reference, such as one in which Earth is moving.

Point out that you can always find a frame of reference in which an object is still and others in which the object moves. There is no correct frame of reference—they are all equally valid.

Develop Critical Thinking

APPLY Describe a puppet show that calls for the puppets to sit on a moving train. The stage set consists of a board with cutouts for windows that is placed behind the puppets.

Ask: How could you convince the audience that the train is moving? *Sample answer: You could paint scenery on a long strip of paper and pull it through the stage set behind the window cutouts. The motion of the pictures would look like scenery passing, because of the reference point of the puppet "observers" inside the train.*

Ongoing Assessment

Describe a frame of reference.

Ask: What is a frame of reference? *the location of an observer of a motion*

READING VISUALS *Answer: To an observer on the bus, an object on the bus appears to be still. The same object will appear to be traveling at the same speed as the bus to an observer on the sidewalk.*

DIFFERENTIATE INSTRUCTION

More Reading Support

D One observer may see a motion differently than another observer. What idea does this statement express? *relative motion*

Inclusion Adapt the relative-motion examples on this page for students with vision impairments. Prompt students to imagine that a person standing on a sidewalk would feel and hear a bus whiz by. To a person on the bus, however, there might be no more sensation of movement than some seat vibrations. These two observers would have completely different perspectives on the motion of the bus.

Ongoing Assessment

CHECK YOUR READING *Sample answer: You observe motion relative to your own position. If you are on a train, for example, you see the ground outside the train moving past you.*

EXPLORE the BIG idea

Revisit "Internet Activity: Relative Motion" on p. 7. Have students explain the reasons for their results.

Reinforce the BIG idea

Have students relate the section to the Big Idea.

Reinforcing Key Concepts, p. 21

1.1 ASSESS & RETEACH

Assess

Section 1.1 Quiz, p. 3

Reteach

Put two marbles randomly on a checkerboard. Have students describe the positions, and then roll the two marbles at different speeds. Emphasize that a good description of motion includes speed. Finally, have students view the rolling of a marble from very different positions. Solicit students' observations. Discuss relative motion.

Technology Resources

Have students visit **ClassZone.com** for reteaching of Key Concepts.

CONTENT REVIEW

CONTENT REVIEW CD-ROM

APPLY In the top picture, the train is moving compared with the camera and the ground. Describe the relative motion of the train, camera, and ground in the bottom picture.

When you ride in a train, a bus, or an airplane, you think of yourself as moving and the ground as standing still. That is, you usually consider the ground as the frame of reference for your motion. If you traveled between two cities, you would say that you had moved, not that the ground had moved under you in the opposite direction.

If you cannot see the ground or objects on it, it is sometimes difficult to tell if a train you are riding in is moving. If the ride is very smooth and you do not look out the window at the scenery, you might never realize you are moving at all.

Suppose you are in a train, and you cannot tell if you are stopped or moving. Outside the window, another train is slowly moving forward. Could you tell which of the following situations is happening?

- Your train is stopped, and the other train is moving slowly forward.
- The other train is stopped, and your train is moving slowly backward.
- Both trains are moving forward, with the other train moving a little faster.
- Your train is moving very slowly backward, and the other train is moving very slowly forward.

Actually, all four of these possibilities would look exactly the same to you. Unless you compared the motion to the motion of something outside the train, such as the ground, you could not tell the difference between these situations.

How does your observation of motion depend on your own motion?

1.1 Review

KEY CONCEPTS

1. What information do you need to describe an object's location?
2. Describe how your position changes as you jump over an object.
3. Give an example of how the apparent motion of an object depends on the observer's motion.

CRITICAL THINKING

4. **Infer** Kyle walks 3 blocks south from his home to school, and Jana walks 2 blocks north from her home to Kyle's home. How far and in what direction is the school from Jana's home?
5. **Predict** If you sit on a moving bus and toss a coin straight up into the air, where will it land?

CHALLENGE

6. **Infer** Jamal is in a car going north. He looks out his window and thinks that the northbound traffic is moving very slowly. Ellen is in a car going south. She thinks the northbound traffic is moving quickly. Explain why Jamal and Ellen have different ideas about the motion of the traffic.

ANSWERS

1. a reference point

2. As a person jumps over an object, both his horizontal and vertical positions change.

3. Sample answer: A person in an airplane will think of her or his seat as not moving, but an observer on the ground will think the seats are moving with the airplane.

4. one block south

5. It should land back in your hand (or directly beneath where you threw it).

6. Jamal and Ellen have different ideas about motion because both compare the northbound traffic with their own motion. From Jamal's point of view, traffic is going in the same direction as he is; therefore, traffic seems to be moving slowly. Ellen sees oncoming cars and thinks they are moving quickly.

COAST GUARD RESCUE

Physics for Rescuers

Performing a rescue operation is often difficult and risky because the person in trouble is in a dangerous situation. Coast Guard Search and Rescue Teams have an especially difficult problem to deal with. As a rescue ship or helicopter approaches a stranded boat, the team must get close enough to help but avoid making the problem worse by colliding with the boat. At the same time, wind, waves, and currents cause changes in the motion of both crafts.

Finding the Problem

A stranded boater fires a flare to indicate his location. The observer on the Coast Guard ship tracks the motion of the flare to its source.

Avoiding Collision

As the boats move closer together, the captain assesses their motion relative to each other. The speeds of the boats must match, and the boats must be close enough that a rope can be thrown across the gap. If the sea is rough, both boats will move up and down, making the proper positioning even more difficult.

Rescue from Above

The helicopter pilot determines where to hover so that the rescue basket lands on target. A mistake could be disastrous for the rescuers as well as the people being rescued.

EXPLORE

1. **PREDICT** Tie a washer to a 30 cm piece of string. Using your hand as a helicopter, lower the rescue washer to a mark on the floor. Turn on a fan to create wind. Predict where you will need to hold the string to land the washer on the mark. Place the fan at a different location and try again. How accurate was your prediction? Does your accuracy improve with practice?
2. **CHALLENGE** Have a partner throw a baseball into the air from behind the corner of a wall. Using the motion of the ball, try to determine the position from which it was thrown. When is it easier—when the ball is thrown in a high arc or lower one?

EXPLORE

PREDICT *Students should improve their accuracy with practice.*
CHALLENGE *thrown in a high arc*

SCIENCE ON THE JOB
Relevance of Science to Non-science Jobs

Set Learning Goal

To understand why rescuers rely on principles of physics to do their job

Present the Science

FINDING POSITIONS Many boats carry a Global Positioning System (GPS) device that enables rescue workers to locate the boat. The GPS device works by transmitting a signal to satellites in orbit around Earth. Each satellite measures a distance between it and the boat. By comparing the distances provided by three or more satellites in different locations, rescuers can determine the boat's position.

HOVERING A helicopter, unlike an airplane, has the ability to hover in place above a target. In order for a helicopter to remain above a moving target, it must match the target's motion.

Discussion Questions

Ask: What are some of the dangers faced by rescuers at sea? *Wind and waves can add forces that make it difficult to predict how a boat will move. Rescuers in helicopters must maintain a balance of forces during the hovering maneuver. There is always the danger of a collision.*

Ask: How does relative motion contribute to these hazards? *Rescuers can adjust for constant relative motion, but unexpected changes in speed make the results unpredictable and increase the possibility of collision.*

Close

Ask: Why does a knowledge of physical science provide a good background for rescuers? *Rescuers need to know about forces and motion, which are explored in physical science.*

1.2 FOCUS

Set Learning Goals

Students will

- Calculate an object's speed.
- Describe an object's velocity.
- Observe through experimentation the relationship between speed and distance.

3-Minute Warm-Up

Display Transparency 4 or copy this exercise on the board:

Suppose you pass a table on which a ball is sitting near the end of a cardboard tube. After a few minutes, you pass the table again. The ball is now near the other end of the tube. What can you say about the motion of the ball between those two times? What can't you say about the motion? *You can say that the ball moved from one end of the tube to another. You can't say anything about the path. The ball may have gone through the tube, or it may have been moved around the tube.*

3-Minute Warm-Up, p. T4

1.2 MOTIVATE

EXPLORE Speed

PURPOSE To introduce the relationship among speed, distance, and time

TIP *10 min.* Select a flat, level area without carpeting.

WHAT DO YOU THINK? *I pushed the ball harder to decrease the time and less hard to increase it. If the time was shorter, the ball had to travel the same distance faster.*

Ongoing Assessment

CHECK YOUR READING *Answer: They are related by time: speed is a measure of how quickly an object changes position.*

KEY CONCEPT

1.2 Speed measures how fast position changes.

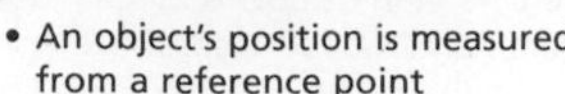

BEFORE, you learned

- An object's position is measured from a reference point
- To describe the position of an object, you can use distance and direction
- An object in motion changes position with time

NOW, you will learn

- How to calculate an object's speed
- How to describe an object's velocity

VOCABULARY

speed p. 16
velocity p. 22
vector p. 22

EXPLORE Speed

How can you measure speed?

PROCEDURE

1. Place a piece of tape on the floor. Measure a distance on the floor 2 m away from the tape. Mark this distance with a second piece of tape.
2. Roll a tennis ball from one piece of tape to the other, timing how long it takes to travel the 2 m.
3. Roll the ball again so that it travels the same distance in less time. Then roll the ball so that it takes more time to travel that distance than it did the first time.

MATERIALS

- tape
- meter stick
- tennis ball
- stopwatch

WHAT DO YOU THINK?

- How did you change the time it took the ball to travel 2 m?
- How did changing the time affect the motion of the ball?

Position can change at different rates.

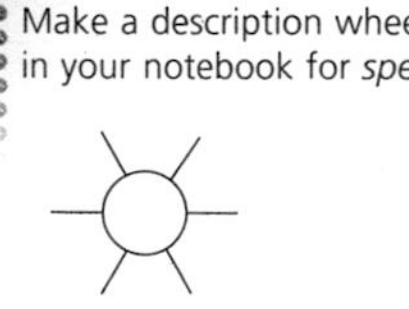

VOCABULARY
Make a description wheel in your notebook for *speed*.

When someone asks you how far it is to the library you can answer in terms of distance or time. You can say it is several blocks, or you can say it is a five-minute walk. When you give a time instead of a distance, you are basing your time estimate on the distance to the library and the person's speed. **Speed** is a measure of how fast something moves or the distance it moves, in a given amount of time. The greater the speed an object has, the faster it changes position.

CHECK YOUR READING How are speed and position related?

RESOURCES FOR DIFFERENTIATED INSTRUCTION

Below Level

UNIT RESOURCE BOOK

- Reading Study Guide A, pp. 24–25
- Decoding Support, p. 49

AUDIO CDS

Advanced

UNIT RESOURCE BOOK

- Challenge and Extension, p. 30

English Learners

UNIT RESOURCE BOOK

Spanish Reading Study Guide, pp. 28–29

AUDIO CDS

- Audio Readings in Spanish
- Audio Readings (English)

C 16 Unit: Motion and Forces

The way in which one quantity changes compared to another quantity is called a rate. Speed is the rate at which the distance an object moves changes compared to time. If you are riding a bike to a movie, and you think you might be late, you increase the rate at which your distance changes by pedaling harder. In other words, you increase your speed.

Calculating Speed

A To calculate speed, you need to know both distance and time measurements. Consider the two bike riders below.

1. The two bikes pass the same point at the same time.
2. After one second, the first bike has traveled four meters, while the second has traveled only two meters. Because the first bike has traveled four meters in one second, it has a speed of four meters per second. The second bike has a speed of two meters per second.
3. If each bike continues moving at the same speed as before, then after two seconds the first rider will have traveled eight meters, while the second one will have traveled only four meters.

Comparing Speed

Objects that travel at different speeds move different distances in the same amount of time.

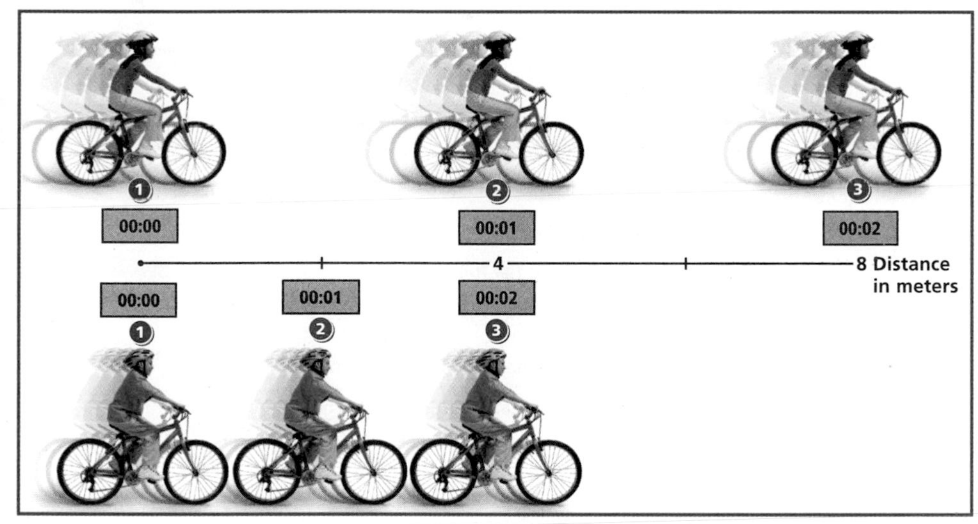

READING VISUALS How far will each rider travel in five seconds?

1.2 INSTRUCT

Teach from Visuals

Help students identify the distance lines for the two bikes. To help students interpret the visual, ask:

- What does each group of numbers in a box represent? *timers that show minutes and seconds*
- How can you tell at a glance which cyclist is pedaling faster? *by observing which bike went farther in the given time (2 seconds)*

Develop Estimation Skills

Make sure students understand how the distance is calculated in step 3. Point out that because the bikes travel at constant speeds, we can sample the speeds in a unit of time, such as one second, and then multiply the base speed by the total number of seconds.

Develop Critical Thinking

PROVIDE EXAMPLES Speed is a rate, that is, a measure of how one quantity (distance) changes compared with another (time). Ask students to give two examples of everyday measures that are rates. *Examples will vary. Heart rate or pulse is a measure of the number of heartbeats per minute, fuel efficiency of cars is measured in miles per gallon, electric power is measured in kilowatts per hour.*

EXPLORE the BIG idea

Revisit "Off the Wall" on p. 7. Have students explain their results.

Ongoing Assessment

READING VISUALS *Answer: The faster rider will travel 20 meters; the slower rider will travel 10 meters.*

DIFFERENTIATE INSTRUCTION

More Reading Support

A Which measurements do you need to calculate speed? *distance and time*

English Learners Students may have difficulty using the words *affect* and *effect* correctly. Explain that *affect* is a verb and *effect* is a noun. A common trick for remembering the difference is that *affect* begins with the letter *a*. A verb is an action, which also begins with the letter *a*. *Affect* is a verb, or a word that denotes an action.

Integrate the Sciences

The fastest-running mammal is the cheetah, a large cat native to Africa. The cheetah can reach a speed of 110 kilometers per hour (about 70 miles per hour) when chasing prey. By contrast, the fastest Olympic runner can reach a speed of about 37 kilometers per hour (about 23 miles per hour). Neither the cheetah nor the runner, however, can maintain that speed for very long.

Ongoing Assessment

Calculate an object's speed.

Ask: Suppose a race is 50 meters, and it takes an athlete 15 seconds to run it. Is that enough information to calculate the runner's speed? Explain your answer. *Yes; the formula for speed is distance divided by time, and both of these measurements are given.*

The runner with the shortest time ran the fastest.

Practice the Math

Answers

1. $S = \frac{d}{t} = \frac{200\ m}{25\ s} = 8\ m/s$
2. $S = \frac{d}{t} = \frac{100\ m}{50\ s} = 2\ m/s$

Racing wheelchairs are specially designed to reach higher speeds than regular wheelchairs.

Speed can be calculated by dividing the distance an object travels by the time it takes to cover the distance. The formula for finding speed is B

$$\text{Speed} = \frac{\text{distance}}{\text{time}} \qquad S = \frac{d}{t}$$

Speed is shown in the formula as the letter *S*, distance as the letter *d*, and time as the letter *t*. The formula shows how distance, time, and speed are related. If two objects travel the same distance, the object that took a shorter amount of time will have the greater speed. Similarly, an object with a greater speed will travel a longer distance in the same amount of time than an object with a lower speed will.

The standard unit for speed is meters per second (m/s). Speed is also given in kilometers per hour (km/h). In the United States, where the English system of measurement is still used, speeds are often given in miles per hour (mi/h or mph). One mile per hour is equal to 0.45 m/s. C

The man participating in the wheelchair race, at left, will win if his speed is greater than the speed of the other racers. You can use the formula to calculate his speed.

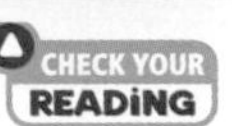
If two runners cover the same distance in different amounts of time, how do their speeds compare?

Calculating Speed

Sample Problem

A wheelchair racer completes a 100-meter course in 20 seconds. What is his speed?

What do you know?	distance = 100 m, time = 20 s
What do you want to find out?	speed
Write the formula:	$S = \frac{d}{t}$
Substitute into the formula:	$S = \frac{100\ m}{20\ s}$
Calculate and simplify:	$S = 5\ m/s$
Check that your units agree:	Unit is m/s. Unit of speed is m/s. Units agree.
Answer:	$S = 5\ m/s$

Practice the Math

1. A man runs 200 m in 25 s. What is his speed?
2. If you travel 100 m in 50 s, what is your speed?

DIFFERENTIATE INSTRUCTION

More Reading Support

B What is the formula for calculating speed? *speed = distance ÷ time*

C What is the standard unit for speed? *meters per second*

Inclusion If you have a student who uses a wheelchair, ask him or her to make a "run" of a predetermined course. Record the time, and measure the course distance. Then ask the student to plug these values into the speed formula introduced on this page to calculate the speed.

Average Speed

Speed is not constant. When you run, you might slow down to pace yourself, or speed up to win a race. At each point as you are running, you have a specific speed. This moment-to-moment speed is called your instantaneous speed. Your instantaneous speed can be difficult to measure; however, it is easier to calculate your average speed over a distance.

In a long race, runners often want to know their times for each lap so that they can pace themselves. For example, an excellent middle school runner might have the following times for the four laps of a 1600-meter race: 83 seconds, 81 seconds, 79 seconds, 77 seconds. The lap times show the runner is gradually increasing her speed throughout the race.

The total time for the four laps can be used to calculate the runner's average speed for the entire race. The total time is 320 seconds (5 min 20 s) for the entire distance of 1600 meters. The runner's average speed is 1600 meters divided by 320 seconds, or 5.0 meters per second.

READING TIP The root of *instantaneous* is *instant,* meaning "moment."

INVESTIGATE Speed and Distance

How does design affect speed?

DESIGN — YOUR OWN — EXPERIMENT

Cars are built in different shapes. How does the shape of the car affect the way it moves? Design your own car, and see how fast it can go.

PROCEDURE

1. Use the clay, film container lids, and toothpicks to design a car that rolls when it is pushed. The car should have a total mass of 150 g or less.
2. Using any or all of the other materials, design an experiment to measure and compare the speed of your car with the speed of someone else's car. Your experiment should be designed so that the design of the car is the only variable being tested. Write up your procedure.
3. Perform the experiment using your car and another student's car. Record the data you need to calculate the speed of both cars.
4. Calculate the speed of each car, and record which car went faster.

WHAT DO YOU THINK?

- What were the constants in your experiment?
- How would you improve your design if you were to repeat the experiment?

SKILL FOCUS
Designing experiments

MATERIALS
- clay
- film container lids
- toothpicks
- beam balance
- board
- books
- string
- straw
- scissors
- stopwatch

TIME
20 minutes

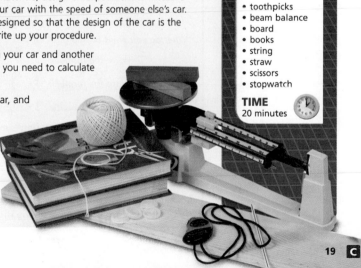

19 C

INVESTIGATE
Speed and Distance

PURPOSE To design a car and an experiment to determine how the car's shape affects the speed at which it moves

TIPS *20 min.*

- To save time, use a kitchen scale to measure the mass of the car. Kitchen scales are an affordable alternative for measuring mass in increments of 10 grams.
- For best comparisons of car design, the masses of the two cars should be as close to each other as possible.
- Tape can be used to mark the end of the race course. Choose a distance from the bottom of the ramp, such as 25 cm, that will make calculations easy.

WHAT DO YOU THINK? *Constants should include mass of car, height of ramp, design of car.*

CHALLENGE *Answers will vary. Improvements might include making the wheels turn more easily, lowering the center of gravity of the car, or changing the mass of the car.*

Datasheet, Speed and Distance, p. 31

Technology Resources

Customize this student lab as needed or look for an alternative. Print rubrics to assess student lab reports.

Lab Generator CD-ROM

DIFFERENTIATE INSTRUCTION

More Reading Support

D What is instantaneous speed? *moment-by-moment speed*

E How can you calculate average speed? *Divide the total distance by total time.*

Real World Example

Japan and many countries in Europe have high-speed passenger trains whose top speeds are advertised to entice travelers. This speed is usually the fastest the train can travel on the best track in open country. It is an instantaneous speed. The train schedules, however, reflect the average speed of the train.

Develop Graphing Skills

Speed is measured by the steepness, or slope, of a line. In mathematics, slope is defined as the change in y-values divided by the change in x-values. Here, slope is calculated by dividing the change in distance by the change in time for a given interval.

- A rising line, or positive slope, indicates that the distance an object travels from its starting point is increasing with time.
- A horizontal line, or 0 slope, indicates that the speed is zero meters per second.

- Math Support, pp. 50
- Math Practice, pp. 51

Ongoing Assessment

Practice the Math

Answers

1. $S = \frac{d}{t} = \frac{40\text{ m} - 40\text{ m}}{40\text{ s} - 20\text{ s}} = \frac{0\text{ m}}{20\text{ s}} = 0\text{ m/s}$

2. $S = \frac{d}{t} = \frac{280\text{ m} - 40\text{ m}}{60\text{ s} - 40\text{ s}} = \frac{240\text{ m}}{20\text{ s}} = 12\text{ m/s}$

Distance-Time Graphs

A convenient way to show the motion of an object is by using a graph that plots the distance the object has traveled against time. This type of graph, called a distance-time graph, shows how speed relates to distance and time. You can use a distance-time graph to see how both distance and speed change with time.

The distance-time graph on page 21 tracks the changing motion of a zebra. At first the zebra looks for a spot to graze. Its meal is interrupted by a lion, and the zebra starts running to escape.

REMINDER
The x-axis and y-axis are arranged as shown:

In a distance-time graph, time is on the horizontal axis, or x-axis, and distance is on the vertical axis, or y-axis.

1. As an object moves, the distance it travels increases with time. This can be seen as a climbing, or rising, line on the graph.
2. A flat, or horizontal, line shows an interval of time where the speed is zero meters per second.
3. Steeper lines show intervals where the speed is greater than intervals with less steep lines.

You can use a distance-time graph to determine the speed of an object. The steepness, or slope, of the line is calculated by dividing the change in distance by the change in time for that time interval.

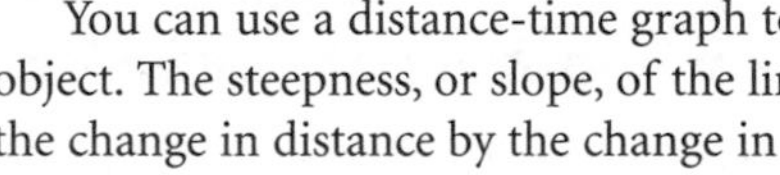

Calculating Speed from a Graph

Sample Problem

How fast is the zebra walking during the first 20 seconds?

What do you know? Reading from the graph:
At time = 0 s, distance = 0 m.
At time = 20 s, distance = 40 m.

What do you want to find out? speed

Write the formula: $S = \frac{d}{t}$

Substitute into the formula: $S = \frac{40\text{ m} - 0\text{ m}}{20\text{ s} - 0\text{ s}}$

Calculate and simplify: $S = \frac{40\text{ m}}{20\text{ s}} = 2\text{ m/s}$

Check that your units agree: Unit is m/s.
Unit of speed is m/s. Units agree.

Answer: $S = 2$ m/s

Practice the Math

1. What is the speed of the zebra during the 20 s to 40 s time interval?
2. What is the speed of the zebra during the 40 s to 60 s interval?

DIFFERENTIATE INSTRUCTION

More Reading Support

F What kind of graph shows how both distance and speed change with time? *a distance-time graph*

G What is the slope of a line? *its steepness*

Advanced Point out that the accuracy of a distance-time graph depends on how frequent the sampling is. Have students describe a way to compile data for a more accurate graph. *Sample answer: sample more frequently, perhaps with the help of technology*

Challenge and Extension, p. 30

Distance-Time Graph

A zebra's speed will change throughout the day, especially if a hungry lion is nearby. You can use a distance-time graph to compare the zebra's speed over different time intervals.

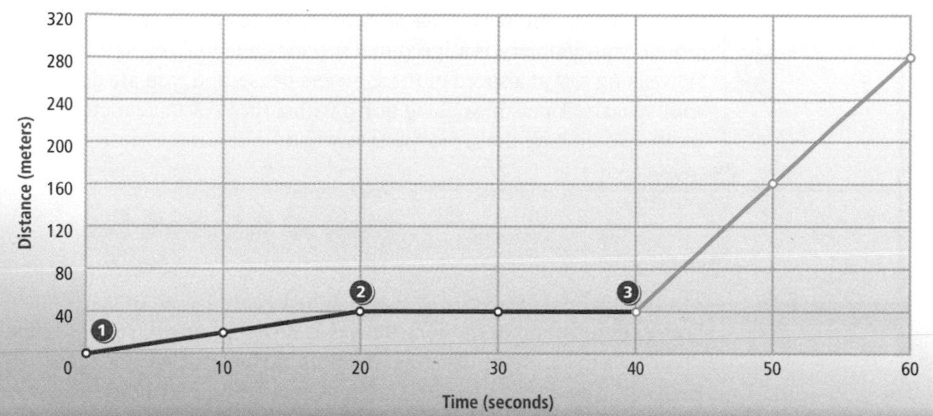

1. When the zebra is walking, its distance from its starting point increases. You can see this motion on the graph as a climbing line.
2. When the zebra stops to graze, it no longer changes its distance from the starting point. Time, however, continues to pass. Therefore, the graph shows a flat, or horizontal, line.
3. As soon as the zebra notices the lion, it stops grazing and starts to run for its life. The zebra is covering a greater distance in each time interval than it was before the chase started, so the line is steeper.

READING VISUALS How do the distances change over each 10-second time interval?

DIFFERENTIATE INSTRUCTION

Below Level Draw attention to the numbered circles in the visual. Ask: What are these numbers used for? *to show order and represent stages in the scenario* Point out that the numbered text tells a story about the zebra. The numbers show the order in which events happen. Demonstrate how the numbers on the graph match up with the numbers in the story.

Teach from Visuals

Have students identify the *x*-axis and *y*-axis and the line plot on the distance-time graph, then ask:

- How does each circled number in the graph relate to the number beside each text block? *The text describes what happened at the corresponding point in the graph.*
- Would average speed give an accurate picture of the zebra's motion during this minute? *no*

 This visual is also available as T6 in the Unit Transparency Book.

Teach Difficult Concepts

A distance-time graph shows how the position of an object changes over time. However, students may think that the path the object takes is the same as the lines shown on the object's distance-time graph. To help students understand how the graph relates to the path of the zebra, try the teacher demo below.

Teacher Demo

Draw the axes of a distance-time graph on the board. Cut out a piece of paper to represent the zebra. Have a student move the paper zebra slowly up the *y*-axis, stop briefly, and then continue at a faster speed. During this process, create the distance-time graph by drawing a line that continually matches the height of the zebra on the *y*-axis as you move your hand horizontally at a constant rate. Compare the distance-time graph to the actual motion of the paper zebra.

Ongoing Assessment

READING VISUALS *Answer: increases, increases, stays the same, stays the same, increases sharply, increases sharply*

Teach Difficult Concepts

The term *vector* may be an entirely new concept to students. Emphasize practical reasons for the distinction between speed and velocity, which is an example of a vector, with the example below.

Your dog Sam often runs away from home. You know that he averages 3 miles per hour on these adventures. Ask: Is that enough information to determine his location when he has been gone for 20 minutes? *no*

A neighbor calls to tell you that she saw Sam making a beeline north. Now do you have enough information to determine his location? *yes, if he hasn't changed direction since then*

Point out that in the first case, Sam could be anywhere within a circle with a radius of one mile from your house. In the second case, you could search along just one straight line north from the neighbor's house to find him.

Ongoing Assessment

Describe an object's velocity.

Ask: If two scooters are moving at the same speed but in opposite directions, is their velocity the same? Explain. *no; they have different directions*

CHECK YOUR READING *Answer: Velocity is speed in a specific direction. Students' examples should include both speed and direction.*

CAPTION The velocity of the ant on the right is greater than that of the ant moving upward and less than the velocity of the ant moving downward.

Velocity includes speed and direction.

Sometimes the direction of motion is as important as its speed. In large crowds, for example, you probably always try to walk in the same direction the crowd is moving and at the same speed. If you walk in even a slightly different direction, you can bump into other people. In a crowd, in other words, you try to walk with the same velocity as the people around you. **Velocity** is a speed in a specific direction. If you say you are walking east at a speed of three meters per second, you are describing your velocity. A person walking north with a speed of three meters per second would have the same speed as you do, but not the same velocity.

? H

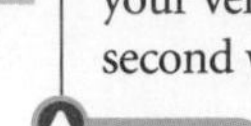

What is velocity? Give an example of a velocity.

Velocity

The picture below shows several ants as they carry leaves along a branch. Each ant's direction of motion changes as it walks along the bends of the branch. As the arrows indicate, each ant is moving in a specific direction. Each ant's velocity is shown by the length and direction of the arrow. A longer arrow means a greater speed in the direction the arrow is pointing. In this picture, for example, the ant moving up the branch is traveling more slowly than the ant moving down the branch.

To determine the velocity of an ant as it carries a leaf, you need to know both its speed and its direction. A change in either speed or direction results in a change in velocity. For example, the velocity of an ant changes if it slows down but continues moving in the same direction. Velocity also changes if the ant continues moving at the same speed but changes direction.

READING TIP

Green arrows show velocity.

A longer arrow indicates a faster speed than a shorter arrow. The direction of the arrow indicates the direction of motion.

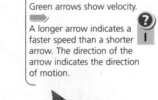

? I

Velocity is an example of a vector. A **vector** is a quantity that has both size and direction. Speed is not a vector because speed is a measure of how fast or slow an object moves, not which direction it moves in. Velocity, however, has a size—the speed—and a direction, so it is a vector quantity.

INFER How does this ant's velocity compare with those of the other ants?

DIFFERENTIATE INSTRUCTION

More Reading Support

H What does velocity have that speed does not have? *a specific direction*

I A vector has a direction. What else does it have? *size*

top view

30 km/h north

30 km/h south

INFER How do the speeds and velocities of these trains compare?

Velocity Versus Speed

Because velocity includes direction, it is possible for two objects to have the same speed but different velocities. If you traveled by train to visit a friend, you might go 30 kilometers per hour (km/h) north on the way there and 30 km/h south on the way back. Your speed is the same both going and coming back, but your velocity is different because your direction of motion has changed.

Another difference between speed and velocity is the way the average is calculated. Your average speed depends on the total distance you have traveled. The average velocity depends on the total distance you are from where you started. Going north, your average speed would be 30 km/h, and your average velocity would be 30 km/h north. After the round-trip ride, your average traveling speed would still be 30 km/h. Your average velocity, however, would be 0 km/h because you ended up exactly where you started.

CHECK YOUR READING Use a Venn diagram to compare and contrast speed and velocity.

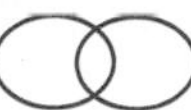

1.2 Review

KEY CONCEPTS

1. How is speed related to distance and time?
2. How would decreasing the time it takes you to run a certain distance affect your speed?
3. What two things do you need to know to describe the velocity of an object?

CRITICAL THINKING

4. **Compare** Amy and Ellie left school at the same time. Amy lives farther away than Ellie, but she and Ellie arrived at their homes at the same time. Compare the girls' speeds.
5. **Calculate** Carlos lives 100 m away from his friend's home. What is his average speed if he reaches his friend's home in 50 s?

CHALLENGE

6. **Synthesize** If you watch a train go by at 20 m/s, at what speed will the people sitting on the train be moving relative to you? Would someone walking toward the back of the train have a greater or lesser speed relative to you? Explain.

ANSWERS

1. Speed is distance divided by time.

2. Your speed would increase.

3. speed and direction of motion

4. Amy traveled faster than Ellie.

5. $S = \frac{d}{t} = \frac{100\ m}{50\ s} = 2\ m/s$

6. 20 m/s; lesser speed because of the frame of reference

Teach from Visuals

To help students interpret the visual on velocity and speed, ask:

- What can you say about the directions of the two trains? *They are opposite.*
- What does the inset show? *two trains going in opposite directions*
- How do the speeds and velocities of the two trains compare? *The speeds are identical, but the velocities are different because the trains are moving in different directions.*

Ongoing Assessment

CHECK YOUR READING *Speed does not include information about direction; velocity does.*

Reinforce the BIG idea

Have students relate the section to the Big Idea.

Reinforcing Key Concepts, p. 32

1.2 ASSESS & RETEACH

Assess

A Section 1.2 Quiz, p. 4

Reteach

On a broad, flat surface, mark with tape a start line and a finish line. Have a volunteer roll a toy car the length of the course. Ask: How could you check the car's speed? Have students time the car, measure the length of the course, and calculate the speed.

Now have two students each roll a car in opposite directions on the course. Ask: How would you find the velocity of each car?

Technology Resources

Have students visit **ClassZone.com** for reteaching of Key Concepts.

CONTENT REVIEW

CONTENT REVIEW CD-ROM

MATH IN SCIENCE
Math Skills Practice for Science

Set Learning Goal
To use units appropriately when performing calculations

Present the Science
An instructive example of not paying attention to units is the failure of the *Mars Climate Orbiter* spacecraft. Because the measurements were not converted from English units into metric units, the spacecraft went too close to the planet and probably burned up in the Martian atmosphere. It never had a chance to enter its observation orbit.

Develop Algebra Skills
Point out the fractional expression in the denominator of a fraction on this page. Emphasize that division by a fraction is equivalent to multiplication by the inverse of the fraction. Present some examples on the board if students aren't following the math on the page.

DIFFERENTIATION TIP Prepare students who have cognitive disabilities to consider units in math formulas by counting different quantities. You could discuss the value of attaching units to types of drinks, foods in packages, sports statistics, and so forth.

Close
Ask: If you were a construction worker, what might happen if you measured your work in the wrong units? *You would probably cause a flaw in construction that either would have to be repaired right away or could cause serious problems after the structure was finished.*

- Math Support, p. 54
- Math Practice, p. 55

Technology Resources
Students can visit **ClassZone.com** for practice with calculations using units.

MATH TUTORIAL

MATH in SCIENCE

MATH TUTORIAL
CLASSZONE.COM
Click on Math Tutorial for more help with units and rates.

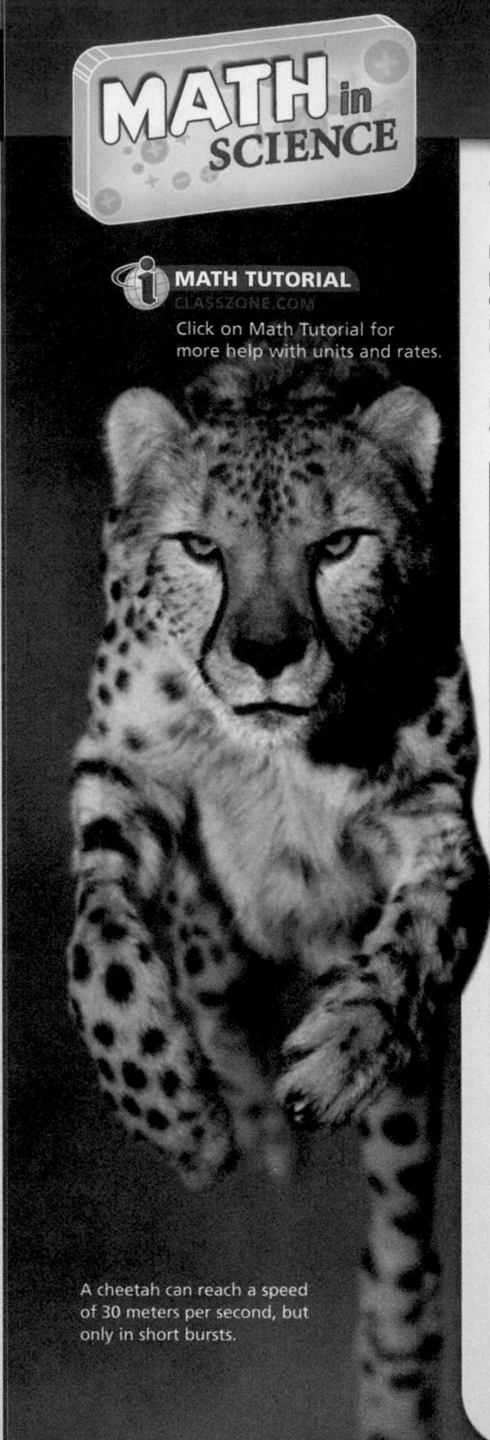

A cheetah can reach a speed of 30 meters per second, but only in short bursts.

SKILL: WORKING WITH UNITS

Time, Distance, and Speed

If someone tells you the store is "five" from the school, you would probably ask, "Five what? Five meters? Five blocks?" You typically describe a distance using standard units of measurement, such as meters, miles, or kilometers. By using units, you help other people understand exactly what your measurement means.

When you work with a formula, the numbers that you substitute into the formula have units. When you calculate with a number, you also calculate with the unit associated with that number.

Example

A cheetah runs at a speed of 30 meters per second. How long does the cheetah take to run 90 meters?

The formula for time in terms of speed and distance is

$$\text{time} = \frac{\text{distance}}{\text{Speed}} \qquad t = \frac{d}{S}$$

(1) Start by substituting the numbers into the formula. Include the units with the numbers.

$$t = \frac{90\text{ m}}{30\text{ m/s}}$$

(2) When the units or calculations include fractions, write out the units as fractions as well:

$$t = \frac{90\text{ m}}{\frac{30\text{ m}}{\text{s}}}$$

(3) Do the calculation and simplify the units by cancellation:

$$t = 90\text{ m} \cdot \frac{\text{s}}{30\text{ m}} = \frac{90}{30} \cdot \frac{\text{m} \cdot \text{s}}{\text{m}} = 3 \cdot \frac{\cancel{\text{m}} \cdot \text{s}}{\cancel{\text{m}}} = 3\text{ s}$$

ANSWER 3 seconds

Note that the answer has a unit of time. Use the units to check that your answer is reasonable. An answer that is supposed to have a unit of time, for example, should not have a unit of distance.

Answer the following questions.

1. How long would it take an object traveling 12 m/s to go 60 m? What unit of time is your answer in?
2. If a car travels 60 km/h, how long would it take the car to travel 300 km? What unit of time is your answer in?
3. If a man walks 3 miles in 1 hour, what is his speed? What unit of speed is your answer in? (Use the formula on page 18.)

CHALLENGE Show that the formula *distance = speed • time* has a unit for distance on both sides of the equal sign.

ANSWERS

1. t = d/S = *60 m/12 m/s = 5 seconds*

2. t = d/S = *300 km/ 60 km/h = 5 hours*

3. *3 miles per hour*

CHALLENGE *Distance has units of length, for example, meters. Speed has units of length per time interval, such as meters per second.*

$$\text{Distance (m)} = \frac{\text{distance (m)}}{\cancel{\text{time (s)}}} \cdot \cancel{\text{time (s)}} = \text{distance (m)}.$$ *Both sides have units of distance.*

KEY CONCEPT

1.3 Acceleration measures how fast velocity changes.

BEFORE, you learned
- Speed describes how far an object travels in a given time
- Velocity is a measure of the speed and direction of motion

NOW, you will learn
- How acceleration is related to velocity
- How to calculate acceleration

VOCABULARY
acceleration p. 25

THINK ABOUT

How does velocity change?

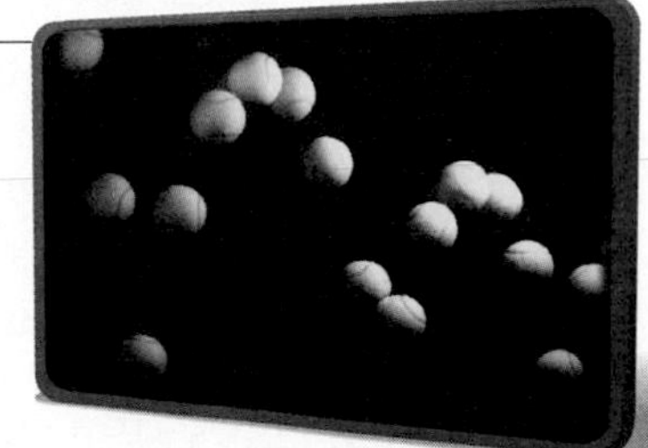

The photograph at right shows the path that a bouncing ball takes. The time between each image of the ball is the same during the entire bounce. Is the ball moving the same distance in each time interval? Is the ball moving the same direction in each time interval?

Speed and direction can change with time.

OUTLINE
Remember to use the blue and red headings in this chapter to help you make notes on acceleration.

I. Main idea
  A. Supporting idea
    1. Detail
    2. Detail
  B. Supporting idea

When you throw a ball into the air, it leaves your hand at a certain speed. As the ball rises, it slows down. Then, as the ball falls back toward the ground, it speeds up again. When the ball hits the ground, its direction of motion changes and it bounces back up into the air. The speed and direction of the ball do not stay the same as the ball moves. The ball's velocity keeps changing.

You can find out how much an object's position changes during a certain amount of time if you know its velocity. In a similar way, you can measure how an object's velocity changes with time. The rate at which velocity changes with time is called **acceleration.** Acceleration is a measure of how quickly the velocity is changing. If velocity does not change, there is no acceleration.

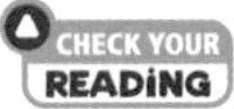
CHECK YOUR READING What is the relationship between velocity and acceleration?

RESOURCES FOR DIFFERENTIATED INSTRUCTION

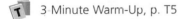

Below Level
UNIT RESOURCE BOOK
- Reading Study Guide A, pp. 35–36
- Decoding Support, p. 49

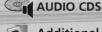

AUDIO CDS

Additional INVESTIGATION, On the Move, A, B, & C, pp. 65–73; Teacher Instructions, pp. 346–347

Advanced
UNIT RESOURCE BOOK
- Challenge and Extension, p. 41
- Challenge Reading, pp. 45–46

English Learners
UNIT RESOURCE BOOK
Spanish Reading Study Guide, pp. 39–40

AUDIO CDS
- Audio Readings in Spanish
- Audio Readings (English)

1.3 FOCUS

Set Learning Goals
Students will
- Explain how acceleration is related to velocity.
- Calculate acceleration.
- Measure acceleration through an experiment.

3-Minute Warm-Up
Display Transparency 5 or copy this exercise on the board:

Decide if these statements are true. If they are not true, correct them.

1. A distance-time graph shows how speed changes with time. *true*
2. Speed is a vector. *Velocity is a vector.*
3. Average speed is total distance multiplied by total time. *Average speed is total distance divided by total time.*

3-Minute Warm-Up, p. T5

1.3 MOTIVATE

THINK ABOUT
PURPOSE To examine the path, velocity, and direction of a bouncing ball

DISCUSS Ask students how the distance the ball travels between time intervals relates to speed. Discuss how speed and direction are changing as the ball bounces. Have students describe situations in which velocity changes. *Examples: a person stopping to open a door; a roller-coaster car moving through a loop*

Answers: No, it does not move the same distance in each time interval; no, it is not moving in the same direction in each time interval.

Ongoing Assessment
CHECK YOUR READING *Answer: Acceleration is the change in velocity over time.*

1.3 INSTRUCT

Address Misconceptions

IDENTIFY Ask: If you ride in a car during acceleration, what happens? If students say "the car speeds up," they may hold the misconception that acceleration only means speeding up, although it means any change in speed or direction.

CORRECT Review the first paragraph on this page. Emphasize that *any* change in velocity is acceleration. Remind students that velocity includes the speed of an object *and* its direction. Mention that "speeding up" is the meaning in English but not the meaning in physics.

REASSESS Ask: Which of the following is an example of acceleration? *all of them*

- a car slowing down from 50 miles per hour to 40 miles per hour
- a car turning a corner
- a car starting up again after stopping

Technology Resources

Visit **ClassZone.com** for background on common student misconceptions.

MISCONCEPTION DATABASE

Integrate the Sciences

Some animals that travel in groups exhibit a very striking form of acceleration: they suddenly change direction as a group. An example includes flocks of birds that suddenly shift direction, totally in sync with each other. This behavior occurs without the guidance of any leader. Computer scientists have studied this behavior and have made computer models of it.

EXPLORE the BIG idea

Revisit "Rolling Along" on p. 7. Have students explain their results.

Ongoing Assessment

Explain how acceleration affects velocity.

Ask: Suppose velocity remains steady. Will there be any acceleration? Explain. *No; acceleration is a change in velocity.*

Answer: When an object accelerates, velocity can increase, decrease, or change direction.

? A

The word *acceleration* is commonly used to mean "speeding up." In physics, however, acceleration refers to any change in velocity. A driver slowing down to stop at a light is accelerating. A runner turning a corner at a constant speed is also accelerating because the direction of her velocity is changing as she turns.

Like velocity, acceleration is a vector, which means it has both size and direction. The direction of the acceleration determines whether an object will slow down, speed up, or turn.

READING TIP

Orange arrows are used to show acceleration.

Remember that green arrows show velocity.

A longer arrow means greater acceleration or velocity.

1. **Acceleration in the Same Direction as Motion** When the acceleration is in the same direction as the object is moving, the speed of the object increases. The car speeds up.

2. **Acceleration in the Opposite Direction of Motion** When the acceleration is opposite to the motion, the speed of the object decreases. The car slows down. Slowing down is also called negative acceleration.

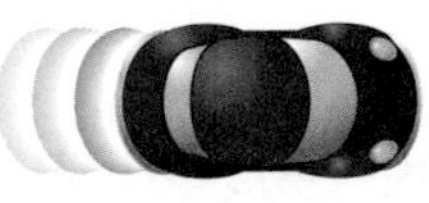

3. **Acceleration at a Right Angle to Motion** When the acceleration is at a right angle to the motion, the direction of motion changes. The car changes the direction in which it is moving by some angle, but its speed does not change.

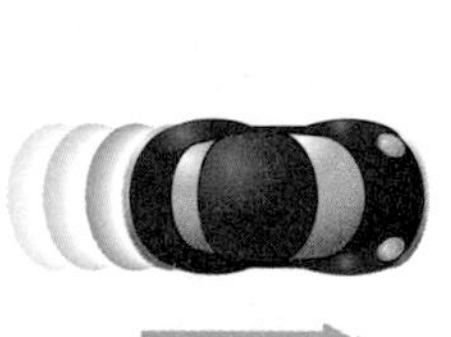

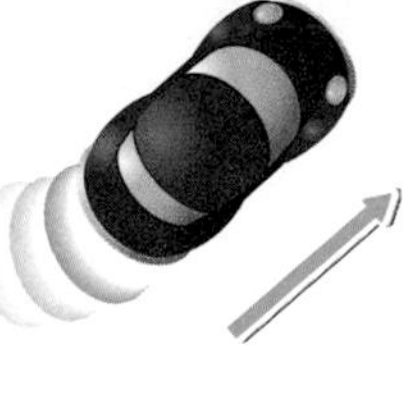

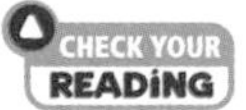

How does acceleration affect velocity? Give examples.

DIFFERENTIATE INSTRUCTION

More Reading Support

A Does *acceleration* mean "speeding up," or does it refer to any possible change in velocity? *any change in velocity*

English Learners English learners may not be familiar with the mathematical concept of right angles on this page. Also, notice the terms *speeding up* and *slowing down* on this page. These idioms may be confusing to English learners who take them literally—in terms of direction (*up* and *down*) instead of acceleration. Explain that the terms refer to the change in velocity.

INVESTIGATE Acceleration

When does an object accelerate?

SKILL FOCUS
Measuring

PROCEDURE

1. Use the template and materials to construct an acceleration measuring tool.
2. Hold the tool in your right hand so that the string falls over the 0 m/s^2 mark. Move the tool in the direction of the arrow. Try to produce both positive and negative acceleration without changing the direction of motion.
3. With the arrow pointing ahead of you, start to walk. Observe the motion of the string while you increase your speed.
4. Repeat step 3, but this time observe the string while slowing down.
5. Repeat step 3 again, but observe the string while walking at a steady speed.

WHAT DO YOU THINK?

- When could you measure an acceleration?
- What was the largest acceleration (positive or negative) that you measured?

CHALLENGE If you moved the acceleration measuring tool backward, how would the measuring scale change?

MATERIALS
- template for tool
- cardboard
- scissors
- glue
- piece of string
- weight

TIME
30 minutes

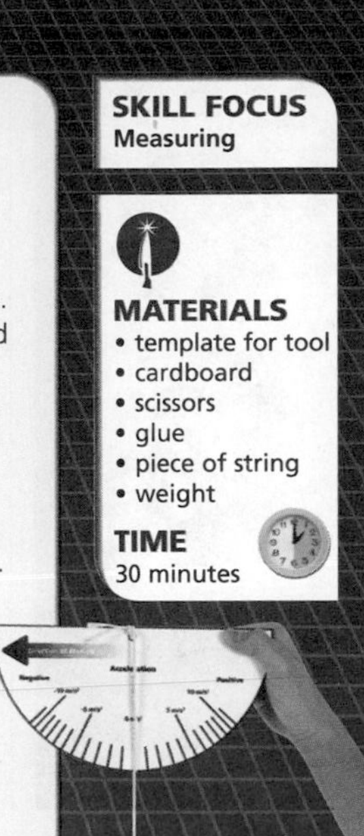

Acceleration can be calculated from velocity and time.

Suppose you are racing a classmate. In one second, you go from standing still to running at six meters per second. In the same time, your classmate goes from standing still to running at three meters per second. How does your acceleration compare with your classmate's acceleration? To measure acceleration, you need to know how velocity changes with time.

- The change in velocity can be found by comparing the initial velocity and the final velocity of the moving object.
- The time interval over which the velocity changed can be measured.

In one second, you increase your velocity by six meters per second, and your friend increases her velocity by three meters per second. Because your velocity changes more, you have a greater acceleration during that second of time than your friend does. Remember that acceleration measures the change in velocity, not velocity itself. As long as your classmate increases her current velocity by three meters per second, her acceleration will be the same whether she is going from zero to three meters per second or from three to six meters per second.

DIFFERENTIATE INSTRUCTION

More Reading Support

B Which two things do you have to know to measure acceleration? *the change in velocity and how long the change took*

Additional Investigation
To reinforce Section 1.3 learning goals, use the following full-period investigation:

Additional INVESTIGATION, On the Move, A, B, & C, pp. 65–73, 346–347 (Advanced students should complete Levels B and C.)

Advanced
Have students who are interested in what happens when large animals accelerate read the following article:

Challenge Reading, pp. 45–46

INVESTIGATE Acceleration

PURPOSE To measure how much acceleration occurs in different types of movements

TIPS *30 min.*

- Use a dark-colored string in the acceleration measuring tool (or color a white string with a black marker) so that it will show up against the background.
- If you have to hold the tool in your left hand, reverse the positive and negative labels and signs in front of the numbers, and point the arrow in the opposite direction.

WHAT DO YOU THINK? *Sample answers: when reversing direction and when speed is changing; students' answers will vary.*

CHALLENGE *You would have to reverse the positive and the negative scale.*

- Template for Tool, p. 42
- Datasheet, Acceleration, p. 43

Technology Resources

Customize this student lab as needed or look for an alternative. Print rubrics to assess student lab reports.

Lab Generator CD-ROM

Metacognitive Strategy

Prompt students to think of ways they can observe velocity and acceleration in their daily lives. Ask whether they think finding examples such as these will improve their understanding of the concepts. Have students write a paragraph and then share their examples in class.

Teacher Demo

Push two wheeled toys simultaneously in parallel paths on a flat surface. Push one toy at a steady speed, but make the other toy gain speed as it rolls.

Ask: Do both toys demonstrate velocity? Explain. *Yes; both are moving at a certain speed in a particular direction.*

Ask: Do both toys demonstrate acceleration? Explain. *No; one toy has a steady velocity, so it has no acceleration.*

Teach Difficult Concepts

Remind students that any value times itself is the value squared. On the board, draw a square with each side labeled *x*. Ask: What is the area of this square? *x times x, or x^2*

Have a volunteer read the paragraph immediately following the acceleration formula on this page. Help students make the connection between the x^2 for area of a square and the unit of acceleration: m/s^2.

Develop Mathematics Skills

- Math Support, p. 52
- Math Practice, p. 53

Mathematics Connection

A common variation on the acceleration formula is $a = \frac{(v_2 - v_1)}{t}$, where v_2 replaces v_{final} and v_1 replaces $v_{initial}$.

Ongoing Assessment

Calculate acceleration.

Ask: Suppose you know the initial velocity of a toy airplane taking off and the final velocity when it is cruising high up in the air. Could you calculate the acceleration? Explain why or why not. *No; you also need the time interval during which the airplane changed velocity.*

Practice the Math

Answers

1. $a = \frac{v_{final} - v_{initial}}{t} = \frac{0.6 \text{ m/s} - 0.5 \text{ m/s}}{1 \text{ s}} = 0.1 \text{ m/s}^2$

2. $a = \frac{v_{final} - v_{initial}}{t} = \frac{0 \text{ m/s} - 10 \text{ m/s}}{20 \text{ s}} = -0.5 \text{ m/s}^2$

Calculating Acceleration

If you know the starting velocity of an object, the final velocity, and the time interval during which the object changed velocity, you can calculate the acceleration of the object. The formula for acceleration is shown below.

$$\text{acceleration} = \frac{\text{final velocity} - \text{initial velocity}}{\text{time}}$$

$$a = \frac{v_{final} - v_{initial}}{t}$$

Remember that velocity is expressed in units of meters per second. The standard units for acceleration, therefore, are meters per second over time, or meters per second per second. This is simplified to meters per second squared, which is written as m/s^2.

As the girl in the photograph at left sleds down the sandy hill, what happens to her velocity? At the bottom of the hill, her velocity will be greater than it was at the top. You can calculate her average acceleration down the hill if you know her starting and ending velocities and how long it took her to get to the bottom. This calculation is shown in the sample problem below.

REMINDER

Remember that velocity is the speed of the object in a particular direction.

Calculating Acceleration

Sample Problem

Ama starts sliding with a velocity of 1 m/s. After 3 s, her velocity is 7 m/s. What is Ama's acceleration?

What do you know? initial velocity = 1 m/s, final velocity = 7 m/s, time = 3 s

What do you want to find out? acceleration

Write the formula: $a = \frac{v_{final} - v_{initial}}{t}$

Substitute into the formula: $a = \frac{7 \text{ m/s} - 1 \text{ m/s}}{3 \text{ s}}$

Calculate and simplify: $a = \frac{6 \text{ m/s}}{3 \text{ s}} = 2 \frac{\text{m/s}}{\text{s}} = 2 \text{ m/s}^2$

Check that your units agree: $\frac{\text{m/s}}{\text{s}} = \frac{\text{m}}{\text{s}} \cdot \frac{1}{\text{s}} = \frac{\text{m}}{\text{s}^2}$

Unit of acceleration is m/s^2. Units agree.

Answer: $a = 2 \text{ m/s}^2$

Practice the Math

1. A man walking at 0.5 m/s accelerates to a velocity of 0.6 m/s in 1 s. What is his acceleration?
2. A train traveling at 10 m/s slows down to a complete stop in 20 s. What is the acceleration of the train?

DIFFERENTIATE INSTRUCTION

More Reading Support

C What values must you plug into the acceleration formula to calculate acceleration? *final velocity, initial velocity, and time*

Below Level Help students understand the expression for acceleration units. On the board write "m/s = velocity," and read it as "meters per second." Write "velocity/s = acceleration," and read it as "velocity per second."

Point out that, because velocity is in units of meters per second, you can substitute those words for velocity in the acceleration expression, and write "meters per second per second = acceleration."

The sledder's final velocity was greater than her initial velocity. If an object is slowing down, on the other hand, the final velocity is less than the initial velocity. Suppose a car going 10 meters per second takes 2 seconds to stop for a red light. In this case, the initial velocity is 10 m/s and the final velocity is 0 m/s. The formula for acceleration gives a negative answer, $-5\ m/s^2$. The negative sign indicates a negative acceleration—that is, an acceleration that decreases the velocity.

RESOURCE CENTER
CLASSZONE.COM
Learn more about acceleration.

CHECK YOUR READING What would be true of the values for initial velocity and final velocity if the acceleration were zero?

Acceleration over Time

Even a very small positive acceleration can lead to great speeds if an object accelerates for a long enough period. In 1998, NASA launched the *Deep Space 1* spacecraft. This spacecraft tested a new type of engine—one that gave the spacecraft an extremely small acceleration. The new engine required less fuel than previous spacecraft engines. However, the spacecraft needed a great deal of time to reach its target velocity.

The acceleration of the *Deep Space 1* spacecraft is less than 2/10,000 of a meter per second per second ($0.0002\ m/s^2$). That may not seem like much, but over 20 months, the spacecraft could increase its speed by 4500 meters per second (10,000 mi/h).

By carefully adjusting both the amount and the direction of the acceleration of *Deep Space 1*, scientists were able to control its flight path. In 2001, the spacecraft successfully flew by a comet, sending back images from about 230 million kilometers (140 million mi) away.

APPLY What makes the new engine technology used by *Deep Space 1* more useful for long-term missions than for short-term ones?

Address Misconceptions

IDENTIFY Ask: What does negative acceleration mean to you? If students answer that it means going backward, then they are confusing negative acceleration with reversing direction.

CORRECT Stage a modified form of the Teacher Demo on page 27: roll one car at a steady speed and make the other car slow down as it rolls. Ask: Which car is demonstrating acceleration? *The car that is slowing down.* Ask: Is the same car demonstrating negative acceleration? *Yes; its velocity is decreasing while it is still going forward.*

REASSESS Write the following scenarios on the board. Say that all demonstrate kinds of acceleration. Ask students which demonstrates negative acceleration.

- a bird shifts direction in flight
- an arrow shoots out of a bow
- an out-of-gas car rolls to a stop

the last one, because the velocity is decreasing

Technology Resources

Visit **ClassZone.com** for background on common student misconceptions.

MISCONCEPTION DATABASE

History of Science

NASA, the space administration agency of the United States, launched *Deep Space 1* (*DS1*) in October 1998. It was powered by ion propulsion. An ion propulsion engine uses a stream of electrically charged particles to move the spacecraft forward in a way similar to that of a jet engine. An ion propulsion engine builds speed by accelerating slowly and steadily. *DS1* flew by an asteroid called Braille in July 1999 and a comet called Borelly in September 2001. The *DS1* mission ended in late 2001.

DIFFERENTIATE INSTRUCTION

More Reading Support

D What happens if an object positively accelerates a little over a long time? *After time has passed, the object will be going very fast.*

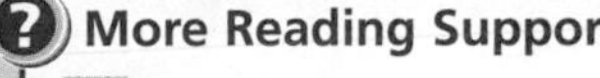

Advanced Ask students to imagine riding on a roller coaster. Ask how acceleration would change throughout the ride (through negative, positive, or vector acceleration). Challenge students to draw cartoons of at least three places in a roller coaster ride that show different kinds of acceleration. Have them label each cartoon to describe briefly what is happening in terms of acceleration.

R Challenge and Extension, p. 41

Ongoing Assessment

CHECK YOUR READING *Answer: The values for initial velocity and final velocity would be the same.*

Teach from Visuals

Make sure students understand how the parts of the velocity-time and distance-time graphs fit together, and then ask:

- Do the graphs represent the same data, different data, or related data? *related data*
- Why does distance change at zero acceleration? *Velocity continues and the boy moves.*

Teach Difficult Concepts

On the velocity-time graph, students see the line fall when velocity slows. They may expect to see a similar drop on the distance-time graph. Point out that the line on a distance-time graph can level out but doesn't fall because it shows the distance already traveled, even if the person in motion slows or stops.

Real World Example

An arrow fired from a bow has a distinctive velocity-time graph shaped like a post. When the arrow leaves the bow, it goes from zero to high velocity in a split second. As the arrow travels through air, its velocity stays the same or very gradually decreases. When the arrow hits a target, its velocity goes from high to zero in a split second.

Develop Graphing Skills

When reading a graph, students often need to estimate intermediate values on one or both axes. Have students scan the distance values in the bottom graph on p. 30. Ask:

- What is the range of numbers? *0 to 60* How much distance is represented by the space between the lines marked 0 and 10? *10 m*
- What is half of 10? *5*
- Where does 5 fall on the *y*-axis? *halfway between the tick marks for 0 and 10*

Velocity-Time Graphs

Velocity-time graphs and distance-time graphs are related. This is because the distance an object travels depends on its velocity. Compare the velocity-time graph on the right with the distance-time graph below it.

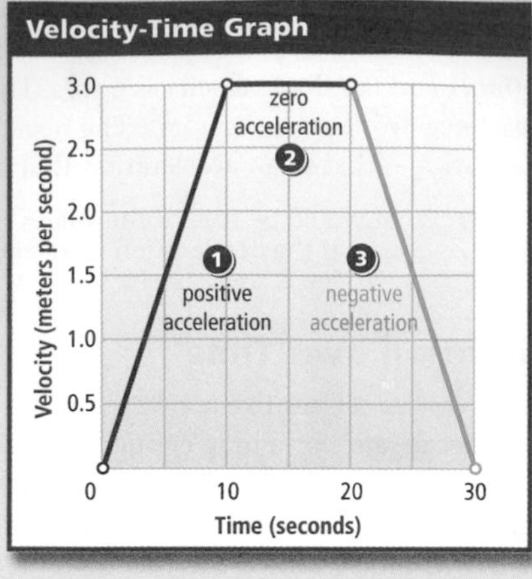

1. As the student starts to push the scooter, his velocity increases. His acceleration is positive, so he moves forward a greater distance with each second that passes.
2. He coasts at a constant velocity. Because his velocity does not change, he has no acceleration, and he continues to move forward the same distance each second.
3. As he slows down, his velocity decreases. His acceleration is negative, and he moves forward a smaller distance with each passing second until he finally stops.

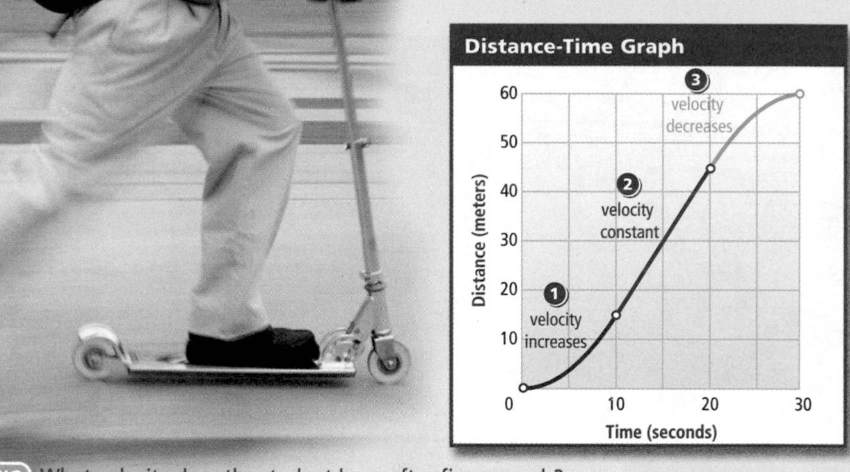

READING VISUALS What velocity does the student have after five seconds? About how far has he moved in that time?

C 30 Unit: Motion and Forces

DIFFERENTIATE INSTRUCTION

Below Level Introduce the visual with the following summary.

All the parts of the page are about the motion of the young man on the scooter. He speeds up, holds this speed, slows down, and stops. The picture tells the story in one way, and the words beside the numbers tell it in another way. The two graphs tell his story in yet another way. They help us analyze acceleration in the motion of the young man and his scooter.

Ongoing Assessment

Answer: 1.5 m/s; about 4 m

Velocity-Time Graphs

Acceleration, like position and velocity, can change with time. Just as you can use a distance-time graph to understand velocity, you can use a velocity-time graph to understand acceleration. Both graphs tell you how something is changing over time. In a velocity-time graph, time is on the horizontal axis, or *x*-axis, and velocity is on the vertical axis, or *y*-axis.

Explore how changing the acceleration of an object changes its motion.

The two graphs on page 30 show a velocity-time graph and a distance-time graph of a student riding on a scooter. He first starts moving and speeds up. He coasts, and then he slows down to a stop.

1. The rising line on the velocity-time graph shows where the acceleration is positive. The steeper the line, the greater the acceleration. The distance-time graph for the same interval is curving upward more and more steeply as the velocity increases.
2. The flat line on the velocity-time graph shows an interval of no acceleration. The distance-time graph has a straight line during this time, since the velocity is not changing.
3. The falling line on the velocity-time graph shows where the acceleration is negative. The same interval on the distance-time graph shows a curve that becomes less and less steep as the velocity decreases. Notice that the overall distance still increases.

Velocity-time graphs and distance-time graphs can provide useful information. For example, scientists who study earthquakes create these graphs in order to study the up-and-down and side-to-side movement of the ground during an earthquake. They produce the graphs from instruments that measure the acceleration of the ground.

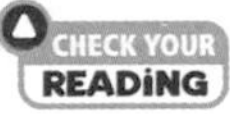

What does a flat line on a velocity-time graph represent?

1.3 Review

KEY CONCEPTS

1. What measurements or observations tell you that a car is accelerating?
2. If an object accelerates in the same direction in which it is moving, how is its speed affected?
3. What measurements do you need in order to calculate acceleration?

CRITICAL THINKING

4. **Calculate** A car goes from 20 m/s to 30 m/s in 10 seconds. What is its acceleration?
5. **Infer** Two runners start a race. After 2 seconds, they both have the same velocity. If they both started at the same time, how do their average accelerations compare?

CHALLENGE

6. **Analyze** Is it possible for an object that has a constant negative acceleration to change the direction in which it is moving? Explain why or why not.

ANSWERS

1. The velocity is changing.

2. The speed increases.

3. starting velocity of an object, final velocity of an object, and time interval during which the object changed velocity

4. $a = \frac{v_{final} - v_{initial}}{t}$

$= \frac{30\ m/s - 20\ m/s}{10s}$

$= \frac{10\ m/s}{10s}$

$= 10\ m/s^2$

5. They are the same.

6. Yes; if the object is still accelerating when the speed reaches 0 m/s, it will start moving in the opposite direction. Once the direction changes, the object will positively accelerate.

Develop Critical Thinking

APPLY Suppose you viewed velocity-time graphs for two cyclists in a race. Because you saw the race, you know that one cyclist built speed rapidly and then coasted to the finish. The other built speed gradually throughout the race. Have students apply their knowledge of graphs and slope to match each cyclist's racing strategy with an appropriate graph. *The graph of the first cyclist would have a steep rise and then a plateau. The graph of the second would show a smoothly rising line.*

Ongoing Assessment

CHECK YOUR READING *Answer: A flat line on a velocity-time graph represents an interval of no acceleration.*

Reinforce the BIG idea

Have students relate the section to the Big Idea.

 Reinforcing Key Concepts, p. 44

1.3 ASSESS & RETEACH

Assess

 Section 1.3 Quiz, p. 5

Reteach

Have students give a real-world example of each type of acceleration:

- positive (speeding up)
- negative (slowing down)
- direction change (changing direction)

Have students review p. 28 and explain why the unit for acceleration is meters per second squared. Students should include the formula for acceleration in their explanation.

Technology Resources

Have students visit **ClassZone.com** for reteaching of Key Concepts.

 **CONTENT REVIEW**

 **CONTENT REVIEW CD-ROM**

CHAPTER INVESTIGATION

Focus

PURPOSE Students will investigate how the steepness of a slope affects acceleration of a marble.

OVERVIEW Students will record the time it takes for a marble to roll down a slope of one book's thickness in multiple trials. They will

- average the times for all the runs
- repeat the trials for a slope of two books' thickness
- use their averaged data to calculate acceleration of the marble on each slope

Lab Preparation

- As an alternative to taping two meter sticks together, students could use a ruler that has a central groove.
- Prior to the investigation, have students read through the investigation, write their hypothesis, and prepare their data tables. Or you may wish to copy and distribute datasheets and rubrics.

 UNIT RESOURCE BOOK, pp. 56–64

 SCIENCE TOOLKIT, F14

Lab Management

- Have one student roll the marble and a partner record the times. Tell students to reverse roles after setting up the ramp on two books.
- Students should release—not push—the marble.
- Make sure students understand that they will have to calculate average times for all the trial data in both columns to complete their data tables. If they need a refresher, review how to calculate an average.
- Verify that students are calculating v_{final} correctly before they attempt to calculate acceleration.

INCLUSION Encourage students with visual impairments to perform tactile tasks, such as rolling the marble.

CHAPTER INVESTIGATION

Acceleration and Slope

OVERVIEW AND PURPOSE When a downhill skier glides down a mountain without using her ski poles, her velocity increases and she experiences acceleration. How would gliding down a hill with a greater slope affect her acceleration? In this investigation you will

- calculate the acceleration of an object rolling down two ramps of different slopes
- determine how the slope of the ramp affects the acceleration of the object

Problem

Write It Up

How does the slope of a ramp affect the acceleration of an object rolling down the ramp?

Hypothesize

Write a hypothesis to explain how changing the slope of the ramp will affect acceleration. Your hypothesis should take the form of an "If . . . , then . . . , because . . ." statement.

MATERIALS
- 2 meter sticks
- masking tape
- marble
- 2 paperback books
- ruler
- stopwatch
- calculator

Procedure

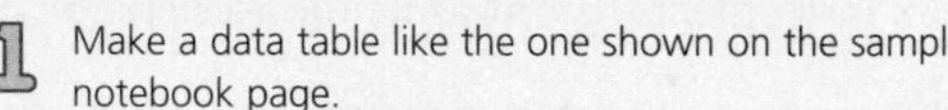

1. Make a data table like the one shown on the sample notebook page.
2. Make a ramp by laying two meter sticks side by side. Leave a small gap between the meter sticks.
3. Use masking tape as shown in the photograph to join the meter sticks. The marble should be able to roll freely along the groove.
4. Set up your ramp on a smooth, even surface, such as a tabletop. Raise one end of the ramp on top of one of the books. The other end of the ramp should remain on the table.
5. Make a finish line by putting a piece of tape on the tabletop 30 cm from the bottom of the ramp. Place a ruler just beyond the finish line to keep your marble from rolling beyond your work area.

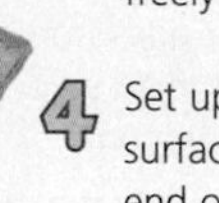

C 32 Unit: Motion and Forces

INVESTIGATION RESOURCES

 CHAPTER INVESTIGATION, Acceleration and Slope
- Level A, pp. 56–59
- Level B, pp. 60–63
- Level C, p. 64

Advanced students should complete Levels B & C.

 Writing a Lab Report, D12–13

Technology Resources

Customize this student lab as needed or look for an alternative. Print rubrics to assess student lab reports.

 **Lab Generator CD-ROM**

6 Test your ramp by releasing the marble from the top of the ramp. Make sure that the marble rolls freely. Do not push on the marble.

7 Release the marble and measure the time it takes for it to roll from the release point to the end of the ramp. Record this time under Column A for trial 1.

8 Release the marble again from the same point, and record the time it takes the marble to roll from the end of the ramp to the finish line. Record this time in Column B for trial 1. Repeat and record three more trials.

9 Raise the height of the ramp by propping it up with both paperback books. Repeat steps 7 and 8.

Observe and Analyze

Write It Up

1. **RECORD OBSERVATIONS** Draw the setup of your procedures. Be sure your data table is complete.
2. **IDENTIFY VARIABLES AND CONSTANTS** Identify the variables and constants in the experiment. List them in your notebook.
3. **CALCULATE**

   Average Time For ramps 1 and 2, calculate and record the average time it took for the marble to travel from the end of the ramp to the finish line.

   Final Velocity For ramps 1 and 2, calculate and record V_{final} using the formula below.

$$v_{final} = \frac{\text{distance from end of ramp to finish line}}{\text{average time from end of ramp to finish line}}$$

   Acceleration For ramps 1 and 2, calculate and record acceleration using the formula below. (**Hint:** Speed at the release of the marble is 0 m/s.)

$$a = \frac{v_{final} - v_{initial}\ \text{(speed at release)}}{\text{average time from release to bottom of ramp}}$$

Conclude

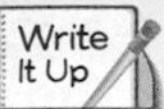

1. **COMPARE** How did the acceleration of the marble on ramp 1 compare with the acceleration of the marble on ramp 2?
2. **INTERPRET** Answer the question posed in the problem.
3. **ANALYZE** Compare your results with your hypothesis. Do your data support your hypothesis?
4. **EVALUATE** Why was it necessary to measure how fast the marble traveled from the end of the ramp to the finish line?
5. **IDENTIFY LIMITS** What possible limitations or sources of error could have affected your results? Why was it important to perform four trials for each measurement of speed?

INVESTIGATE Further

CHALLENGE Design your own experiment to determine how the marble's mass affects its acceleration down a ramp.

Acceleration and Slope

Problem How does the slope of a ramp affect the acceleration of an object rolling down the ramp?

Hypothesize

Observe and Analyze

Table 1. Times for Marble to Travel down Ramp

Height of Ramp (cm)	Trial Number	Column A Time from release to end of ramp	Column B Time from end of ramp to finish line
Ramp 1	1		
	2		
	3		
	4		
	Totals		
		Average	Average

Observe and Analyze

SAMPLE DATA With ramp height of one book (3 cm):

$$v_{final} = \frac{0.3\ m}{2\ s} = 0.15\ m/s;$$

$$\text{Acceleration} = \frac{0.15\ m/s - 0\ m/s}{4.5\ s} = 0.03\ m/s^2.$$

With ramp height of two books (6 cm):

$$v_{final} = \frac{0.3\ m}{1.5\ s} = 0.2\ m/s;$$

$$\text{Acceleration} = \frac{0.2\ m/s - 0\ m/s}{3.5\ s} = 0.06\ m/s^2.$$

1. *See students' drawings and tables.*
2. *Variable: steepness of ramp; constants: marble, length of ramp, distance from the top of the ramp to the finish line*
3. *See students' data.*

Conclude

1. *The higher ramp produced greater acceleration.*
2. *A steeper slope causes greater acceleration.*
3. *Students' answers will vary.*
4. *It allows you to calculate the final velocity.*
5. *Sample answer: Limitations were possible inaccuracy of time measurements and failure to let the marble roll without pushing it. Averaging a number of trials is a way to minimize the effect of errors.*

INVESTIGATE Further

CHALLENGE Students' designs will vary.

Post-Lab Discussion

- Prompt students to think of ways they might increase the accuracy of their data. *Sample answer: Increase the number of trial runs and discard the time measurement that was most different from the others.*
- Ask students what effect adding a second finish line two meters beyond the existing finish line and running more trials would have on their observations about acceleration. *The marble would exhibit negative acceleration between the first and second finish lines.*

BACK TO

the BIG idea

Have students list and explain three ways they have learned to describe motion. *Sample answer: by calculating speed, by describing velocity, and by recognizing and calculating acceleration*

KEY CONCEPTS SUMMARY

SECTION 1.1

Ask: How could you describe the change of position of the jumper shown in the visual? *The jumper starts at point A and stops at point B. You could also measure the distance of her jump in meters.*

SECTION 1.2

Ask: What can you tell about speed from the visual if the bike moves a distance of 3 m? *You could calculate speed from the given data.*

Ask: What can you tell about velocity from the picture if the bike moves a distance of 3 m? *You can calculate speed and observe the bike's direction; that's enough information to figure out velocity.*

SECTION 1.3

Ask: Explain why slowing down is an example of acceleration. *Slowing down is a change in velocity, so it fits the definition of acceleration.*

Review Concepts

- Big Idea Flow Chart, p. T1
- Chapter Outline, pp. T7–T8

1 Chapter Review

the BIG idea

The motion of an object can be described and predicted.

KEY CONCEPTS SUMMARY

1.1 An object in motion changes position.

Position is measured from a reference point.

Motion is measured relative to an observer.

VOCABULARY
position p. 9
reference point p. 10
motion p. 11

1.2 Speed measures how fast position changes.

- Speed is how fast positions change with time.
- Velocity is speed in a specific direction.

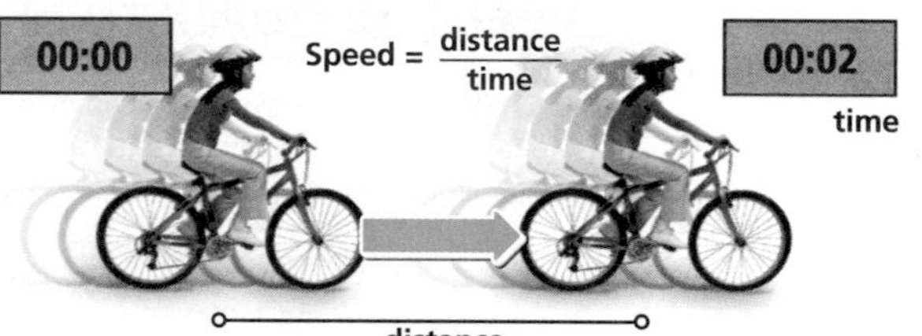

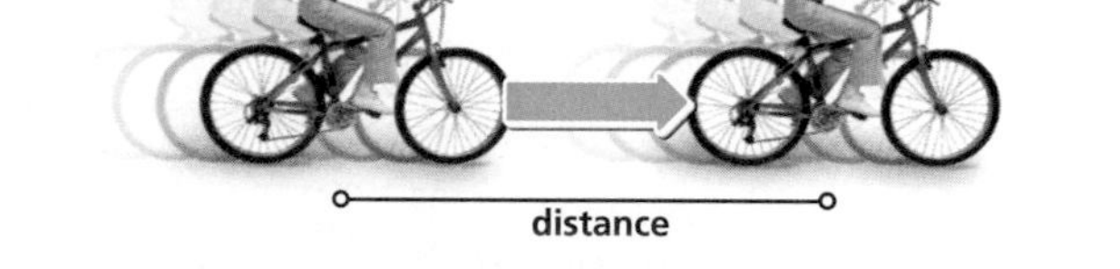

VOCABULARY
speed p. 16
velocity p. 22
vector p. 22

1.3 Acceleration measures how fast velocity changes.

$$\text{acceleration} = \frac{\text{final velocity} - \text{initial velocity}}{\text{time}}$$

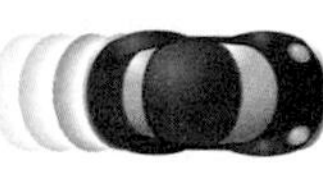

VOCABULARY
acceleration p. 25

C 34 Unit: Motion and Forces

Technology Resources

Have students visit **ClassZone.com** or use the CD-ROM for a cumulative review of concepts.

 CONTENT REVIEW

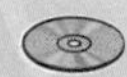 CONTENT REVIEW CD-ROM

Engage students in a whole-class interactive review of Key Concepts. Edit content as you wish.

 POWER PRESENTATIONS

Reviewing Vocabulary

Copy and complete the chart below. If the left column is blank, give the correct term. If the right column is blank, give a brief description.

Term	Description
1.	speed in a specific direction
2.	a change of position over time
3. speed	
4.	an object's location
5. reference point	
6.	the rate at which velocity changes over time
7.	a quantity that has both size and direction

Reviewing Key Concepts

Multiple Choice *Choose the letter of the best answer.*

8. A position describes an object's location compared to
- **a.** its motion
- **b.** a reference point
- **c.** its speed
- **d.** a vector

9. Maria walked 2 km in half an hour. What was her average speed during her walk?
- **a.** 1 km/h
- **b.** 2 km/h
- **c.** 4 km/h
- **d.** 6 km/h

10. A vector is a quantity that has
- **a.** speed
- **b.** acceleration
- **c.** size and direction
- **d.** position and distance

11. Mary and Keisha run with the same constant speed but in opposite directions. The girls have
- **a.** the same position
- **b.** different accelerations
- **c.** different speeds
- **d.** different velocities

12. A swimmer increases her speed as she approaches the end of the pool. Her acceleration is
- **a.** in the same direction as her motion
- **b.** in the opposite direction of her motion
- **c.** at right angles to her motion
- **d.** zero

13. A cheetah can go from 0 m/s to 20 m/s in 2 s. What is the cheetah's acceleration?
- **a.** 5 m/s^2
- **b.** 10 m/s^2
- **c.** 20 m/s^2
- **d.** 40 m/s^2

14. Jon walks for a few minutes, then runs for a few minutes. During this time, his average speed is
- **a.** the same as his final speed
- **b.** greater than his final speed
- **c.** less than his final speed
- **d.** zero

15. A car traveling at 40 m/s slows down to 20 m/s. During this time, the car has
- **a.** no acceleration
- **b.** positive acceleration
- **c.** negative acceleration
- **d.** constant velocity

Short Answer *Write a short answer to each question.*

16. Suppose you are biking with a friend. How would your friend describe your relative motion as he passes you?

17. Describe a situation where an object has a changing velocity but constant speed.

18. Give two examples of an accelerating object.

ASSESSMENT RESOURCES

UNIT ASSESSMENT BOOK
- Chapter Test A, pp. 6–9
- Chapter Test B, pp. 10–13
- Chapter Test C, pp. 14–17
- Alternative Assessment, pp. 18–19

SPANISH ASSESSMENT BOOK
Spanish Chapter Test, pp. 257–260

Technology Resources

Edit test items and answer choices.

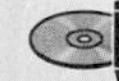

Test Generator CD-ROM

Visit **ClassZone.com** to extend test practice.

Test Practice

Reviewing Vocabulary

1. velocity
2. motion
3. the rate that an object's position changes or the distance it moves in a given amount of time
4. position
5. a location that other locations are compared with
6. acceleration
7. vector

Reviewing Key Concepts

8. b
9. c
10. c
11. d
12. a
13. b
14. c
15. c
16. He would say that you were moving backward compared with him.
17. Sample answer: A woman runs around a corner without changing speed.
18. Sample answer: A bike turning a corner is accelerating. A car coming to a stop at a light is also accelerating.

Thinking Critically

19. Students' answers will vary.

20. See students' drawings.

21. object moving from point A to point C

22. The hare lost the race because its average speed was less than that of the tortoise.

23. Tortoise graph: straight line for the tortoise climbing from 0 to 100 meters over the 40 minutes. Hare graph: steeper slope for part of the time, then a flat line, and then a steep line that climbs to just under 100 meters at the 40-minute mark.

24. Its velocity would be changing constantly.

25. He could time how long it takes the stick to move a certain distance to find its speed. Because the stick and water have the same relative motion, the speed of the stick is the same as the speed of the river.

26. Students' answers will vary.

Using Math Skills in Science

27. $S = \frac{d}{t} = \frac{50\ m}{10\ s} = 5\ m/s$

28. $S = \frac{d}{t} = \frac{10\ m}{5\ s} = 2\ cm/s$

29. $a = \frac{v_{final} - v_{initial}}{t}$

$= \frac{2\ m/s - 7\ m/s}{5\ s}$

$= \frac{-5\ m/s}{5\ s} = -1\ m/s^2$

the BIG idea

30. Velocity will increase.

31. Assumptions should include statements about whether the average velocity will change in the future. If velocity does not change, the car will be 40 km farther east after one hour.

UNIT PROJECTS

Give students the appropriate Unit Project worksheets from the URB for their projects. Both directions and rubrics can be used as a guide.

 Unit Projects, pp. 5–10

Thinking Critically

Use the following graph to answer the next three questions.

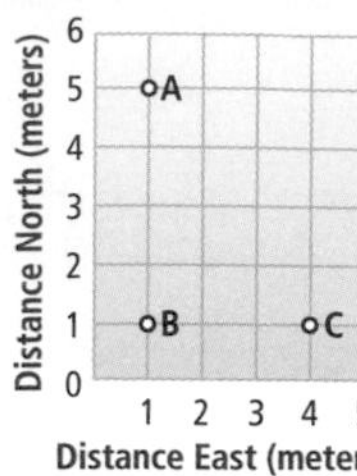

19. **OBSERVE** Describe the location of point A. Explain what you used as a reference point for your location.

20. **COMPARE** Copy the graph into your notebook. Draw two different paths an object could take when moving from point B to point C. How do the lengths of these two paths compare?

21. **ANALYZE** An object moves from point A to point C in the same amount of time that another object moves from point B to point C. If both objects traveled in a straight line, which one had the greater speed?

Read the following paragraph and use the information to answer the next three questions.

In Aesop's fable of the tortoise and the hare, a slow-moving tortoise races a fast-moving hare. The hare, certain it can win, stops to take a long nap. Meanwhile, the tortoise continues to move toward the finish line at a slow but steady speed. When the hare wakes up, it runs as fast as it can. Just as the hare is about to catch up to the tortoise, however, the tortoise wins the race.

22. **ANALYZE** How does the race between the tortoise and the hare show the difference between average speed and instantaneous speed?

23. **MODEL** Assume the racetrack was 100 meters long and the race took 40 minutes. Create a possible distance-time graph for both the tortoise and the hare.

24. **COMPARE** If the racetrack were circular, how would the tortoise's speed be different from its velocity?

25. **APPLY** How might a person use a floating stick to measure the speed at which a river flows?

26. **CONNECT** Describe a frame of reference other than the ground that you might use to measure motion. When would you use it?

Using Math Skills in Science

27. José skated 50 m in 10 s. What was his speed?

28. Use the information in the photograph below to calculate the speed of the ant as it moves down the branch.

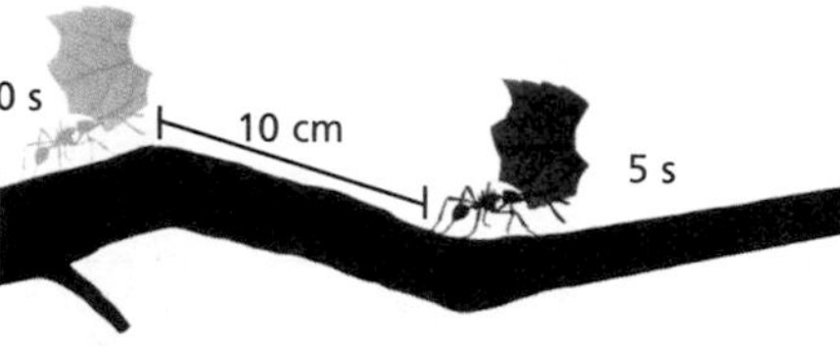

29. While riding her bicycle, Jamie accelerated from 7 m/s to 2 m/s in 5 s. What was her acceleration?

the BIG idea

30. **PREDICT** Look back at the picture at the beginning of the chapter on pages 6–7. Predict how the velocity of the roller coaster will change in the next moment.

31. **WRITE** A car is traveling east at 40 km/h. Use this information to predict where the car will be in one hour. Discuss the assumptions you made to reach your conclusion and the factors that might affect it.

UNIT PROJECTS

If you are doing a unit project, make a folder for your project. Include in your folder a list of the resources you will need, the date on which the project is due, and a schedule to keep track of your progress. Begin gathering data.

MONITOR AND RETEACH

If students have trouble applying the concepts in items 22–24, discuss what a distance-time graph of the race looks like. Use the graph of the tortoise-and-hare race. Draw a possible answer on the board.
Have students calculate the speed of the tortoise for the whole race. Help them calculate the acceleration of the hare in its final spurt in the race. Calculate the hare's speed at the end by taking measurement estimates from the graph. The hare's starting speed after his nap would be 0 m/s. Students may benefit from summarizing sections of the chapter.

 Summarizing the Chapter, pp. 74–75

Standardized Test Practice

For practice on your state test, go to . . . TEST PRACTICE CLASSZONE.COM

Interpreting Graphs

The graph below is a distance-time graph showing a 50-meter race.

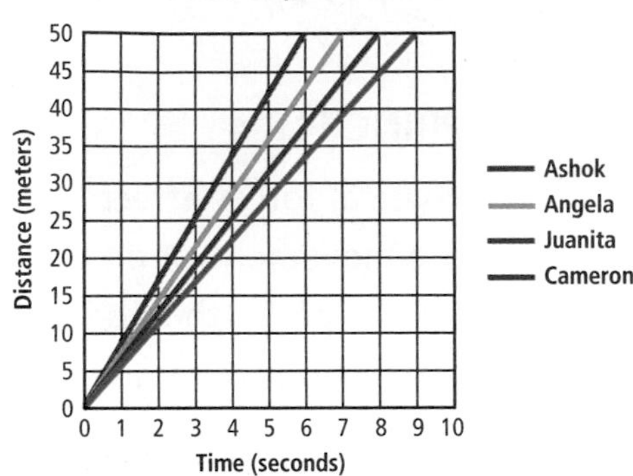

Study the graph and then answer the questions that follow.

1. Which runner reached the finish line first?

a. Ashok
b. Angela
c. Juanita
d. Cameron

2. How far did Juanita run in the first 4 seconds of the race?

a. 5 m
b. 15 m
c. 25 m
d. 35 m

3. How much time passed between the time Angela finished the race and Cameron finished the race?

a. 1 s
b. 2 s
c. 3 s
d. 4 s

4. Which of the following setups would you use to calculate Angela's average speed during the race?

a. $\frac{7 \text{ m}}{50 \text{ s}}$
b. $\frac{7 \text{ s}}{50 \text{ m}}$
c. $\frac{50 \text{ m}}{6 \text{ s}}$
d. $\frac{50 \text{ m}}{7 \text{ s}}$

5. What can you say about the speed of all of the runners?

a. They ran at the same speed.
b. They ran at a steady pace but at different speeds.
c. They sped up as they reached the finish line.
d. They slowed down as they reached the finish line.

Extended Response

Answer the two questions below in detail.

6. Suppose you are biking. What is the difference between your speed at any given moment during your bike ride and your average speed for the entire ride? Which is easier to measure? Why?

7. Suppose you are riding your bike along a path that is also used by in-line skaters. You pass a skater, and another biker passes you, both going in the same direction you're going. You pass a family having a picnic on the grass. Describe your motion from the points of view of the skater, the other biker, and the family.

METACOGNITIVE ACTIVITY

Have students answer the following questions in their **Science Notebook:**

1. What do you think is the most important concept in this chapter? Why?
2. What questions do you still have about motion?
3. Did this chapter change or influence your idea for your Unit Project? How?

Interpreting Graphs

1. a
2. c
3. b
4. d
5. b

Extended Response

6. RUBRIC

4 points for a response that correctly answers both questions and uses the following terms accurately:

- distance
- speed
- average speed

The speed at any given moment during the bike ride would be the very short distance traveled divided by a very short moment in time, or S = d/t. The average speed for the entire race is the entire distance divided by the total time. It is easier to calculate an average speed over a distance because it is easier to measure longer distances and longer times.

3 points for a response that uses two terms accurately
2 points for a response that correctly answers the question and uses one term accurately
1 point for a response that correctly answers the question, but doesn't use the terms

7. RUBRIC

4 points for a response that answers the question and uses the following terms accurately:

- motion
- relative
- frame of reference

My motion would be relative to the position and motion of the people observing me. Motion is described by a frame of reference to the location of the observer of the motion. The skater would think I am fast, the other biker would think I am slow, and the family would think we are all going fast.

3 points for a response that correctly answers the question and uses 2 terms accurately
2 points for a response that correctly answers the question and uses 1 term accurately
1 point for a response that correctly answers the question but doesn't use the terms

CHAPTER 2 Forces

Physical Science UNIFYING PRINCIPLES

PRINCIPLE 1
Matter is made of particles too small to see.

PRINCIPLE 2
Matter changes form and moves from place to place.

PRINCIPLE 3
Energy changes from one form to another, but it cannot be created or destroyed.

PRINCIPLE 4
Physical forces affect the movement of all matter on Earth and throughout the universe.

Unit: Motion and Forces BIG IDEAS

CHAPTER 1 Motion
The motion of an object can be described and predicted.

CHAPTER 2 Forces
Forces change the motion of objects in predictable ways.

CHAPTER 3 Gravity, Friction, and Pressure
Newton's laws apply to all forces.

CHAPTER 4 Work and Energy
Energy is transferred when a force moves an object.

CHAPTER 5 Machines
Machines help people do work by changing the force applied to an object.

CHAPTER 2 KEY CONCEPTS

SECTION 2.1

Forces change motion.

1. A force is a push or a pull.
2. Newton's first law relates force and motion.

SECTION 2.2

Force and mass determine acceleration.

1. Newton's second law relates force, mass, and acceleration.
2. Forces can change the direction of motion.

SECTION 2.3

Forces act in pairs.

1. Newton's third law relates action and reaction forces.
2. Newton's three laws describe and predict motion.

SECTION 2.4

Forces transfer momentum.

1. Objects in motion have momentum.
2. Momentum can be transferred from one object to another.
3. Momentum is conserved.

T The Big Idea Flow Chart is available on p. T9 in the **UNIT TRANSPARENCY BOOK.**

Previewing Content

SECTION

Forces change motion. pp. 41–48

1. A force is a push or a pull.

Force in physics is defined as "a push or a pull." Some forces
- require contact between objects, such as friction; and
- act at a distance, such as gravity and electromagnetic forces.

Net force is the total force that affects an object when multiple forces are combined. The net force depends on both the direction and the size of the individual forces.

To show visually how to measure net force, use the following:

Measuring Net Force

 Identify forces to combine.

 Put tail of second force arrow to tip of first. Maintain length (size) and direction.

 Measure from tail of first force to tip of second force arrow.

Combining forces in the same direction

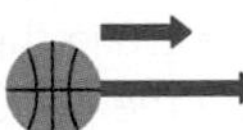  

Combining forces in opposite directions

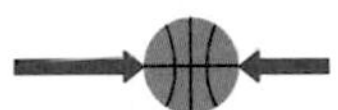  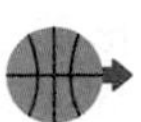

2. Newton's first law relates force and motion.

The key points of Newton's first law are
- objects with no net force acting on them have either constant or zero velocity; and
- force is needed to start or change motion.

Inertia is the resistance of an object to a change in its motion; it is directly proportional to the object's mass.

SECTION

Force and mass determine acceleration. pp. 49–56

1. Newton's second law relates force, mass, and acceleration.

The key points of **Newton's second law** are that the acceleration of an object is
- directly proportional to the force acting on the object;
- inversely proportional to the mass of the object; and
- in the same direction as the net force acting on the object.

Newton's second law is summed up by the equation Force = mass · acceleration, or $F = ma$.

Sample problems show students how to solve for each variable in the equation.

Calculating Mass

Sample Problem

A model rocket is accelerating at 2 m/s². The force on it is 1 N. What is the mass of the rocket?

What do you know? acceleration = 2 m/s², force = 1 N

What do you want to find out? mass

Rearrange the formula: $m = \frac{F}{a}$

Substitute into the formula: $m = \frac{1\ \text{N}}{2\ \text{m/s}^2}$

Calculate and simplify: $m = \frac{1\ \text{N}}{2\ \text{m/s}^2} = \frac{1\ \text{kg} \cdot \text{m/s}^2}{2\ \text{m/s}^2} = 0.5\ \text{kg}$

2. Forces can change the direction of motion.

Force can change the direction of an object without changing its speed if the force acts at right angles to the motion. A force that continuously acts at right angles to an object's motion will pull the object into circular motion. Any force that keeps an object moving in a circle at a constant speed is called a centripetal force. The centripetal force needed to keep an object moving in a circle depends on the mass of the object, the speed of the object, and the radius of the circle.

$$\textbf{centripetal force} = \frac{\textbf{(mass} \cdot \textbf{speed}^2\textbf{)}}{\textbf{radius}}$$

Common Misconceptions

PASSIVE FORCES Many students consider something a force only if there is an action associated with it. Passive forces, such as the force of a table resisting the push from a person, are not seen as force because the table does not move.

 This misconception is addressed on p. 43.

MISCONCEPTION DATABASE
CLASSZONE.COM Background on student misconceptions

FRICTION Students commonly think that if there is no force on an object in motion, it will slow down. In fact, a force is needed to slow the object down. Friction usually slows down moving objects.

TE This misconception is addressed in the Teacher Demo on p. 44.

Previewing Content

SECTION

Forces act in pairs. pp. 57–63

1. **Newton's third law relates action and reaction forces.**
   The key points to **Newton's third law** are that when objects A and B interact,
   - the force of A on B equals the force of B on A; and
   - the forces are opposite in direction.

   In **action/reaction pairs** either force can be considered the action force or the reaction force. The two forces occur simultaneously.
   Example: When you push down on a table, the force from the table's resistance increases instantly to match your force.

   Action/reaction force pairs occur when any two objects interact, not just through contact forces.

   Example: The pull of Earth on a falling baseball is exactly that of the baseball on Earth. Earth is so much more massive, however, that Earth's acceleration from the pull is nearly nothing. The acceleration of the baseball is quite noticeable.

2. **Newton's three laws describe and predict motion.**
   Newton's laws work together to explain changes in the motion of objects, such as a squid moving forward when squirting water backward, or a bird flying higher or changing direction. Newton's laws are also useful in calculating how objects move under the conditions found in everyday life. Scientists such as Albert Einstein have added to our understanding of motion since Newton's time. Under certain conditions, such as extreme speed or extreme gravity, Newton's laws need to be adjusted.

SECTION

Forces transfer momentum. pp. 64–69

1. **Objects in motion have momentum.**
   Momentum can be thought of as inertia for moving objects. It is the tendency of a moving object to keep moving at a constant velocity, and it depends on the mass and velocity of the object.

   momentum = mass · velocity, or $p = mv$

   - Momentum, like velocity, is a vector, so it has both size and direction.
   - Adding the momentum of two objects is similar to adding net forces.

   A force on an object changes the object's momentum. The change in momentum is equal to the force on the object multiplied by the time over which the force is acting.

2. **Momentum can be transferred from one object to another.**
   Momentum is transferred during a **collision.** Colliding objects exert equal and opposite forces on each other while they are in contact. The forces in the collision will change the velocity of each object involved.

3. **Momentum is conserved.**
   In any case where no outside forces are acting on a system, the total momentum of the system will not change, even if the momentum of individual parts of the system changes. This conservation of momentum is most easily seen in collisions. The forces acting are equal and opposite, and they act over the same time period. Therefore, the change in momentum for two colliding objects is equal and opposite, and the total change in momentum is zero.

   In the following example, as the ball that is released strikes the line of balls, momentum transfers to the last ball, causing it to swing out.

Common Misconceptions

ACTION/REACTION PAIRS Action/reaction pairs can be confused with balanced and unbalanced forces. Students may not understand that action/reaction forces act on different objects.

 This misconception is addressed on p. 59.

MISCONCEPTION DATABASE
CLASSZONE.COM Background on student misconceptions

MOMENTUM Some students may think that momentum is the same thing as force. While a moving object can apply a force to another object, that force is not the momentum itself.

 This misconception is addressed on p. 67.

Previewing Labs

Lab Generator CD-ROM
Edit these Pupil Edition labs and generate alternative labs.

EXPLORE the BIG idea

Popping Ping-Pong Balls, p. 39 Students compare the effects of varying force, mass, and acceleration on two objects.	**TIME** 10 minutes **MATERIALS** Ping-Pong ball, flexible ruler, heavier ball
Take Off! p. 39 Students use a model to learn about force pairs.	**TIME** 10 minutes **MATERIALS** long balloon, toy car, 50 cm tape
Internet Activity: Forces, p. 39 Students predict direction and amount of motion.	**TIME** 20 minutes **MATERIALS** computer with Internet access

SECTION  2.1

EXPLORE Changing Motion, p. 41 Students experiment with changing the motion of different objects.	**TIME** 10 minutes **MATERIALS** quarter, book, tennis ball, cup, feather
INVESTIGATE Inertia, p. 46 Students design an experiment to determine the inertia of two balls.	**TIME** 30 minutes **MATERIALS** 2 balls of unknown masses, 1 m string, block, meter stick

SECTION  2.2

EXPLORE Acceleration, p. 49 Students relate the force of gravity on a paper clip to the paper clip's acceleration.	**TIME** 10 minutes **MATERIALS** paper clips, 40 cm string
INVESTIGATE Motion and Force, p. 54 Students hypothesize about the motion of a marble traveling in a circular path when the force is removed.	**TIME** 15 minutes **MATERIALS** newspaper, paper plate, marble, scissors, poster paint, paintbrush

SECTION  2.3

INVESTIGATE Newton's Third Law, p. 58 Students observe action and reaction forces on a spring scale.	**TIME** 15 minutes **MATERIALS** 2 spring scales
CHAPTER INVESTIGATION **Newton's Laws of Motion,** pp. 62–63 Students build a straw bottle rocket and use Newton's laws to improve the rocket's performance.	**TIME** 40 minutes **MATERIALS** 2 straws with different diameters, several plastic bottles (in different sizes), modeling clay, scissors, construction paper, meter stick, 10 cm tape

SECTION 2.4

EXPLORE Collisions, p. 64 Students observe what happens when two balls collide.	**TIME** 10 minutes **MATERIALS** 2 balls of different masses
INVESTIGATE Momentum, p. 66 Students observe changes in momentum when marbles collide.	**TIME** 20 minutes **MATERIALS** 2 rulers, 8 marbles

 **Additional INVESTIGATION,** Newton's First Law, A, B, & C, pp. 138–146; Teacher Instructions, pp. 346–347

Previewing Chapter Resources

	INTEGRATED TECHNOLOGY	LABS AND ACTIVITIES
CHAPTER 2 **Forces**	 **CLASSZONE.COM** • eEdition Plus • EasyPlanner Plus • Misconception Database • Content Review • Test Practice • Simulations • Resource Centers • Internet Activity: Forces • Math Tutorial  **SCILINKS.ORG** SCILINKS 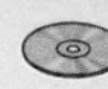 **CD-ROMS** • eEdition • EasyPlanner • Power Presentations • Content Review • Lab Generator • Test Generator 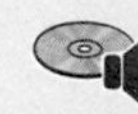 **AUDIO CDS** • Audio Readings • Audio Readings in Spanish	 EXPLORE the Big Idea, p. 39 • Popping Ping-Pong Balls • Take Off! • Internet Activity: Forces  **UNIT RESOURCE BOOK** Unit Projects, pp. 5–10  **Lab Generator CD-ROM** Generate customized labs.
SECTION  **Forces change motion.** pp. 41–48 Time: 2 periods (1 block)  Lesson Plan, pp. 76–77	 **RESOURCE CENTERS,** Inertia; Moving Rocks **UNIT TRANSPARENCY BOOK** • Big Idea Flow Chart, p. T9 • Daily Vocabulary Scaffolding, p. T10 • Note-Taking Model, p. T11 • 3-Minute Warm-Up, p. T12	 • EXPLORE Changing Motion, p. 41 • INVESTIGATE Inertia, p. 46 • Think Science, p. 48  **UNIT RESOURCE BOOK** • Datasheet, Inertia, p. 85 • Additional INVESTIGATION, Newton's First Law, A, B, & C, pp. 138–146
SECTION  **Force and mass determine acceleration.** pp. 49–56 Time: 2 periods (1 block)  Lesson Plan, pp. 87–88	 • **SIMULATION,** Newton's Second Law • **MATH TUTORIAL** **UNIT TRANSPARENCY BOOK** • Daily Vocabulary Scaffolding, p. T10 • 3-Minute Warm-Up, p. T12 	 • EXPLORE Acceleration, p. 49 • INVESTIGATE Motion and Force, p. 54 • Math in Science, p. 56 **UNIT RESOURCE BOOK** • Datasheet, Motion and Force, p. 96 • Math Support and Practice, pp. 125–128
SECTION  **Forces act in pairs.** pp. 57–63 Time: 3 periods (1.5 block)  Lesson Plan, pp. 98–99	 **RESOURCE CENTER,** Newton's Laws of Motion  **UNIT TRANSPARENCY BOOK** • Daily Vocabulary Scaffolding, p. T10 • 3-Minute Warm-Up, p. T13 • "Newton's Three Laws of Motion" Visual, p. T14	 • INVESTIGATE Newton's Third Law, p. 58 • CHAPTER INVESTIGATION, Newton's Laws of Motion, pp. 62–63  **UNIT RESOURCE BOOK** • Datasheet, Newton's Third Law, p. 107 • CHAPTER INVESTIGATION, Newton's Laws of Motion, A, B, & C, pp. 129–137
SECTION  **Forces transfer momentum.** pp. 64–69 Time: 3 periods (1.5 block) Lesson Plan, pp. 109–110	 **RESOURCE CENTER,** Momentum **UNIT TRANSPARENCY BOOK** • Big Idea Flow Chart, p. T9 • Daily Vocabulary Scaffolding, p. T10 • 3-Minute Warm-Up, p. T13 • Chapter Outline, pp. T15–T16	 • EXPLORE Collisions, p. 64 • INVESTIGATE Momentum, p. 66  **UNIT RESOURCE BOOK** Datasheet, Momentum, p. 118

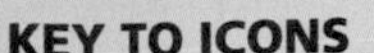  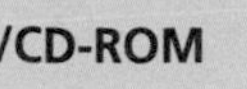 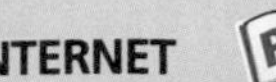  

KEY TO ICONS — CD/CD-ROM · INTERNET · Pupil Edition · Teacher Edition · UNIT RESOURCE BOOK · UNIT TRANSPARENCY BOOK · UNIT ASSESSMENT BOOK · SPANISH ASSESSMENT BOOK · SCIENCE TOOLKIT

READING AND REINFORCEMENT	ASSESSMENT	STANDARDS
 • Magnet Words, B24–25 • Combination Notes, C36 • Daily Vocabulary Scaffolding, H1–8  **UNIT RESOURCE BOOK** • Vocabulary Practice, pp. 122–123 • Decoding Support, p. 124 • Summarizing the Chapter, pp. 147–148  **Audio Readings CD** Listen to Pupil Edition. 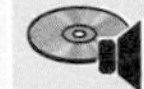 **Audio Readings in Spanish CD** Listen to Pupil Edition in Spanish.	 • Chapter Review, pp. 71–72 • Standardized Test Practice, p. 73  **UNIT ASSESSMENT BOOK** • Diagnostic Test, pp. 20–21 • Chapter Test, A, B, & C, pp. 26–37 • Alternative Assessment, pp. 38–39  Spanish Chapter Test, pp. 261–264  **Test Generator CD-ROM** Generate customized tests.  **Lab Generator CD-ROM** Rubrics for Labs	National Standards A.2–8, A.9.a–c, A.9.e–f, B.2.b–c, E.2–5 See p. 38 for the standards.
 **UNIT RESOURCE BOOK** • Reading Study Guide, A & B, pp. 78–81 • Spanish Reading Study Guide, pp. 82–83 • Challenge and Extension, p. 84 • Reinforcing Key Concepts, p. 86	 Ongoing Assessment, pp. 42–47  Section 2.1 Review, p. 47  **UNIT ASSESSMENT BOOK** Section 2.1 Quiz, p. 22	National Standards A.2–8, A.9.a–c, A.9.e–f, B.2.b, E.2–5
 **UNIT RESOURCE BOOK** • Reading Study Guide, A & B, pp. 89–92 • Spanish Reading Study Guide, pp. 93–94 • Challenge and Extension, p. 95 • Reinforcing Key Concepts, p. 97 • Challenge Reading, pp. 120–121	 Ongoing Assessment, pp. 49–55  Section 2.2 Review, p. 55  **UNIT ASSESSMENT BOOK** Section 2.2 Quiz, p. 23	National Standards A.2–8, A.9.a–c, A.9.e–f,
 **UNIT RESOURCE BOOK** • Reading Study Guide, A & B, pp. 100–103 • Spanish Reading Study Guide, pp. 104–105 • Challenge and Extension, p. 106 • Reinforcing Key Concepts, p. 108	 Ongoing Assessment, pp. 57–61  Section 2.3 Review, p. 61  **UNIT ASSESSMENT BOOK** Section 2.3 Quiz, p. 24	National Standards A.2–8, A.9.a–c, A.9.e–f, B.2.c
 **UNIT RESOURCE BOOK** • Reading Study Guide, A & B, pp. 111–114 • Spanish Reading Study Guide, pp. 115–116 • Challenge and Extension, p. 117 • Reinforcing Key Concepts, p. 119	 Ongoing Assessment, pp. 65–69  Section 2.4 Review, p. 69  **UNIT ASSESSMENT BOOK** Section 2.4 Quiz, p. 25	National Standards A.2–8, A.9.a–c, A.9.e–f,

Previewing Resources for Differentiated Instruction

CHAPTER INVESTIGATION

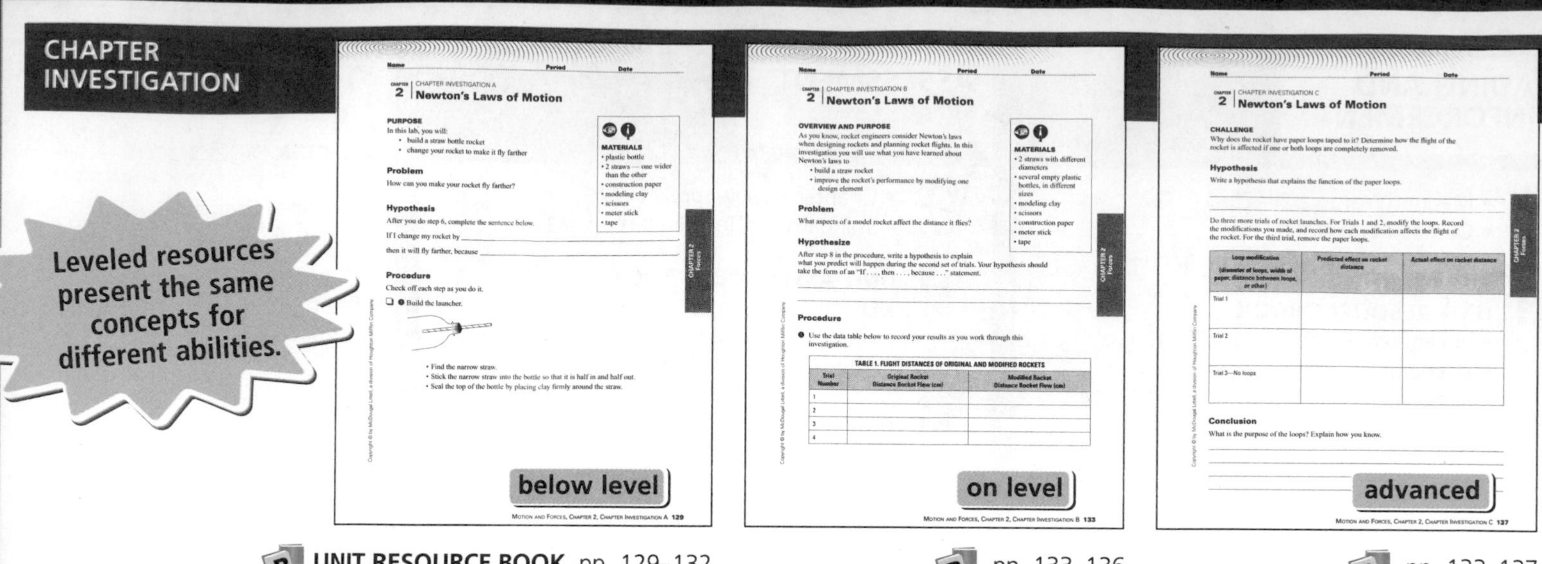

Name Period Date

CHAPTER 2 | CHAPTER INVESTIGATION A
Newton's Laws of Motion

PURPOSE
In this lab, you will:
- build a straw bottle rocket
- change your rocket to make it fly farther

MATERIALS
- plastic bottle
- 2 straws — one wider than the other
- construction paper
- modeling clay
- scissors
- meter stick
- tape

Problem
How can you make your rocket fly farther?

Hypothesis
After you do step 6, complete the sentence below.
If I change my rocket by ____
then it will fly farther, because ____

Procedure
Check off each step as you do it.
❑ 1 Build the launcher.
- Find the narrow straw.
- Stick the narrow straw into the bottle so that it is half in and half out.
- Seal the top of the bottle by placing clay firmly around the straw.

Motion and Forces, Chapter 2, Chapter Investigation A 129

Name Period Date

CHAPTER 2 | CHAPTER INVESTIGATION B
Newton's Laws of Motion

OVERVIEW AND PURPOSE
As you know, rocket engineers consider Newton's laws when designing rockets and planning rocket flights. In this investigation you will use what you have learned about Newton's laws to
- build a straw rocket
- improve the rocket's performance by modifying one design element

MATERIALS
- 2 straws with different diameters
- several empty plastic bottles, in different sizes
- modeling clay
- scissors
- construction paper
- meter stick
- tape

Problem
What aspects of a model rocket affect the distance it flies?

Hypothesize
After step 8 in the procedure, write a hypothesis to explain what you predict will happen during the second set of trials. Your hypothesis should take the form of an "If . . . , then . . . , because . . ." statement.

Procedure
1 Use the data table below to record your results as you work through this investigation.

TABLE 1. FLIGHT DISTANCES OF ORIGINAL AND MODIFIED ROCKETS

Trial Number	Original Rocket Distance Rocket Flew (cm)	Modified Rocket Distance Rocket Flew (cm)
1		
2		
3		
4		

Motion and Forces, Chapter 2, Chapter Investigation B 133

Name Period Date

CHAPTER 2 | CHAPTER INVESTIGATION C
Newton's Laws of Motion

CHALLENGE
Why does the rocket have paper loops taped to it? Determine how the flight of the rocket is affected if one or both loops are completely removed.

Hypothesis
Write a hypothesis that explains the function of the paper loops.

Do three more trials of rocket launches. For Trials 1 and 2, modify the loops. Record the modifications you made, and record how each modification affects the flight of the rocket. For the third trial, remove the paper loops.

Loop modification (diameter of loops, width of paper, distance between loops, or other)	Predicted effect on rocket distance	Actual effect on rocket distance
Trial 1		
Trial 2		
Trial 3—No loops		

Conclusion
What is the purpose of the loops? Explain how you know.

Motion and Forces, Chapter 2, Chapter Investigation C 137

R **UNIT RESOURCE BOOK,** pp. 129–132 **R** pp. 133–136 **R** pp. 133–137

READING STUDY GUIDE

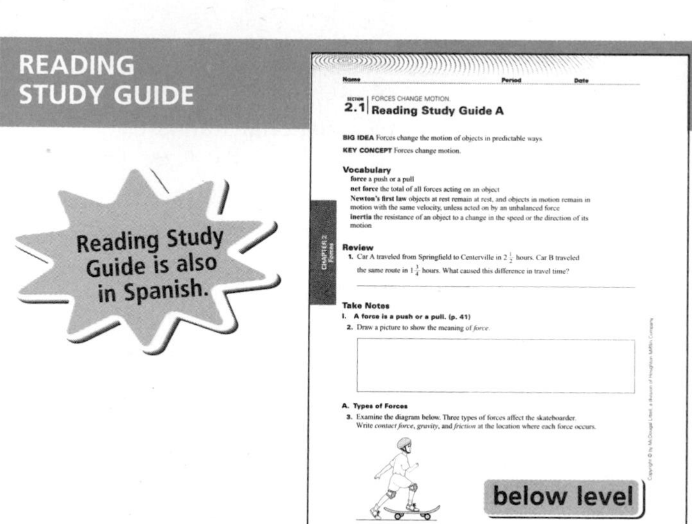

Name Period Date

SECTION 2.1 | FORCES CHANGE MOTION
Reading Study Guide A

BIG IDEA Forces change the motion of objects in predictable ways.
KEY CONCEPT Forces change motion.

Vocabulary
force a push or a pull
net force the total of all forces acting on an object
Newton's first law objects at rest remain at rest, and objects in motion remain in motion with the same velocity, unless acted on by an unbalanced force
inertia the resistance of an object to a change in the speed or the direction of its motion

Review
1. Car A traveled from Springfield to Centerville in $2\frac{1}{2}$ hours. Car B traveled the same route in $1\frac{1}{2}$ hours. What caused this difference in travel time?

Take Notes
I. A force is a push or a pull. (p. 41)
2. Draw a picture to show the meaning of *force*.

A. Types of Forces
3. Examine the diagram below. Three types of forces affect the skateboarder. Write *contact force*, *gravity*, and *friction* at the location where each force occurs.

78 Motion and Forces, Chapter 2, Reading Study Guide A

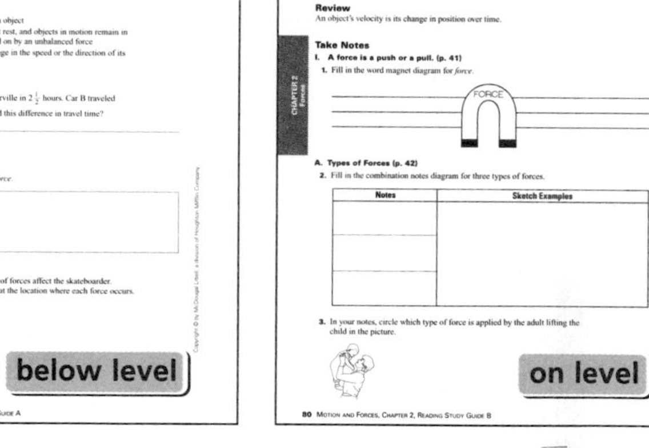

Name Period Date

SECTION 2.1 | FORCES CHANGE MOTION
Reading Study Guide B

BIG IDEA Forces change the motion of objects in predictable ways.
KEY CONCEPT Forces change motion.

Review
An object's velocity is its change in position over time.

Take Notes
I. A force is a push or a pull. (p. 41)
1. Fill in the word magnet diagram for *force*.

A. Types of Forces (p. 42)
2. Fill in the combination notes diagram for three types of forces.

Notes	Sketch Examples

3. In your notes, circle which type of force is applied by the adult lifting the child in the picture.

80 Motion and Forces, Chapter 2, Reading Study Guide B

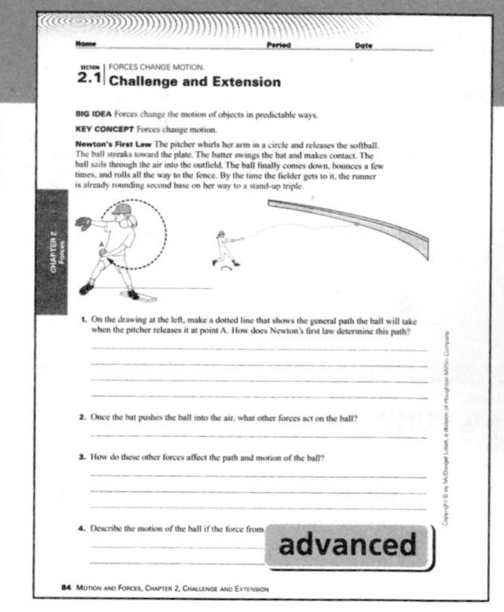

Name Period Date

SECTION 2.1 | FORCES CHANGE MOTION
Challenge and Extension

BIG IDEA Forces change the motion of objects in predictable ways.
KEY CONCEPT Forces change motion.

Newton's First Law The pitcher whirls her arm in a circle and releases the softball. The ball streaks toward the plate. The batter swings the bat and makes contact. The ball sails through the air into the outfield. The ball finally comes down, bounces a few times, and rolls all the way to the fence. By the time the fielder gets to it, the runner is already rounding second base on her way to a stand-up triple.

1. On the drawing at the left, make a dotted line that shows the general path the ball will take when the pitcher releases it at point A. How does Newton's first law determine this path?
2. Once the bat pushes the ball into the air, what other forces act on the ball?
3. How do these other forces affect the path and motion of the ball?
4. Describe the motion of the ball if the force from ...

84 Motion and Forces, Chapter 2, Challenge and Extension

R **UNIT RESOURCE BOOK,** pp. 78–79 **R** pp. 80–81 **R** p. 84

CHAPTER TEST

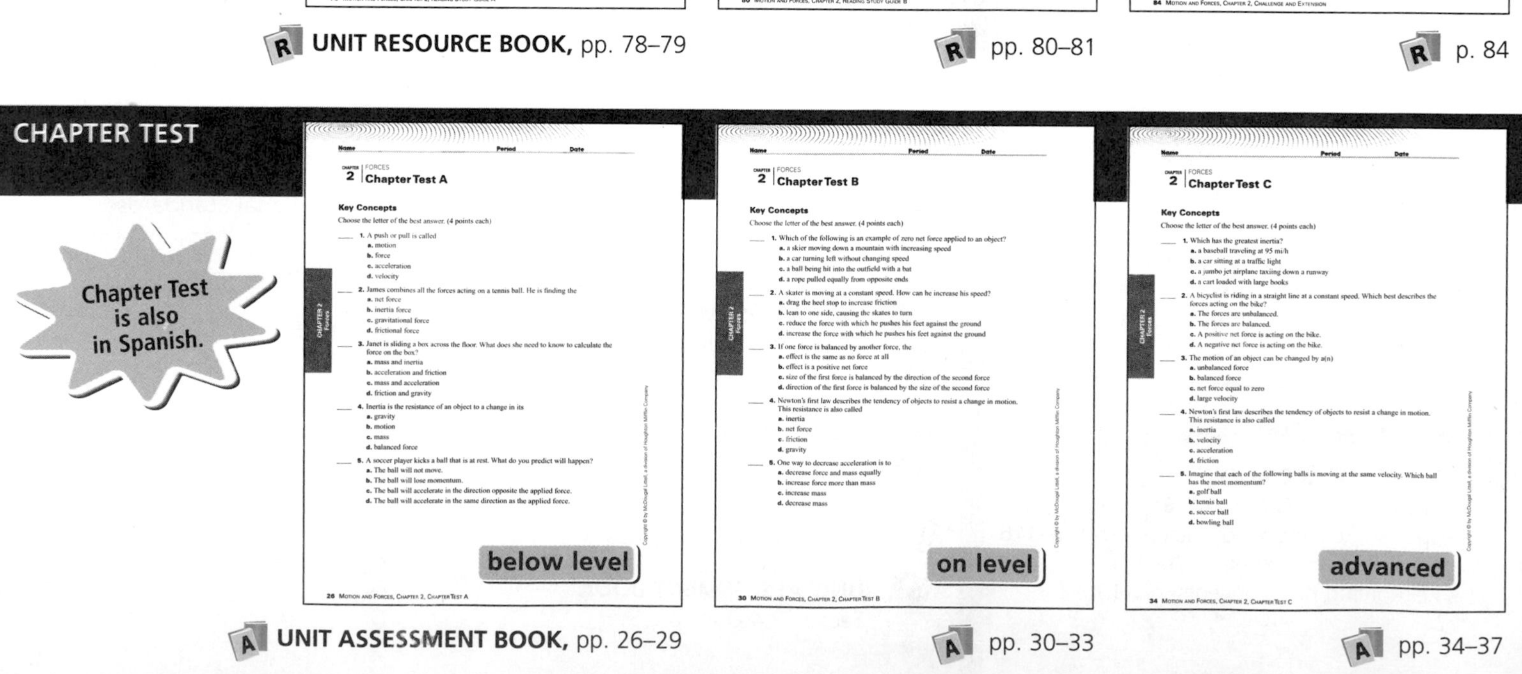

Name Period Date

CHAPTER 2 | FORCES
Chapter Test A

Key Concepts
Choose the letter of the best answer. (4 points each)

___ 1. A push or pull is called
a. motion
b. force
c. acceleration
d. velocity

___ 2. James combines all the forces acting on a tennis ball. He is finding the
a. net force
b. inertia force
c. gravitational force
d. frictional force

___ 3. Janet is sliding a box across the floor. What does she need to know to calculate the force on the box?
a. mass and inertia
b. acceleration and friction
c. mass and acceleration
d. friction and gravity

___ 4. Inertia is the resistance of an object to a change in its
a. gravity
b. motion
c. mass
d. balanced force

___ 5. A soccer player kicks a ball that is at rest. What do you predict will happen?
a. The ball will not move.
b. The ball will lose momentum.
c. The ball will accelerate in the direction opposite the applied force.
d. The ball will accelerate in the same direction as the applied force.

26 Motion and Forces, Chapter 2, Chapter Test A

Name Period Date

CHAPTER 2 | FORCES
Chapter Test B

Key Concepts
Choose the letter of the best answer. (4 points each)

___ 1. Which of the following is an example of zero net force applied to an object?
a. a skier moving down a mountain with increasing speed
b. a car turning left without changing speed
c. a ball being hit into the outfield with a bat
d. a rope pulled equally from opposite ends

___ 2. A skater is moving at a constant speed. How can he increase his speed?
a. drag the heel stop to increase friction
b. lean to one side, causing the skates to turn
c. reduce the force with which he pushes his feet against the ground
d. increase the force with which he pushes his feet against the ground

___ 3. If one force is balanced by another force, the
a. effect is the same as no force at all
b. effect is a positive net force
c. size of the first force is balanced by the direction of the second force
d. direction of the first force is balanced by the size of the second force

___ 4. Newton's first law describes the tendency of objects to resist a change in motion. This resistance is also called
a. inertia
b. net force
c. friction
d. gravity

___ 5. One way to decrease acceleration is to
a. decrease force and mass equally
b. increase force more than mass
c. increase mass
d. decrease mass

30 Motion and Forces, Chapter 2, Chapter Test B

Name Period Date

CHAPTER 2 | FORCES
Chapter Test C

Key Concepts
Choose the letter of the best answer. (4 points each)

___ 1. Which has the greatest inertia?
a. a baseball traveling at 95 mi/h
b. a car sitting at a traffic light
c. a jumbo jet airplane taxiing down a runway
d. a cart loaded with large books

___ 2. A bicyclist is riding in a straight line at a constant speed. Which best describes the forces acting on the bike?
a. The forces are unbalanced.
b. The forces are balanced.
c. A positive net force is acting on the bike.
d. A negative net force is acting on the bike.

___ 3. The motion of an object can be changed by a(n)
a. unbalanced force
b. balanced force
c. net force equal to zero
d. large velocity

___ 4. Newton's first law describes the tendency of objects to resist a change in motion. This resistance is also called
a. inertia
b. velocity
c. acceleration
d. friction

___ 5. Imagine that each of the following balls is moving at the same velocity. Which ball has the most momentum?
a. golf ball
b. tennis ball
c. soccer ball
d. bowling ball

34 Motion and Forces, Chapter 2, Chapter Test C

A **UNIT ASSESSMENT BOOK,** pp. 26–29 **A** pp. 30–33 **A** pp. 34–37

TECHNOLOGY

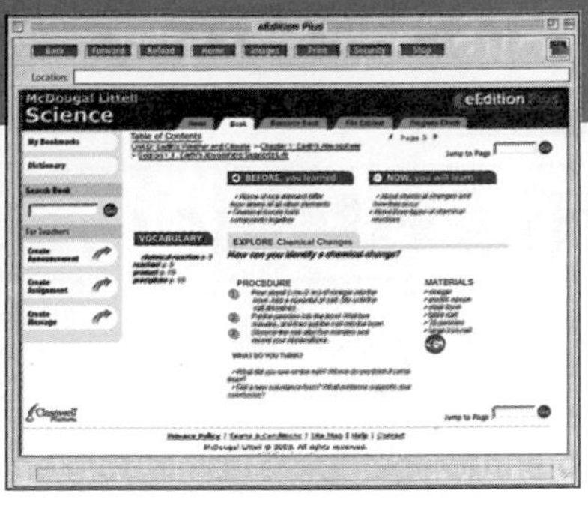

CLASSZONE.COM

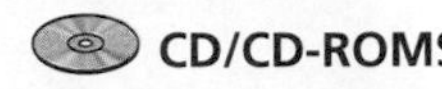
CD/CD-ROMS

CLASSZONE.COM

VISUAL CONTENT

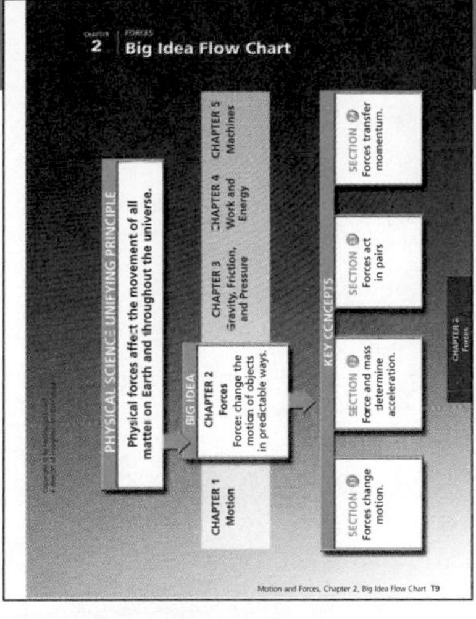

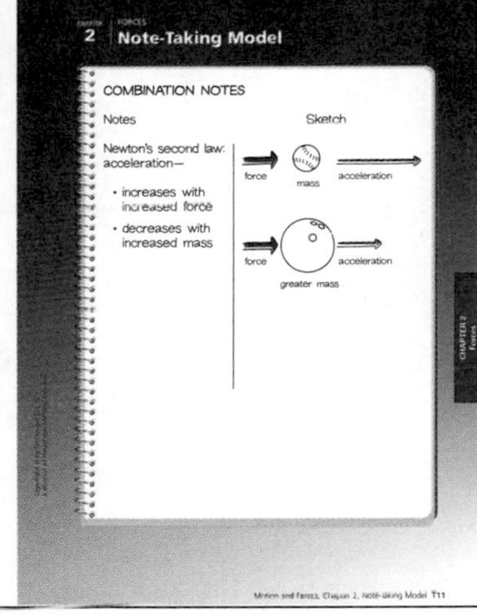

2 Newton's Three Laws of Motion
All three of Newton's laws work together to help describe how an object will move.
First Law
Second Law
Third Law
force of gravity
Force of gravity brings kangaroo back to the ground.
acceleration
With same force, less massive kangaroo has greater acceleration.
action force
reaction force
Kangaroo's legs apply action force.

UNIT TRANSPARENCY BOOK, p. T9

p. T11

p. T14

MORE SUPPORT

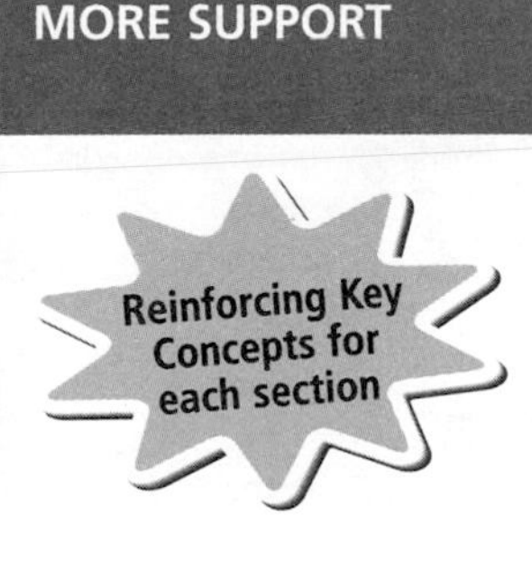

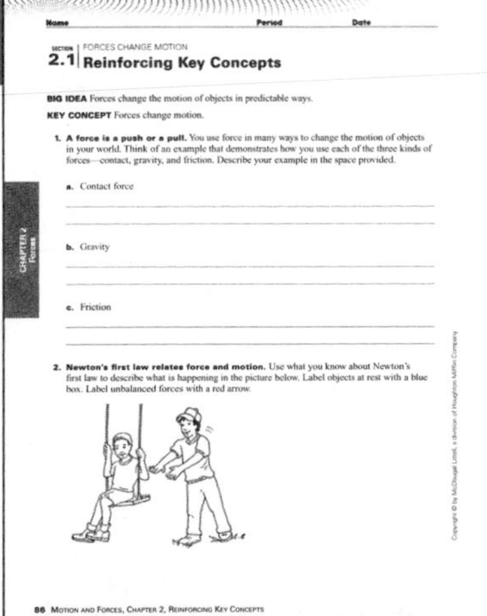

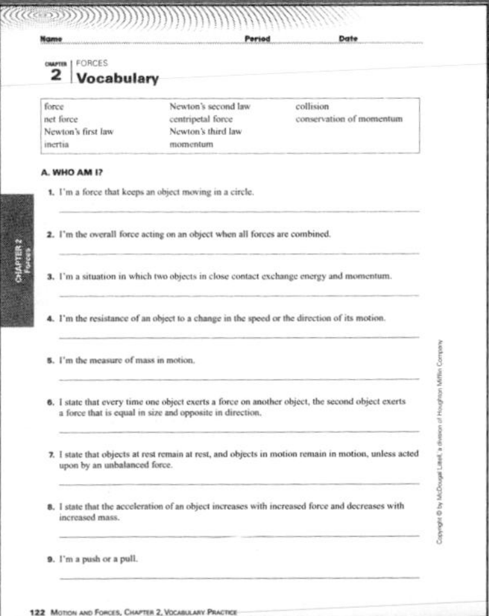

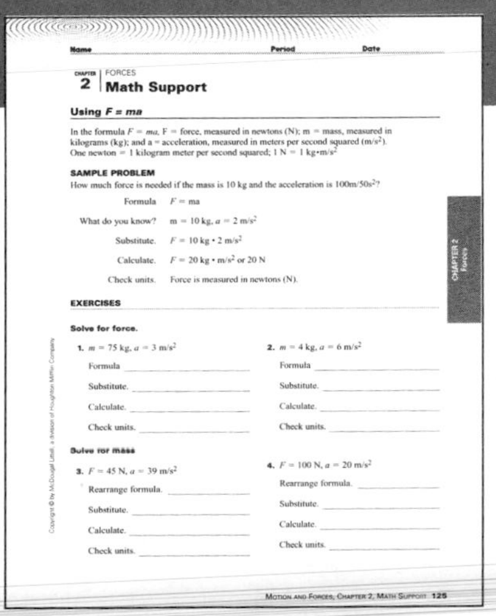

UNIT RESOURCE BOOK, p. 86

pp. 122–123

p. 125

INTRODUCE

the BIG idea

Have students look at the photograph of the tug of war. Discuss how the question in the box links to the Big Idea:

- Why might adding more students to one team make winning easier?
- How do you know one team will move forward and not sideways?
- What do you think of when you hear the word *force?*
- What role does force play in the picture?

National Science Education Standards

Content

B.2.b An object that is not being subjected to a force will continue to move at a constant speed in a straight line.

B.2.c If more than one force acts on an object along a straight line, then the forces will reinforce or cancel one another, depending on their direction and magnitude. Unbalanced forces will cause changes in the speed or direction of an object's motion.

Process

A.2–8 Design and conduct an investigation; use tools to gather and interpret data; use evidence to describe, predict, explain, model; think critically to make relationships between evidence and explanation; recognize different explanations and predictions; communicate scientific procedures and explanations; use mathematics.

A.9.a–c, A.9.e–f Understand scientific inquiry by using different investigations, methods, mathematics, and explanations based on logic, evidence, and skepticism.

E.2–5 Design, implement, and evaluate a product or solution; communicate technological design.

CHAPTER

Forces

the BIG idea

Forces change the motion of objects in predictable ways.

What must happen for a team to win this tug of war?

Key Concepts

SECTION  2.1 **Forces change motion.**
Learn about inertia and Newton's first law of motion.

SECTION  2.2 **Force and mass determine acceleration.**
Learn to calculate force through Newton's second law of motion.

SECTION  2.3 **Forces act in pairs.**
Learn about action forces and reaction forces through Newton's third law of motion.

SECTION  2.4 **Forces transfer momentum.**
Learn about momentum and how it is affected in collisions.

Internet Preview

CLASSZONE.COM

Chapter 2 online resources: Content Review, two Simulations, four Resource Centers, Math Tutorial, Test Practice

INTERNET PREVIEW

CLASSZONE.COM For student use with the following pages:

Review and Practice
- Content Review, pp. 40, 70
- Math Tutorial: Rounding Decimals, p. 56
- Test Practice, p. 73

Activities and Resources
- Internet Activity: Forces, p. 39
- Resource Centers: Inertia, p. 47; Moving Rocks, p. 48; Newton's Laws of Motion, p. 61; Momentum, p. 65
- Simulation: Newton's Second Law, p. 50

Forces **Code: MDL005**

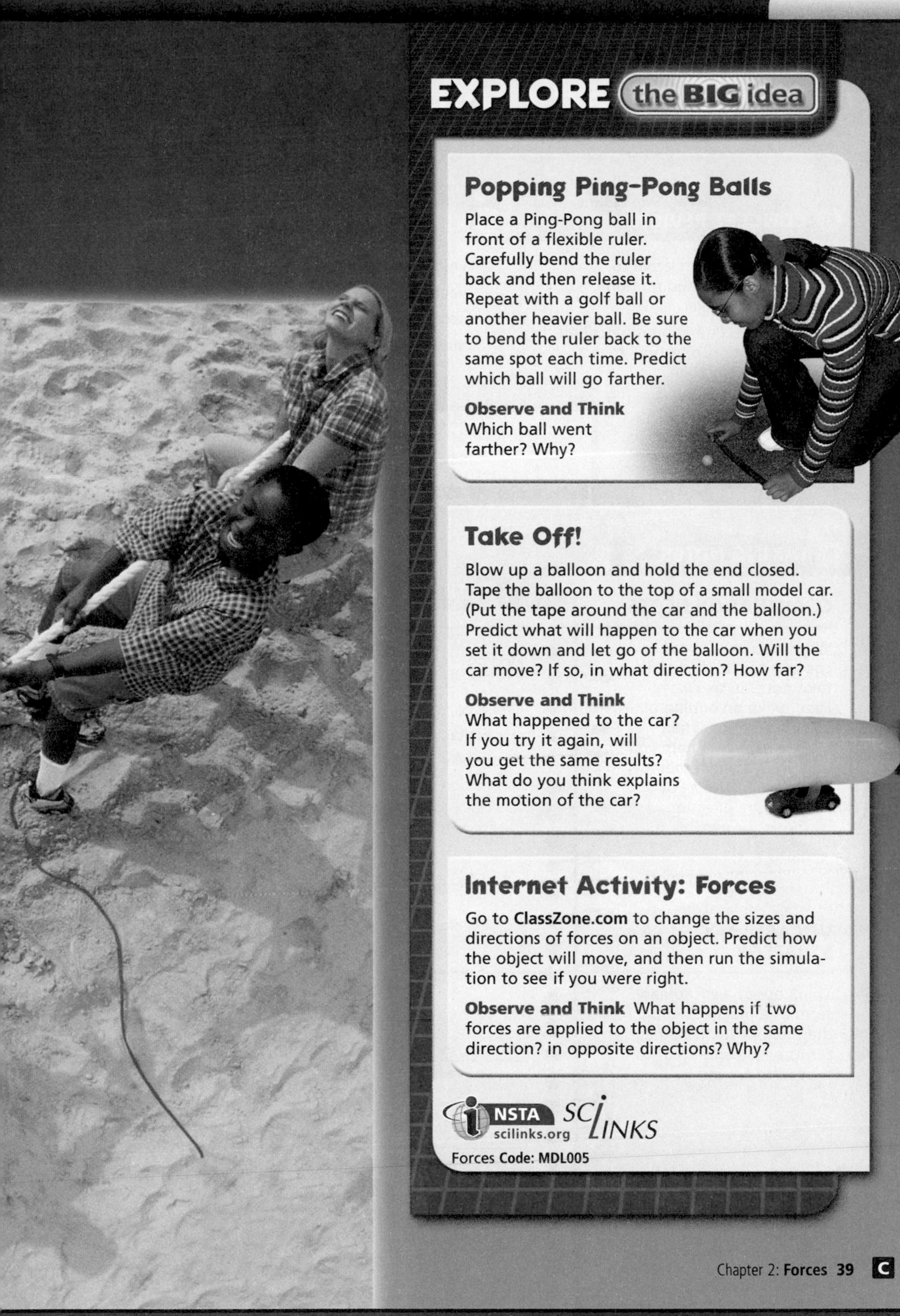

EXPLORE the BIG idea

Popping Ping-Pong Balls

Place a Ping-Pong ball in front of a flexible ruler. Carefully bend the ruler back and then release it. Repeat with a golf ball or another heavier ball. Be sure to bend the ruler back to the same spot each time. Predict which ball will go farther.

Observe and Think
Which ball went farther? Why?

Take Off!

Blow up a balloon and hold the end closed. Tape the balloon to the top of a small model car. (Put the tape around the car and the balloon.) Predict what will happen to the car when you set it down and let go of the balloon. Will the car move? If so, in what direction? How far?

Observe and Think
What happened to the car? If you try it again, will you get the same results? What do you think explains the motion of the car?

Internet Activity: Forces

Go to **ClassZone.com** to change the sizes and directions of forces on an object. Predict how the object will move, and then run the simulation to see if you were right.

Observe and Think What happens if two forces are applied to the object in the same direction? in opposite directions? Why?

NSTA scilinks.org SCILINKS
Forces Code: MDL005

TEACHING WITH TECHNOLOGY

Spreadsheet Several experiments in this chapter lend themselves to analyzing data entered into spreadsheets. See the investigations on pp. 46 and 49. Formulas can be entered into spreadsheets and used to solve problems such as those on pp. 51 and 52.

CBL and Probeware If students have probeware, encourage them to try using a force sensor and a motion detector for "Investigate Inertia" on p. 46.

EXPLORE the BIG idea

These inquiry-based activities are appropriate for use at home or as a supplement to classroom instruction.

Popping Ping-Pong Balls

PURPOSE To introduce students to the relationship of force to mass and acceleration. Students observe that the same force causes objects with different masses to accelerate differently.

TIP *10 min.* Remind students that plastic rulers can break; they should bend the ruler only about an inch.

Answer: the ball with less mass; for a given force, the smaller the mass, the greater the acceleration

REVISIT after p. 53.

Take Off!

PURPOSE To introduce students to the concept of action/reaction force pairs. Students observe the action force, which is the balloon pushing air out, and the reaction force, which is the air pushing on the balloon, moving the car forward.

TIP *10 min.* Students can use different-sized balloons and note differences in speed.

Answer: The car moved in the direction opposite the outflowing air; yes; the force from the air leaving the balloon.

REVISIT after p. 59.

Internet Activity: Forces

PURPOSE To introduce students to the effects of forces of differing size and direction.

TIP *20 min.* Students might try predicting the change in the object's motion before they change the forces.

Answer: The object moves in that direction; the object moves in the direction of the greater force; forces are combined using both size and direction.

REVISIT after p. 45.

PREPARE

CONCEPT REVIEW

Activate Prior Knowledge

- Make a transparency of a piece of graph paper and display it on an overhead projector. Place a marble or a Superball on the transparency.
- Show the marble or ball in motion.
- Ask students to describe how the action demonstrates each principle listed in the concept review.

TAKING NOTES

Combination Notes

Combining pictures with notes will help students connect abstract concepts with concrete examples. The two-column format allows students to write their notes in one column and draw their pictures in the second.

Vocabulary Strategy

Students might think that there must be an attraction between the magnet word and the words around it. Emphasize to students that the magnet shows a connection, but attraction is not a requirement.

Vocabulary and Note-Taking Resources

- Vocabulary Practice, pp. 122–123
- Decoding Support, p. 124

- Daily Vocabulary Scaffolding, p. T10
- Note-Taking Model, p. T11

- Magnet Words, B24–25
- Combination Notes, C36
- Daily Vocabulary Scaffolding, H1–8

CHAPTER 2

Getting Ready to Learn

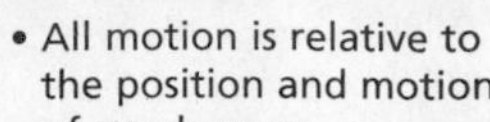

CONCEPT REVIEW

- All motion is relative to the position and motion of an observer.
- An object's motion is described by position, direction, speed, and acceleration.
- Velocity and acceleration can be measured.

VOCABULARY REVIEW

velocity p. 22
vector p. 22
acceleration p. 25
mass *See Glossary.*

CONTENT REVIEW
CLASSZONE.COM
Review concepts and vocabulary.

TAKING NOTES

COMBINATION NOTES

When you read about a concept for the first time, take notes in two ways. First, make an outline of the information. Then make a sketch to help you understand and remember the concept. Use arrows to show the direction of forces.

VOCABULARY STRATEGY

Think about a vocabulary term as a **magnet word** diagram. Write the other terms or ideas related to that term around it.

See the Note-Taking Handbook on pages R45–R51.

SCIENCE NOTEBOOK

C 40 Unit: Motion and Forces

CHECK READINESS

Administer the Diagnostic Test to determine students' readiness for new science content and their mastery of requisite math skills.

 Diagnostic Test, pp. 20–21

Technology Resources

Students needing content and math skills should visit **ClassZone.com**.

- CONTENT REVIEW
- MATH TUTORIAL

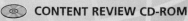

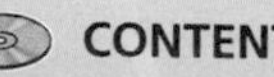 CONTENT REVIEW CD-ROM

KEY CONCEPT

2.1 Forces change motion.

BEFORE, you learned

- The velocity of an object is its change in position over time
- The acceleration of an object is its change in velocity over time

NOW, you will learn

- What a force is
- How unbalanced forces change an object's motion
- How Newton's first law allows you to predict motion

VOCABULARY

force p. 41
net force p. 43
Newton's first law p. 45
inertia p. 46

EXPLORE Changing Motion

How can you change an object's motion?

PROCEDURE

1. Choose an object from the materials list and change its motion in several ways, from
   - not moving to moving
   - moving to not moving
   - moving to moving faster
   - moving to moving in a different direction
2. Describe the actions used to change the motion.
3. Experiment again with another object. First, decide what you will do; then predict how the motion of the object will change.

MATERIALS

- quarter
- book
- tennis ball
- cup
- feather

WHAT DO YOU THINK?
In step 3, how were you able to predict the motion of the object?

A force is a push or a pull.

Think about what happens during an exciting moment at the ballpark. The pitcher throws the ball across the plate, and the batter hits it high up into the stands. A fan in the stands catches the home-run ball. In this example, the pitcher sets the ball in motion, the batter changes the direction of the ball's motion, and the fan stops the ball's motion. To do so, each must use a **force,** or a push or a pull.

REMINDER
Motion is a change in position over time.

You use forces all day long to change the motion of objects in your world. You use a force to pick up your backpack, to open or close a car door, and even to move a pencil across your desktop. Any time you change the motion of an object, you use a force.

2.1 FOCUS

Set Learning Goals

Students will

- Describe forces and how unbalanced forces change an object's motion.
- Explain how Newton's first law allows them to predict motion.
- Explain how the inertia of an object affects its motion.
- Design an experiment to investigate inertia.

3-Minute Warm-Up

Display Transparency 12 or copy this exercise on the board:

Decide if these statements are true. If they are not true, correct them.

1. Speed includes direction, while velocity does not. *Velocity includes direction, while speed does not.*
2. A moving object covers the same distance in less time if its velocity is greater. *true*
3. Acceleration measures only change in speed. *Acceleration measures change in velocity over time.*

3-Minute Warm-Up, p. T12

2.1 MOTIVATE

EXPLORE Changing Motion

PURPOSE To introduce the concept that force is needed to change motion

TIP *10 min.* Have groups compare two or three objects of their choice.

WHAT DO YOU THINK? *The force and direction with which you push determine the speed and direction of the object's movement.*

RESOURCES FOR DIFFERENTIATED INSTRUCTION

Below Level

UNIT RESOURCE BOOK
- Reading Study Guide A, pp. 78–79
- Decoding Support, p. 124

AUDIO CDS

Additional INVESTIGATION,
Newton's First Law, A, B, & C, pp. 138–146; Teacher Instructions, pp. 346–347

Advanced

UNIT RESOURCE BOOK
Challenge and Extension, p. 84

English Learners

UNIT RESOURCE BOOK
Spanish Reading Study Guide, pp. 82–83

AUDIO CDS
- Audio Readings in Spanish
- Audio Readings (English)

2.1 INSTRUCT

Teach from Visuals

Remind students that the length of the arrows shows the size of the force. To help students interpret the visual of the skater, ask:

- What do the red arrows represent in the picture of the skater? *a force in action*
- What objects are gravity, friction, and contact forces acting upon? *Gravity is acting upon on the skater; friction, on the wheels and ground; contact force, on the wheels and ground. Some students might mention friction between the skater and the air.*
- Would the forces shown be similar or different for a person walking? *similar*

Develop Critical Thinking

APPLY Have students discuss examples of contact forces in the following situations:

- turning a page
- pulling a chair
- brushing hair

For each example, ask them to visualize, demonstrate, or explain the size and direction of the contact forces.

Sample answer: hand applying a contact force to the page; hand applying contact force to the chair and contact force between the ground and the chair legs; brush applying contact force to the hair

Ongoing Assessment

CHECK YOUR READING *Answer: It is when one object pushes or pulls another object by touching it; the force your shoe and the ground exert on each other when you walk.*

Types of Forces

A variety of forces are always affecting the motion of objects around you. For example, take a look at how three kinds of forces affect the skater in the photograph on the left.

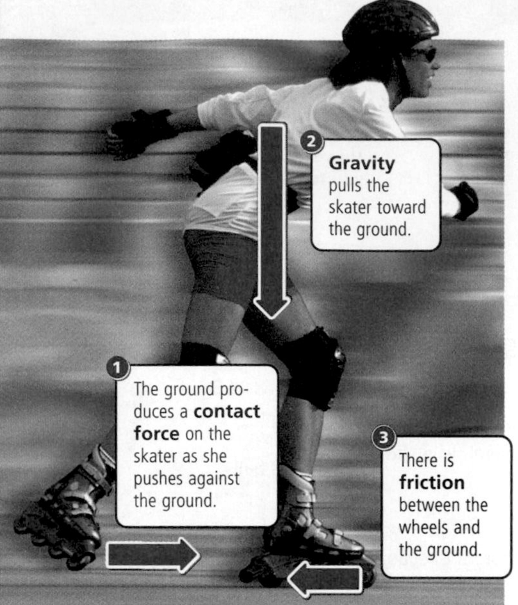

2 Gravity pulls the skater toward the ground.
1 The ground produces a contact force on the skater as she pushes against the ground.
3 There is friction between the wheels and the ground.

1. **Contact Force** When one object pushes or pulls another object by touching it, the first object is applying a contact force to the second. The skater applies a contact force as she pushes against the ground. The ground applies a contact force that pushes the skater forward.
2. **Gravity** Gravity is the force of attraction between two masses. Earth's gravity is pulling on the skater, holding her to the ground. The strength of the gravitational force between two objects depends on their masses. For example, the pull between you and Earth is much greater than the pull between you and a book.
3. **Friction** Friction is a force that resists motion between two surfaces that are pressed together. (? A) Friction between the surface of the ground and the wheels of the skates exerts a force that resists the skater's forward motion.

You will learn more about gravity and friction in Chapter 3. In this chapter, most of the examples involve contact forces. You use contact forces constantly. Turning a page, pulling a chair, using a pencil to write, pushing your hair away from your eyes—all involve contact forces.

CHECK YOUR READING What is a contact force? Give an example of a contact force.

Size and Direction of Forces

Like velocity, force is a vector. That means that force has both size and direction. For example, think about what happens when you try to make a shot in basketball. To get the ball through the hoop, you must apply the right amount of force to the ball and aim the force in the right direction. If you use too little force, the ball will not reach the basket. If you use too much force, the ball may bounce off the backboard and into your opponent's hands.

In the illustrations in this book, red arrows represent forces. The direction of an arrow shows the direction of the force, and the length of the arrow indicates the amount, or size, of the force. A blue box represents mass.

READING TIP

Red arrows are used to show force.

Blue boxes show mass.

DIFFERENTIATE INSTRUCTION

More Reading Support

A What does it mean to say that friction resists motion? *It slows motion.*

B Force has size and what else? *direction*

English Learners Some students may lack background knowledge of the baseball and basketball terminology on pp. 41 and 42.

C 42 Unit: **Motion and Forces**

Balanced and Unbalanced Forces

Considering the size and the direction of all the forces acting on an object allows you to predict changes in the object's motion. The overall force acting on an object when all the forces are combined is called the **net force.**

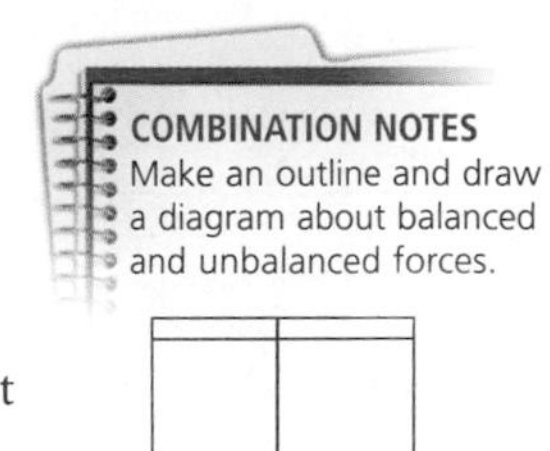

COMBINATION NOTES
Make an outline and draw a diagram about balanced and unbalanced forces.

C If the net force on an object is zero, the forces acting on the object are balanced. Balanced forces have the same effect as no force at all. That is, the motion of the object does not change. For example, think about the forces on the basketball when one player attempts a shot and another blocks it. In the photograph below on the left, the players are pushing on the ball with equal force but from opposite directions. The forces on the ball are balanced, and so the ball does not move.

D Only an unbalanced force can change the motion of an object. If one of the basketball players pushes with greater force than the other player, the ball will move in the direction that player is pushing. The motion of the ball changes because the forces on the ball become unbalanced.

It does not matter whether the ball started at rest or was already moving. Only an unbalanced force will change the ball's motion.

READING VISUALS **COMPARE** Compare the net force on the balls in these two photographs. Which photograph shows a net force of zero?

Address Misconceptions

IDENTIFY Ask: Does an object always move when a force acts on it? If students answer yes, they may hold the misconception that movement always accompanies a force.

CORRECT Place a book on a desk. Ask students what forces are acting on it. Explain that an upward force from the table balances the downward pull of gravity. Even though the book does not move, there are forces acting on it.

REASSESS Ask: If you are holding a book so that it does not move, what can be said about the force of gravity on the book and the force of your hand on the book? *The forces are present but balanced.*

Technology Resources

Visit **ClassZone.com** for background on common student misconceptions.

MISCONCEPTION DATABASE

Teach from Visuals

Students can add diagrams of shooting a basketball to their notes.

- Encourage students to add the direction of force. Ask what happens if you push the ball with the right amount of force but in the wrong direction. *The ball misses the basket.*
- Challenge students to sketch other examples of direction and size of forces, such as force on a bowling ball resulting in a gutter ball, a strike, or a split.

Ongoing Assessment

Describe how unbalanced forces change an object's motion.

Ask: When you lift and turn a page, how do unbalanced forces change its motion? *To lift the page, you apply an unbalanced force upward. When you let it go, gravity becomes the unbalanced force pushing downward.*

Answer: the left-hand picture

DIFFERENTIATE INSTRUCTION

More Reading Support

C If forces on an object are balanced, what is the net force? *zero*

D What kind of force changes the motion of an object? *unbalanced*

Advanced Have students predict the angle at which a ball at rest will move when pushed with equal force from two directions. Ask what happens if equal forces on the ball are at right angles. *Equal forces send the object moving at a 45-degree angle.* Have students experiment with different angles. Challenge them to draw diagrams showing the different results.

R Challenge and Extension, p. 84

Teach Difficult Concepts

Because friction slows down moving objects, students may think that a continuous force is needed to keep an object moving. To help students understand that a force is needed to stop a moving object, you might try the following demonstration.

Teacher Demo

This demonstration can be used to address the misconception cited on p. 37B.

- Spin a wheel of any type.
- Ask: What keeps the wheel in motion after you take your hand away? *No force has made it stop moving.*
- Ask: What will stop the wheel? *a force, probably friction*

History of Science

Galileo developed hypotheses about events that were impossible to observe in nature; he applied knowledge from related observable events. For example, he thought that if you started a ball rolling, it would move forever if no friction existed. By observing how a ball rolled along a series of ramps, he was able to deduce that without a force such as friction, a rolling ball would roll forever. The ability to hypothesize about what cannot be observed using knowledge of what can be observed enables scientists like Galileo and Newton to describe unobservable events in nature accurately.

Ongoing Assessment

CHECK YOUR READING *Answer: The object continues moving in the same direction with the same speed; first sentence of paragraph 2.*

Forces on Moving Objects

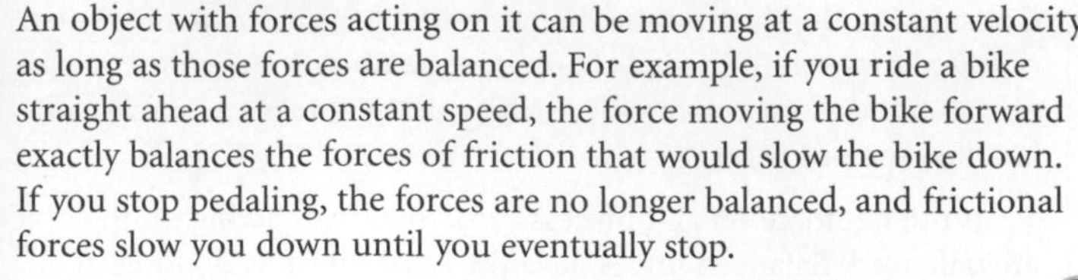

An object with forces acting on it can be moving at a constant velocity as long as those forces are balanced. For example, if you ride a bike straight ahead at a constant speed, the force moving the bike forward exactly balances the forces of friction that would slow the bike down. If you stop pedaling, the forces are no longer balanced, and frictional forces slow you down until you eventually stop.

Balanced forces cannot change an object's speed or its direction. An unbalanced force is needed to change an object's motion. E

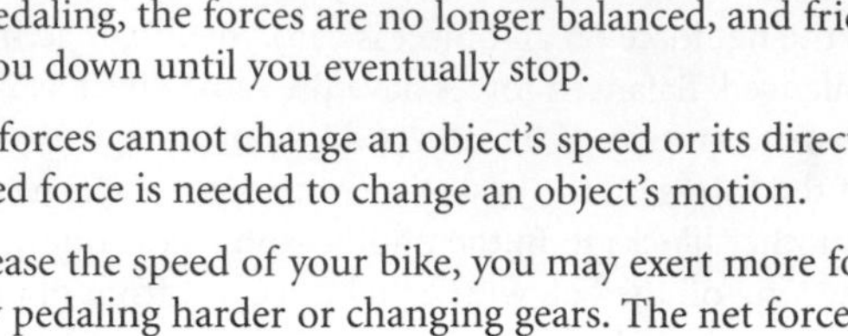

- To increase the speed of your bike, you may exert more forward force by pedaling harder or changing gears. The net force moves the bike ahead faster.
- To turn your bike, you apply an unbalanced force by leaning to one side and turning the handlebars.
- To stop the bike, you use the extra force of friction that your bike brakes provide.

CHECK YOUR READING What happens to a moving object if all the forces on it are balanced? Which sentence above tells you?

Newton's first law relates force and motion.

In the mid-1600s, the English scientist Sir Isaac Newton studied the effects of forces on objects. He formulated three laws of motion that are still helping people describe and predict the motions of objects today. Newton's ideas were built on those of other scientists, in particular the Italian scientist Galileo Galilei (gal-uh-LEE-oh gal-uh-LAY). Both Galileo and Newton overturned thinking that had been accepted since the times of the ancient Greek philosophers.

The ancient Greeks had concluded that it was necessary to apply a continuous force to keep an object in motion. For example, if you set a book on a table and give the book a quick push, the book slides a short way and then stops. To keep the book moving, you need to keep pushing it. The Greeks reasoned that the book stops moving because you stop pushing it.

Galileo's Thought Experiment

READING TIP
Contrast the last sentence of this paragraph with the last sentence of the previous paragraph.

In the early 1600s, Galileo suggested a different way of interpreting such observations. He imagined a world without friction and conducted a thought experiment in this ideal world. He concluded that, in the absence of friction, a moving object will continue moving even if there is no force acting on it. In other words, it does not take a force to keep an object moving; it takes a force—friction—to stop an object that is already moving.

DIFFERENTIATE INSTRUCTION

More Reading Support

E What happens to a moving object if balanced forces act on it? What will it do if there is no force? *In both cases, it keeps moving with same velocity.*

Below Level Use the table to help students understand the conflicting ideas about force and motion.

Greeks	Galileo
Force keeps an object in motion.	Force is needed to change motion.
Force causes continuing movement.	Force is not required to keep an object moving.

Objects at rest and objects in motion both resist changes in motion. That is, objects at rest tend to stay at rest, and objects that are moving tend to continue moving unless a force acts on them. Galileo reasoned there was no real difference between an object that is moving at a constant velocity and an object that is standing still. An object at rest is simply an object with zero velocity.

CHECK YOUR READING How were Galileo's ideas about objects in motion different from the ideas of the ancient Greeks?

Newton's First Law

F Newton restated Galileo's conclusions as his first law of motion. **Newton's first law** states that objects at rest remain at rest, and objects in motion remain in motion with the same velocity, unless acted upon by an unbalanced force. You can easily observe the effects of unbalanced forces, both on the ball at rest and the ball in motion, in the pictures below.

Newton's First Law

Objects at rest remain at rest, and objects in motion remain in motion with the same velocity, unless acted upon by an unbalanced force.

An Object at Rest

An object at rest **(the ball)** remains at rest unless acted upon by an unbalanced force **(from the foot)**.

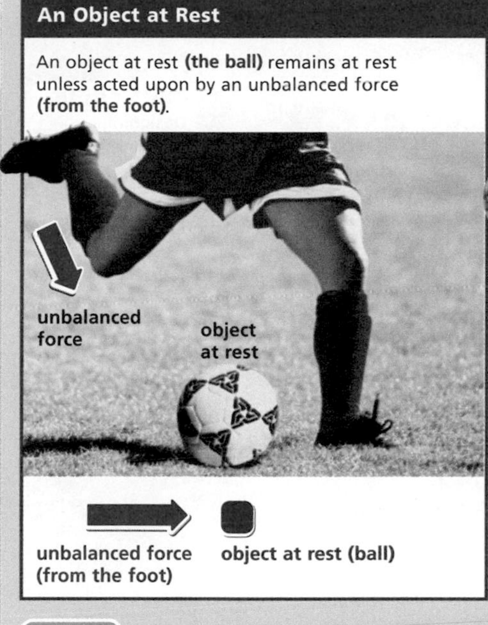

An Object in Motion

An object in motion **(the ball)** remains in motion with the same velocity, unless acted upon by an unbalanced force **(from the hand)**.

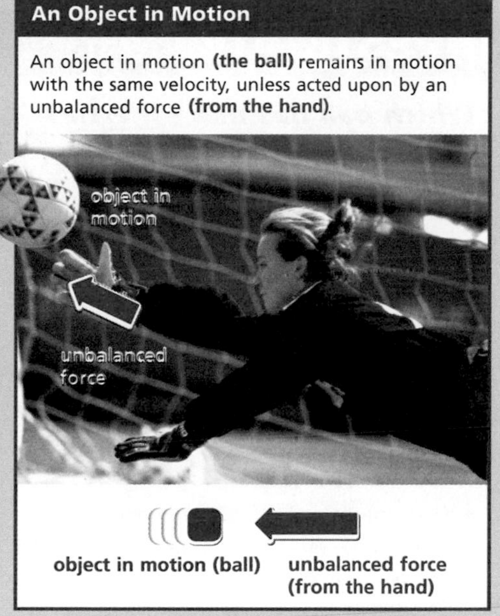

READING VISUALS What will happen to the ball's motion in each picture? Why?

DIFFERENTIATE INSTRUCTION

More Reading Support

F State Newton's first law in your own words. *Answer: Unmoving objects stay still. A moving object moves in the same direction unless a force changes its motion.*

Additional Investigation To reinforce Section 2.1 learning goals, use the following full-period investigation:

R **Additional INVESTIGATION,** Newton's First Law, A, B, & C, pp. 138–146, 346–347

Advanced Have students research Galileo's experiments with inclined planes and inertia. Students should then design a recreation of Galileo's inclined plane experiment, or design another experiment that demonstrates the same principle.

Teach from Visuals

To help students interpret the visuals of kids playing soccer, ask:

- What symbol is used to show an object in motion? *a blue box with motion lines to one side of it*
- How might the symbol be different in a third picture after the goalie hits the ball? *The motion lines would be drawn on the right side of the box instead of on the left.*

Real World Example

Consider this dramatic example of a force changing the motion of a 10,000 kg satellite. On Earth, any force that an astronaut could apply by pushing on the satellite would be balanced by the force of friction. In space, however, there is no friction to balance the push. Even a small unbalanced force on the satellite can cause it to start moving or change its motion, although the change would be extremely small due to the satellite's large mass.

EXPLORE the BIG idea

Revisit "Internet Activity: Forces" on p. 39. Have students explain their results.

Ongoing Assessment

Explain how Newton's first law allows you to predict motion.

Ask: What unbalanced forces change the motion of a volleyball that is hit hard over the net? How will its velocity change? *The contact force of a hand changes the ball's direction, and may increase its velocity.*

CHECK YOUR READING *Answer: Galileo: friction slows down motion; neither objects at rest nor objects with constant velocity are being acted upon by a net force; it takes force to change velocity. Greeks: an object stays in motion only if force is continuously applied; no motion means no force is present.*

READING VISUALS *Answer: The ball's motion will change. In the picture on the left, it will start moving. In the picture on the right, it will probably change both speed and direction.*

INVESTIGATE Inertia

PURPOSE To design an experiment to determine the inertia of two objects

TIPS *30 min.* Allow students a few minutes to explore, then suggest the following:

- Think of ways to apply equal forces to both balls.
- Change the velocity of the ball from both rest and motion.

WHAT DO YOU THINK? *The ball with greater mass has greater inertia. The variable is the mass of the balls. Constants should include the amount of force applied and may include the standard of measuring change in motion. The ball with greater mass has greater inertia, and it is more difficult to change its motion.*

 Datasheet, Inertia, p. 85

Technology Resources

Customize this student lab as needed or look for an alternative. Print rubrics to assess student lab reports.

 **Lab Generator CD-ROM**

You will find many examples of Newton's first law around you. For instance, if you throw a stick for a dog to catch, you are changing the motion of the stick. The dog changes the motion of the stick by catching it and by dropping it at your feet. You change the motion of a volleyball when you spike it, a tennis racket when you swing it, a paintbrush when you make a brush stroke, and an oboe when you pick it up to play or set it down after playing. In each of these examples, you apply a force that changes the motion of the object.

Inertia

VOCABULARY
Make a magnet word diagram for *inertia* in your notebook.

Inertia (ih-NUR-shuh) is the resistance of an object to a change in the speed or the direction of its motion. Newton's first law, which describes the tendency of objects to resist changes in motion, is also called the law of inertia. Inertia is closely related to mass. When you measure the mass of an object, you are also measuring its inertia. You know from experience that it is easier to push or pull an empty box than it is to push or pull the same box when it is full of books. Likewise, it is easier to stop or to turn an empty wagon than to stop or turn a wagon full of sand. In both of these cases, it is harder to change the motion of the object that has more mass.

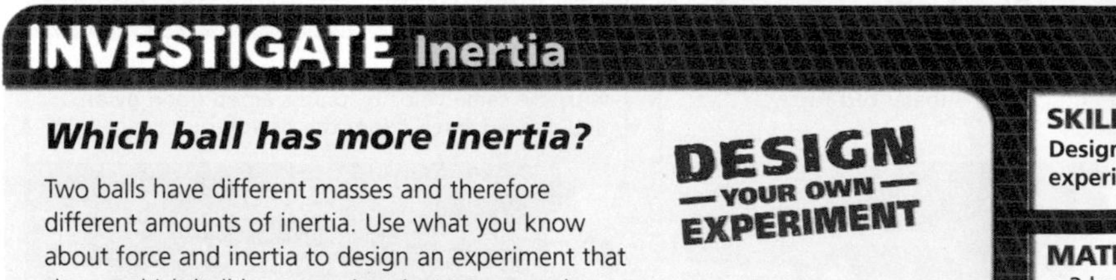

INVESTIGATE Inertia

Which ball has more inertia?

DESIGN — YOUR OWN — EXPERIMENT

Two balls have different masses and therefore different amounts of inertia. Use what you know about force and inertia to design an experiment that shows which ball has more inertia. Your procedure cannot include lifting the balls, weighing the balls, or touching the balls with your hands.

SKILL FOCUS
Designing experiments

MATERIALS
- 2 balls of unknown masses
- string
- block
- meter stick

TIME
30 minutes

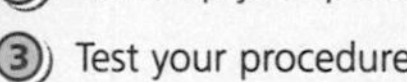

PROCEDURE

1. Figure out how to use the meter stick or other materials to compare the inertia of the two balls.
2. Write up your procedure.
3. Test your procedure.

WHAT DO YOU THINK?

- What were the results of your experiment? Did it work? Why or why not?
- What was the variable? What were the constants?
- How does your experiment demonstrate the property of inertia?

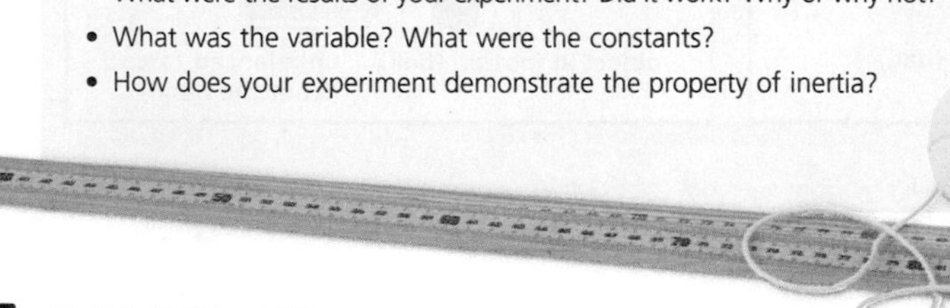

Metacognitive Strategy

Ask students to write a paragraph about one thing that posed a problem while they were designing the procedure. Have them describe their solution to this problem.

Teaching with Technology

If students have probeware, encourage then to use a force sensor and motion detector in their experiments.

Ongoing Assessment

Explain how the inertia of an object affects its motion.

Ask: How would the inertia of each object used in the Explore on p. 41 affect its motion? *the greater the mass, the more difficult it is to change its motion*

DIFFERENTIATE INSTRUCTION

More Reading Support

G What is Newton's first law also called? *the law of inertia* How do you measure inertia? *by measuring the mass of an object*

English Learners Help English learners understand that on p. 46, *resist* and *resistance* are verb and noun forms of the same word. English learners may also be confused by the ways in which certain English words are used in different contexts, as in, "it is harder to change the motion of the object that has more mass" (p. 46). *Harder* describes the degree of difficulty, not how firm or solid an object is. Help students recognize when a tactile word is being used to describe something abstract.

Inertia is the reason that people in cars need to wear seat belts. A moving car has inertia, and so do the riders inside it. When the driver applies the brakes, an unbalanced force is applied to the car. Normally, the bottom of the seat applies an unbalanced force—friction—which slows the riders down as the car slows. If the driver stops the car suddenly, however, this force is not exerted over enough time to stop the motion of the riders. Instead, the riders continue moving forward with most of their original speed because of their inertia.

RESOURCE CENTER
CLASSZONE.COM

Find out more about inertia.

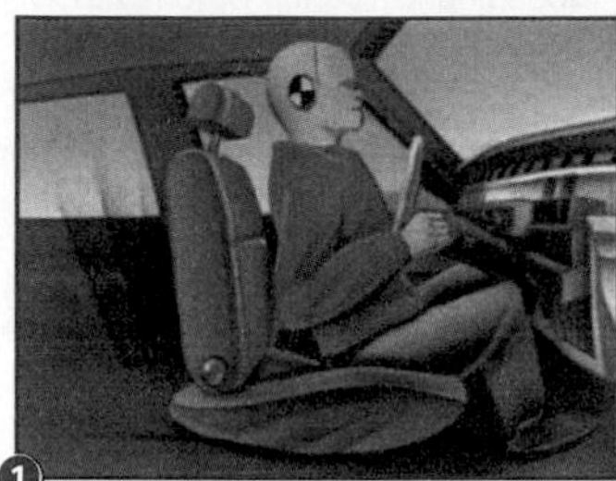

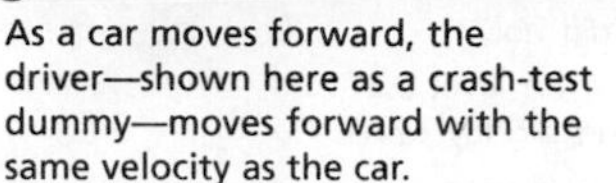

1 As a car moves forward, the driver—shown here as a crash-test dummy—moves forward with the same velocity as the car.

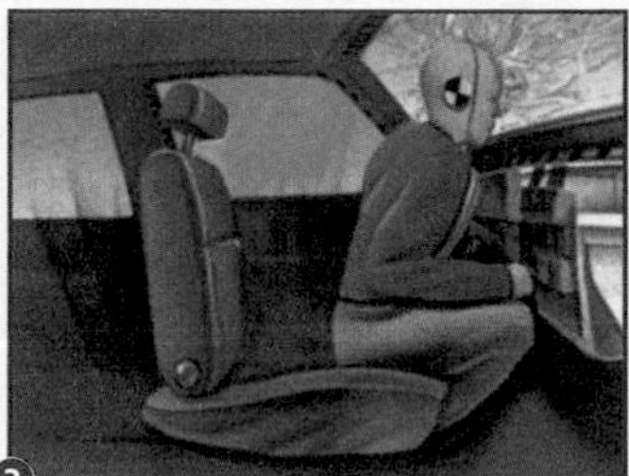

2 When the driver hits the brakes, the car stops. If the stop is sudden and the driver is not wearing a seat belt, the driver keeps moving forward.

3 Finally, the windshield applies an unbalanced force that stops the driver's forward motion.

If the driver is wearing a seat belt, the seat belt rather than the windshield applies the unbalanced force that stops the driver's forward motion. The force from the seat belt is applied over a longer time, so the force causes less damage. In a collision, seat belts alone are sometimes not enough to stop the motion of drivers or passengers. Air bags further cushion people from the effects of inertia in an accident.

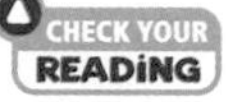

CHECK YOUR READING If a car makes a sudden stop, what happens to a passenger riding in the back seat who is not wearing a seat belt?

2.1 Review

KEY CONCEPTS

1. Explain the difference between balanced and unbalanced forces.
2. What is the relationship between force and motion described by Newton's first law?
3. What is inertia? How is the inertia of an object related to its mass?

CRITICAL THINKING

4. **Infer** Once a baseball has been hit into the air, what forces are acting upon it? How can you tell that any forces are acting upon the ball?
5. **Predict** A ball is at rest on the floor of a car moving at a constant velocity. What will happen to the ball if the car swerves suddenly to the left?

CHALLENGE

6. **Synthesize** What can the changes in an object's position tell you about the forces acting on that object? Describe an example from everyday life that shows how forces affect the position of an object.

ANSWERS

1. Balanced forces do not change an object's motion; unbalanced forces exert a net force and can change an object's motion.

2. An object's state of motion does not change unless acted upon by an unbalanced force.

3. an object's resistance to having its state of motion changed; the greater its mass, the greater its inertia

4. gravity (friction with the air may also be mentioned), as shown by change in speed and direction

5. It will continue straight ahead (to the right side of the car).

6. If changes in an object's position demonstrate a change in speed or direction, the object has an unbalanced force acting upon it.

Ongoing Assessment

CHECK YOUR READING *Answer: The passenger will continue to move forward and will hit the back of the front seat.*

Reinforce the BIG idea

Have students relate the section to the Big Idea.

Reinforcing Key Concepts, p. 86

2.1 ASSESS & RETEACH

Assess

Section 2.1 Quiz, p. 22

Reteach

Have students examine the picture of the bicycle on p. 44. Ask for volunteers to diagram the bicycle's motion. Explain the following concepts and have volunteers add force/mass symbols to the diagram.

- Balanced forces applied by both hands keep the handlebars straight.
- Pedaling applies an unbalanced force.
- Steering applies an unbalanced force.
- When you brake, friction slows the turning wheels.
- If you coast on a level surface, the bicycle slows due to friction.

Technology Resources

Have students visit **ClassZone.com** for reteaching of Key Concepts.

CONTENT REVIEW

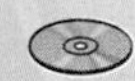

CONTENT REVIEW CD-ROM

THINK SCIENCE
Scientific Methods of Thinking

Set Learning Goal
To evaluate hypotheses by checking them against observations

Present the Science
The sliding rock phenomenon on this desert floor has been observed since the early 1900s, but no one has ever seen the rocks move. Ask students to discuss what the evidence of sliding is. *Rocks are found in new locations; tracks in the clay show movement.*

Guide the Activity
- Remind students that to evaluate means to judge a statement based on criteria.
- Students should determine whether each observation supports each hypothesis.
- Point out that if even one observation does not support the hypothesis, students should recheck the observation or revise the hypothesis.
- Remind students to use visual clues.

COOPERATIVE LEARNING STRATEGY
Divide the class into groups of three to six students. In each group, assign a facilitator, a recorder, and a reporter. Assign each group one hypothesis to evaluate. The facilitator ensures that everyone has a chance to respond. The recorder writes the group's consensus. The reporter presents each observation to the class.

Close
Ask: Why does a hypothesis have to be checked against each observation? *One observation can disprove the entire hypothesis.*

Technology Resources

Students can visit **ClassZone.com** for links to Racetrack Playa pictures, animations, and more information.

RESOURCE CENTER

Think SCIENCE

SKILL: EVALUATING HYPOTHESES

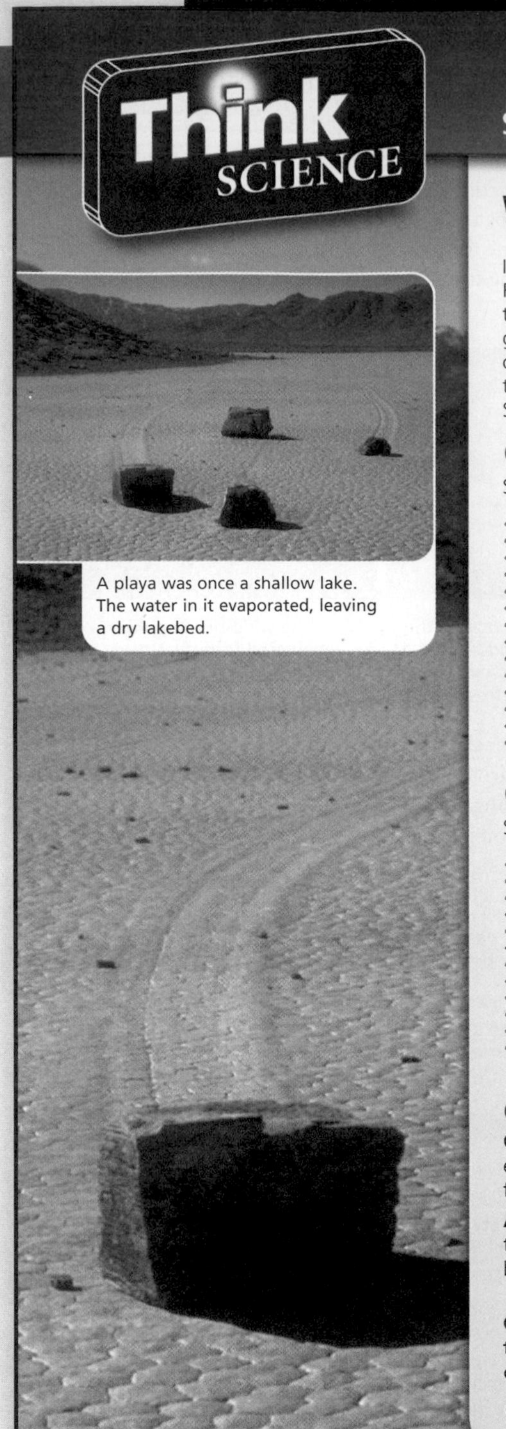

A playa was once a shallow lake. The water in it evaporated, leaving a dry lakebed.

Why Do These Rocks Slide?
In Death Valley, California, there is a dry lakebed known as Racetrack Playa. Rocks are mysteriously moving across the ground there, leaving tracks in the clay. These rocks can have masses as great as 320 kilograms (corresponding to 700 lb). No one has ever observed the rocks sliding, even though scientists have studied their tracks for more than 50 years. What force moves these rocks? Scientists do not yet know.

Observations
Scientists made these observations.

a. Some rocks left trails that are almost parallel.
b. Some rocks left trails that took abrupt turns.
c. Sometimes a small rock moved while a larger rock did not.
d. Most of the trails are on level surfaces. Some trails run slightly uphill.
e. The temperature in that area sometimes drops below freezing.

This rock made a U-turn.

Hypotheses
Scientists formed these hypotheses about how the rocks move.

- When the lakebed gets wet, it becomes so slippery that gravity causes the rocks to slide.
- When the lakebed gets wet, it becomes so slippery that strong winds can move the rocks.
- When the lakebed gets wet and cold, a sheet of ice forms and traps the rocks. Strong winds move both the ice sheet and the trapped rocks.

Evaluate Each Hypothesis
On Your Own Think about whether all the observations support each hypothesis. Some facts may rule out some hypotheses. Some facts may neither support nor contradict a particular hypothesis.

As a Group Decide which hypotheses are reasonable. Discuss your thinking and conclusions in a small group, and list the reasonable hypotheses.

CHALLENGE What further observations would you make to test any of these hypotheses? What information would each observation add?

RESOURCE CENTER CLASSZONE.COM — Learn more about the moving rocks.

ANSWERS
Hypothesis 1 is not reasonable and is not supported by observations a, b, and d. Gravity does not cause rocks to slide in parallel lines, turn, or move uphill.
Hypothesis 2 is reasonable. Observation a does not support or weaken it, and b–e support it: winds can change direction, move smaller rocks, cause rocks to move uphill, and move rocks on ice.
Hypothesis 3 is reasonable and supported by all observations: if frozen in ice, rocks can move together; winds can make rocks change direction, move small rocks, move rocks uphill; ice will form when temperature is below freezing.
CHALLENGE Sample answer: record movement and temperatures; this might prove a relationship between movement and freezing temperatures

KEY CONCEPT

Force and mass determine acceleration.

BEFORE, you learned

- Mass is a measure of inertia
- The motion of an object will not change unless the object is acted upon by an unbalanced force

NOW, you will learn

- How Newton's second law relates force, mass, and acceleration
- How force works in circular motion

VOCABULARY

Newton's second law p. 50
centripetal force p. 54

EXPLORE Acceleration

How are force and acceleration related?

PROCEDURE

1. Tie a paper clip to each end of a long string. Hook two more paper clips to one end.
2. Hold the single paper clip in the middle of a smooth table; hang the other end of the string over the edge. Let go and observe.
3. Add one more paper clip to the hanging end and repeat the experiment. Observe what happens. Repeat.

MATERIALS

- paper clips
- string

WHAT DO YOU THINK?

- What happened each time that you let go of the single paper clip?
- Explain the relationship between the number of hanging paper clips and the motion of the paper clip on the table.

Newton's second law relates force, mass, and acceleration.

Suppose you are eating lunch with a friend and she asks you to pass the milk container. You decide to slide it across the table to her. How much force would you use to get the container moving? You would probably use a different force if the container were full than if the container were empty.

If you want to give two objects with different masses the same acceleration, you have to apply different forces to them. You must push a full milk container harder than an empty one to slide it over to your friend in the same amount of time.

REMINDER
Acceleration is a change in velocity over time.

CHECK YOUR READING What three concepts are involved in Newton's second law?

RESOURCES FOR DIFFERENTIATED INSTRUCTION

Below Level

UNIT RESOURCE BOOK

- Reading Study Guide A, pp. 89–90
- Decoding Support, p. 124

Advanced

UNIT RESOURCE BOOK

- Challenge and Extension, p. 95
- Challenge Reading, pp. 120–121

English Learners

UNIT RESOURCE BOOK

Spanish Reading Study Guide, pp. 93–94

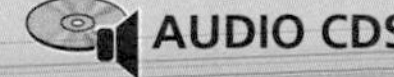

- Audio Readings in Spanish
- Audio Readings (English)

2.2 FOCUS

Set Learning Goals

Students will

- Explain how Newton's second law relates force, mass, and acceleration.
- Describe how force works in circular motion.
- Hypothesize about how circular motion is affected by force.

3-Minute Warm-Up

Display Transparency 12 or copy this exercise on the board:

Draw a force/mass diagram for the following information:

Two identical disks are on a table. Disk A is pushed from the left with a small force. Disk B is pushed from the left with a force twice as large as disk A and pushed from the right with a force the same size as disk A. How does the motion of disk A compare to the motion of disk B? *It is the same.*

Combining forces in the same direction

Combining forces in opposite directions

3-Minute Warm-Up, p. T12

2.2 MOTIVATE

EXPLORE Acceleration

PURPOSE To introduce the concept that force and acceleration are interrelated

TIP *10 min.* The paper clips must be the same size, so that the only independent variable is the number of paper clips.

WHAT DO YOU THINK? *The more paper clips on the other end of the thread, the faster the motion of the paper clip on the table.*

Ongoing Assessment

Answer: force, mass, acceleration

2.2 INSTRUCT

Teach from Visuals

To help students interpret the visuals depicting Newton's Second Law, ask:

- In the picture on the left, what stays constant? What changes? *Mass stays constant; force and acceleration change.*
- In the picture on the right, what stays constant? What changes? *Force stays constant; mass and acceleration change.*

Ongoing Assessment

Explain how Newton's second law relates force, mass, and acceleration.

Ask: An apple and a bowling ball are pushed with the same force. Which one will accelerate more? Why? *the apple, because it has less mass*

READING VISUALS *Answer: The upper arrows show size and direction of force, and the bottom arrows show size and direction of acceleration.*

Newton's Second Law

Explore Newton's second law.

Newton studied how objects move, and he noticed some patterns. He observed that the acceleration of an object depends on the mass of the object and the size of the force applied to it. **Newton's second law** states that the acceleration of an object increases with increased force and decreases with increased mass. The law also states that the direction in which an object accelerates is the same as the direction of the force.

The photographs below show Newton's second law at work in a supermarket. The acceleration of each shopping cart depends upon two things:

- the size of the force applied to the shopping cart
- the mass of the shopping cart

In the left-hand photograph, the force on the cart changes, while the mass of the cart stays the same. In the right-hand photograph, the force on the cart stays the same, while the mass of the cart varies. Notice how mass and force affect acceleration.

Newton's Second Law

The acceleration of an object increases with increased force, decreases with increased mass, and is in the same direction as the force.

Increasing Force Increases Acceleration

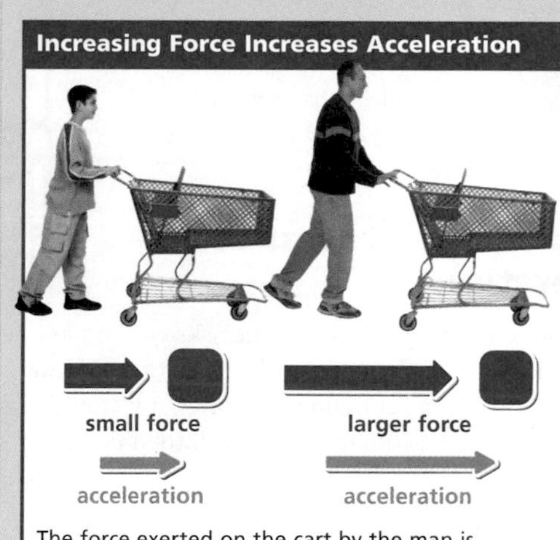

The force exerted on the cart by the man is greater than the force exerted on the same cart by the boy, so the acceleration is greater.

Increasing Mass Decreases Acceleration

small mass

acceleration

larger mass

acceleration

The mass of the full cart is greater than the mass of the empty cart, and the boy is pushing with the same force, so the acceleration is less.

READING VISUALS What do the arrows in these diagrams show?

DIFFERENTIATE INSTRUCTION

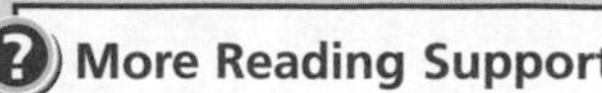

More Reading Support

A What happens to the acceleration of an object when the force on it increases? *It increases.*

English Learners English learners may have difficulty distinguishing between comparative *(-er)* and superlative *(-est)* adjectives. When an adjective ends in *-er,* it is comparing two people, places, things, or ideas (as *greater* is in the captions on this page). When an adjective ends in *-est,* it is comparing three or more people, places, things, or ideas.

Force Equals Mass Times Acceleration

Newton was able to describe the relationship of force, mass, and acceleration mathematically. You can calculate the force, the mass, or the acceleration if you know two of the three factors. The mathematical form of Newton's second law, stated as a formula, is

$$\textbf{Force} = \textbf{mass} \cdot \textbf{acceleration}$$
$$F = ma$$

To use this formula, you need to understand the unit used to measure force. In honor of Newton's contribution to our understanding of force and motion, the standard unit of force is called the newton (N). Because force equals mass times acceleration, force is measured in units of mass (kilograms) times units of acceleration (meters per second per second). A newton is defined as the amount of force that it takes to accelerate one kilogram (1 kg) of mass one meter per second per second (1 m/s^2). So 1 N is the same as 1 kg • m/s^2.

REMINDER

Meters per second per second is the same as *m/s^2*, which can be read "meters per second squared."

CHECK YOUR READING If the same force is applied to two objects of different mass, which object will have the greater acceleration?

The mathematical relationship of force, mass, and acceleration allow you to solve problems about how objects move. If you know the mass of an object and the acceleration you want to achieve, you can use the equation to find the force you need to exert to produce that acceleration. Use Newton's second law to find the force that is needed to accelerate the shopping cart in the sample problem.

Calculating Force

Sample Problem

What force is needed to accelerate a 10 kg shopping cart 3 m/s^2?

What do you know?	mass = 10 kg, acceleration = 3 m/s^2
What do you want to find out?	Force
Write the formula:	$F = ma$
Substitute into the formula:	$F = 10\text{ kg} \cdot 3\text{ m/s}^2$
Calculate and simplify:	$F = 10\text{ kg} \cdot \frac{3\text{m}}{\text{s}^2} = 30\text{ kg} \cdot \text{m/s}^2$
Check that your units agree:	Unit is kg • m/s^2. Unit of force is newton, which is also kg • m/s^2. Units agree.
Answer:	$F = 30$ N

Practice the Math

1. If a 5 kg ball is accelerating 1.2 m/s^2, what is the force on it?
2. A person on a scooter is accelerating 2 m/s^2. If the person has a mass of 50 kg, how much force is acting on that person?

Integrate the Sciences

Many athletes train with weights to improve their performance. Strengthening their muscles allows athletes to apply more force to objects they use in their sports. Increased force means increased acceleration. A stronger baseball player can throw and hit the ball farther, a stronger swimmer can move through the water more quickly, and a stronger discuss thrower can throw the discuss farther.

Develop Algebra Skills

- Math Support, p. 125
- Math Practice, p. 126

Ongoing Assessment

CHECK YOUR READING *Answer: the object with less mass*

Practice the Math *Answers:*

1. $F = ma = 5\text{ kg} \cdot 1.2\text{ m/s}^2$
$= 6\text{ kg} \cdot \text{m/s}^2 = 6\text{ N}$
2. $F = ma = 50\text{ kg} \cdot 2\text{ m/s}^2$
$= 100\text{ kg} \cdot \text{m/s}^2 = 100\text{ N}$

DIFFERENTIATE INSTRUCTION

More Reading Support

B What is the standard unit of force? *newton*

C What unit is used for mass when calculating force or acceleration? *kilogram*

Advanced Have students describe a scenario that involves force, mass, and acceleration. Using this scenario, have them write related additional problems to solve using $F = ma$. Each set of problems should include at least one problem for each variable. Review how to rearrange the equation to solve for any of the variables. Have students exchange problems and solve.

R Challenge and Extension, p. 95

Develop Algebra Skills

Remind students that formulas can be rearranged to find any variable in the formula.

- Isolate the desired variable on one side of the equation.
- To isolate the desired variable, "undo" any process done to the variable in the formula. Multiplication and division are opposite processes, as are addition and subtraction.
- To solve for C in the formula $F = (9C/5) + 32$, subtract 32 from both sides of the equation. Divide both sides by 9, then multiply by 5. The formula for C is $5(F - 32)/9$.

Teacher Demo

Use a toy with wheels to demonstrate the difference between velocity and acceleration. Pull the toy at a steady speed for 6 m while a student measures the amount of time it takes to pull the toy over each 2 m section. Ask students to calculate the speed of the toy in each section. These values should agree closely. Ask students how much the speed changed from one section to another; in other words, what was the acceleration? Students should be able to see that the acceleration of the toy is zero, even though it moves constantly.

Teach from Visuals

To help students interpret the visual of people pulling a plane and the information in the caption, ask:

- What was the average velocity of the plane while it was pulled? *$v = d/t = 3.7\ m/6.74\ s = 0.55\ m/s$*
- If the people pulled a car with the same amount of force they used on the plane, how would the car's acceleration compare with the plane's? Why? *It would be greater because its mass is less.*

Ongoing Assessment

Practice the Math *Answers:*

1. *$a = F/m = 4500N/30{,}000\ kg = 0.15\ m/s^2$*
2. *$a = F/m = 3N/6\ kg = 0.5\ m/s^2$*

This team of 20 people pulled a 72,000-kilogram (159,000 lb) Boeing 727 airplane 3.7 meters (12 ft) in 6.74 seconds.

The photograph above shows people who are combining forces to pull an airplane. Suppose you knew the mass of the plane and how hard the people were pulling. How much would the plane accelerate? The sample problem below shows how Newton's second law helps you calculate the acceleration.

Calculating Acceleration

Sample Problem

If a team pulls with a combined force of 9000 N on an airplane with a mass of 30,000 kg, what is the acceleration of the airplane?

What do you know? mass = 30,000 kg, force = 9000 N

What do you want to find out? acceleration

(? D) *Rearrange the formula:* $a = \frac{F}{m}$

Substitute into the formula: $a = \frac{9000\ N}{30{,}000\ kg}$

(? E) *Calculate and simplify:* $a = \frac{9000\ N}{30{,}000\ kg} = \frac{9000\ kg \cdot m/s^2}{30{,}000\ kg} = 0.3\ m/s^2$

Check that your units agree: Unit is m/s^2.
Unit for acceleration is m/s^2.
Units agree.

Answer: $a = 0.3\ m/s^2$

Practice the Math

1. Half the people on the team decide not to pull the airplane. The combined force of those left is 4500 N, while the airplane's mass is still 30,000 kg. What will be the acceleration?
2. A girl pulls a wheeled backpack with a force of 3 N. If the backpack has a mass of 6 kg, what is its acceleration?

52 Unit: **Motion and Forces**

DIFFERENTIATE INSTRUCTION

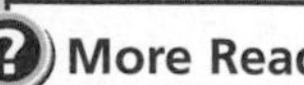

More Reading Support

D What formula is used to calculate acceleration? $a = \frac{F}{m}$

E A mass is 2 kg. What other information do you need to calculate acceleration? *force*

Below Level Have students draw a graphic organizer that shows the steps used to solve problems for each form of $F = ma$. The graphic organizer should show what formula is used to solve for each variable and a sample problem of each type. Have students use their organizers to solve the practice problems on pp. 52 and 53.

Mass and Acceleration

Mass is also a variable in Newton's second law. If the same force acts on two objects, the object with less mass will have the greater acceleration. For instance, if you push a soccer ball and a bowling ball with equal force, the soccer ball will have a greater acceleration.

If objects lose mass, they can gain acceleration if the force remains the same. When a rocket is first launched, most of its mass is the fuel it carries. As the rocket burns fuel, it loses mass. As the mass continually decreases, the acceleration continually increases.

APPLY This NASA launch rocket accelerates with enough force to lift about 45 cars off the ground. As the rocket loses fuel, will it accelerate more or less? Why?

Calculating Mass

Sample Problem

A model rocket is accelerating at 2 m/s². The force on it is 1 N. What is the mass of the rocket?

What do you know? acceleration = 2 m/s², force = 1 N

What do you want to find out? mass

Rearrange the formula: $m = \frac{F}{a}$

Substitute into the formula: $m = \frac{1\text{ N}}{2\text{ m/s}^2}$

Calculate and simplify: $m = \frac{1\text{ N}}{2\text{ m/s}^2} = \frac{1\text{ kg} \cdot \text{m/s}^2}{2\text{ m/s}^2} = 0.5\text{ kg}$

Check that your units agree: Unit is kg.
Unit of mass is kg. Units agree.

Answer: $m = 0.5$ kg

Practice the Math

1. Another model rocket is accelerating at a rate of 3 m/s² with a force of 1 N. What is the mass of the rocket?
2. A boy pushes a shopping cart with a force of 10 N, and the cart accelerates 1 m/s². What is the mass of the cart?

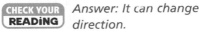

Forces can change the direction of motion.

F

Usually, we think of a force as either speeding up or slowing down the motion of an object, but force can also make an object change direction. If an object changes direction, it is accelerating. Newton's second law says that if you apply a force to an object, the direction in which the object accelerates is the same as the direction of the force. You can change the direction of an object without changing its speed. For example, a good soccer player can control the motion of a soccer ball by applying a force that changes the ball's direction but not its speed.

CHECK YOUR READING How can an object accelerate when it does not change speed?

Real World Example

A fully loaded tractor-trailer and a passenger car are going 55 miles per hour. If both drivers apply their brakes at the same time, the truck will travel farther than the car before it stops. While the truck's brakes apply about 16 times the force of the car's brakes, the truck is also about 27 times as massive as the car. Therefore, the truck's acceleration when braking is only about 60% of the car's acceleration. This is why truck drivers are advised to keep a greater following distance than passenger car drivers are.

EXPLORE the BIG idea

Revisit "Popping Ping-Pong Balls" on p. 39. Have students explain their results.

Ongoing Assessment

CHECK YOUR READING *Answer: It can change direction.*

PHOTO CAPTION Answer: It will accelerate more because mass becomes less.

Practice the Math *Answers:*

1. *$m = F/a = 1\text{N}/3\text{ m/s}^2 = 1/3$ kg $= 0.3$ kg*
2. *$m = F/a = 10\text{N}/1\text{ m/s}^2 = 10$ kg*

DIFFERENTIATE INSTRUCTION

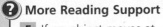

More Reading Support

F If an object moves at a constant speed, but it accelerates, what changes? *direction*

Inclusion Students with learning disabilities might have difficulty rearranging formulas to solve for all possible variables. For these students, list all possible forms of a formula, so that they can choose the one needed rather than having to derive it. For example, for the force formula, list $F = ma$, $a = \frac{F}{m}$, and $m = \frac{F}{a}$. Be sure students understand that all of these formulas present the same information.

INVESTIGATE Motion and Force

PURPOSE To hypothesize about the circular motion of a ball when the force on it changes

TIPS *15 min.*

- Have water and paper towels on hand for cleaning up paint.
- To anchor a lightweight plate, place an object in the center of the plate. Make sure that the object will not interfere with the motion of the ball.
- Revise any hypotheses that are not supported by experimental results, then repeat the experiment.

WHAT DO YOU THINK? *Students should determine if observations supported their hypotheses. The marble moved in a straight line in the direction it was moving when it left the plate; the marble was affected by friction.*

CHALLENGE *The marble will roll in the same direction, will leave the plate at a different speed, and will roll a different distance before it stops.*

 Datasheet, Motion and Force, p. 96

Technology Resources

Customize this student lab as needed or look for an alternative. Print rubrics to assess student lab reports.

 **Lab Generator CD-ROM**

INVESTIGATE Motion and Force

What affects circular motion?

PROCEDURE

1. Spread newspaper over your work surface. Place the paper plate down on the newspaper.
2. Practice rolling the marble around the edge of the plate until you can roll it around completely at least once.
3. Cut out a one-quarter slice of the paper plate. Put a dab of paint on the edge of the plate where the marble will leave it. Place the plate back down on the newspaper.
4. Hypothesize: How will the marble move once it rolls off the plate? Why?
5. Roll the marble all the way around the paper plate into the cut-away section and observe the resulting motion as shown by the trail of paint.

WHAT DO YOU THINK?

- Did your observations support your hypothesis?
- What forces affected the marble's motion after it left the plate?

CHALLENGE How will changing the speed at which you roll the marble change your results? Repeat the activity to test your prediction.

SKILL FOCUS
Hypothesizing

MATERIALS
- newspaper
- paper plate
- marble
- scissors
- poster paint
- paintbrush

TIME
15 minutes

Centripetal Force

When you were younger, you may have experimented with using force to change motion. Perhaps you and a friend took turns swinging each other in a circle. If you remember this game, you may also remember that your arms got tired because they were constantly pulling your friend as your friend spun around. It took force to change the direction of your friend's motion. Without that force, your friend could not have kept moving in a circle.

Any force that keeps an object moving in a circle is known as a **centripetal force** (sehn-TRIHP-ih-tuhl). This force points toward the center of the circle. Without the centripetal force, the object would go flying off in a straight line. When you whirl a ball on a string, what keeps the ball moving in a circle? The force of the string turns the ball, changing the ball's direction of motion. When the string turns, so does the ball. As the string changes direction, the force from the string also changes direction. The force is always pointing along the string toward your hand, the center of the circle. The centripetal force on the whirling ball is the pull from the string. If you let go of the string, the ball would fly off in the direction it was headed when you let go.

VOCABULARY
Remember to make a magnet word diagram for *centripetal force.*

G

CHECK YOUR READING How does centripetal force change the motion of an object?

Metacognitive Strategy

Ask students to discuss with a partner any additions or modifications they made to the procedure of the lab.

Teaching with Technology

If students have probeware, they may wish to use a motion detector for this activity.

Ongoing Assessment

CHECK YOUR READING *Answer: It consistently changes its direction.*

DIFFERENTIATE INSTRUCTION

More Reading Support

G In what direction does centripetal force point? *toward the center of the circle*

Alternative Assessment Have students prepare diagrams that show how centripetal force affects their daily lives. Each diagram should show motion in a circle and arrows indicating force. Examples include a car on a curved ramp and amusement park rides that spin.

Advanced Have students who are interested in learning how forces affect the motion of a boomerang read:

 Challenge Reading, pp. 120–121

centripetal force
top view
Centripetal force
The force that keeps the female skater moving in a circle is the pull exerted by her partner. The diagram shows the direction of the centripetal force.

Circular Motion and Newton's Second Law

Suppose the male skater shown above spins his partner faster. Her direction changes more quickly than before, so she accelerates more. To get more acceleration, he must apply more force. The same idea holds for a ball you whirl on a string. You have to pull harder on the string when you whirl the ball faster, because it takes more centripetal force to keep the ball moving at the greater speed.

You can apply the formula for Newton's second law even to an object moving in a circle. If you know the size of the centripetal force acting upon the object, you can find its acceleration. A greater acceleration requires a greater centripetal force. A more massive object requires a greater centripetal force to have the same circular speed as a less massive object. But no matter what the mass of an object is, if it moves in a circle, its force and acceleration are directed toward the center of the circle.

 **How does increasing the centripetal force on an object affect its acceleration?**

2.2 Review

KEY CONCEPTS

1. If the force acting upon an object is increased, what happens to the object's acceleration?
2. How does the mass of an object affect its acceleration?
3. What force keeps an object moving in a circle? In what direction does this force act?

CRITICAL THINKING

4. **Infer** Use Newton's second law to determine how much force is being applied to an object that is traveling at a constant velocity.
5. **Calculate** What force is needed to accelerate an object 5 m/s^2 if the object has a mass of 10 kg?

CHALLENGE

6. **Synthesize** Carlos pushes a 3 kg box with a force of 9 N. The force of friction on the box is 3 N in the opposite direction. What is the acceleration of the box? **Hint:** Combine forces to find the net force.

ANSWERS

1. Acceleration increases.

2. Assuming force is constant, the greater the mass, the less the acceleration.

3. centripetal force; toward the center of the circle

4. No net force is applied. If a force were applied, the object would change velocity, and thus change acceleration.

5. $F = ma = 10 \text{ kg} \cdot 5 \text{ m/s}^2 = 50 \text{ N}$

6. net force $= 9N - 3N = 6N$
$a = F/m = 6N/3 \text{ kg} = 2 \text{ m/s}^2$

Real World Example

If possible, borrow a centrifuge from a local medical facility to show students. Explain that a tube in a centrifuge can spin several thousand times a minute. The motion separates the contents of the tube according to their density. Ask students to predict where the parts of blood end up in the tube after it is placed in a centrifuge. The denser parts—red and white blood cells—move to the outer part of the circle. Plasma is in the inner part because it is less dense.

Ongoing Assessment

Describe how force works in circular motion.

Ask: What is the centripetal force that keeps the Moon in orbit around Earth? *gravity*

CHECK YOUR READING *Answer: Its acceleration increases.*

Reinforce the BIG idea

Have students relate the section to the Big Idea.

 Reinforcing Key Concepts, p. 97

2.2 ASSESS & RETEACH

Assess

 Section 2.2 Quiz, p. 23

Reteach

Tie a heavy cord about 1 m long to a bucket. Place an inch of water in the bucket. Swing it around in a circle. Ask students why the water remains in the bucket. *Centripetal force acts on the water and keeps it in the bucket.*

Technology Resources

Have students visit **ClassZone.com** for reteaching of Key Concepts.

 **CONTENT REVIEW**

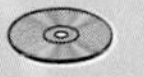 **CONTENT REVIEW CD-ROM**

MATH IN SCIENCE
Math Skills Practice for Science

Set Learning Goal

To analyze the number of significant figures in sets of data and in calculations using these data

Present the Science

Significant figures apply to measurements only, not to counting numbers. For example, 35 cm has two significant figures, but significant figures are not considered with a counted amount, such as 35 coins. Thus, the product of a measurement and a counted number has the same number of significant figures as the measurement factor. For example, the mass of 12 buttons, each of which has a mass of 5.43 g, would be 65.2 g.

Develop Skills in Estimation

Provide students with several measuring devices. Examples might be meter sticks or rulers, balances, and measuring cups or other graduated containers. Tell students that significant measurements can be made by estimating one place farther than the smallest measurement on the instrument. For example, if a meter stick is calibrated so that the smallest unit on it is centimeters, length can be measured to millimeters by estimating between the centimeter units. Have pairs of students use the instruments to measure several objects.

Close

Ask: A rectangular garden is 2.5 m by 4.23 m. If the area of the garden is found by multiplying these figures together, how many significant figures will be in the area? *two*

- Math Support, p. 127
- Math Practice, p. 128

Technology Resources

Students can visit **ClassZone.com** for practice working with rounding decimals.

MATH TUTORIAL

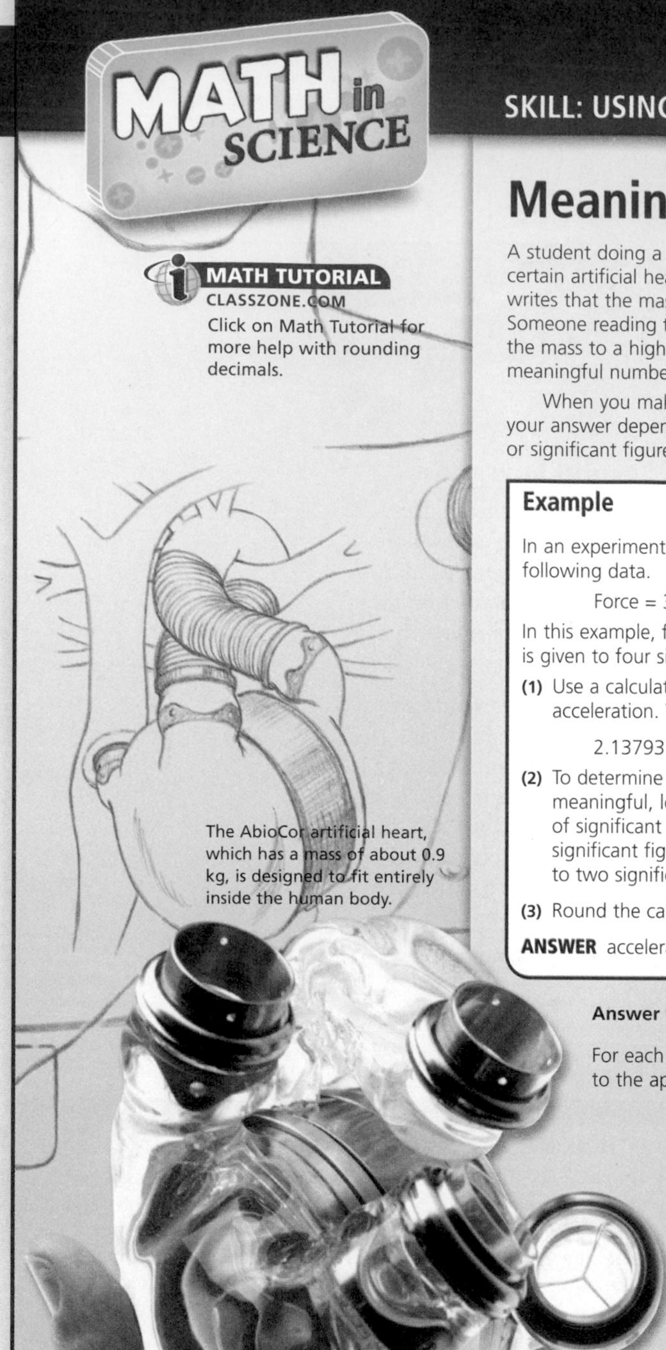

SKILL: USING SIGNIFICANT FIGURES

Meaningful Numbers

MATH TUTORIAL
CLASSZONE.COM
Click on Math Tutorial for more help with rounding decimals.

A student doing a science report on artificial hearts reads that a certain artificial heart weighs about 2 pounds. The student then writes that the mass of the artificial heart is 0.907185 kilograms. Someone reading this report might think that the student knows the mass to a high precision, when actually he knows it only to one meaningful number.

When you make calculations, the number of digits to include in your answer depends in part on the number of meaningful digits, or significant figures, in the numbers you are working with.

The AbioCor artificial heart, which has a mass of about 0.9 kg, is designed to fit entirely inside the human body.

Example

In an experiment to find acceleration, a scientist might record the following data.

Force = 3.1 N   mass = 1.450 kg

In this example, force is given to two significant figures, and mass is given to four significant figures.

(1) Use a calculator and the formula $a = F/m$ to find the acceleration. The display on the calculator shows

2.1379310345

(2) To determine how many of the digits in this answer are really meaningful, look at the measurement with the least number of significant figures. In this example, force is given to two significant figures. Therefore, the answer is meaningful only to two significant figures.

(3) Round the calculated number to two digits.

ANSWER acceleration = 2.1 m/s^2

Answer the following questions.

For each pair of measurements, calculate the acceleration to the appropriate number of digits.

1. Force = 3.100 N   mass = 3.1 kg

2. Force = 2 N   mass = 4.2 kg

3. Force = 1.21 N   mass = 1.1000 kg

CHALLENGE Suppose a scientist measures a force of 3.25 N and a mass of 3.3 kg. She could round the force to two significant figures and then divide, or she could divide and then round the answer. Compare these two methods. Which method do you think is more accurate?

ANSWERS

1. $a = F/m = 3.100$ N/3.1 kg = 1.000 N/kg = 1.0 m/s^2

2. $a = F/m = 2$ N/4.2 kg = 0.5 N/kg = 0.5 m/s^2

3. $a = F/m = 1.21$ N/1.1000 kg = 1.10 N/kg = 1.10 m/s^2

CHALLENGE *Using 3.25 N: $a = F/m = 3.25$ N/3.3 kg = 0.98 m/s^2. Using 3.3 N: $a = F/m = 3.3$ N/3.3 kg = 1.0 m/s^2. In general, the solution rounded off after dividing is more accurate.*

KEY CONCEPT

Forces act in pairs.

BEFORE, you learned

- A force is a push or a pull
- Increasing the force on an object increases the acceleration
- The acceleration of an object depends on its mass and the force applied to it

NOW, you will learn

- How Newton's third law relates action/reaction pairs of forces
- How Newton's laws work together

VOCABULARY

Newton's third law p. 57

THINK ABOUT

How do jellyfish move?

Jellyfish do not have much control over their movements. They drift with the current in the ocean. However, jellyfish do have some control over their up-and-down motion. By squeezing water out of its umbrella-like body, the jellyfish shown here applies a force in one direction to move in the opposite direction. If the water is forced downward, the jellyfish moves upward. How can a person or an object move in one direction by exerting a force in the opposite direction?

Newton's third law relates action and reaction forces.

COMBINATION NOTES
In your notebook, make an outline and draw a diagram about Newton's third law.

Newton made an important observation that explains the motion of the jellyfish. He noticed that forces always act in pairs. **Newton's third law** states that every time one object exerts a force on another object, the second object exerts a force that is equal in size and opposite in direction back on the first object. As the jellyfish contracts its body, it applies a downward force on the water. The water applies an equal force back on the jellyfish. It is this equal and opposite force on the jellyfish that pushes it up. This is similar to what happens when a blown-up balloon is released. The balloon pushes air out the end, and the air pushes back on the balloon and moves it forward.

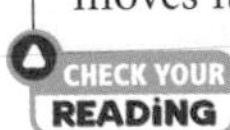

CHECK YOUR READING What moves the jellyfish through the water?

2.3 FOCUS

Set Learning Goals

Students will

- Explain how Newton's third law relates action/reaction pairs of forces.
- Describe how Newton's laws work together.
- Investigate how action and reaction forces compare.

3-Minute Warm-Up

Display Transparency 13 or copy this exercise on the board:

A worker pushes a 5-kilogram box with a force of 10 N. What is the acceleration of the box? Use the formula F = ma to find the answer. *a = F/m = 10 N/5 kg = 2 m/s²*

3-Minute Warm-Up, p. T13

2.3 MOTIVATE

THINK ABOUT

PURPOSE To understand action/reaction forces as a way of causing motion

DISCUSS Point out that the movement of the jellyfish is in the opposite direction of the force it exerts. Have students give parallel examples based on human movement, such as a strong sneeze forcing a person's head back.

Ongoing Assessment

CHECK YOUR READING *Answer: As the jellyfish applies a downward force on the water, the water applies an equal force on the jellyfish.*

RESOURCES FOR DIFFERENTIATED INSTRUCTION

Below Level

UNIT RESOURCE BOOK

- Reading Study Guide A, pp. 100–101
- Decoding Support, p. 124

AUDIO CDS

Advanced

UNIT RESOURCE BOOK

Challenge and Extension, p. 106

English Learners

UNIT RESOURCE BOOK

Spanish Reading Study Guide, pp. 104–105

AUDIO CDS

- Audio Readings in Spanish
- Audio Readings (English)

2.3 INSTRUCT

INVESTIGATE Newton's Third Law

PURPOSE To observe pairs of action and reaction forces

TIP *15 min.* Hold the scales securely and pull only hard enough to provide a reasonable reading.

WHAT DO YOU THINK? *Whether one partner pulled or both partners pulled, the force was the same on both spring scales. The scales show equal and opposite action and reaction.*

CHALLENGE *Sample answer: First law: fasten a spring scale to an object; pull on the object so that its speed or direction changes; the force needed for this motion can be read on the scale. Second law: fasten a spring scale to an object; pull the object by the scale, noting that, as force increases, acceleration increases.*

Datasheet, Newton's Third Law, p. 107

Technology Resources

Customize this student lab as needed or look for an alternative. Print rubrics to assess student lab reports.

Lab Generator CD-ROM

Metacognitive Strategy

Ask students to explain what they would do differently if they repeated the investigation.

Ongoing Assessment

CHECK YOUR READING *Answer: the rocket's engine pushes exhaust downward/exhaust pushes back on rocket; force of toe on table/force of table on toe; force of hand on table/force of table on hand*

Action and Reaction Pairs

A

The force that is exerted on an object and the force that the object exerts back are known together as an action/reaction force pair. One force in the pair is called the action force, and the other is called the reaction force. For instance, if the jellyfish pushing on the water is the action force, the water pushing back on the jellyfish is the reaction force. Likewise, if the balloon pushing the air backward is the action force, the air pushing the balloon forward is the reaction force.

You can see many examples of action and reaction forces in the world around you. Here are three:

- You may have watched the liftoffs of the space shuttle on television. When the booster rockets carrying the space shuttle take off, their engines push fuel exhaust downward. The exhaust pushes back on the rockets, sending them upward.
- When you bang your toe into the leg of a table, the same amount of force that you exert on the table is exerted back on your toe.
- Action and reaction forces do not always result in motion. For example, if you press down on a table, the table resists the push with the same amount of force, even though nothing moves.

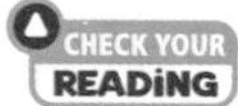

Identify the action/reaction forces in each example described above.

INVESTIGATE Newton's Third Law

How do action and reaction forces compare?

SKILL FOCUS
Observing

MATERIALS
2 spring scales

TIME
15 minutes

PROCEDURE

1. With a partner, hook the two spring scales together.
2. Pull gently on your spring scale while your partner holds but does not pull on the other scale.
3. Observe and record the amount of force that is shown on your scale and on your partner's scale.
4. Both of you pull together. Observe the force shown on each scale.

WHAT DO YOU THINK?

- What happened to your partner's force as your force increased?
- What happened when you both pulled?
- Explain why you think what you observed in each case happened.

CHALLENGE Can you think of a way to use the scales to show Newton's first or second law?

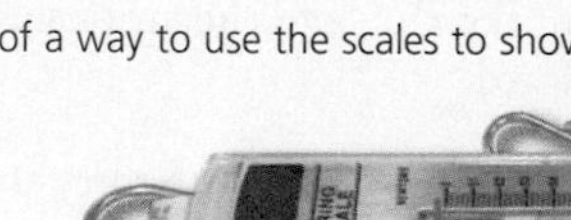
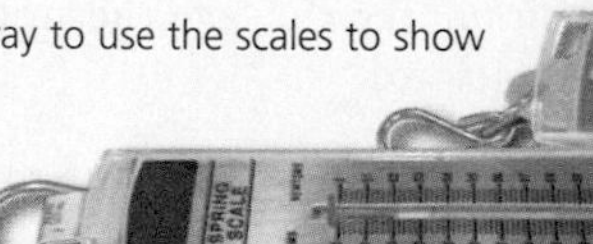

DIFFERENTIATE INSTRUCTION

More Reading Support

A If the action force is the force of scissors on paper, what is the reaction force? *the force of paper on scissors*

English Learners English learners might not recognize sentences beginning with *If* that do not also contain *then*. Point out the sentence on p. 58 that begins with *If*. Explain that *then* is implied and that a cause-and-effect relationship is at work.

Action and Reaction Forces Versus Balanced Forces

Because action and reaction forces are equal and opposite, they may be confused with balanced forces. Keep in mind that balanced forces act on a single object, while action and reaction forces act on different objects.

B **Balanced Forces** If you and a friend pull on opposite sides of a backpack with the same amount of force, the backpack doesn't move, because the forces acting on it are balanced. In this case, both forces are exerted on one object—the backpack.

C **Action and Reaction** As you drag a heavy backpack across a floor, you can feel the backpack pulling on you with an equal amount of force. The action force and the reaction force are acting on two different things—one is acting on the backpack, and the other is acting on you.

The illustration below summarizes Newton's third law. The girl exerts an action force on the boy by pushing him. Even though the boy is not trying to push the girl, an equal and opposite reaction force acts upon the girl, causing her to move as well.

Newton's Third Law

When one object exerts a force on another object, the second object exerts an equal and opposite force on the first object.

① **One Skater Pushes**

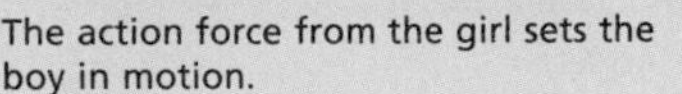

The action force from the girl sets the boy in motion.

② **Both Skaters Move**

Even though the boy does not do anything, the reaction force from him sets the girl in motion as well.

READING VISUALS How does the direction of the force on the girl relate to her motion?

Address Misconceptions

IDENTIFY Ask: If one object has two equal but opposite forces acting on it, what type of forces are they? If students answer that the forces are action/reaction pairs, they may hold the misconception that action/reaction forces act on a single object.

CORRECT Make two lists on the board. Label one "Balanced forces" and the other "Action/reaction." Ask students to give examples of force pairs. List each in the correct column after students agree whether or not the forces act on a single object.

REASSESS Ask: What type of force pair is present when someone steps from a boat onto a dock, and the boat moves away from the dock? Explain your answer. *An action/reaction force pair; the force on the boat is from the shoe, the force on the shoe is from the boat.*

Technology Resources

Visit **ClassZone.com** for background on common student misconceptions.

MISCONCEPTION DATABASE

EXPLORE the BIG idea

Revisit "Take Off!" on p. 39. Have students explain their results.

Ongoing Assessment

Explain how Newton's third law relates action/reaction pairs of forces.

Ask: If a reaction force is 5 N, what is the action force? *5 N*

READING VISUALS *Answer: The girl's motion is in the same direction as the force acting on her. This force is equal and opposite to the direction of the force she applied.*

DIFFERENTIATE INSTRUCTION

More Reading Support

B What type of force pair acts on only one object? *balanced*

C What type of force pair acts on different objects? *action/reaction*

Advanced

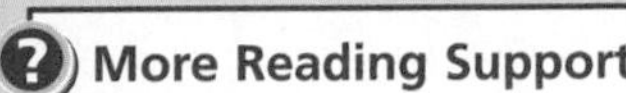

R Challenge and Extension, p. 106

Teach from Visuals

To help students understand the visual demonstrating Newton's three laws of motion, ask:

- Would a lighter animal, such as a cat, require as much force to jump as high? Explain your answer in terms of Newton's first law. *No; less mass means less inertia to overcome.*
- Would a dog be able to leap 30 ft, as a kangaroo can? Explain your answer in terms of Newton's third law. *No; a dog does not have legs as powerful as those of a kangaroo. A dog could not produce as much of an action force, so the reaction force would also be less.*

The visual "Newton's Three Laws of Motion" is available as T14 in the Unit Transparency Book.

Real World Example

Have students apply Newton's three laws of motion in comparing how they jump with how a kangaroo jumps. Gravity acts on the students as it does on the kangaroo, or their inertia would keep them rising upward after they jumped. Even though a student's mass is less than that of a kangaroo, a student's acceleration after jumping would be less because the force applied by the kangaroo is much greater. Because the action force that the student applies on the ground is much less than the force a kangaroo applies, the reaction force is also much less, so the kangaroo jumps much farther.

Ongoing Assessment

READING VISUALS *Answer: gravity on the kangaroo; force of the kangaroo on the ground; force of the ground on the kangaroo*

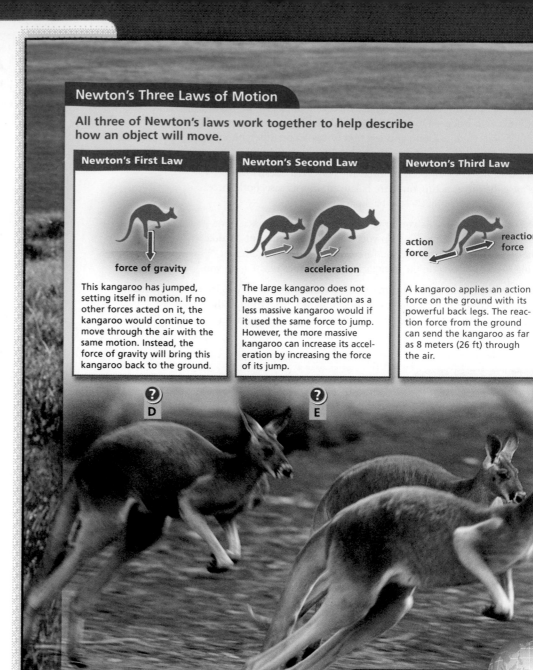

Newton's Three Laws of Motion

All three of Newton's laws work together to help describe how an object will move.

Newton's First Law

force of gravity

This kangaroo has jumped, setting itself in motion. If no other forces acted on it, the kangaroo would continue to move through the air with the same motion. Instead, the force of gravity will bring this kangaroo back to the ground.

Newton's Second Law

acceleration

The large kangaroo does not have as much acceleration as a less massive kangaroo would if it used the same force to jump. However, the more massive kangaroo can increase its acceleration by increasing the force of its jump.

Newton's Third Law

action force

reaction force

A kangaroo applies an action force on the ground with its powerful back legs. The reaction force from the ground can send the kangaroo as far as 8 meters (26 ft) through the air.

D

E

Common Name: Red kangaroo
Scientific Name: *Macropus rufus*
Home: Australia
Top Speed: 65 km/h (40 mi/h)
Maximum Leap: 8 m (26 ft)

AUSTRALIA

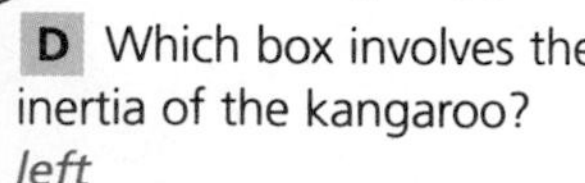

DIFFERENTIATE INSTRUCTION

More Reading Support

D Which box involves the inertia of the kangaroo? *left*

E Which box relates force, mass, and acceleration? *middle*

Newton's three laws describe and predict motion.

Newton's three laws can explain the motion of almost any object, including the motion of animals. The illustrations on page 60 show how all three of Newton's laws can be used to describe how kangaroos move. The three laws are not independent of one another; they are used together to explain the motion of objects.

RESOURCE CENTER CLASSZONE.COM

Find out more about Newton's laws of motion.

You can use the laws of motion to explain how other animals move as well. For example, Newton's laws explain why a squid moves forward while squirting water out behind it. These laws also explain that a bird is exerting force when it speeds up to fly away or when it changes its direction in the air.

You can also use Newton's laws to make predictions about motion. If you know the force acting upon an object, then you can predict how that object's motion will change. For example, if you want to send a spacecraft to Mars, you must be able to predict exactly where Mars will be by the time the spacecraft reaches it. You must also be able to control the force on your spacecraft so that it will arrive at the right place at the right time.

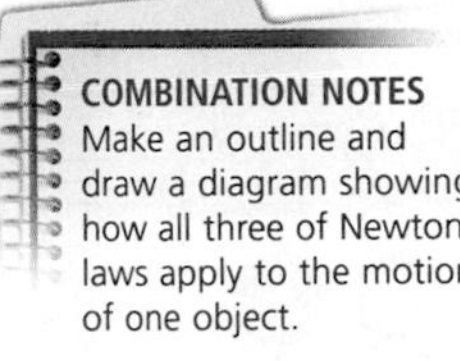

COMBINATION NOTES

Make an outline and draw a diagram showing how all three of Newton's laws apply to the motion of one object.

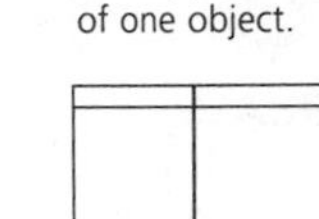

Knowing how Newton's three laws work together can also help you win a canoe race. In order to start the canoe moving, you need to apply a force to overcome its inertia. Newton's second law might affect your choice of canoes, because a less massive canoe is easier to accelerate than a more massive one. You can also predict the best position for your paddle in the water. If you want to move straight ahead, you push backward on the paddle so that the canoe moves forward. Together, Newton's laws can help you explain and predict how the canoe, or any object, will move.

2.3 Review

KEY CONCEPTS

1. Identify the action/reaction force pair involved when you catch a ball.
2. Explain the difference between balanced forces and action/reaction forces.
3. How do Newton's laws of motion apply to the motion of an animal, such as a cat that is running?

CRITICAL THINKING

4. **Apply** A man pushes on a wall with a force of 50 N. What are the size and the direction of the force that the wall exerts on the man?
5. **Evaluate** Jim will not help push a heavy box. He says, "My force will produce an opposite force and cancel my effort." Evaluate Jim's statement.

CHALLENGE

6. **Calculate** Suppose you are holding a basketball while standing still on a skateboard. You and the skateboard have a mass of 50 kg. You throw the basketball with a force of 10 N. What is your acceleration before and after you throw the ball?

ANSWERS

1. force of the ball on your hand, force of your hand on the ball

2. Balanced forces act on one object, and each cancels the effect of the other. Action and reaction forces act on two different objects.

3. force exerted by cat overcomes cat's inertia; force cat needs to accelerate depends on its mass; when running, cat applies backward force on ground, ground applies forward force on cat

4. 50 N in direction opposite to his force on the wall

5. It is incorrect; the reaction force doesn't act on the box, it acts on Jim.

6. before: no change in velocity = 0 m/s²; after: a = F/m = 10 N/50 kg = 0.2 m/s²

Ongoing Assessment

Describe how Newton's laws work together.

Ask: Use Newton's three laws to explain how you take a step. *Sample answer: Force is needed to overcome your inertia. Applying force accelerates your foot. The action force of the foot on the ground equals the reaction force of the ground on the foot.*

Reinforce  the BIG idea

Have students relate the section to the Big Idea.

 Reinforcing Key Concepts, p. 108

2.3 ASSESS & RETEACH

Assess

 Section 2.3 Quiz, p. 24

Reteach

Stand on a skateboard or wear a pair of roller skates. Have students hypothesize how you could move the skateboard without touching any part of the room or any student. *Sample answer: Throw an object in the direction opposite the direction you want to move.*

Technology Resources

Have students visit **ClassZone.com** for reteaching of Key Concepts.

 CONTENT REVIEW

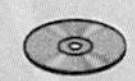 CONTENT REVIEW CD-ROM

CHAPTER INVESTIGATION

Focus

PURPOSE To learn how Newton's laws can be applied to make rockets fly farther

OVERVIEW Students create a straw bottle rocket and modify it to test how the modification affects flight. Students will find that the distance of flight increases

- as the size of the bottle increases;
- as the mass of the straws decreases;
- as the amount of clay decreases; and
- as the size of the paper loops increases.

Lab Preparation

- Have students cut strips in advance to allow more time to investigate.
- Have extra plastic bottles in a variety of sizes on hand in case cracks develop from squeezing.
- Prior to the investigation, have students read through the investigation and prepare their data tables. Or you may wish to copy and distribute datasheets and rubrics.

UNIT RESOURCE BOOK, pp. 129–137

SCIENCE TOOLKIT, F14

Lab Management

- Have different groups of students try different modifications, so the class can address all the ways in which a rocket's flight could be affected.
- Students can work in pairs and take turns launching the rocket and recording the distance it flies. Remind students to exert the same force in order to achieve consistent results.
- Remind students to measure the distance the rocket flies consistently. For example, always measure from the launching point (tip of the straw) to the place where the rocket lands (tip of the straw).

SAFETY Emphasize that students should always point projectiles away from other students.

CHAPTER INVESTIGATION

Newton's Laws of Motion

OVERVIEW AND PURPOSE As you know, rocket engineers consider Newton's laws when designing rockets and planning rocket flights. In this investigation you will use what you have learned about Newton's laws to

- build a straw rocket
- improve the rocket's performance by modifying one design element

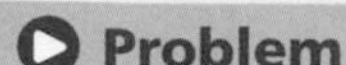

Problem

What aspects of a model rocket affect the distance it flies?

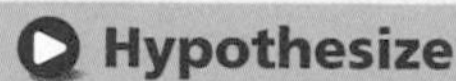

Hypothesize

After step 8 in the procedure, write a hypothesis to explain what you predict will happen during the second set of trials. Your hypothesis should take the form of an "If . . . , then . . . , because . . ." statement.

MATERIALS
- 2 straws with different diameters
- several plastic bottles, in different sizes
- modeling clay
- scissors
- construction paper
- meter stick
- tape

Procedure

1. Make a data table like the one shown on the sample notebook page.

2. Insert the straw with the smaller diameter into one of the bottles. Seal the mouth of the bottle tightly with modeling clay so that air can escape only through the straw. This is the rocket launcher.

straw
clay

3. Cut two thin strips of paper, one about 8 cm long and the other about 12 cm long. Connect the ends of the strips to make loops.

4. To create the rocket, place the straw with the larger diameter through the smaller loop and tape the loop to the straw at one end. Attach the other loop to the other end of the straw in the same way. Both loops should be attached to the same side of the straw to stabilize your rocket in flight.

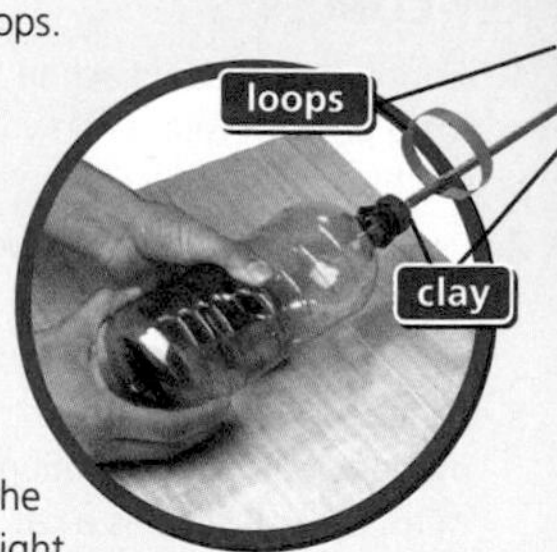

INVESTIGATION RESOURCES

CHAPTER INVESTIGATION, Newton's Laws of Motion
- Level A, pp. 129–132
- Level B, pp. 133–136
- Level C, p. 137

Advanced students should complete Levels B & C.

Writing a Lab Report, D12–13

Technology Resources

Customize this student lab as needed or look for an alternative. Print rubrics to assess student lab reports.

Lab Generator CD-ROM

5 Use a small ball of modeling clay to seal the end of the straw near the smaller loop.

6 Slide the open end of the rocket over the straw on the launcher. Place the bottle on the edge of a table so that the rocket is pointing away from the table.

7 Test launch your rocket by holding the bottle with two hands and squeezing it quickly. Measure the distance the rocket lands from the edge of the table. Practice the launch several times. Remember to squeeze with equal force each time.

8 Launch the rocket four times. Keep the amount of force you use constant. Measure the distance the rocket travels each time, and record the results in your data table.

9 List all the variables that may affect the distance your rocket flies. Change the rocket or launcher to alter one variable. Launch the rocket and measure the distance it flies. Repeat three more times, and record the results in your data table.

Observe and Analyze

Write It Up

1. **RECORD OBSERVATIONS** Draw a diagram of both of your bottle rockets. Make sure your data table is complete.
2. **IDENTIFY VARIABLES** What variables did you identify, and what variable did you modify?

Conclude

1. **COMPARE** How did the flight distances of the original rocket compare with those of the modified rocket?
2. **ANALYZE** Compare your results with your hypothesis. Do the results support your hypothesis?
3. **IDENTIFY LIMITS** What possible limitations or errors did you experience or could you have experienced?
4. **APPLY** Use Newton's laws to explain why the rocket flies.
5. **APPLY** What other real-life example can you think of that demonstrates Newton's laws?

INVESTIGATE Further

CHALLENGE Why does the rocket have paper loops taped to it? Determine how the flight of the rocket is affected if one or both loops are completely removed. Hypothesize about the function of the paper loops and design an experiment to test your hypothesis.

Newton's Laws of Motion

Problem What aspects of a model rocket affect the distance it flies?

Hypothesize

Observe and Analyze

Table 1. Flight Distances of Original and Modified Rocket

	Original Rocket	Modified Rocket
Trial Number	Distance Rocket Flew (cm)	Distance Rocket Flew (cm)
1		
2		
3		
4		

Conclude

Observe and Analyze

SAMPLE DATA Original rocket: 200 cm, 175 cm, 185 cm, 195 cm; modified rocket with larger bottle: 250 cm, 225 cm, 230 cm, 240 cm

1. *See students' diagrams.*
2. *Variables include the size of the bottle, the length and diameter of the straws, the amount of modeling clay, the size of the loops, and the distance between the loops.*

Conclude

1. *Student answers will vary.*
2. *Student answers will vary, depending on students' results and original hypotheses.*
3. *Possible limits and errors include inconsistent force, inconsistent measurements, and changing more than one variable.*
4. *Newton's first law: The unbalanced force of the air pushing on the straw changes the straw's motion. Newton's second law: The acceleration of the straw should be greater with a larger bottle because it produces more force; a straw with smaller mass or less clay accelerates more easily than a straw with greater mass or more clay. Newton's third law: As the straw forces air backward, air forces the straw forward.*
5. *Sample answer: rowing a boat or using a bow and arrow*

INVESTIGATE Further

CHALLENGE The loops stabilize the rocket. An experiment might include adding additional loops and then removing the loops one at a time and observing the results.

Post-Lab Discussion

- On the board, list the different modifications made. Have groups of students write whether each modification made the rocket fly farther, the same distance, or a shorter distance. Keep a tally from period to period to show that larger data samples show trends more accurately.
- Ask students to work in groups and have each group come to a consensus on the questions. Then have the class as a whole come to a consensus.

2.4 FOCUS

Set Learning Goals

Students will

- Calculate momentum.
- Explain how momentum is affected by collisions.
- Investigate what happens to momentum when objects collide.

3-Minute Warm-Up

Display Transparency 13 or copy this exercise on the board:

Match each definition with the correct term.

Definitions

1. resistance of an object to a change in its speed or direction *b*
2. a force that is exerted on an object and the force that the object exerts back *c*
3. a push or a pull *d*

Terms

a. acceleration
b. inertia
c. action and reaction force pair
d. force

 3-Minute Warm-Up, p. T13

2.4 MOTIVATE

EXPLORE Collisions

PURPOSE To observe that energy and momentum are transferred during a collision

TIP *10 min.* For easily observed results, use balls that vary considerably in mass.

WHAT DO YOU THINK? *Greater speeds before the collision yielded greater speeds and more motion after the collision; a ball at rest moved after the collision, and the other ball moved more slowly; motion after the collision depended on the mass of the balls.*

KEY CONCEPT

2.4 Forces transfer momentum.

BEFORE, you learned

- A force is a push or a pull
- Newton's laws help to describe and predict motion

NOW, you will learn

- What momentum is
- How to calculate momentum
- How momentum is affected by collisions

VOCABULARY

momentum p. 64
collision p. 66
conservation of momentum p. 67

EXPLORE Collisions

What happens when objects collide?

PROCEDURE

1. Roll the two balls toward each other on a flat surface. Try to roll them at the same speed. Observe what happens. Experiment by changing the speeds of the two balls.
2. Leave one ball at rest, and roll the other ball so that it hits the first ball. Observe what happens. Then repeat the experiment with the balls switched.

MATERIALS

2 balls of different masses

WHAT DO YOU THINK?

- How did varying the speed of the balls affect the motion of the balls after the collision?
- What happened when one ball was at rest? Why did switching the two balls affect the outcome?

Objects in motion have momentum.

If you throw a tennis ball at a wall, it will bounce back toward you. What would happen if you could throw a wrecking ball at the wall at the same speed that you threw the tennis ball? The wall would most likely break apart. Why would a wrecking ball have a different effect on the wall than the tennis ball?

A moving object has a property that is called momentum. **Momentum** (moh-MEHN-tuhm) is a measure of mass in motion; the momentum of an object is the product of its mass and its velocity. At the same velocity, the wrecking ball has more momentum than the tennis ball because the wrecking ball has more mass. However, you could increase the momentum of the tennis ball by throwing it faster.

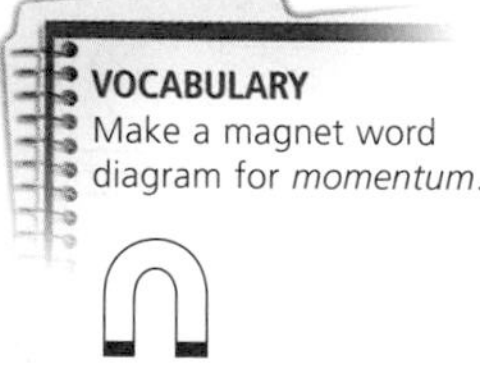

VOCABULARY
Make a magnet word diagram for *momentum*.

RESOURCES FOR DIFFERENTIATED INSTRUCTION

Below Level

UNIT RESOURCE BOOK

- Reading Study Guide A, pp. 111–112
- Decoding Support, p. 124

Advanced

UNIT RESOURCE BOOK

Challenge and Extension, p. 117

English Learners

UNIT RESOURCE BOOK

Spanish Reading Study Guide, pp. 115–116

- Audio Readings in Spanish
- Audio Readings (English)

Momentum is similar to inertia. Like inertia, the momentum of an object depends on its mass. Unlike inertia, however, momentum takes into account how fast the object is moving. A wrecking ball that is moving very slowly, for example, has less momentum than a fast-moving wrecking ball. With less momentum, the slower-moving wrecking ball would not be able to do as much damage to the wall.

REMINDER
Inertia is the resistance of an object to changes in its motion.

To calculate an object's momentum, you can use the following formula:

$$\text{momentum} = \text{mass} \cdot \text{velocity}$$
$$p = mv$$

In this equation, p stands for momentum, m for mass, and v for velocity. In standard units, the mass of an object is given in kilograms (kg), and velocity is given in meters per second (m/s). Therefore, the unit of momentum is the kilogram-meter per second (kg • m/s). Notice that the unit of momentum combines mass, length, and time.

Explore momentum.

Like force, velocity, and acceleration, momentum is a vector—it has both a size and a direction. The direction of an object's momentum is the same as the direction of its velocity. You can use speed instead of velocity in the formula as long as you do not need to know the direction of motion. As you will read later, it is important to know the direction of the momentum when you are working with more than one object.

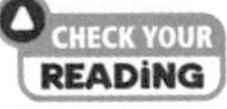
How do an object's mass and velocity affect its momentum?

Calculating Momentum

Sample Problem

What is the momentum of a 1.5 kg ball moving at 2 m/s?

What do you know? mass = 1.5 kg, velocity = 2 m/s

What do you want to find out? momentum

Write the formula: $p = mv$

Substitute into the formula: p = 1.5 kg • 2 m/s

Calculate and simplify: p = 3 kg • m/s

Check that your units agree: Unit is kg • m/s.
Unit of momentum is kg • m/s. Units agree.

Answer: p = 3 kg • m/s

1. A 3 kg ball is moving with a velocity of 1 m/s. What is the ball's momentum?
2. What is the momentum of a 0.5 kg ball moving 0.5 m/s?

DIFFERENTIATE INSTRUCTION

More Reading Support

A Compare and contrast the inertia and the momentum of an object. *Both depend on mass, but momentum depends on velocity, while inertia does not.*

English Learners English learners often employ memorization as a tool in learning new concepts. When using definitions, formulas, or laws, be sure the language is always consistent. This will help students remember key ideas and important vocabulary. Remind students of the difference between *affect* and *effect*. *Affect* is a verb, as in, "the speed of the balls affects the motion of the balls" (in the Explore on p. 64). *Effect* is a noun, as in "a different effect."

2.4 INSTRUCT

History of Science

René Descartes (1596–1650), a French mathematician and philosopher, first used the term *momentum* to mean "the amount of motion."

- Through experimental observations, Descartes discovered that this amount of motion could be calculated by multiplying the mass of an object and its speed.
- Descartes thought that momentum is always positive. His theory could not explain how momentum is conserved when two objects with equal momentum moving from opposite directions collide and stop moving. The Dutch scientist Christiaan Huygens (1629–1695) solved this problem by stating that the "amount of motion" can be either positive or negative when two objects are moving relative to each other. If an object is moving in the opposite direction, the "amount of motion," or momentum, is considered to be negative.

Ongoing Assessment

Describe how forces act in collisions.

Ask: In terms of forces, what happens when a tennis ball collides with a racket? *The ball and the racket exert equal and opposite forces on each other.*

Define and calculate momentum.

Ask: What is the momentum of a 2 kg rock moving at 3 m/s? *p = mv = 2 kg · 3 m/s = 6 kg m/s*

CHECK YOUR READING *Answer: An increase in either mass or velocity increases momentum; a decrease in either mass or velocity decreases momentum.*

Practice the Math *Answers:*

1. *p = mv = 3 kg · 1 m/s = 3 kg m/s*
2. *p = mv = 0.5 kg · 0.5 m/s = 0.25 kg m/s*

INVESTIGATE Momentum

PURPOSE To observe what happens when objects collide

TIPS *20 min.* Allow students a few minutes to explore, then suggest the following:

- If the marbles do not roll easily, try another ruler or use the groove in the ruler.
- The setup should be far enough from the edge of the table that marbles do not roll onto the floor.
- Make sure you work on a flat, level table.

WHAT DO YOU THINK? *If one marble collides with the row of marbles, one marble moves away from the row. Equal numbers of marbles move for two and three marbles, also. Momentum transfers to the last marbles in the line.*

CHALLENGE *Sample hypothesis: The number of marbles that collide with the stationary marbles equals the number of marbles that move from the end of the row because the marbles are the same mass. Designs might include a similar experiment using different sizes of marbles.*

 Datasheet, Momentum, p. 118

Technology Resources

Customize this student lab as needed or look for an alternative. Print rubrics to assess student lab reports.

 **Lab Generator CD-ROM**

INVESTIGATE Momentum

What happens when objects collide?

PROCEDURE

1. Set up two parallel rulers separated by one centimeter. Place a line of five marbles, each touching the next, in the groove between the rulers.
2. Roll a marble down the groove so that it collides with the line of marbles, and observe the results.
3. Repeat your experiment by rolling two and then three marbles at the line of marbles. Observe the results.

WHAT DO YOU THINK?

- What did you observe when you rolled the marbles?
- Why do you think the marbles moved the way they did?

CHALLENGE Use your answers to write a hypothesis explaining your observations. Design your own marble experiment to test this hypothesis. Do your results support your hypothesis?

SKILL FOCUS
Observing

MATERIALS
- 2 rulers
- 8 marbles

TIME
20 minutes

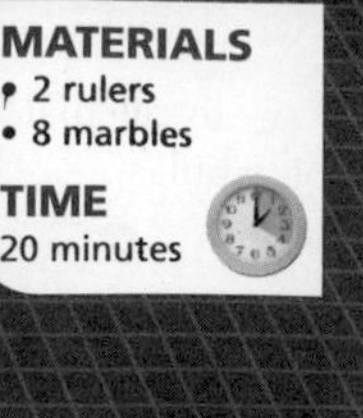

Momentum can be transferred from one object to another.

If you have ever ridden in a bumper car, you have experienced collisions. A **collision** is a situation in which two objects in close contact exchange energy and momentum. As another car bumps into the back of yours, the force pushes your car forward. Some of the momentum of the car behind you is transferred to your car. At the same time, the car behind you slows because of the reaction force from your car. You gain momentum from the collision, and the other car loses momentum. The action and reaction forces in collisions are one way in which objects transfer momentum.

If two objects involved in a collision have very different masses, the one with less mass has a greater change in velocity. For example, consider what happens if you roll a tennis ball and a bowling ball toward each other so that they collide. Not only will the speed of the tennis ball change, but the direction of its motion will change as it bounces back. The bowling ball, however, will simply slow down. Even though the forces acting on the two balls are the same, the tennis ball will be accelerated more during the collision because it has less mass.

 How can a collision affect the momentum of an object?

Metacognitive Strategy

Ask students to explain what in the procedure or in the results surprised them about the investigation.

Ongoing Assessment

CHECK YOUR READING *Answer: When two objects collide, they exchange energy and transfer momentum.*

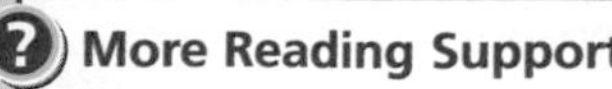

DIFFERENTIATE INSTRUCTION

More Reading Support

B If a bowling ball and a tennis ball collide, which ball will have the greater change in velocity? *the tennis ball*

Below Level Have students model transferring momentum. Have several student volunteers stand in a line with each student facing the back of the student in front of him. Each student places his right hand on the right shoulder of the person in front of him with his elbow locked in place. Have the student in the back of the line gently push forward on the shoulder of the student in front of him, and observe how far along the line the push could be felt. Repeat the activity with different amounts of force to the push. Caution students to push gently.

Momentum is conserved.

During a collision between two objects, each object exerts a force on the other. The colliding objects make up a system—a collection of objects that affect one another. As the two objects collide, the velocity and the momentum of each object change. However, as no other forces are acting on the objects, the total momentum of both objects is unchanged by the collision. This is due to the conservation of momentum. The principle of **conservation of momentum** states that the total momentum of a system of objects does not change, as long as no outside forces are acting on that system.

READING TIP

A light blue-green arrow shows the momentum of an individual object.

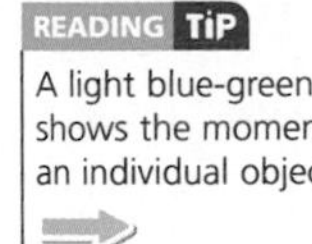

A dark blue-green arrow shows the total momentum.

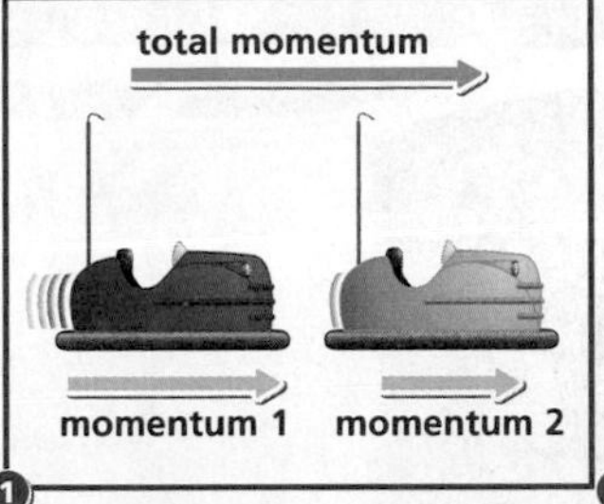

1 **Before the collision** The momentum of the first car is greater than the momentum of the second car. Their combined momentum is the total momentum of the system.

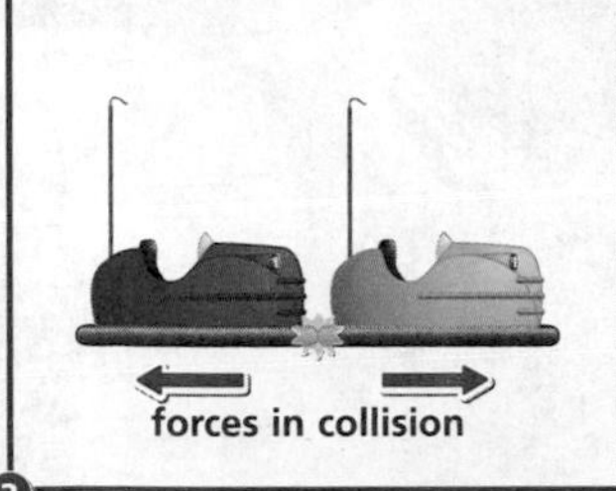

2 **During the collision** The forces on the two cars are equal and opposite, as described by Newton's third law. Momentum is transferred from one car to the other during the collision.

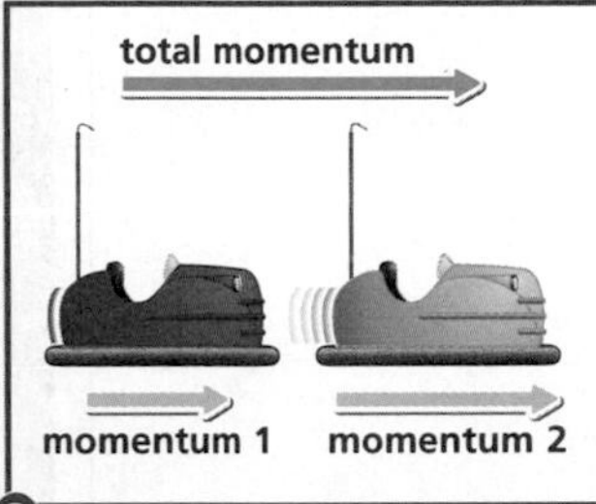

3 **After the collision** The momentum lost by one car was gained by the other car. The total momentum of the system remains the same as it was before the collision.

How much an object's momentum changes when a force is applied depends on the size of the force and how long that force is applied. Remember Newton's third law—during a collision, two objects are acted upon by equal and opposite forces for the same length of time. This means that the objects receive equal and opposite changes in momentum, and the total momentum does not change.

You can find the total momentum of a system of objects before a collision by combining the momenta of the objects. Because momentum is a vector, like force, the direction of motion is important. To find the total momentum of objects moving in the same direction, add the momenta of the objects. For two objects traveling in opposite directions, subtract one momentum from the other. Then use the principle of conservation of momentum and the equation for momentum to predict how the objects will move after they collide.

READING TIP

The plural of *momentum* is *momenta*.

CHECK YOUR READING What is meant by "conservation of momentum"? What questions do you have about the application of this principle?

Address Misconceptions

IDENTIFY Ask: Is there any difference between force and momentum? If students answer no, they hold the misconception that momentum is the same thing as force.

CORRECT Move a toy car through the air to demonstrate that momentum is not the same as force. Ask if the car applies any forces. Students may suggest a force on the air or your hand. Discuss the force exerted on surrounding air particles. Ask if the car would apply any force if it were moving by itself in empty space where there is nothing for the car to hit. Help students realize that the moving car applies a force only when it interacts with another object, such as an air particle; it does not apply a force simply because it is moving. However, the car will have momentum because of its motion.

REASSESS Ask: What measurements are necessary to calculate force? to calculate momentum? *mass and acceleration; mass and velocity*

Technology Resources

Visit **ClassZone.com** for background on common student misconceptions.

MISCONCEPTION DATABASE

Develop Critical Thinking

APPLY Ask: What is the overall momentum of a system made up of a 1 kg ball moving south at 2 m/s and a 1.5 kg ball moving north at the same speed? *1 kg · 2 m/s = 2 kg · m/s south; 1.5 kg · 2 m/s = 3 kg m/s north; opposite, so subtract: 3 kg · m/s − 2 kg m/s = 1 kg m/s north*

Ongoing Assessment

Explain how momentum is affected by collisions.

Ask: What happens to the momentum of a bowling ball when it strikes the pins? *some momentum transfers to pins, but total momentum remains the same*

CHECK YOUR READING *Answer: The amount of momentum a system of objects has does not change as long as there are no outside forces acting on the system.*

DIFFERENTIATE INSTRUCTION

More Reading Support

C If the momentum of a system before a collision is 5 kg m/s, what is the momentum of the system after the collision? *5 kg m/s*

Advanced Have students relate common uses of the word *conservation.* Examples of uses of this term include environmental conservation; conservation of works of art; and other scientific uses, such as conservation of mass or energy. In each case, what is conserved is not wasted or used up.

R Challenge and Extension, p. 117

Teach from Visuals

To help students understand the visual of a crash test, ask:

- In addition to the bending metal, what else uses some of the energy in the collision? *Sample answer: thermal energy caused by friction between parts of the cars or between the cars and the pavement*
- What role does friction between the pavement and the cars play in the conservation of momentum in this crash? *Because friction acts as an outside force on the cars and slows their velocity, it reduces the amount of momentum the two cars have after the crash.*

Real World Example

In most situations students are familiar with, momentum does not appear to be conserved. Outside forces such as friction transfer momentum to Earth or other objects that do not move noticeably due to their large mass. In some systems, such as balls in a game of pool, this loss of momentum is not very noticeable. If a pool ball hits a ball at rest, the initial forward momentum is shared by the two balls after the collision. Momentum is conserved not only in the forward direction, but also in the sideways direction. The two will move to opposite sides of the table. Ultimately, the momentum of each ball is completely transferred to the table by friction, and the balls come to a stop.

Ongoing Assessment

CHECK YOUR READING *Answer: In the two types of collisions, movement is different but momentum is conserved.*

Two Types of Collisions

When bumper cars collide, they bounce off each other. Most of the force goes into changing the motion of the cars. The two bumper cars travel separately after the collision, just as they did before the collision. The combined momentum of both cars after the collision is the same as the combined momentum of both cars before the collision.

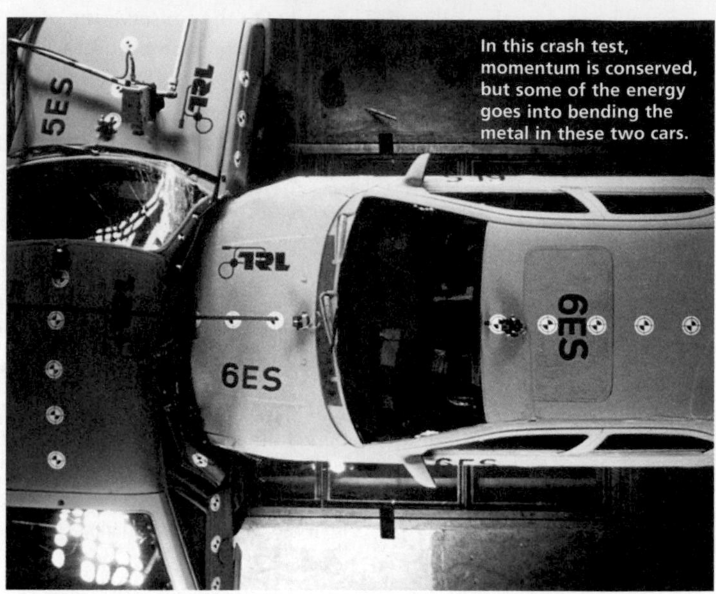

In this crash test, momentum is conserved, but some of the energy goes into bending the metal in these two cars.

When two cars collide during a crash test, momentum is also conserved during the collision. Unlike the bumper cars, however, which separate, the two cars shown in the photograph above stick and move together after the collision. Even in this case, the total momentum of both cars together is the same as the total momentum of both cars before the collision. Before the crash shown in the photograph, the yellow car had a certain momentum, and the blue car had no momentum. After the crash, the two cars move together with a combined momentum equal to the momentum the yellow car had before the collision.

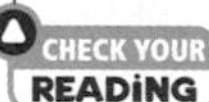

CHECK YOUR READING Compare collisions in which objects separate with collisions in which objects stick together.

Momentum and Newton's Third Law

Collisions are not the only events in which momentum is conserved. In fact, momentum is conserved whenever the only forces acting on objects are action/reaction force pairs. Conservation of momentum is really just another way of looking at Newton's third law.

DIFFERENTIATE INSTRUCTION

More Reading Support

D Is momentum conserved in a collision where the objects stick together? *yes*

Below Level Have students working alone or in pairs use the note-taking strategy explained on p. 40 to summarize what they know about momentum. Remind them that the heads in the student text can be part of their outlines. Diagrams should include arrows to represent momentum before and after collisions.

When a firefighter turns on a hose, water comes out of the nozzle in one direction, and the hose moves back in the opposite direction. You can explain why by using Newton's third law. The water is forced out of the hose. A reaction force pushes the hose backward. You can also use the principle of conservation of momentum to explain why the hose moves backward:

- Before the firefighter turns on the water, the hose and the water are not in motion, so the hose/water system has no momentum.
- Once the water is turned on, the water has momentum in the forward direction.
- For the total momentum of the hose and the water to stay the same, the hose must have an equal amount of momentum in the opposite direction. The hose moves backward.

Firefighters must apply a force to the water hose to prevent it from flying backward when the water comes out.

If the hose and the water are not acted on by any other forces, momentum is conserved. Water is pushed forward, and the hose is pushed backward. However, the action and reaction force pair acting on the hose and the water are not usually the only forces acting on the hose/water system, as shown in the photograph above. There the firefighters are holding the hose steady.

The force the firefighters apply is called an outside force because it is not being applied by the hose or the water. When there is an outside force on a system, momentum is not conserved. Because the firefighters hold the hose, the hose does not move backward, even though the water has a forward momentum.

CHECK YOUR READING Under what condition is momentum not conserved? What part of the paragraph above tells you?

2.4 Review

KEY CONCEPTS

1. How does increasing the speed of an object change its momentum?
2. A car and a truck are traveling at the same speed. Which has more momentum? Why?
3. Give two examples showing the conservation of momentum. Give one example where momentum is not conserved.

CRITICAL THINKING

4. **Predict** A performing dolphin speeds through the water and hits a rubber ball originally at rest. Describe what happens to the velocities of the dolphin and the ball.
5. **Calculate** A 50 kg person is running at 2 m/s. What is the person's momentum?

CHALLENGE

6. **Apply** A moving train car bumps into another train car with the same mass. After the collision, the two cars are coupled and move off together. How does the final speed of the two train cars compare with the initial speed of the moving train cars before the collision?

ANSWERS

1. Momentum increases.

2. The truck has more momentum because it has more mass.

3. Sample answer: Momentum is conserved when a bat hits a ball or when two cars collide. Momentum is not conserved when an inflated balloon releases air but is held in place.

4. The dolphin loses momentum and velocity; the ball gains momentum and velocity.

5. $p = mv = 50\ kg \cdot 2\ m/s = 100\ kg \cdot m/s$

6. The total momentum is conserved. The mass of the two cars together is twice the mass of the original moving car, so the velocity will be half the original velocity.

Ongoing Assessment

CHECK YOUR READING *Answer: According to the second sentence, momentum is not conserved if outside forces act on the system.*

Reinforce the BIG idea

Have students relate the section to the Big Idea.

 Reinforcing Key Concepts, p. 119

2.4 ASSESS & RETEACH

Assess

 Section 2.4 Quiz, p. 25

Reteach

Use this demonstration to emphasize the importance of outside forces on the conservation of momentum. Place a long piece of waxed paper on a desk. Roll two tennis balls at the same speed so that they collide on the waxed paper. Have students observe that the speed of the balls after the collision is nearly the same as their speed before the collision. Roll the balls again, so that they collide on sandpaper or a piece of textured fabric. The resulting speed will be less because of increased friction.

Technology Resources

Have students visit **ClassZone.com** for reteaching of Key Concepts.

 **CONTENT REVIEW**

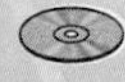 **CONTENT REVIEW CD-ROM**

BACK TO

the BIG idea

Have students explain why each of the following words in the Big Idea statement is important: *change, motion,* and *predictable. Answer: Change implies that a force alters motion; changing motion is an important result of forces acting on an object; predictable indicates that forces always change motion in the same way if under the same conditions.*

KEY CONCEPTS SUMMARY

SECTION 2.1

Ask: In the picture on the left, in which direction will the ball accelerate? *to the right*

Ask: In the second picture, what would you need to do to stop the blue box from accelerating? *Place a force on the box that pushes to the right and is equal to the force pushing the box to the left.*

SECTION 2.2

Ask: If the box has a mass of 20 kg and a force of 10 N acting on it, what is the box's acceleration? *$a = F/m = 10\ N/20\ kg = 0.5\ m/s^2$*

Ask: How could you modify the force or the mass to make the box accelerate faster? *decrease box's mass or increase force*

SECTION 2.3

Ask: If the girl pushes on the boy with a force of 5 N, with what force does the boy push back on her? *5 N*

Ask: If the girl were pushing against a wall instead, why wouldn't the wall move? *Forces from ground and wall prevent its moving.*

SECTION 2.4

A 1 kg ball moving at a speed of 3 m/s hits a 2 kg ball at rest. Ask: If they stick together, at what speed will they move? *1 m/s*

Review Concepts

- Big Idea Flow Chart, p. T9
- Chapter Outline, pp. T15–T16

2 Chapter Review

the BIG idea

Forces change the motion of objects in predictable ways.

KEY CONCEPTS SUMMARY

2.1 Forces change motion.

Newton's first law
Objects at rest remain at rest, and objects in motion remain in motion with the same velocity, unless acted upon by an unbalanced force.

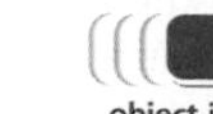

VOCABULARY
force p. 41
net force p. 43
Newton's first law p. 45
inertia p. 46

2.2 Force and mass determine acceleration.

Newton's second law
The acceleration of an object increases with increased force and decreases with increased mass, and is in the same direction as the force.

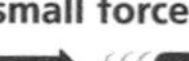
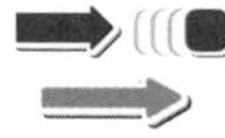
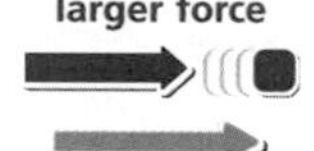

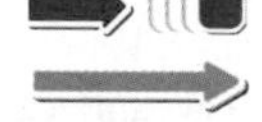
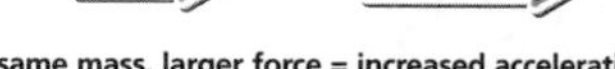

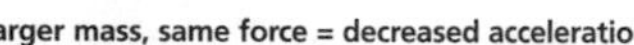

VOCABULARY
Newton's second law p. 50
centripetal force p. 54

2.3 Forces act in pairs.

Newton's third law
When one object exerts a force on another object, the second object exerts an equal and opposite force on the first object.

VOCABULARY
Newton's third law p. 57

2.4 Forces transfer momentum.

- Momentum is a property of a moving object.
- Forces in collisions are equal and opposite.
- Momentum is conserved in collisions.

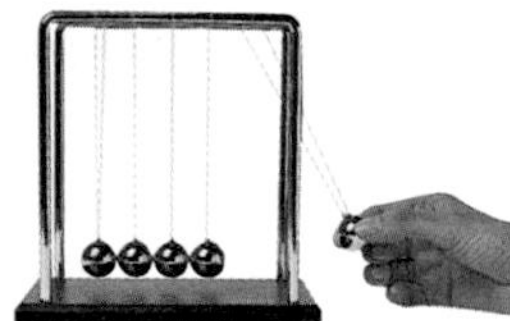

VOCABULARY
momentum p. 64
collision p. 66
conservation of momentum p. 67

Technology Resources

Have students visit **ClassZone.com** or use the CD-ROM for a cumulative review of concepts.

CONTENT REVIEW

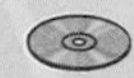
CONTENT REVIEW CD-ROM

Engage students in a whole-class interactive review of Key Concepts. Edit content as you wish.

POWER PRESENTATIONS

Reviewing Vocabulary

Copy and complete the chart below. If the left column is blank, give the correct term. If the right column is blank, give an example from real life.

Term	Example from Real Life
1. acceleration	
2. centripetal force	
3.	The pull of a handle on a wagon
4. inertia	
5. mass	
6. net force	
7. Newton's first law	
8. Newton's second law	
9.	When you're walking, you push backward on the ground, and the ground pushes you forward with equal force.
10. momentum	

Reviewing Key Concepts

Multiple Choice *Choose the letter of the best answer.*

11. Newton's second law states that to increase acceleration, you
- **a.** increase force
- **b.** decrease force
- **c.** increase mass
- **d.** increase inertia

12. What units are used to measure force?
- **a.** kilograms
- **b.** meters
- **c.** newtons
- **d.** seconds

13. A wagon is pulled down a hill with a constant velocity. All the forces on the wagon are
- **a.** balanced
- **b.** unbalanced
- **c.** increasing
- **d.** decreasing

14. An action force and its reaction force are
- **a.** equal in size and direction
- **b.** equal in size and opposite in direction
- **c.** different in size but in the same direction
- **d.** different in size and in direction

15. John pulls a box with a force of 4 N, and Jason pulls the box from the opposite side with a force of 3 N. Ignore friction. Which of the following statements is true?
- **a.** The box moves toward John.
- **b.** The box moves toward Jason.
- **c.** The box does not move.
- **d.** There is not enough information to determine if the box moves.

16. A more massive marble collides with a less massive one that is not moving. The total momentum after the collision is equal to
- **a.** zero
- **b.** the original momentum of the more massive marble
- **c.** the original momentum of the less massive marble
- **d.** twice the original momentum of the more massive marble

Short Answer *Write a short answer to each question.*

17. List the following objects in order, from the object with the least inertia to the object with the most inertia: feather, large rock, pencil, book. Explain your reasoning.

18. During a race, you double your velocity. How does that change your momentum?

19. Explain how an object can have forces acting on it but not be accelerating.

20. A sea scallop moves by shooting jets of water out of its shell. Explain how this works.

Reviewing Vocabulary

Sample answers:

1. *a ball speeding up as it falls to the ground*
2. *the pull of a string on a yo-yo as the yo-yo spins in a circle*
3. *force*
4. *the difficulty experienced moving a heavy table*
5. *a brick*
6. *the combined force of two people pushing on a box*
7. *a still ball moves when kicked*
8. *pulling harder to accelerate a full wagon*
9. *Newton's third law of motion*
10. *the mass and velocity of an automobile*

Reviewing Key Concepts

11. *a*
12. *c*
13. *a*
14. *b*
15. *a*
16. *b*
17. *Feather, pencil, book, large rock; the more mass an object has, the more inertia it will have.*
18. *It also doubles.*
19. *The forces are balanced.*
20. *According to Newton's third law, the scallop pushes water out of its shell in one direction (the action force) and the water pushes on the shell in the opposite direction (the reaction force) to move the animal backward.*

ASSESSMENT RESOURCES

UNIT ASSESSMENT BOOK
- Chapter Test A, pp. 26–29
- Chapter Test B, pp. 30–33
- Chapter Test C, pp. 34–37
- Alternative Assessment, pp. 38–39

SPANISH ASSESSMENT BOOK
Spanish Chapter Test, pp. 261–264

Technology Resources

Edit test items and answer choices.

Test Generator CD-ROM

Visit **ClassZone.com** to extend test practice.

Test Practice

Thinking Critically

21. The forces are balanced because the ball is not moving.

22. Student diagrams should show balanced forces; the downward force of gravity balances an upward force from the string.

23. When ball 1 hits ball 2, ball 2 exerts a force on ball 1, stopping its motion.

24. The velocity of the first ball became zero. The momentum from the first ball was transferred from the first ball down the balls to the last ball.

25. no change in motion

26. change in motion

27. no change in motion

28. The net force on the baseball must be three times the net force on the tennis ball.

Using Math Skills in Science

29. $5\ kg \cdot 2\ m/s^2 = 10\ N$

30. $10\ N/5\ m/s^2 = 2\ kg$

31. $5\ N/10\ kg = 0.5\ m/s^2$

32. $40\ kg \cdot 0.5\ m/s = 20\ kg \cdot m/s$

the BIG idea

33. Sample answer: You need to know the mass of the people on both sides of the rope and the force with which they can pull. Newton's first law states that the rope will not move unless the forces on it are unbalanced. If the force is unbalanced, the rope will accelerate toward the greater force.

34. Student answers will vary.

35. Student answers will vary.

UNIT PROJECTS

Students should have begun designing their models or multimedia presentations by this time. Remind them to continue researching as needed. Encourage them to try different solutions to problems they encounter.

 Unit Projects, pp. 5–10

Thinking Critically

Use the information in the photographs below to answer the next four questions.

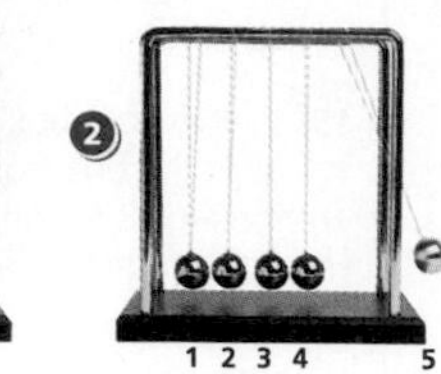

The photographs above show a toy called Newton's Cradle. In the first picture (1), ball 1 is lifted and is being held in place.

21. Are the forces on ball 1 balanced? How do you know?

22. Draw a diagram showing the forces acting on ball 2. Are these forces balanced?

In the second picture (2), ball 1 has been let go.

23. Ball 1 swung down, hit ball 2, and stopped. Use Newton's laws to explain why ball 1 stopped.

24. Use the principle of conservation of momentum to explain why ball 5 swung into the air.

Copy the chart below. Write what will happen to the object in each case.

Cause	Effect
25. Balanced forces act on an object.	
26. Unbalanced forces act on an object.	
27. No force acts on an object.	

28. INFER A baseball is three times more massive than a tennis ball. If the baseball and the tennis ball are accelerating equally, what can you determine about the net force on each?

Using Math Skills in Science

Complete the following calculations.

29. What force should Lori apply to a 5 kg box to give it an acceleration of 2 m/s^2?

30. If a 10 N force accelerates an object 5 m/s^2, how massive is the object?

31. Ravi applies a force of 5 N to a wagon with a mass of 10 kg. What is the wagon's acceleration?

32. Use the information in the photograph on the right to calculate the momentum of the shopping cart.

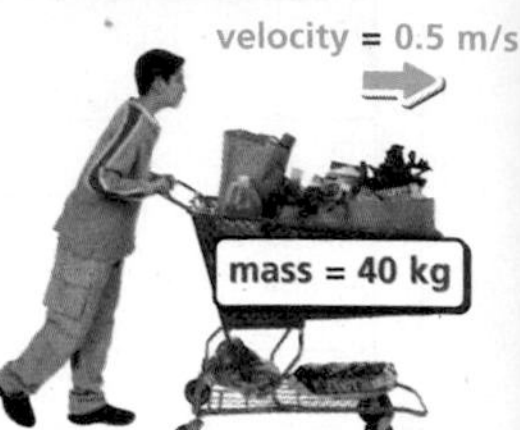

the BIG idea

33. PREDICT Look again at the tug of war pictured on pages 38–39. Describe what information you need to know to predict the outcome of the game. How would you use that information and Newton's laws to make your prediction?

34. WRITE Pick an activity you enjoy, such as running or riding a scooter, and describe how Newton's laws apply to that activity.

35. SYNTHESIZE Think of a question you have about Newton's laws that is still unanswered. What information do you need in order to answer the question? How might you find the information?

UNIT PROJECTS

If you need to do an experiment for your unit project, gather the materials. Be sure to allow enough time to observe results before the project is due.

MONITOR AND RETEACH

If students have trouble applying the concepts in items 21–24, they should review the diagram on p. 60 showing Newton's three laws in relation to a kangaroo jump. They can create a similar diagram for the canoe example on p. 61. **Part 1** should show an unbalanced force applied to the canoe. **Part 2** should show how an increase in mass or force affects acceleration (include a formula). **Part 3** should show equal and opposite forces on you and the paddle.

Students may benefit from summarizing sections of the chapter.

 Summarizing the Chapter, pp. 147–148

Standardized Test Practice

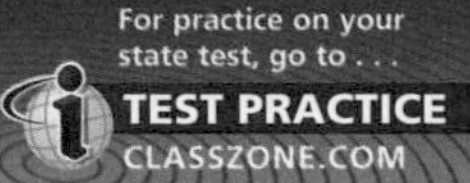
For practice on your state test, go to . . .
TEST PRACTICE
CLASSZONE.COM

Analyzing Data

To test Newton's second law, Jodie accelerates blocks of ice across a smooth, flat surface. The table shows her results. (For this experiment, you can ignore the effects of friction.)

Accelerating Blocks of Ice							
Mass (kg)	1.0	1.5	2.0	2.5	3.0	3.5	4.0
Acceleration (m/s²)	4.0	2.7	2.0	1.6	1.3	1.1	1.0

Study the data table and then answer the questions that follow.

1. The data show that as mass becomes greater, acceleration
a. increases
b. decreases
c. stays the same
d. cannot be predicted

2. From the data, you can tell that Jodie was applying a force of
a. 1 N **c.** 3 N
b. 2 N **d.** 4 N

3. If Jodie applied less force to the ice blocks, the accelerations would be
a. greater **c.** the same
b. less **d.** inconsistent

4. If Jodie applied a force of 6 N to the 2 kg block of ice, the acceleration would be
a. 2 m/s^2 **c.** 3 m/s^2
b. 4 m/s^2 **d.** 5 m/s^2

5. The average mass of the ice blocks she pushed was
a. 1.5 kg **c.** 3 kg
b. 2.5 kg **d.** 4 kg

6. If Jodie used a 3.25 kg block in her experiment, the force would accelerate the block somewhere between
a. 1.0 and 1.1 m/s^2
b. 1.1 and 1.3 m/s^2
c. 1.3 and 1.6 m/s^2
d. 1.6 and 2.0 m/s^2

Extended Response

Answer the two questions in detail. Include some of the terms shown in the word box. Underline each term you use in your answer.

Newton's second law	velocity
mass	inertia
gravity	balanced forces
centripetal force	unbalanced forces

7. Tracy ties a ball to a string and starts to swing the ball around her head. What forces are acting on the ball? What happens if the string breaks?

8. Luis is trying to pull a wagon loaded with rocks. What can he do to increase the wagon's acceleration?

METACOGNITIVE ACTIVITY

Have students answer the following questions in their **Science Notebooks:**

1. What about forces did you find most challenging to understand?
2. What questions do you still have about force and motion?
3. How have you solved a problem while working on your Unit Project?

Analyzing Data

1. b *4. c*
2. d *5. b*
3. b *6. b*

Extended Response

7. RUBRIC

4 points for a response that correctly answers both questions and uses the following terms accurately:

- centripetal force
- inertia
- velocity
- unbalanced force(s)

Sample answer: As Tracy spins the string, the centripetal force from the string acts on the ball to pull it toward the center of the circle. This unbalanced force continually changes the velocity of the ball by changing its direction. If the string were to break, the ball would fly off in a straight line because its inertia would resist change in velocity. (Students may also mention gravity.)

3 points correctly answers both questions and uses three terms accurately
2 points correctly answers one question and uses two terms accurately
1 point correctly answers one question or uses one term accurately

8. RUBRIC

4 points for a response that uses the following terms accurately:

- unbalanced force(s)
- mass
- Newton's second law
- acceleration

Sample answer: According to Newton's second law, decreasing mass will increase acceleration for the same force. To decrease the mass, Luis can remove rocks. Applying a greater unbalanced force will also increase the wagon's acceleration. To increase the force, Luis can pull harder or ask a friend to help pull.

3 points uses three terms accurately
2 points uses two terms accurately
1 point uses one term accurately

CHAPTER

Gravity, Friction, and Pressure

Physical Science UNIFYING PRINCIPLES

PRINCIPLE 1
Matter is made of particles too small to see.

PRINCIPLE 2
Matter changes form and moves from place to place.

PRINCIPLE 3
Energy changes from one form to another, but it cannot be created or destroyed.

PRINCIPLE 4
Physical forces affect the movement of all matter on Earth and throughout the universe.

Unit: Motion and Forces BIG IDEAS

CHAPTER 1
Motion
The motion of an object can be described and predicted.

CHAPTER 2
Forces
Forces change the motion of objects in predictable ways.

CHAPTER 3
Gravity, Friction, and Pressure
Newton's laws apply to all forces.

CHAPTER 4
Work and Energy
Energy is transferred when a force moves an object.

CHAPTER 5
Machines
Machines help people do work by changing the force applied to an object.

CHAPTER 3 KEY CONCEPTS

SECTION 3.1 

Gravity is a force exerted by masses.
1. Masses attract each other.
2. Gravity keeps objects in orbit.

SECTION 3.2 

Friction is a force that opposes motion.
1. Friction occurs when surfaces slide against each other.
2. Motion through fluids produces friction.

SECTION 3.3 

Pressure depends on force and area.
1. Pressure describes how a force is spread over an area.
2. Pressure acts in all directions in fluids.
3. Pressure in fluids depends on depth.

SECTION 3.4

Fluids can exert a force on objects.
1. Fluids can exert an upward force on objects.
2. The motion of a fluid affects its pressure.
3. Forces can be transmitted through fluids.

The Big Idea Flow Chart is available on p. T17 in the **UNIT TRANSPARENCY BOOK**.

Previewing Content

SECTION

Gravity is a force exerted by masses. pp. 77–84

1. Masses attract each other.

Gravity is the force objects exert on each other because of their mass. It attracts any two masses anywhere in the universe. The strength of the gravitational force is proportional to the product of the masses divided by the distance between them squared.

Greater mass results in greater force.

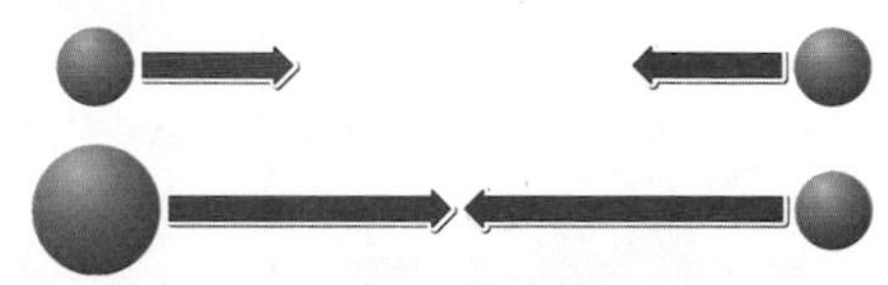

Greater distance results in smaller force.

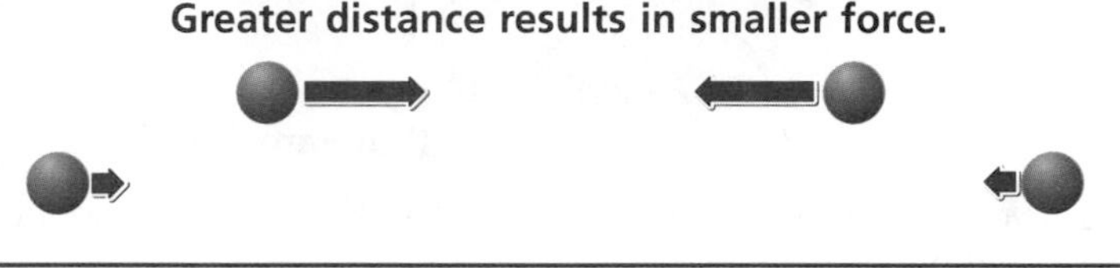

- Gravitational acceleration is symbolized by *g* and equals 9.8 m/s^2 at Earth's surface. Any object falling in a vacuum, no matter how massive, has this acceleration. The force of gravity, *F*, equals *mg* at Earth's surface.
- **Mass** and **weight** are not synonymous. Mass is the amount of matter something contains. Weight is the effect of gravity on the object.

2. Gravity keeps objects in orbit.

An **orbit** is an elliptical path that one object takes around another object. An orbital path is the result of the speed of the orbiting body and the gravitational pull between the two objects.

The speed an object must have to escape the gravitational pull of another body, such as a spacecraft leaving a planet, is called escape velocity. Speeds lower than the escape velocity will result in an orbit.

A spacecraft and its contents in orbit are in free fall. The environment is such that an astronaut can't feel gravity.

SECTION

Friction is a force that opposes motion. pp. 85–90

1. Friction occurs when surfaces slide against each other.

Friction is a force that resists the movement of two surfaces that are in contact with each other. Several factors determine the amount of friction between two surfaces.

While Object Moves

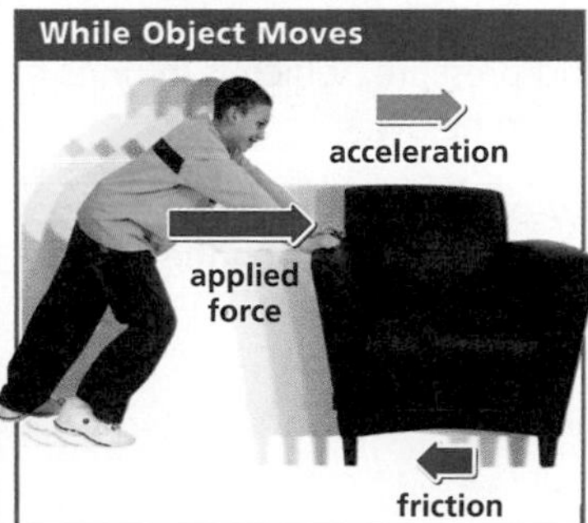

More Weight

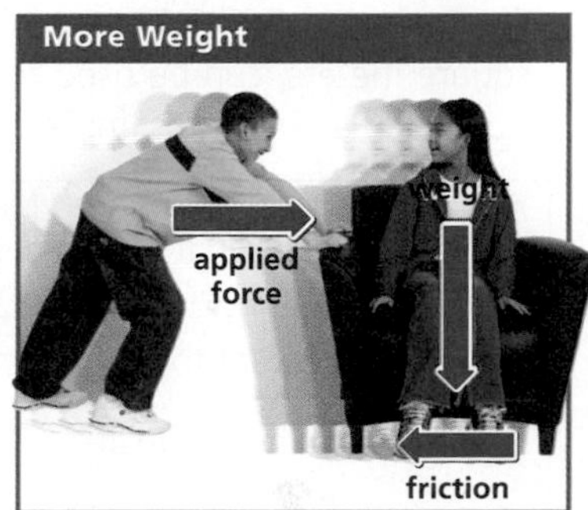

- The type of surface determines the amount of friction. Generally, smooth surfaces have less friction than rough surfaces.
- The motion of the surfaces affects friction. It takes more force to start an object moving (static friction) than it does to keep one moving (sliding friction).
- As the force pressing the surfaces together increases, friction increases. This force is often called the normal force because it is perpendicular, or normal, to the surface.

Friction between surfaces produces heat.

2. Motion through fluids produces friction.

A **fluid** is a substance that flows easily, such as liquids and gases. As an object moves through a fluid, its surface moves against particles in the fluid, causing friction.

When the fluid is air, the friction caused by a moving object is called **air resistance.**

- The amount of air resistance is based on the surface area of the object and the speed at which it moves.
- When an object falls through air at a speed at which air resistance balances gravity, the object reaches its maximum speed, called terminal velocity.

Common Misconceptions

WEIGHT AND MASS Students frequently think that *weight* and *mass* are terms that can be used interchangeably. In fact, mass is the amount of matter an object contains, and it remains the same no matter where the object is. Weight depends on the force of gravity exerted upon an object.

 This misconception is addressed on p. 80.

MISCONCEPTION DATABASE
CLASSZONE.COM Background on student misconceptions

FRICTION AND SURFACES Because students generally observe that rougher surfaces have greater friction, students may think that roughness is the sole cause of friction. In fact, friction is more complicated than just roughness, and is due to the interaction between surfaces.

 This misconception is addressed in Teach Difficult Concepts on p. 86.

Previewing Content

SECTION

3.3 Pressure depends on force and area. pp. 91–97

1. Pressure describes how a force is spread over an area.
Pressure measures how much force is acting on a certain area.
- It increases when force stays the same but acts on a smaller area.
- It increases when area stays the same but force increases.

The equation $P = F/A$, where force is in newtons and area is in square meters, can be used to find pressure, which is in units of **pascals,** Pa. See the sample problem below.

> A winter hiker weighing 500 N is wearing snowshoes that cover an area of 0.2 m^2. What pressure does the hiker exert on the snow?
> What do you know? Area = 0.2 m^2, Force = 500 N
> What do you want to find out? Pressure
>
> Write the formula: $P = \frac{F}{A}$
>
> Substitute into the formula: $P = \frac{500 \text{ N}}{0.2 \text{ m}^2}$
>
> Calculate and simplify: $P = 2500 \frac{\text{N}}{\text{m}^2} = 2500 \text{ N/m}^2$

2. Pressure acts in all directions in fluids.
The particles in a fluid move constantly and rapidly. They collide with objects that come in contact with the fluid, applying pressure to the surface of the object. The amount of pressure exerted depends on the density and the depth of the fluid.
Air exerts pressure on all objects in air. It is more dense at lower elevations and less dense at higher elevations. The denser the air, the more pressure it exerts.
Because water is denser than air, it exerts more pressure on objects in it.

3. Pressure in fluids depends on depth.
The pressure that a fluid exerts depends on depth and density of the fluid. At sea level, air exerts a pressure called atmospheric pressure. Air has weight.
- The more air above you, the greater the weight of that air.
- Air at higher elevations weighs less.
- Air at lower elevations is more compressed, therefore denser, and weighs more.

Water has a greater density than air, and therefore exerts more pressure on objects than air.

SECTION

3.4 Fluids can exert a force on objects. pp. 98–103

1. Fluids can exert an upward force on objects.
On Earth, objects are subject to forces from all directions, but these forces might not be balanced. The difference in water pressure at different depths produces an upward force, called **buoyant force,** which is illustrated below.

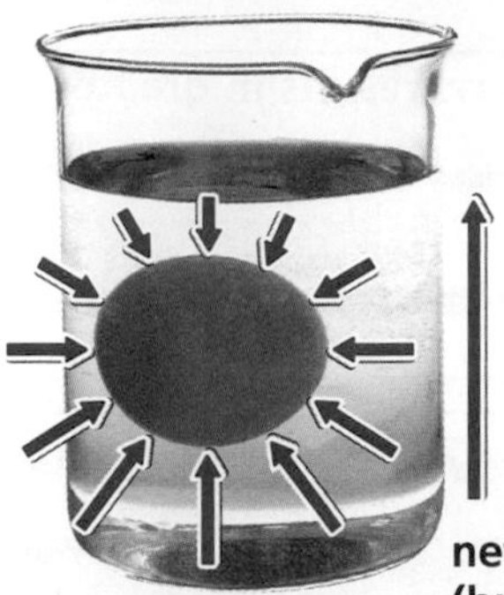

For a particular object, this force directly relates to the amount of fluid the object replaces.

Density is the amount of matter per unit of volume, $D = m/V$, where mass is commonly in grams and volume is commonly in cubic centimeters.

Because of buoyancy, a less dense material will float on another, denser material.

2. The motion of a fluid affects its pressure.
Bernoulli's principle states that, as speed of a fluid increases, the pressure inside the fluid decreases.

3. Forces can be transmitted through fluids.
Pascal's principle states that, when outside pressure is applied to a fluid in a container, that pressure is transmitted equally throughout the entire fluid.
Hydraulic machines use liquids to transmit forces. Gases would be less effective because they change volume when force is applied.

Common Misconceptions

BUOYANCY Students might think that ships and boats float because the materials in them are less dense than water. The overall density of the ship or boat is less than that of water only if you include the air it contains.

This misconception is addressed on p. 99.

MISCONCEPTION DATABASE
CLASSZONE.COM Background on student misconceptions

Previewing Labs

Lab Generator CD-ROM
Edit these Pupil Edition labs and generate alternative labs.

EXPLORE the BIG idea

Let It Slide, p. 75 Students examine the effect of friction on motion by using a ramp with various surfaces.	**TIME** 10 minutes **MATERIALS** board, books, small object, materials of different textures to place on the ramp, such as sandpaper or waxed paper
Under Pressure, p. 75 Students are introduced to the relationship of pressure to the amount of gas. They compare bottles of carbonated drinks.	**TIME** 10 minutes **MATERIALS** 2 plastic bottles of carbonated soft drink
Internet Activity: Gravity, p. 75 Students are introduced to the gravitational attraction of masses.	**TIME** 20 minutes **MATERIALS** computer with Internet access

SECTION 3.1  **EXPLORE Downward Acceleration,** p. 77 Students predict and compare the acceleration of different falling objects.	**TIME** 10 minutes **MATERIALS** golf ball, Ping-Pong ball
INVESTIGATE Gravity, p. 82 Students predict how gravity affects falling objects and test their ideas with cups of water.	**TIME** 15 minutes **MATERIALS** pencil, paper cup, water, large dishpan
SECTION 3.2  **INVESTIGATE Friction in Air,** p. 88 Students design an experiment to show how the shape of an object affects how it falls.	**TIME** 30 minutes **MATERIALS** 3 identical sheets of paper
SECTION 3.3  **EXPLORE Pressure,** p. 91 Students observe imprints on Styrofoam to find out how surface area affects pressure.	**TIME** 10 minutes **MATERIALS** sharpened pencil, Styrofoam board, book
CHAPTER INVESTIGATION **Pressure in Fluids,** pp. 96–97 Students vary depth and volume of water to determine what factors affect pressure.	**TIME** 40 minutes **MATERIALS** nail; 2 plastic bottles, small and large with tops cut off; ruler; plastic container; meter stick; coffee can
SECTION 3.4  **EXPLORE Forces in Liquid,** p. 98 Students observe water exerting a force on paper clips.	**TIME** 10 minutes **MATERIALS** 3 pieces of string, pencil, 8 paper clips, cup full of water
INVESTIGATE Bernoulli's Principle, p. 100 Students observe how the speed of air affects air pressure.	**TIME** 15 minutes **MATERIALS** pen, ruler, 2 clear straws, clear plastic cup filled with water, food coloring

 **Additional INVESTIGATION,** What Floats Your Boat? A, B, & C, pp. 210–218; Teacher Instructions, pp. 346–347

Previewing Chapter Resources

	INTEGRATED TECHNOLOGY	LABS AND ACTIVITIES
Chapter 3 **Gravity, Friction, and Pressure**	 **CLASSZONE.COM** • eEdition Plus • EasyPlanner Plus • Misconception Database • Content Review • Test Practice • Visualization • Simulation • Resource Centers • Internet Activity: Gravity • Math Tutorial  **SCILINKS.ORG** SCILINKS  **CD-ROMS** • eEdition • EasyPlanner • Power Presentations • Content Review • Lab Generator • Test Generator  **AUDIO CDS** • Audio Readings • Audio Readings in Spanish	 EXPLORE the Big Idea, p. 75 • Let It Slide • Under Pressure • Internet Activity: Gravity  **UNIT RESOURCE BOOK** Unit Projects, pp. 5–10  **Lab Generator CD-ROM** Generate customized labs.
SECTION  3.1 **Gravity is a force exerted by masses.** pp. 77–84 Time: 2 periods (1 block)  Lesson Plan, pp. 149–150	 • **RESOURCE CENTER,** Gravitational Lenses • **VISUALIZATION,** Gravity in a Vacuum  **UNIT TRANSPARENCY BOOK** • Big Idea Flow Chart, p. T17 • Daily Vocabulary Scaffolding, p. T18 • Note-Taking Model, p. T19 • 3-Minute Warm-Up, p. T20 • "Orbits" Visual, p. T22	 • EXPLORE Downward Acceleration, p. 77 • INVESTIGATE Gravity, p. 82 • Extreme Science, p. 84  **UNIT RESOURCE BOOK** Datasheet, Gravity, p. 158
SECTION  3.2 **Friction is a force that opposes motion.** pp. 85–90 Time: 2 periods (1 block)  Lesson Plan, pp. 160–161	 • **RESOURCE CENTER,** Friction, Forces, and Surfaces • **MATH TUTORIAL**  **UNIT TRANSPARENCY BOOK** • Daily Vocabulary Scaffolding, p. T18 • 3-Minute Warm-Up, p. T20	 • INVESTIGATE Friction in Air, p. 88 • Math in Science, p. 90  **UNIT RESOURCE BOOK** • Datasheet, Friction in Air, p. 169 • Math Support, p. 199 • Math Practice, p. 200
SECTION  3.3 **Pressure depends on force and area.** pp. 91–97 Time: 3 periods (1.5 blocks) Lesson Plan, pp. 171–172	 **SIMULATION,** Fluids and Pressure  **UNIT TRANSPARENCY BOOK** • Daily Vocabulary Scaffolding, p. T18 • 3-Minute Warm-Up, p. T21	 • EXPLORE Pressure, p. 91 • CHAPTER INVESTIGATION, Pressure in Fluids, pp. 96–97  **UNIT RESOURCE BOOK** • Math Support & Practice, pp. 197–198 • CHAPTER INVESTIGATION, Pressure in Fluids, A, B, & C, pp. 201–209
SECTION  3.4 **Fluids can exert a force on objects.** pp. 98–103 Time: 3 periods (1.5 blocks)  Lesson Plan, pp. 181–182	 **UNIT TRANSPARENCY BOOK** • Big Idea Flow Chart, p. T17 • Daily Vocabulary Scaffolding, p. T18 • 3-Minute Warm-Up, p. T21 • Chapter Outline, pp. T23–T24	 • EXPLORE Forces in Liquid, p. 98 • INVESTIGATE Bernoulli's Principle, p. 100  **UNIT RESOURCE BOOK** • Datasheet, Bernoulli's Principle, p. 190 • Additional INVESTIGATION, What Floats Your Boat?, A, B, & C, pp. 210–218

KEY TO ICONS

 CD/CD-ROM
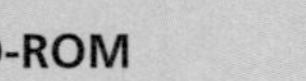 Teacher Edition
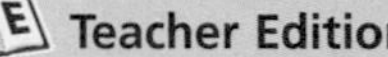
 UNIT TRANSPARENCY BOOK
 SPANISH ASSESSMENT BOOK
INTERNET
 Pupil Edition
 UNIT RESOURCE BOOK
 UNIT ASSESSMENT BOOK
SCIENCE TOOLKIT

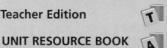

READING AND REINFORCEMENT	ASSESSMENT	STANDARDS
 • Four Square, B22–23 • Supporting Main Ideas, C42 • Daily Vocabulary Scaffolding, H1–8  **UNIT RESOURCE BOOK** • Vocabulary Practice, pp. 194–195 • Decoding Support, p. 196 • Summarizing the Chapter, pp. 219–220 **Audio Readings CD** Listen to Pupil Edition. 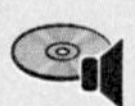 **Audio Readings in Spanish CD** Listen to Pupil Edition in Spanish.	• Chapter Review, pp. 105–106 • Standardized Test Practice, p. 107 **UNIT ASSESSMENT BOOK** • Diagnostic Test, pp. 40–41 • Chapter Test, A, B, & C, pp. 46–57 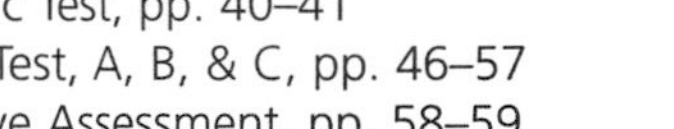 • Alternative Assessment, pp. 58–59  Spanish Chapter Test, pp. 265–268  **Test Generator CD-ROM** Generate customized tests.  **Lab Generator CD-ROM** Rubrics for Labs	National Standards A.2–8, A.9.a–c, A.9.e–f, B.1.a, D.3.c, E.2–5, G.1.a–b See p. 74 for the standards.
 **UNIT RESOURCE BOOK** • Reading Study Guide, A & B, pp. 151–154 • Spanish Reading Study Guide, pp. 155–156 • Challenge and Extension, p. 157 • Reinforcing Key Concepts, p. 159	 Ongoing Assessment, pp. 77–83  Section 3.1 Review, p. 83  **UNIT ASSESSMENT BOOK** Section 3.1 Quiz, p. 42	National Standards A.2–8, A.9.a–c, A.9.e–f, D.3.c, G.1.b
 **UNIT RESOURCE BOOK** • Reading Study Guide, A & B, pp. 162–165 • Spanish Reading Study Guide, pp. 166–167 • Challenge and Extension, p. 168 • Reinforcing Key Concepts, p. 170	Ongoing Assessment, pp. 85, 87, 89  Section 3.2 Review, p. 89  **UNIT ASSESSMENT BOOK** Section 3.2 Quiz, p. 43	National Standards A.2–8, A.9.a–c, A.9.e–f, E.2–5, G.1.b
 **UNIT RESOURCE BOOK** • Reading Study Guide, A & B, pp. 173–176 • Spanish Reading Study Guide, pp. 177–178 • Challenge and Extension, p. 179 • Reinforcing Key Concepts, p. 180	 Ongoing Assessment, pp. 92–93, 95  Section 3.3 Review, p. 95  **UNIT ASSESSMENT BOOK** Section 3.3 Quiz, p. 44	National Standards A.2–8, A.9.a–c, A.9.e–f, G.1.b
 **UNIT RESOURCE BOOK** • Reading Study Guide, A & B, pp. 183–186 • Spanish Reading Study Guide, pp. 187–188 • Challenge and Extension, p. 189 • Reinforcing Key Concepts, p. 191 • Challenge Reading, pp. 192–193	 Ongoing Assessment, pp. 99–102  Section 3.4 Review, p. 103  **UNIT ASSESSMENT BOOK** Section 3.4 Quiz, p. 45	National Standards A.2–8, A.9.a–c, A.9.e, B.1.a, G.1.a–b

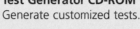

Previewing Resources for Differentiated Instruction

CHAPTER INVESTIGATION

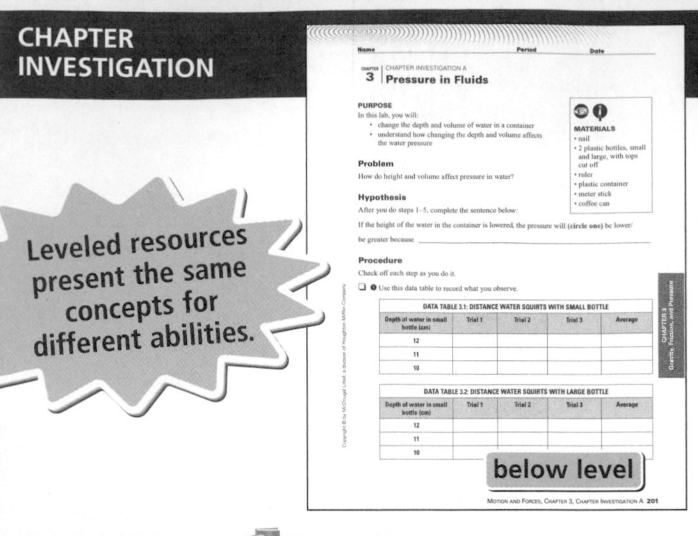

Leveled resources present the same concepts for different abilities.

Name Period Date

CHAPTER 3 | CHAPTER INVESTIGATION A

Pressure in Fluids

PURPOSE

MATERIALS

Problem

How do height and volume affect pressure in water?

Hypothesis

Procedure

DATA TABLE 3.1: DISTANCE WATER SQUIRTS WITH SMALL BOTTLE

DATA TABLE 3.2: DISTANCE WATER SQUIRTS WITH LARGE BOTTLE

below level

MOTION AND FORCES, CHAPTER 3, CHAPTER INVESTIGATION A 201

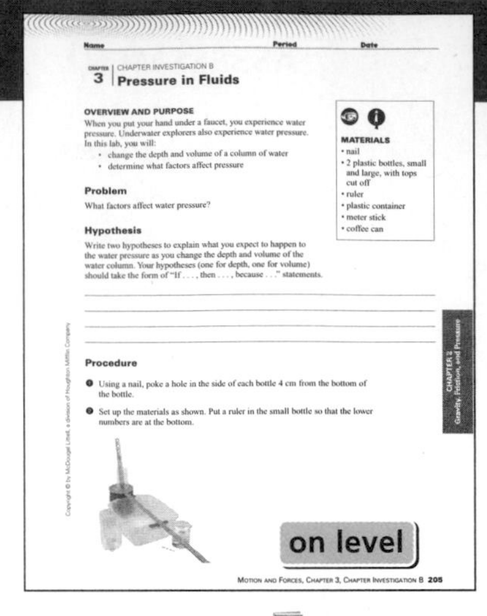

Name Period Date

CHAPTER 3 | CHAPTER INVESTIGATION B

Pressure in Fluids

OVERVIEW AND PURPOSE

MATERIALS

Problem

What factors affect water pressure?

Hypothesis

Procedure

on level

MOTION AND FORCES, CHAPTER 3, CHAPTER INVESTIGATION B 205

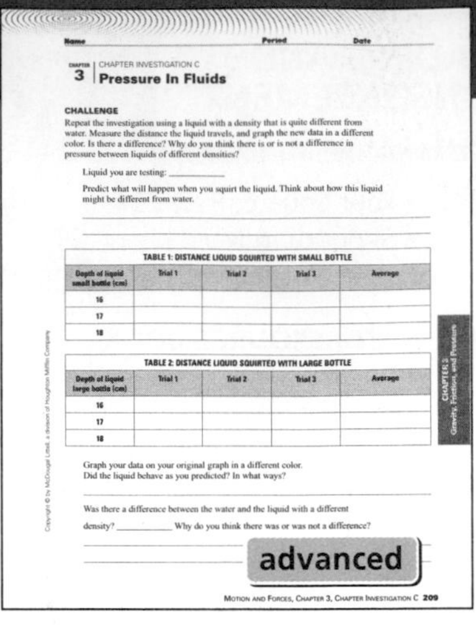

Name Period Date

CHAPTER 3 | CHAPTER INVESTIGATION C

Pressure in Fluids

CHALLENGE

TABLE 1: DISTANCE LIQUID SQUIRTED WITH SMALL BOTTLE

TABLE 2: DISTANCE LIQUID SQUIRTED WITH LARGE BOTTLE

advanced

MOTION AND FORCES, CHAPTER 3, CHAPTER INVESTIGATION C 209

UNIT RESOURCE BOOK, pp. 201–204 | pp. 205–208 | pp. 205–209

READING STUDY GUIDE

Reading Study Guide is also in Spanish.

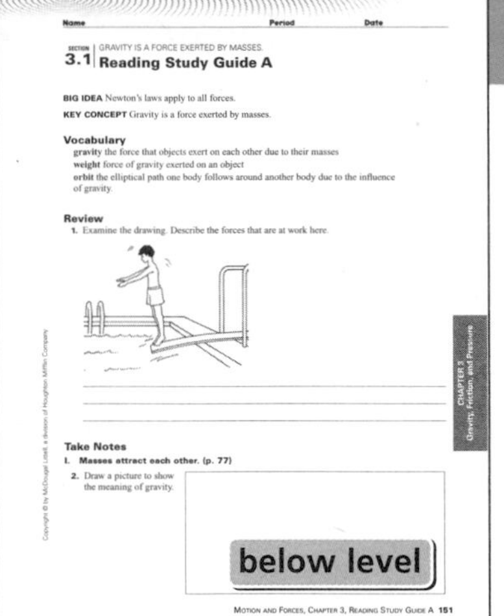

Name Period Date

SECTION 3.1 | GRAVITY IS A FORCE EXERTED BY MASSES.

Reading Study Guide A

BIG IDEA Newton's laws apply to all forces.

KEY CONCEPT Gravity is a force exerted by masses.

Vocabulary

Review

Take Notes

I. Masses attract each other. (p. 77)

below level

MOTION AND FORCES, CHAPTER 3, READING STUDY GUIDE A 151

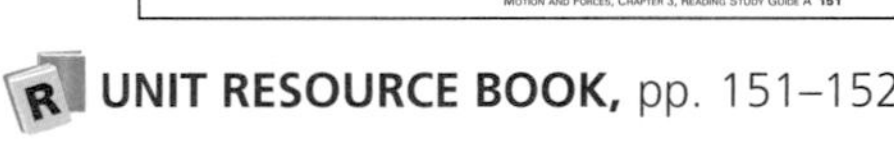

UNIT RESOURCE BOOK, pp. 151–152

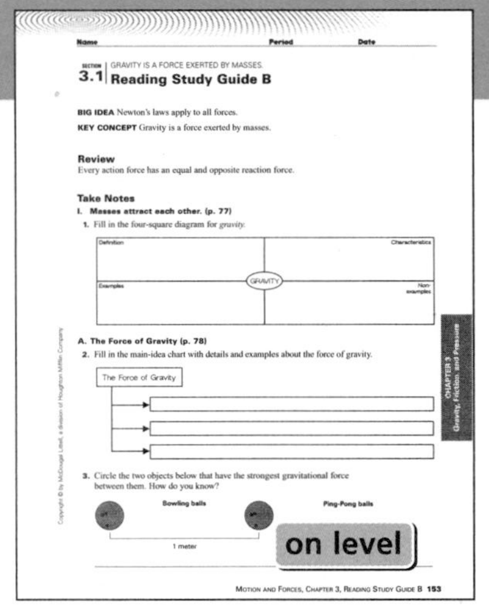

Name Period Date

SECTION 3.1 | GRAVITY IS A FORCE EXERTED BY MASSES.

Reading Study Guide B

BIG IDEA Newton's laws apply to all forces.

KEY CONCEPT Gravity is a force exerted by masses.

Review

Every action force has an equal and opposite reaction force.

Take Notes

I. Masses attract each other. (p. 77)

A. The Force of Gravity (p. 78)

on level

MOTION AND FORCES, CHAPTER 3, READING STUDY GUIDE B 153

pp. 153–154

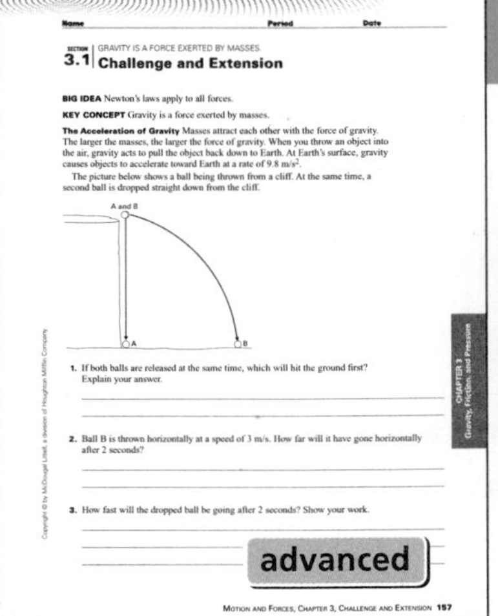

Name Period Date

SECTION 3.1 | GRAVITY IS A FORCE EXERTED BY MASSES.

Challenge and Extension

BIG IDEA Newton's laws apply to all forces.

KEY CONCEPT Gravity is a force exerted by masses.

advanced

MOTION AND FORCES, CHAPTER 3, CHALLENGE AND EXTENSION 157

p. 157

CHAPTER TEST

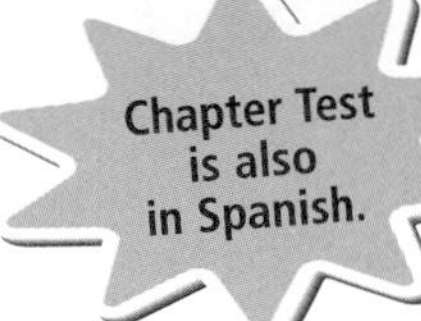

Chapter Test is also in Spanish.

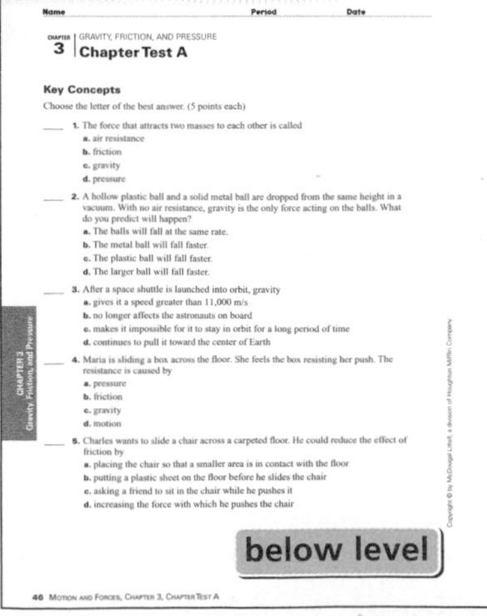

Name Period Date

CHAPTER 3 | GRAVITY, FRICTION, AND PRESSURE

Chapter Test A

Key Concepts

Choose the letter of the best answer. (5 points each)

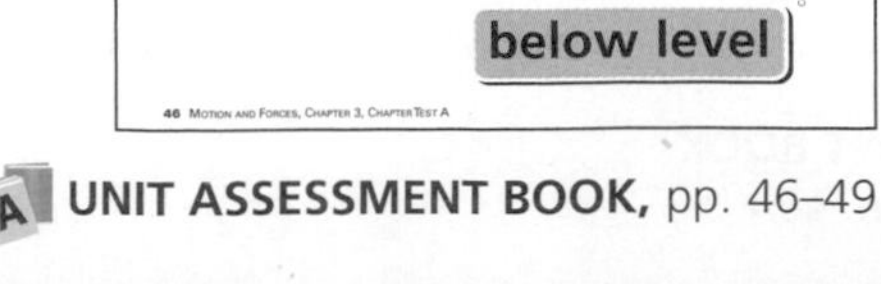

below level

46 MOTION AND FORCES, CHAPTER 3, CHAPTER TEST A

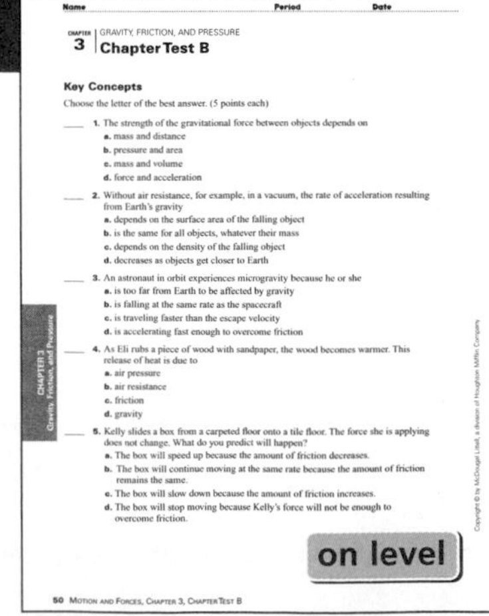

Name Period Date

CHAPTER 3 | GRAVITY, FRICTION, AND PRESSURE

Chapter Test B

Key Concepts

Choose the letter of the best answer. (5 points each)

on level

50 MOTION AND FORCES, CHAPTER 3, CHAPTER TEST B

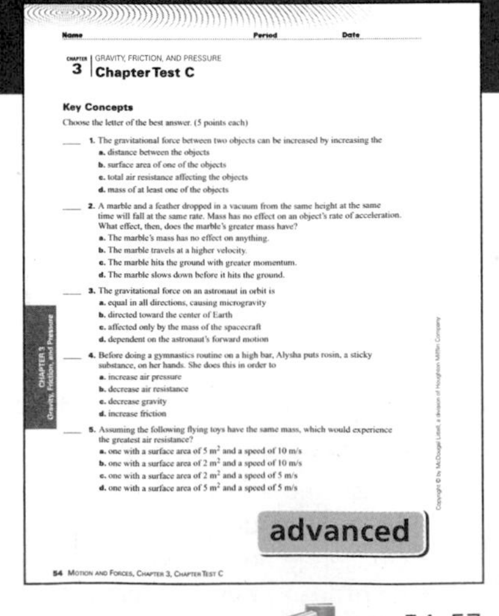

Name Period Date

CHAPTER 3 | GRAVITY, FRICTION, AND PRESSURE

Chapter Test C

Key Concepts

Choose the letter of the best answer. (5 points each)

advanced

54 MOTION AND FORCES, CHAPTER 3, CHAPTER TEST C

UNIT ASSESSMENT BOOK, pp. 46–49 | pp. 50–53

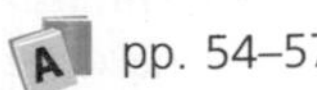

pp. 54–57

TECHNOLOGY

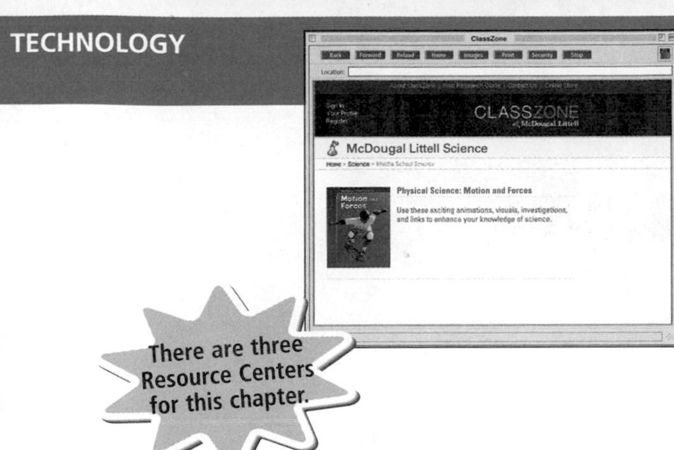
There are three Resource Centers for this chapter.

McDOUGAL LITTELL
AUDIO READINGS
McDOUGAL LITTELL
LAB GENERATOR
Customize and edit labs with this easy-to-use CD-ROM
• Searchable database of all labs from the program
• Additional lab options
• Template for creating your own labs
• Rubrics and other resources

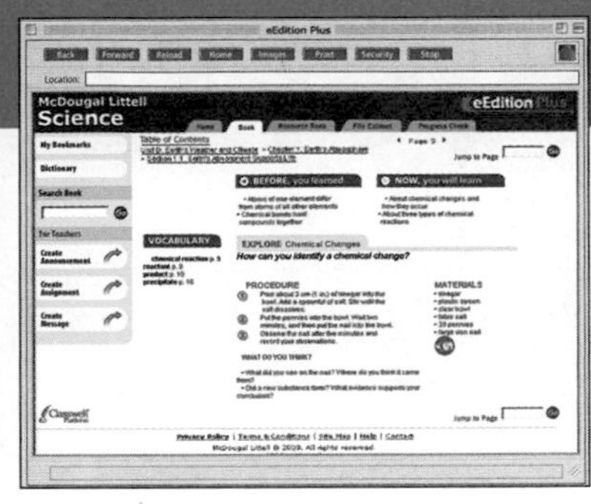

CLASSZONE.COM

CD/CD-ROMS

CLASSZONE.COM

VISUAL CONTENT

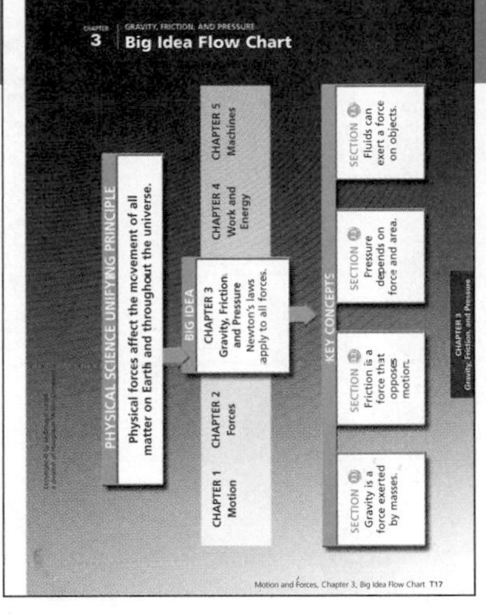
3 Big Idea Flow Chart

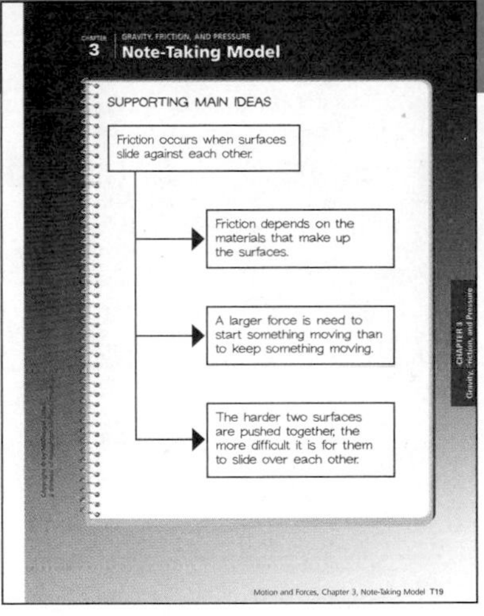
3 Note-Taking Model
SUPPORTING MAIN IDEAS
Friction occurs when surfaces slide against each other.
Friction depends on the materials that make up the surfaces.
A larger force is need to start something moving than to keep something moving.
The harder two surfaces are pushed together, the more difficult it is for them to slide over each other.

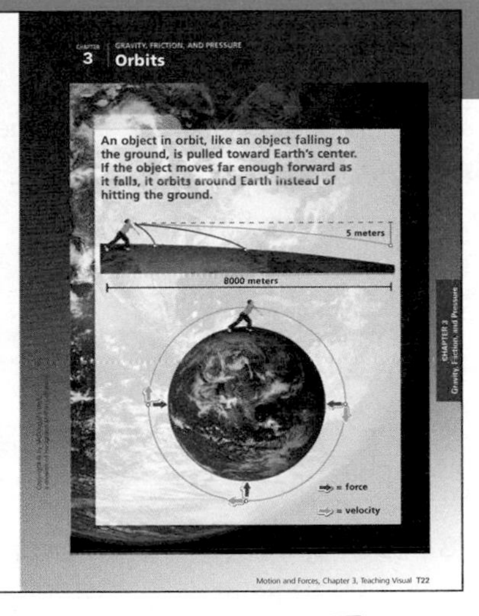
3 Orbits
An object in orbit, like an object falling to the ground, is pulled toward Earth's center. If the object moves far enough forward as it falls, it orbits around Earth instead of hitting the ground.

UNIT TRANSPARENCY BOOK, p. T17 p. T19 p. T22

MORE SUPPORT

Reinforcing Key Concepts for each section

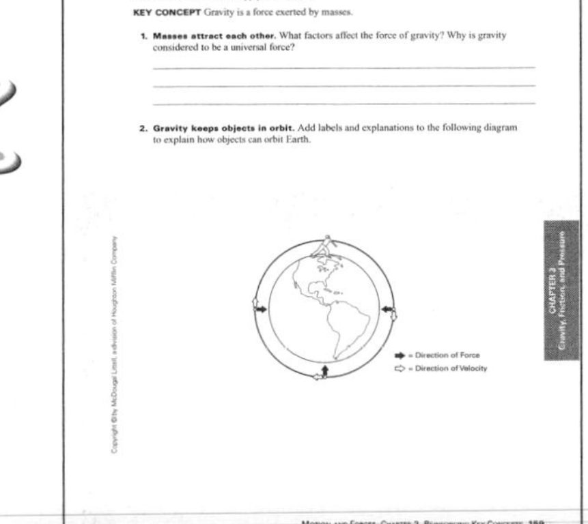
3.1 Reinforcing Key Concepts

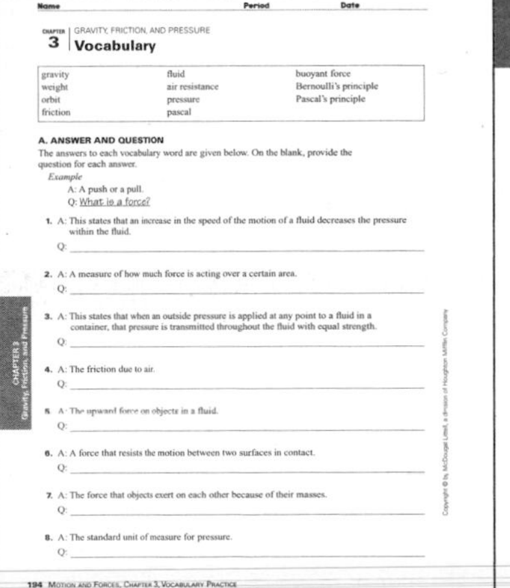
3 Vocabulary

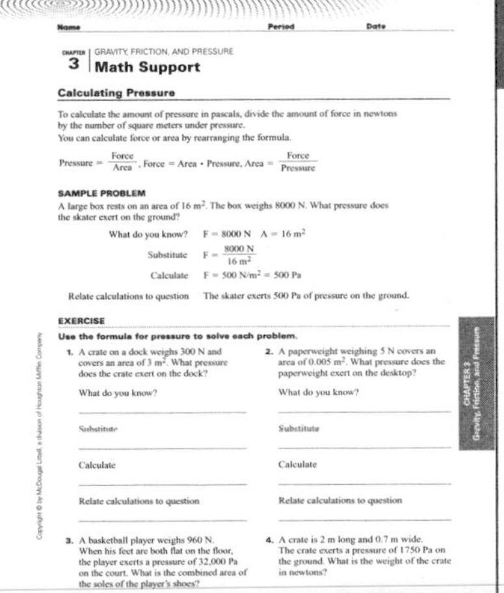
3 Math Support

UNIT RESOURCE BOOK, p. 159

pp. 194–195 p. 197

CHAPTER

Gravity, Friction, and Pressure

INTRODUCE

the BIG idea

Have students look at the photograph of the snowboarder and describe the scene. Have them discuss how the question in the box on p. 75 links to the Big Idea:

What effect do each of the forces acting on the snowboarder have on his motion?

National Science Education Standards

Content

B.1.a Substances have certain properties, including density.

D.3.c Gravity is the force that keeps planets in orbit around the Sun and governs the rest of the motion in the solar system. Gravity alone holds us to Earth's surface and explains the phenomena of the tides.

Process

A.2–8 Design and conduct an investigation; use tools to gather and interpret data; use evidence to describe, predict, explain, model; think critically to make relationships between evidence and explanation; recognize different explanations and predictions; communicate scientific procedures and explanations; use mathematics.

A.9.a–c, A.9.e–f Understand scientific inquiry by using different investigations, methods, mathematics, and explanations based on logic, evidence and skepticism.

E.2–5 Design, implement, and evaluate a solution; communicate technological design.

G.1.a–b Science as human endeavor

CHAPTER

Gravity, Friction, and Pressure

the BIG idea

Newton's laws apply to all forces.

Key Concepts

SECTION

3.1 Gravity is a force exerted by masses.
Learn about gravity, weight, and orbits.

SECTION

3.2 Friction is a force that opposes motion.
Learn about friction and air resistance.

SECTION

3.3 Pressure depends on force and area.
Learn about pressure and how forces act on objects in fluids.

SECTION

3.4 Fluids can exert a force on objects.
Learn how fluids apply forces to objects and how forces are transmitted through fluids.

Internet Preview

CLASSZONE.COM

Chapter 3 online resources: Content Review, Simulation, two Visualizations, three Resource Centers, Math Tutorial, Test Practice

INTERNET PREVIEW

CLASSZONE.COM For student use with the following pages:

Review and Practice
- Content Review, pp. 76, 104
- Math Tutorial: Creating a Line Graph, p. 90
- Test Practice, 107

Activities and Resources
- Internet Activity: Gravity, p. 75
- Visualization, p. 79
- Simulation, p. 93
- Resource Centers: Gravitational Lenses, p. 84; Friction, Forces, and Surfaces, p. 86

Pressure **Code: MDL006**

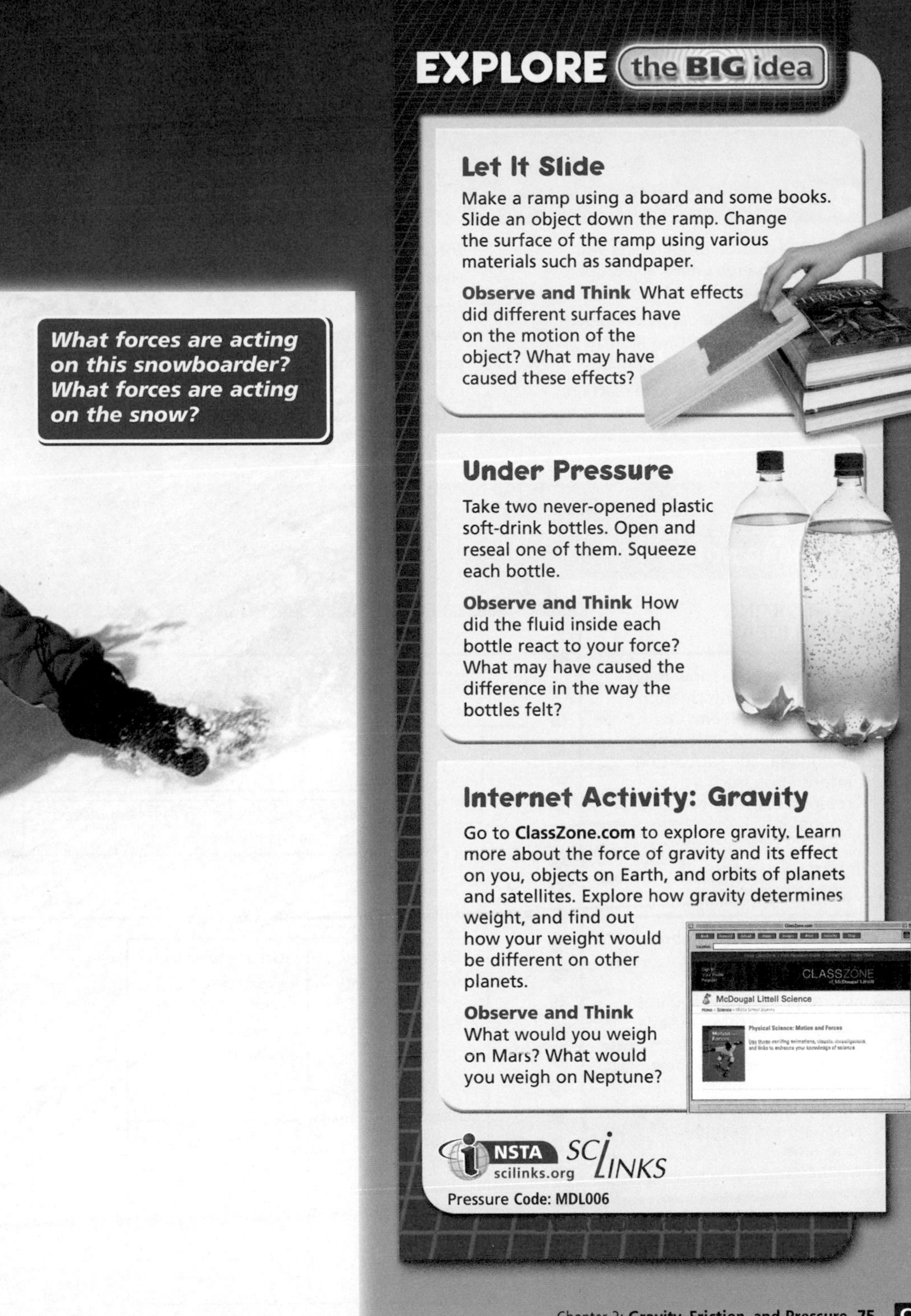

EXPLORE the BIG idea
Let It Slide
Make a ramp using a board and some books. Slide an object down the ramp. Change the surface of the ramp using various materials such as sandpaper.
Observe and Think What effects did different surfaces have on the motion of the object? What may have caused these effects?
What forces are acting on this snowboarder? What forces are acting on the snow?
Under Pressure
Take two never-opened plastic soft-drink bottles. Open and reseal one of them. Squeeze each bottle.
Observe and Think How did the fluid inside each bottle react to your force? What may have caused the difference in the way the bottles felt?
Internet Activity: Gravity
Go to ClassZone.com to explore gravity. Learn more about the force of gravity and its effect on you, objects on Earth, and orbits of planets and satellites. Explore how gravity determines weight, and find out how your weight would be different on other planets.
Observe and Think What would you weigh on Mars? What would you weigh on Neptune?
CLASSZONE
McDougal Littell Science
NSTA scilinks.org SCILINKS
Pressure Code: MDL006
Chapter 3: Gravity, Friction, and Pressure 75 C

TEACHING WITH TECHNOLOGY

CBL and Probeware Use an accelerometer probe to measure the acceleration of the paper falling in the Investigate on p. 88. Students should compare their qualitative and quantitative results.

Video Camera Students can film the investigations on pp. 82 and 88 that involve falling objects. Playing the videotape in slow motion will help students analyze their observations.

EXPLORE the BIG idea

These inquiry-based activities are appropriate for use at home or as a supplement to classroom instruction.

Let It Slide

PURPOSE To introduce students to the effect of friction on motion. Students slide an object down a ramp a few times, changing the surface of the ramp each time, to see the effects of friction.

TIP *10 min.* One surface should provide a lot of friction and one a little friction, such as waxed paper. The object should slide, not roll, down the ramp.

Answer: The rougher the surface, the more the motion slowed. A rough surface "holds back" motion of the object.

REVISIT after p. 86.

Under Pressure

PURPOSE To introduce students to the relationship of pressure to amount of gas. Students compare closed and opened bottles of carbonated drinks.

TIP *10 min.* Check that the cap is tightly fastened on the bottle that has been opened to avoid spills from squeezing the bottle.

Answer: The liquid did not change in volume, but the gas is compressed more in the closed bottle. There was less gas in the opened bottle, so the molecules could be pushed closer together.

REVISIT after p. 93.

Internet Activity: Gravity

PURPOSE To introduce students to the effect of mass on gravitational attraction. Students choose masses and see how they interact.

TIP *20 min.* Students can predict the gravitational attraction of two masses before they try the activity to determine it.

REVISIT after p. 78.

PREPARE

CONCEPT REVIEW

Activate Prior Knowledge

- Ask students to describe a situation that illustrates each of Newton's three laws.
- Have students use each law to explain the importance of using restraints on amusement-park rides.

TAKING NOTES

Supporting Main Ideas

Students will find these charts analogous to an outline because of the way they organize main points and supporting information. The chart provides a meaningful way to organize information, especially for visual learners.

Vocabulary Strategy

As students start the chapter, have each student write a four square diagram for gravity. Use their individual diagrams to compile a class diagram for the concept. Discuss why individual diagrams contain different information.

Vocabulary and Note-Taking Resources

- Vocabulary Practice, pp. 194–195
- Decoding Support, p. 196

- Daily Vocabulary Scaffolding, p. T18
- Note-Taking Model, p. T19

- Four Square, B22–23
- Supporting Main Ideas, C42
- Daily Vocabulary Scaffolding, H1–8

CHAPTER 3

Getting Ready to Learn

CONCEPT REVIEW

- The motion of an object will not change unless acted upon by an unbalanced force.
- The acceleration of an object depends on force and mass.
- For every action force there is an equal and opposite reaction.

VOCABULARY REVIEW

force p. 41
Newton's first law p. 45
Newton's second law p. 50
Newton's third law p. 57
density *See Glossary.*

CONTENT REVIEW
CLASSZONE.COM
Review concepts and vocabulary.

TAKING NOTES

SUPPORTING MAIN IDEAS

Make a chart to show main ideas and the information that supports them. Copy the main ideas. Below each main idea, add supporting information, such as reasons, explanations, and examples.

VOCABULARY STRATEGY

Write each new vocabulary term in the center of a **four square** diagram. Write notes in the squares around each term. Include a definition, some characteristics, and some examples of the term. If possible, write some things that are not examples of the term.

See the Note-Taking Handbook on pages R45–R51.

SCIENCE NOTEBOOK

Force of gravity depends on mass and distance.
→ More mass = more gravitational force
→ More distance = less gravitational force

Definition	Characteristics
force of gravity acting on an object	• changes if gravity changes • measured in newtons
Examples	**Nonexamples**
A 4 kg bowling ball weighs 39 N.	Mass in kg is not a weight.

WEIGHT

CHECK READINESS

Administer the Diagnostic Test to determine students' readiness for new science content and their mastery of requisite math skills.

Diagnostic Test, pp. 40–41

Technology Resources

Students needing content and math skills should visit **ClassZone.com**.

- **CONTENT REVIEW**
- **MATH TUTORIAL**

CONTENT REVIEW CD-ROM

KEY CONCEPT

Gravity is a force exerted by masses.

BEFORE, you learned

- Every action force has an equal and opposite reaction force
- Newton's laws are used to describe the motions of objects
- Mass is the amount of matter an object contains

NOW, you will learn

- How mass and distance affect gravity
- What keeps objects in orbit

VOCABULARY

gravity p. 77
weight p. 79
orbit p. 80

EXPLORE Downward Acceleration

How do the accelerations of two falling objects compare?

PROCEDURE

1. Make a prediction: Which ball will fall faster?
2. Drop both balls from the same height at the same time.
3. Observe the balls as they hit the ground.

WHAT DO YOU THINK?

- Were the results what you had expected?
- How did the times it took the two balls to hit the ground compare?

MATERIALS

- golf ball
- Ping-Pong ball

Masses attract each other.

When you drop any object—such as a pen, a book, or a football—it falls to the ground. As the object falls, it moves faster and faster. The fact that the object accelerates means there must be a force acting on it. The downward pull on the object is due to gravity. **Gravity** is the force that objects exert on each other because of their masses. You are familiar with the force of gravity between Earth and objects on Earth.

VOCABULARY Create a four square diagram for *gravity* in your notebook.

Gravity is present not only between objects and Earth, however. Gravity is considered a universal force because it acts between any two masses anywhere in the universe. For example, there is a gravitational pull between the Sun and the Moon. Even small masses attract each other. The force of gravity between dust and gas particles in space helped form the solar system.

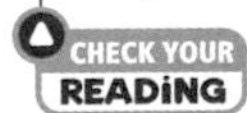
Why is gravity considered a universal force?

RESOURCES FOR DIFFERENTIATED INSTRUCTION

Below Level

UNIT RESOURCE BOOK
- Reading Study Guide A, pp. 151–152
- Decoding Support, p. 196

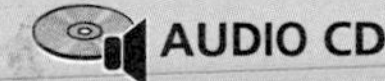

Advanced

UNIT RESOURCE BOOK
Challenge and Extension, p. 157

English Learners

UNIT RESOURCE BOOK
Spanish Reading Study Guide, pp. 155–156

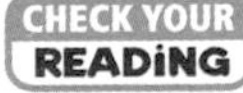
AUDIO CDS

- Audio Readings in Spanish
- Audio Readings (English)

3.1 FOCUS

Set Learning Goals

Students will

- Describe how mass and distance affect gravity.
- Explain what keeps objects in orbit.
- Investigate through experimentation how gravity affects falling objects.

3-Minute Warm-Up

Display Transparency 20 or copy this exercise on the board:

Match each definition with a term.

Definitions

1. the rate at which velocity changes with time *b*
2. a measure of mass in motion *e*
3. the force that keeps an object moving in a circle *a*

Terms

a. centripetal force
b. acceleration
c. mass
d. contact force
e. momentum

 3-Minute Warm-Up, p. T20

3.1 MOTIVATE

EXPLORE Downward Acceleration

PURPOSE To introduce the relationship between gravity and mass

TIP *10 min.* If students have difficulty seeing which ball hits first, have them listen to the balls hitting the floor.

WHAT DO YOU THINK? *Both balls will hit the floor at the same time. Gravity accelerates all masses equally.*

Ongoing Assessment

CHECK YOUR READING *Answer: Gravity is present everywhere in the universe.*

3.1 INSTRUCT

Real World Example

Although building codes and architects seek to ensure that buildings are stable, some basic designs are more stable than others. One of the most stable structures is the geodesic dome, like the one at the Epcot Center in Florida. Its rounded shape excels at resisting damage from strong winds and heavy snows. One of its greatest advantages is that its shape spreads the force of gravity down the sides. The compression that gravity would otherwise cause is evenly distributed. Another example of a stable dome is the igloo, used as a temporary dwelling by the Inuit people.

Teach Difficult Concepts

If all objects were single points, there would be no difficulty finding the distance between any two of them. The center of mass of an object is the point in space where an object (or system of objects) behaves as if all the mass were concentrated at that point. For example, a large sphere can be treated like it is a point-sized mass located at the center of the sphere. When finding the gravitational force between two objects, the distance between them is the distance between each object's center of mass.

EXPLORE the BIG idea

Revisit "Internet Activity: Gravity" on p. 75. Have students explain their results.

Ongoing Assessment

Describe how mass and distance affect gravity.

Ask: Which has more gravitational attraction, a pair of students standing one meter apart or the same students standing three meters apart? *those standing one meter apart*

CHECK YOUR READING *Answer: As mass increases, gravity increases. As distance increases, gravity decreases.*

SUPPORTING MAIN IDEAS Support the main ideas about the force of gravity with details and examples.

The Force of Gravity

If there is a force between all masses, why are you not pulled toward your desk by the desk's gravity when you walk away from it? Remember that the net force on you determines how your motion changes. The force of gravity between you and the desk is extremely small compared with other forces constantly acting on you, such as friction, the force from your muscles, Earth's gravity, and the gravitational pull from many other objects. The strength of the gravitational force between two objects depends on two factors, mass and distance.

A **The Mass of the Objects** The more mass two objects have, the greater the force of gravity the masses exert on each other. If one of the masses is doubled, the force of gravity between the objects is doubled.

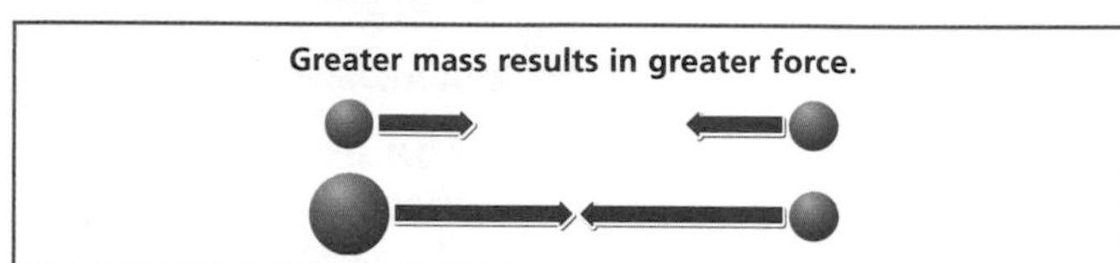
Greater mass results in greater force.

The Distance Between the Objects As distance between the objects increases, the force of gravity decreases. If the distance is doubled, the force of gravity is one-fourth as strong as before.

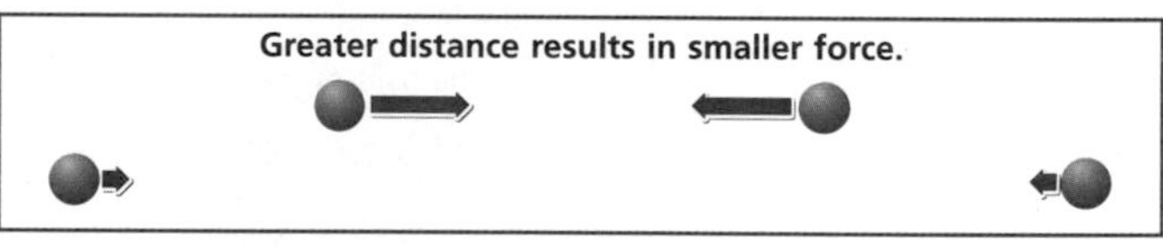
Greater distance results in smaller force.

CHECK YOUR READING
How do mass and distance affect the force of gravity?

Gravity on Earth

The force of gravity acts on both masses equally, even though the effects on both masses may be very different. Earth's gravity exerts a downward pull on a dropped coin. Remember that every action force has an equal and opposite reaction force. The coin exerts an equal upward force on Earth. Because the coin has an extremely small mass compared with Earth, the coin can be easily accelerated. Earth's acceleration due to the force of the coin is far too small to notice because of Earth's large mass.

The acceleration due to Earth's gravity is called g and is equal to 9.8 m/s^2 at Earth's surface. You can calculate the force of gravity on an object using the object's mass and this acceleration. The formula that expresses Newton's second law is $F = ma$. If you use g as the acceleration, the formula for calculating the force due to gravity on a mass close to Earth's surface becomes $F = mg$.

DIFFERENTIATE INSTRUCTION

More Reading Support

A Two objects are exerting force on each other. If one of the masses is doubled, what happens to the force of gravity between the objects? *It is also doubled.*

English Learners This chapter contains many hypothetical sentences that start with *if* and *when* phrases, such as, "*When* you drop any object . . . it falls," (p. 77). Help students understand these abstract ideas by eliminating *when* and separating the sentence into two separate parts. ("You drop an object. It falls.") Point out the cause and effect between the parts of the original sentence.

C 78 Unit: Motion and Forces

Acceleration Due to Gravity

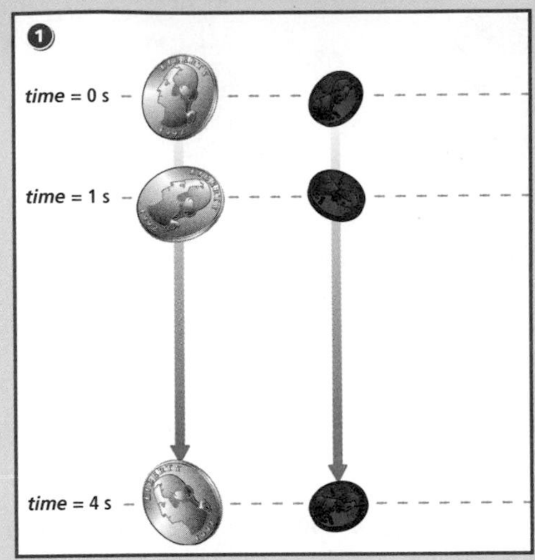

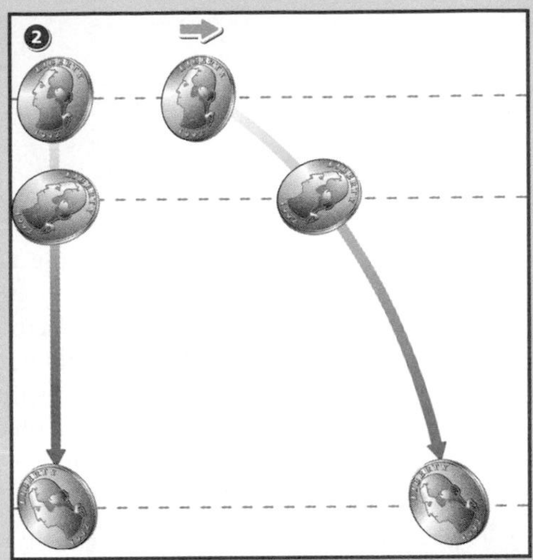

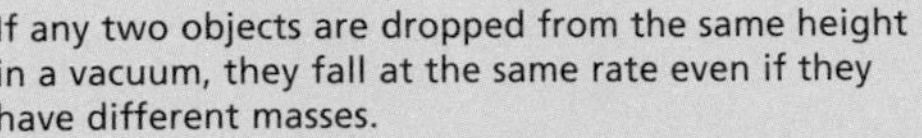

If any two objects are dropped from the same height in a vacuum, they fall at the same rate even if they have different masses.

If an object has a velocity in the horizontal direction when it falls, the horizontal velocity does not change its downward acceleration.

In a vacuum—that is, where there is no air—all falling objects have the same acceleration.

1. The quarter falls at the same rate as the penny when they are dropped together. Because the quarter has more mass, gravity exerts more force on it. But greater mass also means more inertia, so the greater force does not produce a larger acceleration. Objects with different masses fall with the same acceleration.

2. A coin that is dropped falls at the same rate as one that is thrown forward. Horizontal velocity does not affect acceleration due to gravity. Because gravity is directed downward, it changes only the downward velocity of the coin, not its forward velocity.

VISUALIZATION CLASSZONE.COM

Explore how objects fall at the same rate in a vacuum.

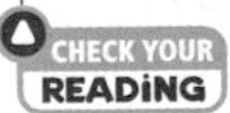
CHECK YOUR READING Compare the times it takes two objects with different masses to fall from the same height.

Weight and Mass

While weight and mass are related, they are not the same properties. Mass is a measure of how much matter an object contains. **Weight** is the force of gravity on an object. Mass is a property that an object has no matter where it is located. Weight, on the other hand, depends on the force of gravity acting on that object.

Teach from Visuals

To help students interpret the visual on acceleration of falling coins:

- Tell students that gravity has a greater pull on the quarter because the quarter has greater mass. Ask: Why do the two coins accelerate at the same rate? *The quarter has greater inertia.*
- Ask: If you dropped a ball from a height of two meters and threw an identical ball horizontally from the same height, which ball will reach the ground first? *They will reach the ground at the same time.*

History of Science

People always have wondered why things move in the ways they do. The ancient Greek philosopher Aristotle proposed that all solid objects eventually fall to Earth simply because it is natural for them to fall toward the center of Earth. He also said that any object in the sky—such as a planet, star, or comet—constantly circles Earth because it naturally moves in a circle. Many of Aristotle's ideas survived up to the 16th century, when scientists such as Galileo and Newton developed different ideas about motion and gravity. These scientists showed how the force of gravity could explain not only why a rock falls to Earth but also why the Moon circles Earth.

Ongoing Assessment

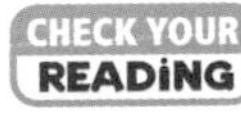
CHECK YOUR READING *Answer: They fall in the same time when there is no air.*

DIFFERENTIATE INSTRUCTION

More Reading Support

B What is the measure of the amount of matter an object contains? *mass*

Advanced The gravitational attraction between two masses is inversely proportional to the square of the distance between them. *Inversely proportional* means that, as one value increases, the other value decreases. Have students explain why, if two masses that are four meters apart are moved to twelve meters apart, the new gravitational attraction will be 1/9 of the original attraction. $\frac{12\ m}{4\ m}$ *= 3 times as far apart;* $\left(\frac{1}{3}\right)^2 = \frac{1}{9}$.

Challenge and Extension, p. 157

Address Misconceptions

IDENTIFY Ask: If a student reading a beam balance determines that a rock is 0.5 kg, what is the student measuring? If students answer weight, they may hold the misconception that mass and weight can be used interchangeably.

CORRECT Explain to students that kilograms are commonly used to describe weight because most measurements are taken on Earth. Ask students what would happen to the weight of the rock if it were taken to the Moon, which has 1/6 of Earth's gravity. *It would weigh less.* Ask them if the amount of matter—its mass—would change. *no*

REASSESS Ask: On Earth, a person who weighs 300 N has a mass of about 31 kg. What would that person's weight be on the Moon? *50 N* What would her mass be? *31 kg*

Technology Resources

Visit **ClassZone.com** for background on common student misconceptions.

MISCONCEPTION DATABASE

History of Science

By the middle 1500s, scientific observations were changing the way people understood astronomical orbits. In Denmark, Tycho Brahe made the most accurate observations of the stars and planets up to that time. The German astronomer Johannes Kepler used Brahe's data to show that the planets don't travel in perfect circles but in ellipses. In the 1600s, Galileo turned his telescope toward the sky. His observations, including observations of moons orbiting Jupiter, led him to champion the theory that Earth orbits the Sun. Regardless of the evidence Galileo offered, many people still did not accept his ideas.

Ongoing Assessment

Explain what keeps objects in orbit.

Ask: What two factors keep an object in orbit? *its speed and gravity*

When you use a balance, you are measuring the mass of an object. A person with a mass of 50 kilograms will balance another mass of 50 kilograms whether she is on Earth or on the Moon. Traveling to the Moon would not change how much matter a person is made of.

When you use a spring scale, such as a bathroom scale, to measure the weight of an object, however, you are measuring how hard gravity is pulling on an object. The Moon is less massive than Earth, and its gravitational pull is one-sixth that of Earth's. A spring scale would show that a person who has a weight of 490 newtons (110 lb) on Earth would have a weight of 82 newtons (18 lb) on the Moon.

Gravity keeps objects in orbit.

Sir Isaac Newton hypothesized that the force that pulls objects to the ground—gravity—also pulls the Moon in its orbit around Earth. An **orbit** is the elliptical path one body, such as the Moon, follows around another body, such as Earth, due to the influence of gravity. The centripetal force keeping one object in orbit around another object is due to the gravitational pull between the two objects.

In the case of the Moon's orbit, the centripetal force is the gravitational pull between the Moon and Earth. Similarly, Earth is pulled around the Sun by the gravitational force between Earth and the Sun.

READING TIP

An ellipse is shaped as shown below. A circle is a special type of ellipse.

You can think of an object orbiting Earth as an object that is falling around Earth rather than falling to the ground. Consider what happens to the ball in the illustration on page 81. A dropped ball will fall about five meters during the first second it falls. Throwing the ball straight ahead will not change that falling time. What happens as you throw faster and faster?

Earth is curved. This fact is noticeable only over very long distances. For every 8000 meters you travel, Earth curves downward about 5 meters. If you could throw a ball at 8000 meters per second, it would fall to Earth in such a way that its path would curve the same amount that Earth curves. Since the ball would fall along the curve of Earth, the ball would never actually land on the ground. The ball would be in orbit.

DIFFERENTIATE INSTRUCTION

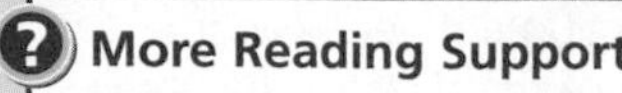

More Reading Support

C Does a spring scale measure mass or weight? *weight*

D What is the centripetal force of the Moon's orbit? *the gravitational attraction of Earth*

Inclusion Tie a metal paper clip to one end of a 30-cm piece of string. Tape the other end of the string to a desk. Have students hold the paper clip straight up in the air. Bring a magnet close to the paper clip, so that it is not touching the paper clip but suspends the clip in the air. Point out that the force of the magnet is stronger than the force of gravity on the paper clip. This activity is suitable for any student; students with a visual impairment can do the activity using touch.

Orbits

An object in orbit, like an object falling to the ground, is pulled toward Earth's center. If the object moves far enough forward as it falls, it orbits around Earth instead of hitting the ground.

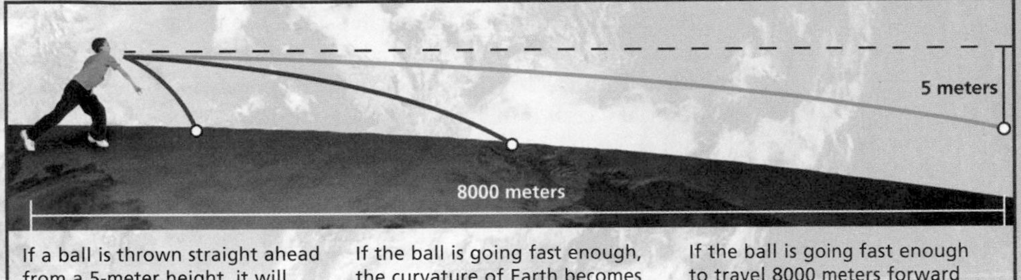

If a ball is thrown straight ahead from a 5-meter height, it will drop 5 meters in the first second it falls. At low speeds, the ball will hit the ground after 1 second.

If the ball is going fast enough, the curvature of Earth becomes important. While the ball still drops 5 meters in the first second, it must fall farther than 5 meters to hit the ground.

If the ball is going fast enough to travel 8000 meters forward as it drops downward 5 meters, it follows the curvature of Earth. The ball will fall around Earth, not into it.

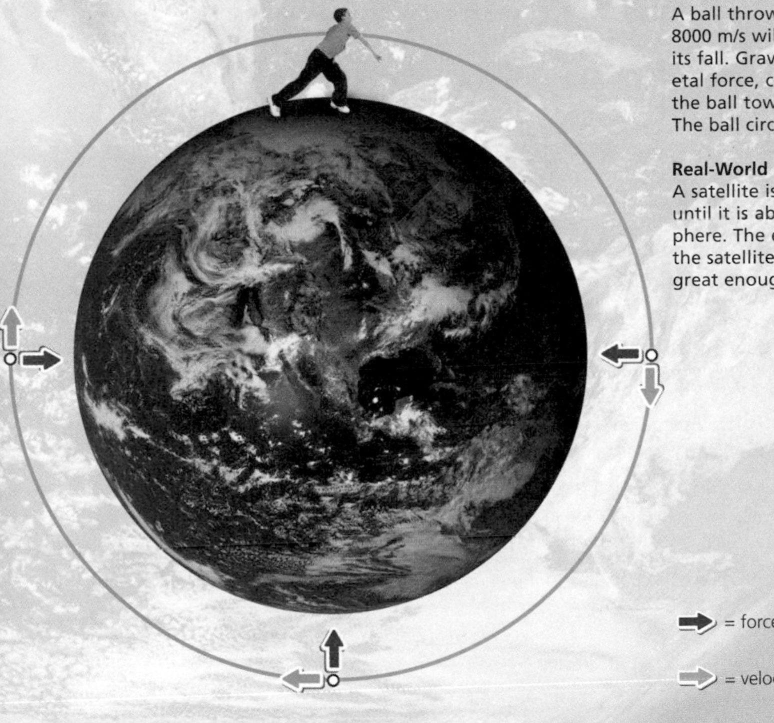

A ball thrown horizontally at 8000 m/s will not hit Earth during its fall. Gravity acts as a centripetal force, continually pulling the ball toward Earth's center. The ball circles Earth in an orbit.

Real-World Application
A satellite is launched upward until it is above Earth's atmosphere. The engine then gives the satellite a horizontal speed great enough to keep it in orbit.

= force

= velocity

READING VISUALS Compare the direction of the velocity with the direction of the force for an object in a circular orbit.

Teach from Visuals

Explain to students that the visual "Orbits" represents a thought experiment. While it is not possible for a person to throw a ball at the speeds described, students can use what they know about throwing a ball to imagine the results if a ball could be thrown at extremely high speeds. Another imaginary factor in the visual is the size of the boy. People are not five meters high. Five meters is closer to the height of a single-story building. Students also may need to be told that the image of the boy is not shown at the same scale as the image of Earth in either picture.

 The visual "Orbits" is available as T22 in the Unit Transparency Book.

Teach Difficult Concepts

You may need to spend extra time on the idea that the velocity of an object in a circular orbit and the force acting on it have different directions. Emphasize that the motion of the object is a combination of its velocity and the centripetal force from gravity.

Ongoing Assessment

READING VISUALS *Answer: The force and the velocity are at right angles to each other.*

DIFFERENTIATE INSTRUCTION

Advanced Have students design and perform an experiment that shows the effects of horizontal velocity on the falling time of an object. Students should include a discussion of which variables need to be held constant and which will change.

INVESTIGATE Gravity

PURPOSE Predict how gravity affects falling objects and check the prediction

TIPS *15 min.*

- Have paper towels available to clean up spills immediately.
- Students should hold the cup straight when they drop it.
- Step 4 can be repeated if needed.

WHAT DO YOU THINK? *In step 3, the water came out the hole in a horizontal stream that fell into the pan. In step 5, all the water in the cup fell with the cup and none came out the hole. The cup and water fell at the same rate.*

CHALLENGE *The first time, the cup was not falling. Gravity pulled on the water. The second time both were falling—the cup could not produce a reaction force on water since the water was not applying a force to the falling cup. Water does apply a force to a still cup.*

 Datasheet, Gravity, p. 158

Technology Resources

Customize this student lab as needed or look for an alternative. Print rubrics to assess student lab reports.

 **Lab Generator CD-ROM**

Metacognitive Strategy

Ask students to write a paragraph about how working with a partner might help them complete the activity.

Ongoing Assessment

CHECK YOUR READING *Answer: Answers might include that gravity on a spaceship is so close to what it is on Earth's surface.*

Spacecraft in Orbit

The minimum speed needed to send an object into orbit is approximately 8000 meters per second. At this speed, the path of a falling object matches the curve of Earth's surface. If you launch a spacecraft or a satellite at a slower speed, it will eventually fall to the ground.

A spacecraft launched at a greater speed can reach a higher orbit than one launched at a lower speed. The higher the orbit, the weaker the force from Earth's gravity. The force of gravity is still very strong, however. If a craft is in a low orbit—about 300 kilometers (190 mi)—Earth's gravitational pull is about 91 percent of what it is at Earth's surface. The extra distance makes a difference in the force of only about 9 percent.

If a spacecraft is launched with a speed of 11,000 meters per second or more, it is moving too fast to go into an orbit. Instead, the spacecraft will ultimately escape the pull of Earth's gravity altogether. The speed that a spacecraft needs to escape the gravitational pull of an object such as a planet or a star is called the escape velocity. A spacecraft that escapes Earth's gravity will go into orbit around the Sun unless it is also going fast enough to escape the Sun's gravity.

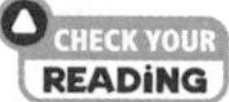 Did any facts in the text above surprise you? If so, which surprised you and why?

INVESTIGATE Gravity

How does gravity affect falling objects?

SKILL FOCUS Predicting

PROCEDURE

1. Carefully use the pencil to punch a hole that is the width of the pencil in the side of the cup, about one-third of the way up from the bottom.
2. Holding your finger over the hole, fill the cup three-fourths full of water.
3. Hold the cup above the dishpan. Predict what will happen if you remove your finger from the hole. Remove your finger and observe what happens.
4. With your finger over the hole, refill the cup to the same level as in step 2. Predict how the water will move if you hold the cup 50 cm above the dishpan and drop the cup and its contents straight down into the pan.
5. Drop the cup and observe what happens to the water while the cup is falling.

MATERIALS
- pencil
- paper cup
- water
- dishpan

TIME 15 minutes

WHAT DO YOU THINK?
- What happened to the water in step 3? in step 5?
- How did gravity affect the water when you dropped the cup?

CHALLENGE Why did the water behave differently the second time?

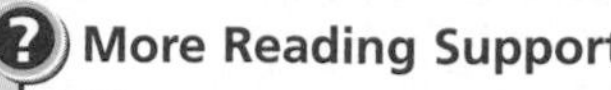

DIFFERENTIATE INSTRUCTION

More Reading Support

E Will a launched spacecraft always go into orbit? *No; if the speed is great enough, it will escape Earth's gravity.*

Alternative Assessment Have students make a poster that shows time-lapse drawings of what happens to the water in steps 3 and 5 of the Investigate.

People in Orbit

Astronaut Mae Jemison is shown here working in a microgravity environment.

When an elevator you are riding in accelerates downward, you may feel lighter for a short time. If you were standing on a scale during the downward acceleration, the scale would show that you weighed less than usual. Your mass would not have changed, nor would the pull of gravity. What would cause the apparent weight loss?

When the elevator is still, the entire force of your weight presses against the scale. When the elevator accelerates downward, you are not pressing as hard on the scale, because the scale is also moving downward. Since the scale measures how hard you are pushing on it, you appear to weigh less. If you and the scale were in free fall—a fall due entirely to gravity—the scale would fall as fast as you did. You would not press against the scale at all, so you would appear to be weightless.

A spacecraft in orbit is in free fall. Gravity is acting on the astronauts and on the ship—without gravity, there could be no orbit. However, the ship and the astronauts are falling around Earth at the same rate. While astronauts are in orbit, their weight does not press against the floor of the spacecraft. The result is an environment, called a microgravity environment, in which objects behave as if there were no gravity. People and objects simply float as if they were weightless.

CHECK YOUR READING Why do astronauts float when they are in orbit?

3.1 Review

KEY CONCEPTS

1. What effect would increasing the mass of two objects have on the gravitational attraction between them?
2. What effect would decreasing the distance between objects have on their gravitational attraction to each other?
3. How does gravity keep the Moon in orbit around Earth?

CRITICAL THINKING

4. **Compare** How does the size of the force exerted by Earth's gravity on a car compare with the size of the force the car exerts on Earth?
5. **Apply** What would be the effect on the mass and the weight of an object if the object were taken to a planet with twice the gravity of Earth?

CHALLENGE

6. **Synthesize** Precision measurements of the acceleration due to gravity show that the acceleration is slightly different in different locations on Earth. Explain why the force of gravity is not exactly the same everywhere on Earth's surface. **Hint:** Think about the details of Earth's surface.

ANSWERS

1. The gravitational attraction would increase.

2. The gravitational attraction would increase.

3. Gravity acts as a centripetal force. A combination of this centripetal force and the velocity of the Moon keep the Moon in orbit.

4. They are equal and opposite.

5. Its weight would be twice as much, but its mass would remain the same.

6. Places that have more mass between the object and the center of Earth, such as a mountain, would have slightly greater gravity.

Ongoing Assessment

CHECK YOUR READING *Answer: The astronauts and the spaceship are falling at the same rate. They are in free fall.*

Reinforce the BIG idea

Have students relate the section to the Big Idea.

Reinforcing Key Concepts, p. 159

3.1 ASSESS & RETEACH

Assess

Section 3.1 Quiz, p. 42

Reteach

Have students list on the board twenty classroom objects of different masses. Assume each of the items is positioned one meter from another item. Have students sequence the items from greatest to smallest amount of gravitational attraction exerted.

Technology Resources

Have students visit **ClassZone.com** for reteaching of Key Concepts.

CONTENT REVIEW

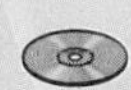
CONTENT REVIEW CD-ROM

EXTREME SCIENCE
Fun and Motivating Science

Set Learning Goal
To understand that light can be bent by extremely massive objects

Present the Science
The bending of light from a distant, bright object by a massive object is called gravitational lensing because the intervening object acts as a lens, focusing the image of the distant, bright object to new locations. If the bright object, the massive object, and Earth are not in a straight line, the light paths travel different distances around the massive object. Because the distances from all objects is so great in space, the size of the massive object and that of the bright object can be considered to be points in space. If the distant object varies in brightness, the brightness of the different images will change at different times. Astronomers can use the time differences to calculate the differences in path lengths. The distance to the bright object can then be calculated.

Discussion Questions
Ask: Why do you think that the bending of light shown here is called gravitational lensing? *Lenses bend light. In this concept, gravity of an extremely massive object bends light.*

Ask: What is the importance of Einstein's prediction? *Sample answer: After Einstein predicted this bending of light, scientists were aware of the concept and recognized it when they observed it.*

Close
Ask: Would it be accurate to state that every very large object could bend light? *No; a large object might not have a large mass, and gravity depends on mass.*

Technology Resources
Students can visit **ClassZone.com** to find out more about gravitational lenses.

RESOURCE CENTER

EXTREME SCIENCE

GRAVITY IN THE EXTREME

Bending Light
You know that gravity can pull objects toward each other, but did you know that gravity can also affect light? Very extreme sources of gravity cause the normally straight path of a light beam to bend.

Going in Circles
Although Earth is massive, the effects of its gravity on light are not noticeable. However, scientists can model what a familiar scene might look like with an extreme source of gravity nearby. The image to the left shows how the light from the Seattle Space Needle could be bent almost into circles if an extremely small yet extremely massive object, such as a black hole, were in front of it.

Seeing Behind Galaxies
How do we know that gravity can bend light? Astronomers, who study space, have seen the phenomenon in action. If a very bright but distant object is behind a very massive one, such as a large galaxy, the mass of the galaxy bends the light coming from the distant object. This effect, called gravitational lensing, can produce multiple images of the bright object along a ring around the massive galaxy. Astronomers have observed gravitational lensing in their images.

Facts About Bending Light

- Gravitational lensing was predicted by Albert Einstein in the early 1900s, but the first example was not observed until 1979.
- The masses of distant galaxies can be found by observing their effect on light.

Seeing Quadruple
This gravitational lens is called the Einstein Cross. The four bright objects that ring the central galaxy are all images of the same very bright yet very distant object that is located 20 times farther away than the central galaxy.

EXPLORE

1. **INFER** Why are you unable to notice the gravitational bending of light by an object such as a large rock?
2. **CHALLENGE** Look at the photographs in the Resource Center. Find the multiple images of the distant objects and the more massive object bending the light from them.

RESOURCE CENTER CLASSZONE.COM Find out more information about gravitational lenses.

C 84

EXPLORE
INFER *The large rock does not have enough mass to produce noticeable bending of light.*

KEY CONCEPT

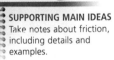

3.2 Friction is a force that opposes motion.

BEFORE, you learned
- Gravity is the attractive force masses exert on each other
- Gravity increases with greater mass and decreases with greater distance
- Gravity is the centripetal force keeping objects in orbit

NOW, you will learn
- How friction affects motion
- About factors that affect friction
- About air resistance

VOCABULARY

friction p. 85
fluid p. 88
air resistance p. 89

THINK ABOUT

What forces help you to walk?

As a person walks, she exerts a backward force on the ground. A reaction force moves her forward. But some surfaces are harder to walk on than others. Ice, for example, is harder to walk on than a dry surface because ice is slippery. How can different surfaces affect your ability to walk?

Friction occurs when surfaces slide against each other.

Have you ever pushed a heavy box across the floor? You probably noticed that it is easier to push the box over some surfaces than over others. You must apply a certain amount of force to the box to keep it moving. The force that acts against your pushing force is called friction. **Friction** is a force that resists the motion between two surfaces in contact.

When you try to slide two surfaces across each other, the force of friction resists the sliding motion. If there were no friction, the box would move as soon as you applied any force to it. Although friction can make some tasks more difficult, most activities, including walking, would be impossible without it. Friction between your feet and the ground is what provides the action and reaction forces that enable you to walk.

SUPPORTING MAIN IDEAS
Take notes about friction, including details and examples.

3.2 FOCUS

Set Learning Goals

Students will
- Describe how friction affects motion.
- List the factors that affect friction.
- Explain air resistance.
- Design an experiment to investigate how the shape of an object affects how it falls.

3-Minute Warm-Up

Display Transparency 20 or copy this exercise on the board:

Gravity is pulling the Moon toward the center of Earth. Write a brief paragraph explaining why the Moon does not crash into Earth. *Gravity is pulling at right angles to the motion of the Moon. The force changes the direction in which the Moon is moving, rather than pulling the Moon downward to Earth's surface. The Moon is in orbit.*

 3-Minute Warm-Up, p. T20

3.2 MOTIVATE

THINK ABOUT

PURPOSE To introduce the concept of friction

DISCUSS Ask students to list previous experiences they have had walking on different surfaces. List them on the board and ask students to group the experiences according to ease of walking.

Examples: On slick surfaces, such as ice, it is more difficult to exert a backward force on the ground. Thus, the reaction force that moves you forward is also less. It is easier to exert a backward force on rougher surfaces.

Ongoing Assessment

Describe how friction affects motion.

Ask: Is it more difficult to move a box across a tile floor or across carpet? *across carpet*

RESOURCES FOR DIFFERENTIATED INSTRUCTION

Below Level

UNIT RESOURCE BOOK
- Reading Study Guide A, pp. 162–163
- Decoding Support, p. 196

 **AUDIO CDS**

Advanced

UNIT RESOURCE BOOK
Challenge and Extension, p. 168

English Learners

UNIT RESOURCE BOOK
Spanish Reading Study Guide, pp. 166–167

 **AUDIO CDS**
- Audio Readings in Spanish
- Audio Readings (English)

3.2 INSTRUCT

Teach Difficult Concepts

Because friction is generally greater between rough surfaces, students may not realize that friction is complicated by more factors than surface roughness. Friction is produced by the interaction of atoms or molecules on surfaces as they resist sliding over each other. In some cases, smoothing two surfaces actually produces greater friction between them.

Real World Example

Friction can have both positive and negative effects in machines. If a moving part of a machine rubs against another part of the machine or against something existing in the environment, the resulting friction can cause damage. For example, an electric drill can overheat because of friction between motor parts. Drill bits can become so hot from the friction between the bit and the material it is drilling through that it burns the user's skin. In a sander, the friction is necessary for the machine to smooth a surface and, in fact, to work at all.

EXPLORE the BIG idea

Revisit "Let It Slide" on p. 75. Have students explain the reasons for their results.

Learn more about friction, forces, and surfaces.

REMINDER

Remember that balanced forces on an object do not change the object's motion.

Forces and Surfaces

If you look down from a great height, such as from the window of an airplane, a flat field appears to be smooth. If you were to walk in the field, however, you would see that the ground has many bumps and holes. In the same way, a flat surface such as a piece of plastic may look and feel smooth. However, if you look at the plastic through a strong microscope, you see that it has tiny bumps and ridges. Friction depends on how these bumps and ridges on one surface interact with and stick to the bumps and ridges on other surfaces. There are several factors that determine the friction between two surfaces.

Types of Surfaces Friction between two surfaces depends on the materials that make up the surfaces. Different combinations of surfaces produce different frictional forces. A rubber hockey puck sliding across ice has a smaller frictional force on it than the same puck sliding across a wooden floor. The friction between rubber and ice is less than the friction between rubber and wood.

Motion of the Surfaces You need a larger force to start something moving than you do to keep something moving. If you have ever tried to push a heavy chair, you may have noticed that you had to push harder and harder until the chair suddenly accelerated forward.

As you apply a force to push a chair or any other object that is not moving, the frictional force keeping it from sliding increases so the forces stay balanced. However, the frictional force has a limit to how

Friction and Motion

Before Object Moves

When an object is standing still, there is a maximum force needed to overcome friction and start it moving. Any force less than this will be exactly balanced by the force of friction, and the object will not move.

While Object Moves

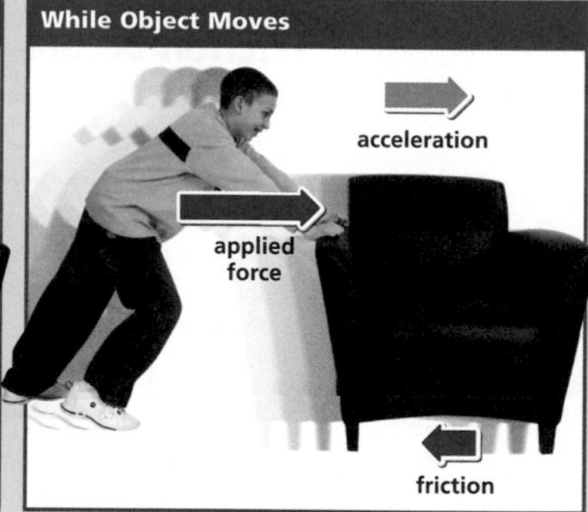

Once the object is moving, the frictional force remains constant. This constant force is less than the maximum force needed to start the object moving.

DIFFERENTIATE INSTRUCTION

A Why might a wooden hockey puck on a wooden floor have a different frictional force than a rubber puck? *Different surface combinations produce different frictional forces.*

English Learners Help students with the complex sentences in this section by breaking each down into smaller sentences and ideas. Then help students make the connections between the phrases in the more complex sentence.

large it can be. When your force is greater than this limit, the forces on the chair are no longer balanced, and the chair moves. The frictional force remains at a new lower level once the chair is moving.

Force Pressing the Surfaces Together The harder two surfaces are pushed together, the more difficult it is for the surfaces to slide over each other. When an object is placed on a surface, the weight of the object presses on that surface. The surface exerts an equal and opposite reaction force on the object. This reaction force is one of the factors that determines how much friction there is.

If you push a chair across the floor, there will be a certain amount of friction between the chair and the floor. Increasing the weight of the chair increases the force pushing the surfaces together. The force of friction between the chair and the floor is greater when a person is sitting in it than when the chair was empty.

Friction depends on the total force pressing the surfaces together, not on how much area this force acts over. Consider a rectangular cardboard box. It can rest with its smaller or larger side on the floor. The box will have the same force from friction regardless of which side sits on the floor. The larger side has more area in contact with the floor than the smaller side, but the weight of the box is more spread out on the larger side.

CHECK YOUR READING What factors influence frictional force? Give two examples.

Friction and Weight

Less Weight

The force of friction depends on the total force pushing the surfaces together. Here the weight of the chair is the force pressing the surfaces together.

More Weight

The weight of the chair increases when someone sits in it. The force of friction is now greater than when the chair was empty.

Teach from Visuals

To help students interpret the visual on friction and weight, ask:

How does increasing the weight in the chair influence force? *Increasing the weight of the chair increases the force of friction between the chair and the floor.*

Develop Critical Thinking

CONNECT To completely understand friction, students should relate the concepts to practical situations.

- Ask: What are some everyday examples illustrating that friction increases because surfaces press together with greater force? *pressing down on a piece of sandpaper that is used to sand wood, filling tractor tires with water or lime to add weight for extra traction, or using a finger to hold down strings while tying a knot*
- Ask: What are some everyday examples illustrating that friction decreases because the texture of the surfaces changes? *tires wearing down, ice forming on a roadway or sidewalk, or using polished metal or plastic to make a slide*

Ongoing Assessment

List the factors that affect friction.

Ask: How could you decrease friction between a dishcloth and a plate? *by changing the surface texture or the force pressing the surfaces together*

CHECK YOUR READING *Answer: types of surface, motion of the surfaces, force pressing the surfaces together; examples will vary*

DIFFERENTIATE INSTRUCTION

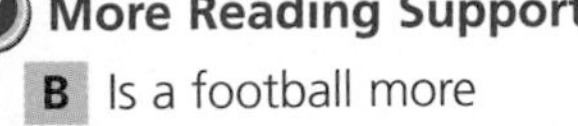

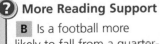

More Reading Support

B Is a football more likely to fall from a quarterback's hand if he holds it tightly or loosely? *loosely*

Below Level Have students make a ramp using a sturdy binder or a board and a stack of textbooks. Have them lay a large hardware nut or washer at the top of the ramp and adjust the height until the nut slides down. Have them place several other materials on the surface of the ramp, such as waxed paper, sandpaper, or cloth, and again observe what happens when the nut or washer is placed at the top. Ask students to explain their observations, which should include that the object will slide more easily on some surfaces because friction is less.

INVESTIGATE
Friction in Air

PURPOSE Design an experiment to determine the relationship between the shape of an object and how it falls.

TIPS *30 min.*

- Emphasize that the pieces of paper must be identical.
- Students should write a procedure and have it approved before they follow it.

WHAT DO YOU THINK? *Paper that has been crumpled will fall more quickly than paper that has not been crumpled. The more tightly it is crumpled, the faster it will fall. Students should explain if results supported their hypothesis. A small, round shape has less air resistance than a flat shape has.*

CHALLENGE *Answers might include air currents in the room or the humidity of the air. Check students' plans for testing.*

 Datasheet, Friction in Air, p. 169

Technology Resources

Customize this student lab as needed or look for an alternative. Print rubrics to assess student lab reports.

 **Lab Generator CD-ROM**

Integrate the Sciences

Some scientists study how much friction and heat earthquakes produce. During an earthquake, large moving pieces that form Earth's crust, called tectonic plates, collide, move apart, or slide past each other. The friction from this sliding should produce a large amount of heat. However, in many cases, scientists do not see the increases in temperature that they expect based on their current understanding of friction.

Friction produces sparks between a match head and a rough surface. The heat from friction eventually lights the match.

Friction and Heat

Friction between surfaces produces heat. You feel heat produced by friction when you rub your hands together. As you rub, friction causes the individual molecules on the surface of your hands to move faster. As the individual molecules in an object move faster, the temperature of the object increases. The increased speed of the molecules on the surface of your hands produces the warmth that you feel.

The heat produced by friction can be intense. The friction that results from striking a match against a rough surface produces enough heat to ignite the flammable substance on the head of the match. In some machines, such as a car engine, too much heat from friction can cause serious damage. Substances such as oil are often used to reduce friction between moving parts in machines. Without motor oil, a car's engine parts would overheat and stop working.

Motion through fluids produces friction.

As you have seen, two objects falling in a vacuum fall with the same acceleration. Objects falling through air, however, have different accelerations. This difference occurs because air is a fluid. A **fluid** is a substance that can flow easily. Gases and liquids are fluids.

INVESTIGATE Friction in Air

How does the shape of an object affect how it falls?

DESIGN — YOUR OWN — EXPERIMENT

Write a hypothesis that explains how shape affects the speed of falling objects. Design an experiment that tests your hypothesis.

PROCEDURE

1. Figure out how you can use the three sheets of paper to test your hypothesis. Remember to control all other variables, including the mass of the paper.
2. Write up your procedure.
3. Conduct your experiment.

WHAT DO YOU THINK?

- What were the results of your experiment?
- Did the results support your hypothesis? Explain your answer.
- Write a statement that summarizes your findings.

CHALLENGE What other variable might affect falling time? How could you test it?

SKILL FOCUS Designing experiments

MATERIALS 3 identical sheets of paper

TIME 30 minutes

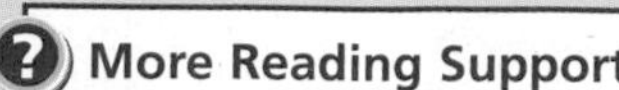

DIFFERENTIATE INSTRUCTION

More Reading Support

C How does friction affect temperature? *Temperature increases.*

D Is the amount of heat produced by friction always the same? *no*

Advanced Vehicle tires are designed to increase friction between the tire and the surface over which it moves. Have students investigate different materials, tread depths, and tread patterns used in tires to increase friction. Challenge students to explain the advantages and disadvantages of each tread choice.

 Challenge and Extension, p. 168

When an object moves through a fluid, it pushes the molecules of the fluid out of the way. At the same time, the molecules of the fluid exert an equal and opposite force on the object that slows it down. This force resisting motion through a fluid is a type of friction that is often called drag. Friction in fluids depends on the shape of the moving object. Objects can be designed either to increase or reduce the friction caused by a fluid. Airplane designs, for example, improve as engineers find ways to reduce drag.

When the force of air resistance equals the force from gravity, a skydiver falls at a constant speed.

The friction due to air is often called **air resistance.** Air resistance differs from the friction between solid surfaces. Air resistance depends on surface area and the speed of an object in the following ways:

- An object with a larger surface area comes into contact with more molecules as it moves than an object with a smaller surface area. This increases the air resistance.
- The faster an object moves through air, the more molecules it comes into contact with in a given amount of time. As the speed of the object increases, air resistance increases.

When a skydiver jumps out of a plane, gravity causes the skydiver to accelerate toward the ground. As the skydiver falls, his body pushes against the air. The air pushes back—with the force of air resistance. As the skydiver's speed increases, his air resistance increases. Eventually, air resistance balances gravity, and the skydiver reaches terminal velocity, which is the final, maximum velocity of a falling object. When the skydiver opens his parachute, air resistance increases still further, and he reaches a new, slower terminal velocity that enables him to land safely.

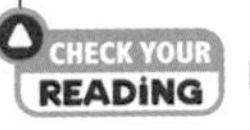
CHECK YOUR READING How do speed and surface area affect air resistance?

3.2 Review

KEY CONCEPTS

1. How does friction affect forward motion? Give an example.
2. Describe two ways to change the frictional force between two solid surfaces.
3. How does air resistance affect the velocity of a falling object?

CRITICAL THINKING

4. **Infer** What two sources of friction do you have to overcome when you are walking?
5. **Synthesize** If you push a chair across the floor at a constant velocity, how does the force of friction compare with the force you exert? Explain.

CHALLENGE

6. **Synthesize** If you push a book against a wall hard enough, it will not slide down even though gravity is pulling it. Use what you know about friction and Newton's laws of motion to explain why the book does not fall.

ANSWERS

1. It opposes forward motion. According to Newton's first law, a moving bicycle would continue to move forward without being pedaled if friction did not stop the motion.

2. Answers might include changing the amount of force, pressing the surfaces together, and changing the textures of the surfaces.

3. Air resistance increases until it balances gravity. At that point, an object reaches its maximum velocity.

4. You must overcome friction between your feet and the ground and air resistance.

5. The force will be equal to the force of friction because the forces are balanced (constant velocity).

6. The hand causes force that increases the force of friction between the book and the wall. Friction balances gravity.

Ongoing Assessment

Explain air resistance.

Ask: Which will fall more slowly, a feather or a button of the same mass? Why? *the feather, because it has more surface area*

CHECK YOUR READING *Answer: Increasing speed and increasing surface area both increase air resistance.*

Reinforce the BIG idea

Have students relate the section to the Big Idea.

 Reinforcing Key Concepts, p. 170

3.2 ASSESS & RETEACH

Assess

 Section 3.2 Quiz, p. 43

Reteach

Provide student groups with two pieces of wood, such as 15-cm lengths of board. Have them write a plan to minimize the amount of friction between the boards. *Suggestions might include sanding the wood or coating it with a friction-reducing material, such as grease or enamel.*

Write an outline on the board, using the section's blue heads as main topics. Invite students to fill in subtopics. Have them use red heads, topic sentences of paragraphs, and vocabulary terms.

Technology Resources

Have students visit **ClassZone.com** for reteaching of Key Concepts.

 CONTENT REVIEW

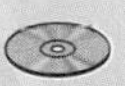 CONTENT REVIEW CD-ROM

MATH IN SCIENCE
Math Skills Practice for Science

Set Learning Goal

To use data and graphs to relate terminal velocity to time and mass

Present the Science

Smoke jumpers are highly skilled firefighters who parachute into areas surrounding forest fires in remote places that are not otherwise accessible.

A parachuter or skydiver has a terminal velocity based on the amount of surface area exposed, and the mass.

To maximize terminal velocity, the skydiver would minimize surface area by keeping arms close to the body and legs close together, aiming the body slightly downward. To minimize terminal velocity, the skydiver would spread out arms and legs. A smoke jumper would probably want to minimize terminal velocity so as to have better control over the landing site.

Develop Graphing Skills

It is standard to plot the independent variable on the horizontal axis of a graph and the dependent variable on the vertical axis.

Have students interpret the graph in the Example. *The graph shows that velocity remained constant after 10 seconds. Therefore, the diver reached terminal velocity at 10 seconds.*

Close

Ask: Which of the data will you graph on the horizontal axis? on the vertical axis? *extra mass; terminal velocity*

- Math Support, p. 199
- Math Practice, p. 200

Technology Resources

Students can visit **ClassZone.com** for practice in graphing.

MATH TUTORIAL

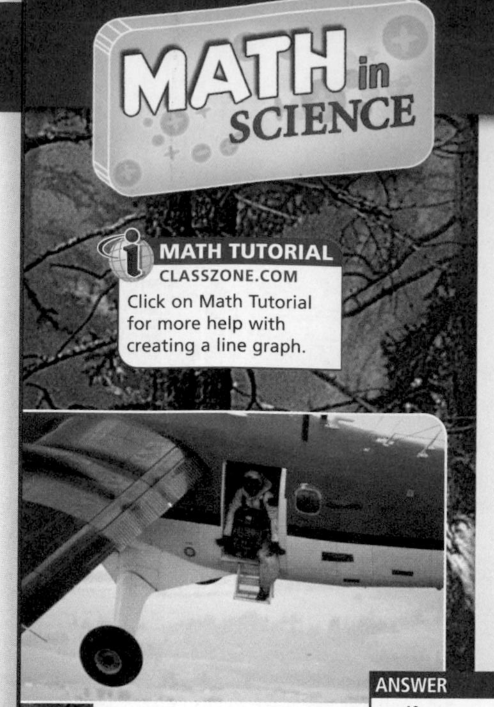

MATH in SCIENCE
MATH TUTORIAL
CLASSZONE.COM
Click on Math Tutorial for more help with creating a line graph.

Smoke jumpers parachute into burning forests in order to contain the flames.

SKILL: CREATING A LINE GRAPH

Smoke Jumpers in Action

Scientists often use graphs as a way to present data. Sometimes information is easier to understand when it is presented in graphic form.

Example

Smoke jumpers are firefighters who parachute down into a forest that is on fire. Suppose you measured how the velocity of a smoke jumper changed as he was free-falling, and recorded the following data:

Time (s)	0	2	4	6	8	10	12	14	16	18
Velocity (m/s)	0	18	29	33	35	36	36	36	36	36

Follow these steps to make a line graph of the data in the table.

(1) For both variables, decide the scale that each box on your graph will represent and what range you will show for each variable. For the above time data you might choose a range of 0 to 18 s, with each interval representing 2 s. For velocity, a range of 0 to 40 m/s with intervals of 5 m/s each is reasonable.

(2) Determine the dependent and independent variables. In this example, the velocity depends on the falling time, so velocity is the dependent variable.

(3) Plot the independent variable along the horizontal axis, or *x*-axis. Plot the dependent variable along the vertical axis, or *y*-axis. Connect the points with a smooth line.

ANSWER

Velocity (m/s): 0, 5, 10, 15, 20, 25, 30, 35, 40
Time (s): 2, 4, 6, 8, 10, 12, 14, 16, 18

Use the data below to answer the following questions.

Suppose a smoke jumper varied the mass of his equipment over 5 jumps, and you measured his different terminal velocities as follows:

Extra Mass (kg)	0	5	10	15	20
Terminal Velocity (m/s)	36	37	38	39	40

1. Identify the independent and dependent variables.
2. Choose the scales and intervals you would use to graph the data. **Hint:** Your velocity range does not have to start at 0 m/s.
3. Plot your graph.

CHALLENGE How do different scales give different impressions of the data? Try comparing several different scales for the same data.

C 90

ANSWERS

1. The independent variable is extra mass, and the dependent variable is terminal velocity.

2. Scales and intervals will vary, but intervals for each axis should be equal.

3. The graph will be a straight line going from (0, 36) to (20, 40).

CHALLENGE *Different scales can make the line look more or less steep.*

KEY CONCEPT

Pressure depends on force and area.

BEFORE, you learned

- Frictional forces oppose motion when surfaces resist sliding
- Frictional force depends on the surface types and the total force pushing them together
- Air resistance is a type of friction on objects moving through air

NOW, you will learn

- How pressure is determined
- How forces act on objects in fluids
- How pressure changes in fluids

VOCABULARY

pressure p. 91
pascal p. 92

EXPLORE Pressure

How does surface area affect pressure?

PROCEDURE

1. Place the pencil flat on the Styrofoam board. Balance the book on top of the pencil. After 5 seconds, remove the book and the pencil. Observe the Styrofoam.
2. Balance the book on top of the pencil in an upright position as shown. After 5 seconds, remove the book and the pencil. Observe the Styrofoam.

MATERIALS

- sharpened pencil
- Styrofoam board
- book

WHAT DO YOU THINK?

- How did the effect on the Styrofoam change from step 1 to step 2?
- What do you think accounts for any differences you noted?

Pressure describes how a force is spread over an area.

VOCABULARY
Create a four square diagram for *pressure* in your notebook.

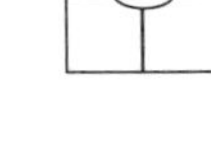

Pressure is a measure of how much force is acting on a certain area. In other words, pressure describes how concentrated a force is. When a cat lies down on your lap, all the force of the cat's weight is spread out over a large area of your lap. If the cat stands up, however, all the force from the cat's weight is concentrated into its paws. The pressure the cat exerts on you increases when the cat stands up in your lap.

While the increased pressure may make you feel as if there is more force on you, the force is actually the same. The cat's weight is simply pressing on a smaller area. How you feel a force when it is pressing on you depends on both the force and the area over which it is applied.

3.3 FOCUS

Set Learning Goals

Students will

- Explain how pressure is determined.
- Describe how forces act on objects in fluids.
- Describe pressure changes in fluids.

3-Minute Warm-Up

Display Transparency 21 or copy this exercise on the board:

Write a paragraph about what happens when you try to walk on ice. Compare walking on ice to walking on a sidewalk in the summer. *Students' paragraphs should discuss the role that friction between surfaces plays in walking.*

3-Minute Warm-Up, p. T21

3.3 MOTIVATE

EXPLORE Pressure

PURPOSE To introduce how surface area affects pressure

TIP *10 min.* A small piece of Styrofoam of about five square centimeters is adequate. Students should balance the book with their hand and not press down on it.

WHAT DO YOU THINK? *The total force remained the same. The effect was greater in step 2 because the force was exerted over a smaller area.*

RESOURCES FOR DIFFERENTIATED INSTRUCTION

Below Level

UNIT RESOURCE BOOK

- Reading Study Guidel A, pp. 173–174
- Decoding Support, p. 196

AUDIO CDS

Advanced

UNIT RESOURCE BOOK
Challenge and Extension, p. 179

English Learners

UNIT RESOURCE BOOK
Spanish Reading Study Guide, pp. 177–178

AUDIO CDS

- Audio Readings in Spanish
- Audio Readings (English)

3.3 INSTRUCT

Teacher Demo

Use shoes with different sized heels to demonstrate how pressure changes when area changes and force stays constant. Wear shoes such as athletic shoes and stand on a piece of soft wood, Styrofoam, or corrugated cardboard. Switch to shoes with increasingly narrower heels to show how pressure increases, as shown by increased indentations, as the area of the heel decreases. You can use a student volunteer, but it is important that one person performs the entire activity so that force remains constant throughout. You must use a new piece of Styrofoam or cardboard each time.

Develop Algebra Skills

Review with students the importance of using the correct formula and correct units when solving problems.

- Show students how to use the pressure formula to solve for force or area. Remind them that multiplication and division are opposite operations, and they "undo" each other. Emphasize that an operation must be done to both sides of the equation.
- Units can be especially tricky when students cannot easily see their relationship within the problem. The relationship among square meters, newtons, and pascals is not obvious. Remind students that area must be in square meters for these problems.

- Math Support, p. 197
- Math Practice, p. 198

Ongoing Assessment

Explain how to determine pressure.

Ask: What two quantities do you need to know to calculate pressure? *force and area*

Practice the Math *Answers:*

1. $P = \frac{F}{A} = \frac{500\ N}{0.075\ m^2} = 6670\ Pa$

2. $F = PA = 2000\ \frac{N}{m^2} \cdot 20\ m^2 = 40{,}000\ N$

One way to increase pressure is to increase force. If you press a wall with your finger, the harder you press, the more pressure you put on the wall. But you can also increase the pressure by decreasing the area. When you push a thumbtack into a wall, you apply a force to the thumbtack. The small area of the sharp point of the thumbtack produces a much larger pressure on the wall than the area of your finger does. The greater pressure from the thumbtack can pierce the wall, while the pressure from your finger alone cannot.

The following formula shows exactly how pressure depends on force and area:

$$\text{Pressure} = \frac{\text{Force}}{\text{Area}} \qquad P = \frac{F}{A}$$

READING TIP Notice that when a unit, such as pascal or newton, is named for a person, the unit is not capitalized but its abbreviation is.

In this formula, P is the pressure, F is the force in newtons, and A is the area over which the force is exerted, measured in square meters (m^2). The unit for pressure is the **pascal** (Pa). One pascal is the pressure exerted by one newton (1 N) of force on an area of one square meter (1 m^2). That is, one pascal is equivalent to one N/m^2. B

Sometimes knowing pressure is more useful than knowing force. For example, many surfaces will break or crack if the pressure on them is too great. A person with snowshoes can walk on top of snow, while a person in hiking boots will sink into the snow.

COMPARE How does the pressure from her snowshoes compare to the pressure from her boots?

Calculating Pressure

Sample Problem

A winter hiker weighing 500 N is wearing snowshoes that cover an area of 0.2 m². What pressure does the hiker exert on the snow?

What do you know? Area = 0.2 m^2, Force = 500 N

What do you want to find out? Pressure

Write the formula: $P = \frac{F}{A}$

Substitute into the formula: $P = \frac{500\ N}{0.2\ m^2}$

Calculate and simplify: $P = 2500\ \frac{N}{m^2} = 2500\ N/m^2$

Check that your units agree: Unit is N/m^2. Unit of pressure is Pa, which is also N/m^2. Units agree.

Answer: $P = 2500$ Pa

Practice the Math

1. If a winter hiker weighing 500 N is wearing boots that have an area of 0.075 m^2, how much pressure is exerted on the snow?
2. A pressure of 2000 Pa is exerted on a surface with an area of 20 m^2. What is the total force exerted on the surface?

DIFFERENTIATE INSTRUCTION

More Reading Support

A If force increases, what happens to pressure? *It increases.*

B What unit is used for pressure? *pascal*

English Learners English learners may not have prior knowledge of the snowshoes mentioned on this page, elevation on p. 94, and ice skating on p. 95. The Chapter Investigation on p. 96 asks students to write hypotheses in the form of "If . . . then . . . because . . ." statements. This concept should be taught directly to English learners. Offer examples to illustrate how the sentence structure works. For example, "If you climb a mountain, then the air pressure will decrease, because air is less dense at higher elevations."

Pressure acts in all directions in fluids.

Explore how a fluid produces pressure.

Fluids are made of loosely connected particles that are too small to see. These particles are in constant, rapid motion. The motion is random, which means particles are equally likely to move in any direction. Particles collide with—or crash into—one another and into the walls of a container holding the fluid. The particles also collide with any objects in the fluid.

As particles collide with an object in the fluid, they apply a constant force to the surfaces of the object. This force produces a pressure against the surfaces that the particles come in contact with. A fluid contains many particles, each moving in a different direction, and the force from each particle can be exerted in any direction. Therefore, the pressure exerted by the fluid acts on an object from all directions.

The diver in the picture below experiences a constant pressure from the particles—or molecules—in the water. Water molecules are constantly hitting her body from all directions. The collisions on all parts of her body produce a net force on the surface of her body.

CHECK YOUR READING How does understanding particle motion help you understand fluid pressure?

Pressure in Fluids

Randomly moving water molecules collide with a diver. The net force from the many collisions produces the pressure on the diver.

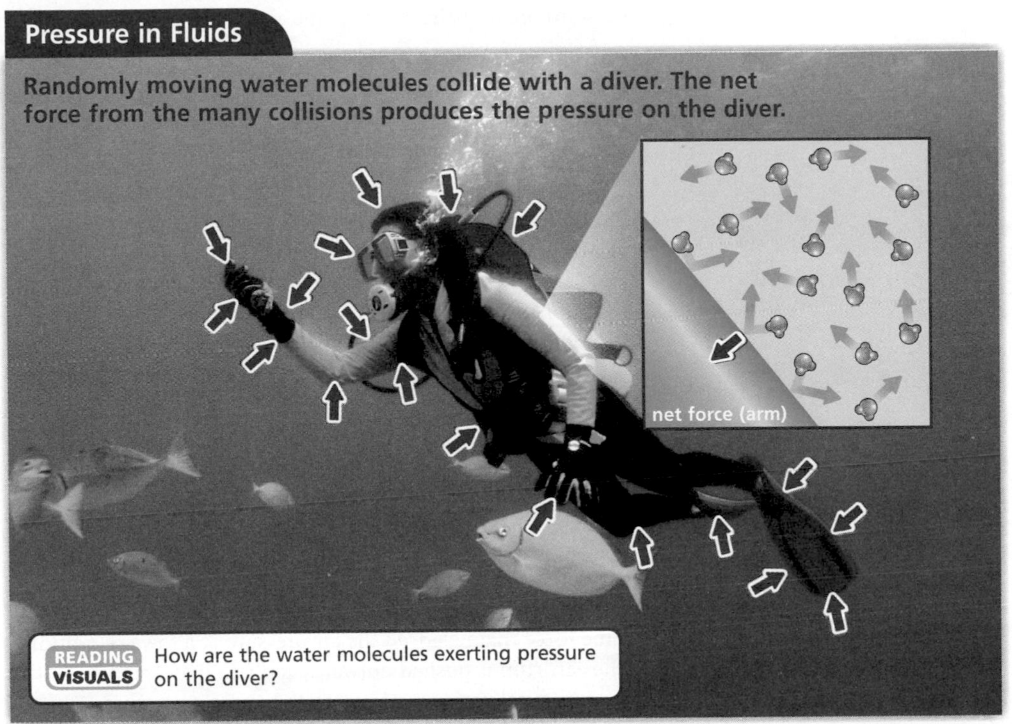

READING VISUALS How are the water molecules exerting pressure on the diver?

DIFFERENTIATE INSTRUCTION

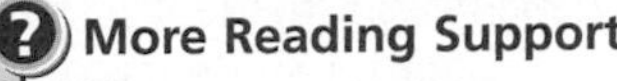
More Reading Support

C What is the result of collisions of particles with an object? *pressure on the object*

D In which direction does pressure act on an object in a fluid? *in all directions*

Alternative Assessment Have students make a diagram of the heart or the ear and show where and why pressure increases or decreases.

Integrate the Sciences

Pressure exerted by fluids has many biological implications.

Blood exerts pressure against the walls of blood vessels. The pressure in arteries can be determined by using a blood pressure cuff and a stethoscope. Blood pressure is reported as a fraction. The top number is the pressure of blood when the heart is beating. The lower number measures the pressure of blood when the heart is resting.

Differences in air pressure and the pressure of air in the middle ear can cause ear pain. This pressure difference occurs when the eustachian tube does not open enough to equalize the pressures. This problem most often occurs when air pressure is much less than usual, such as in an airplane.

EXPLORE the BIG idea

Revisit "Under Pressure" on p. 75. Have students explain the reasons for their results.

Ongoing Assessment

Describe how forces act on objects in fluids.

Ask: A balloon is floating in the air. In which direction is pressure acting on the balloon? *in all directions*

CHECK YOUR READING *Answer: A fluid contains many particles, and the forces from each particle can be exerted in any direction. Therefore, fluid pressure acts on an object from all directions.*

READING VISUALS *Answer: The water molecules are hitting the diver from all directions. The collisions produce a net force which produces pressure on the diver.*

Develop Critical Thinking

APPLY One practical use of air pressure is scuba diving. Pressurized gases are placed into tanks that divers use as a source of air for breathing when diving.

- Remind students that pressure is caused by collisions with an object, such as the walls of a container. Ask: What happens to the pressure of gases as they are forced into the tanks? *Pressure increases because the more particles, the more collisions. In this case, the gas particles hit the container more frequently.*
- Certain parts of the body are compressible. That is, these parts become smaller in volume when pressure is increased, such as when a diver goes deeper into water. These parts are the middle ear, the sinuses, and the lungs. Ask: What do these body parts have in common that would make them affected by increased pressure? *They all contain some air.*
- The air in a scuba tank exerts a pressure that is about 200 times greater than normal atmospheric pressure. Ask: Why does a scuba tank need a regulator to control the pressure of the air that is breathed? *If the pressure were not reduced to an appropriate level, it might damage lung tissue.*

Teach Difficult Concepts

Relate friction and pressure to an everyday example. Ask: If you are going on a long road trip, why might the tires on your car have a greater pressure when you reach your destination than they did when you began your journey? *Friction between the tires and the road heats the tires, raising the pressure.*

Pressure in fluids depends on depth.

The pressure that a fluid exerts depends on the density and the depth of the fluid. Imagine that you have a tall cylinder sitting on the palm of your hand. As you fill the cylinder with water, the force of the water's weight exerts more and more pressure on your hand. The force of the water's weight increases as you put in more water.

Suppose you had two identical cylinders of water sitting on your hand. The cylinders would push with twice the weight of a single cylinder, but the force would be spread over twice the area. Therefore, the pressure would still be the same. The pressure does not depend on the total volume of the fluid, only on the depth and density.

Pressure in Air

Although you do not notice the weight of air, air exerts pressure on you at all times. At sea level, air exerts a pressure on you equal to about 100,000 pascals. This pressure is called atmospheric pressure and is referred to as one atmosphere. At this pressure, every square centimeter of your body experiences a force of ten newtons (2.2 lb). You do not notice it pushing your body inward, however, because the materials in your body provide an equal outward pressure that balances the air pressure.

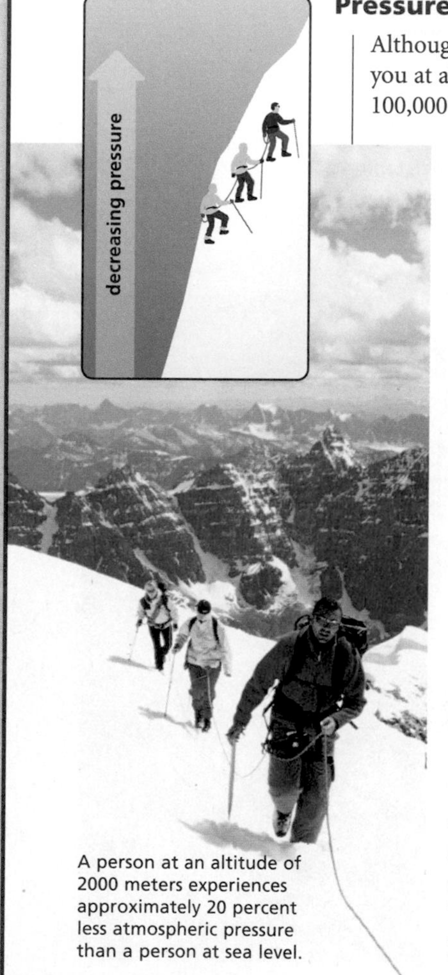

A person at an altitude of 2000 meters experiences approximately 20 percent less atmospheric pressure than a person at sea level.

Changing Elevation Air has weight. The more air there is above you, the greater the weight of that air. As you climb a mountain, the column of air above you is shorter and weighs less, so the pressure of air on you at higher elevations is less than one atmosphere. ? E

Changing Density The air at the top of a column presses down on the air below it. The farther down the column, the more weight there is above to press downward. Air at lower elevations is more compressed, and therefore denser, than air at higher elevations.

Effects on Pressure Pressure is exerted by individual molecules colliding with an object. In denser air, there are more molecules—and therefore more collisions. An increase in the number of collisions results in an increase in the force, and therefore pressure, exerted by the air.

As you travel up a mountain, the air pressure on you decreases. For a short time, the pressure on the inside surface of your eardrum may continue to push out with the same force that balanced the air pressure at a lower elevation. The eardrum is pushed outward, and you may feel pain until your internal pressure adjusts to the new air pressure.

DIFFERENTIATE INSTRUCTION

More Reading Support

E Will the pressure at the top of a mountain be greater than one atmosphere or less than one atmosphere? *less*

Advanced The fact that pressure depends on the depth of a fluid, not the total volume, can be difficult for students to accept. Have advanced students create a visual that will help convince other students that pressure depends on depth.

R Challenge and Extension, p. 179

Pressure in Water

Unlike air molecules, water molecules are already very close together. The density of water does not change very much with depth. However, the deeper you go underwater, the more water there is above you. The weight of that water above you produces the water pressure acting on your body. Just as air pressure increases at lower elevations, water pressure increases with greater water depth.

A deep-diving whale at 1000 meters below the surface experiences about 34 times more pressure than a turtle diving to a depth of 20 meters (65 ft).

Water exerts more pressure on you than air does because water has a greater density than air. Therefore, the change in weight of the column of water above you as you dive is greater for each meter that you descend than it is in air. There is a greater difference in pressure if you dive ten meters farther down in the ocean than if you walked ten meters down a mountain. In fact, ten meters of water above you applies about as much pressure on you as the entire atmosphere does.

If you were to dive 1000 meters (3300 ft) below the surface of the ocean, the pressure would be nearly 100 times greater than pressure from the atmosphere. The force of this pressure would collapse your lungs unless you were protected by special deep-sea diving equipment. As scientists explore the ocean to greater depths, new underwater vehicles are designed that can withstand the increase in water pressure. Some whales, however, can dive to a depth of 1000 meters without being injured. As these whales dive to great depths, their lungs are almost completely collapsed by the pressure. However, the whales have adapted to the collapse—they store most of their oxygen intake in their muscles and blood instead of within their lungs.

CHECK YOUR READING Why is water pressure greater than air pressure?

3.3 Review

KEY CONCEPTS

1. How is pressure related to force and surface area?
2. Describe the way in which a fluid exerts pressure on an object immersed in it.
3. How does changing elevation affect air pressure? How does changing depth affect water pressure?

CRITICAL THINKING

4. **Calculate** If a board with an area of 3 m^2 has a 12 N force exerted on it, what is the pressure on the board?
5. **Infer** What might cause a balloon blown up at a low altitude to burst if it is taken to a higher altitude?

CHALLENGE

6. **Synthesize** During cold winters, ice can form on small lakes and ponds. Many people enjoy skating on the ice. Occasionally, a person skates on thin ice and breaks through it. Why do rescue workers lie flat on the ice instead of walking upright when reaching out to help rescue a skater?

ANSWERS

1. *Pressure describes how force is spread over an area.*
2. *Pressure is exerted in all directions.*
3. *Increasing elevation decreases air pressure. As water becomes deeper, its pressure increases.*
4. $P = F/A = 12\ N/ 3\ m^2 = 4\ Pa$
5. *As the balloon is taken to a higher altitude, outside pressure decreases. The balloon expands because the inside pressure becomes greater than the outside air pressure. If the balloon expands enough, it will burst.*
6. *The weight (force) of the person causes less pressure if exerted over a larger area. The area over which the force is exerted is much greater for a person lying on the ice than for the same person standing up on it.*

Ongoing Assessment

Describe pressure changes in fluids.

Ask: Will a diver experience greater pressure 20 m below the surface of the water or 30 m below the surface? *30 m*

CHECK YOUR READING *Answer: Water is denser than air.*

Reinforce the BIG idea

Have students relate the section to the Big Idea.

Reinforcing Key Concepts, p. 180

3.3 ASSESS & RETEACH

Assess

Section 3.3 Quiz, p. 44

Reteach

Have students solve problems in which they are given two values of pressure, area, and force and are asked to find the value of the unknown variable. Have them write the correct form of the equation and show their work.

Sample problems

1. A box with a weight of 25 N is placed on a brick with an area of 0.12 m^2. How much pressure does the box exert on the brick? $P = F/A = 25\ N/0.12\ m^2 = 208\ Pa$
2. What force is exerted on 14 m^2 if the pressure is 300 Pa? $F = PA = 300\ Pa \cdot 14\ m^2 = 4200\ N$
3. A force of 500 N produces a pressure of 125 Pa. Over what area is the force applied? $A = F/P = 500\ N/125\ Pa = 4\ m^2$

Technology Resources

Have students visit **ClassZone.com** for reteaching of Key Concepts.

CONTENT REVIEW

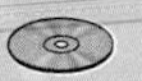

CONTENT REVIEW CD-ROM

CHAPTER INVESTIGATION

Focus

PURPOSE Students will learn how water pressure changes when depth and volume of the water change.

OVERVIEW Students will determine how much pressure is produced by varying volumes and depths of water. Students will determine that

- Pressure is the same for a larger volume
- Pressure is greater for a greater depth

Lab Preparation

- Ask students to bring plastic bottles of different sizes from home several days before doing the lab. For consistency, use only bottles with smooth sides.
- To save class time and for student safety, carefully cut the tops off the bottles ahead of time with scissors or a sharp knife.
- Prior to the investigation, have students read through the investigation and prepare their data tables. Or you may wish to copy and distribute datasheets and rubrics.

UNIT RESOURCE BOOK, pp. 201–209

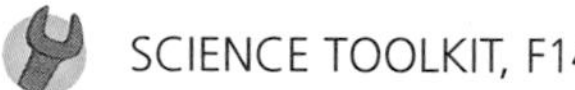
SCIENCE TOOLKIT, F14

Lab Management

- Advise students to allow for all depths down to four centimeters when making their data tables.
- Warn students not to squeeze the bottles when they put their finger over the hole. Squeezing the bottle will give a volume reading that is too high.
- Have rags available for cleaning up any spills.

SAFETY Students should use caution in cutting the bottles and making the holes. Clean up any spills immediately.

INCLUSION Use colored tape to mark off the meter stick in measurements of 10 cm to help students see the numbers.

CHAPTER INVESTIGATION

Pressure in Fluids

OVERVIEW AND PURPOSE When you put your hand under a faucet, you experience water pressure. Underwater explorers also experience water pressure. In this investigation you will

- change the depth and volume of a column of water
- determine what factors affect pressure

Problem

What factors affect water pressure?

Hypothesize

Write two hypotheses to explain what you expect to happen to the water pressure as you change the depth and volume of the water column. Your hypotheses (one for depth, one for volume) should take the form of "If . . . , then . . . , because . . ." statements.

MATERIALS
- nail
- 2 plastic bottles, small and large, with tops cut off
- ruler
- plastic container
- meter stick
- coffee can
- water

Procedure

1. Create a data table like the one shown on the sample notebook page.
2. Using a nail, poke a hole in the side of each bottle 4 cm from the bottom of the bottle.
3. Set up the materials as shown on the left. Put a ruler in the small bottle so that the lower numbers are at the bottom.
4. Put your finger over the hole so no water will squirt out. Add or remove water (by lifting your finger off the hole) so that the water level is exactly at the 12 cm mark.
5. Release your finger from the hole, while your partner reads the exact mark where the water hits the meter stick. Cover the hole immediately after your partner reads the distance the water squirted. Record the distance on the line for this depth in your table.

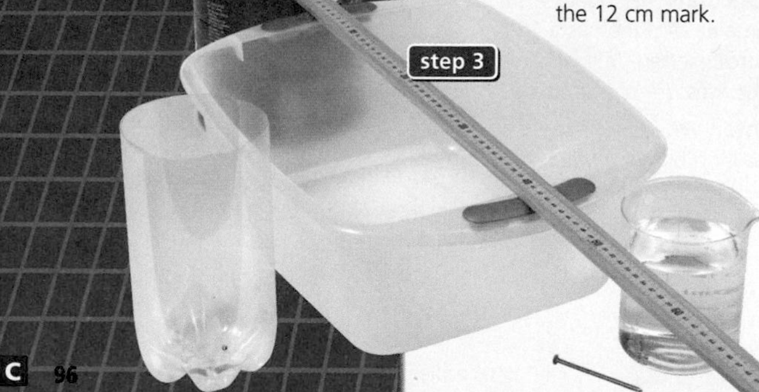

C 96

INVESTIGATION RESOURCES

CHAPTER INVESTIGATION, Pressure in Fluids
- Level A, pp. 201–204
- Level B, pp. 205–208
- Level C, p. 209

Advanced students should complete Levels B & C.

Writing a Lab Report, D12–13

Technology Resources

Customize this student lab as needed or look for an alternative. Print rubrics to assess student lab reports.

Lab Generator CD-ROM

6 Add or remove water so that the water level is now exactly at the 11 cm mark. Repeat step 5.

7 Continue adding, removing, and squirting water at each whole centimeter mark until no more water squirts from the bottle.

8 Repeat steps 4–7 two more times for a total of three trials.

9 Repeat steps 4–8 using the large bottle.

Observe and Analyze

Write It Up

1. **RECORD OBSERVATIONS** Be sure that your data table is complete.
2. **GRAPH** Construct a graph showing distance versus depth. Draw two curves, one for the small bottle and one for the large bottle. Use different colors for the two curves.
3. **IDENTIFY VARIABLES AND CONSTANTS** List the variables and constants for the experiment using the small bottle and the experiment using the large bottle.
4. **ANALYZE** Is the depth greater when the bottle is more full or more empty? When did the water squirt farther, when the bottle was more full or more empty?
5. **ANALYZE** Did the water squirt farther when you used the small or the large bottle?

Conclude

Write It Up

1. **INTERPRET** Answer the question posed in the problem.
2. **ANALYZE** Examine your graph and compare your results with your hypotheses. Do your results support your hypotheses?
3. **INFER** How does depth affect pressure? How does volume affect pressure?
4. **IDENTIFY LIMITS** What possible limitations or errors did you experience or could you have experienced with this investigation?
5. **APPLY** Dams store water for irrigation, home use, and hydroelectric power. Explain why dams must be constructed so that they are much thicker at the bottom than at the top.
6. **APPLY** Have you ever dived to the bottom of a swimming pool to pick up a coin? Describe what you felt as you swam toward the bottom.

INVESTIGATE Further

CHALLENGE Repeat the investigation using a liquid with a density that is quite different from water. Measure the distance the liquid travels, and graph the new data in a different color. Is there a difference? Why do you think there is or is not a difference in pressure between liquids of different densities?

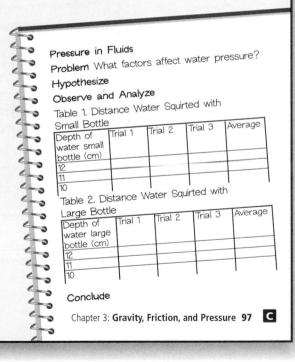

Pressure in Fluids

Problem What factors affect water pressure?

Hypothesize

Observe and Analyze

Table 1. Distance Water Squirted with Small Bottle

Depth of water small bottle (cm)	Trial 1	Trial 2	Trial 3	Average
12				
11				
10				

Table 2. Distance Water Squirted with Large Bottle

Depth of water large bottle (cm)	Trial 1	Trial 2	Trial 3	Average
12				
11				
10				

Conclude

Observe and Analyze

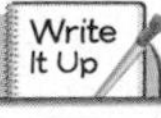

SAMPLE DATA Depth of Water, Small Bottle, 11 cm; Distance, 12 cm, 11.5 cm, 14 cm; Distance Average, 12.5 cm; Depth of Water, Large Bottle, 11 cm; Distance, 11 cm, 12.5 cm, 11.5 cm; Distance Average, 11.2 cm

2. *Graphs should show depth as the independent variable on the horizontal axis and distance on the vertical axis. The curve will be the same for both bottles.*
3. *For both, the independent variable is depth, and the dependent variable is distance. Volume at the same depth is the independent variable. Constants include water temperature, height of the bottle, location of the meter stick, and fluid used.*
4. *When the bottle was more full, the water squirted farther and the depth was greater.*
5. *The water squirted the same distance from the larger bottle.*

Conclude

1. *Depth affects water pressure.*
2. *Student answers will vary.*
3. *Greater depth results in greater pressure. Volume does not affect pressure.*
4. *Sample answers: distance on the meter stick could be difficult to determine; placement of the bottles on the can was not the same.*
5. *Water pressure is greater at the bottom.*
6. *Pressure increases as you get closer to the bottom.*

INVESTIGATE Further

SAFETY Do not let students use liquids that are flammable or toxic, or have hazardous vapors.

CHALLENGE Students' results will depend on the viscosity of the liquid they choose. However, if the hole is large enough, the results should be similar to the original experiment.

Post-Lab Discussion

- Ask students to compare their results to those of others. They should note that the curves produced are similar but might not be exactly the same. Ask them to compare the results with the results attained when different bottles were used.
- Discuss how the results of this investigation could be used in designing a swimming pool.

3.4 FOCUS

Set Learning Goals

Students will

- Explain how fluids apply forces to objects.
- Describe how the motion of a fluid affects the pressure it exerts.
- Explain how forces are transmitted through fluids.
- Observe through experimentation how the speed of air affects air pressure.

3-Minute Warm-Up

Display Transparency 21 or copy this exercise on the board:

Decide if these statements are true. If not true, correct them.

1. A person exerts the same force on a floor whether she is lying on it or standing on it. *true*
2. A force is applied over two square centimeters. The same force is applied over four square centimeters. The amount of force per square centimeter is the same for both. *The amount of force per square centimeter differs.*
3. If $F = mg$, then m can be found by multiplying g by F. *If $F = mg$, then m can be found by dividing F by g.*

3-Minute Warm-Up, p. T21

3.4 MOTIVATE

EXPLORE Forces in Liquid

PURPOSE To observe how water applies a force on objects

TIP *10 min.* Use large paper clips so that students can easily tie string to them.

WHAT DO YOU THINK? *The paper clips in water seemed lighter, and they seemed out of balance. The water was applying an upward force on them.*

KEY CONCEPT

3.4 Fluids can exert a force on objects.

BEFORE, you learned

- Pressure depends on force and area
- Pressure acts in all directions in fluids
- Density is mass divided by volume

NOW, you will learn

- How fluids apply forces to objects
- How the motion of a fluid affects the pressure it exerts
- How forces are transmitted through fluids

VOCABULARY

buoyant force p. 98
Bernoulli's principle p. 100
Pascal's principle p. 102

EXPLORE Forces in Liquid

How does water affect weight?

PROCEDURE

1. Tie a piece of string to the middle of the pencil. Tie 4 paper clips to each end of the pencil as shown.
2. Move the middle string along the pencil until the paper clips are balanced and the pencil hangs flat.
3. While keeping the pencil balanced, slowly lower the paper clips on one end of the pencil into the water. Observe what happens.

MATERIALS

- 3 pieces of string
- pencil
- 8 paper clips
- cup full of water

WHAT DO YOU THINK?

- How did the water affect the balance between the two sets of paper clips?
- Did the water exert a force on the paper clips? Explain.

Fluids can exert an upward force on objects.

If you drop an ice cube in air, it falls to the floor. If you drop the ice cube into water, it may sink a little at first, but the cube quickly rises upward until it floats. You know that gravity is pulling downward on the ice, even when it is in the water. If the ice cube is not sinking, there must be some force balancing gravity that is pushing upward on it.

The upward force on objects in a fluid is called **buoyant force,** or buoyancy. Buoyancy is why ice floats in water. Because of buoyant force, objects seem lighter in water. For example, it is easier to lift a heavy rock in water than on land because the buoyant force pushes upward on the rock, reducing the net force you need to lift it.

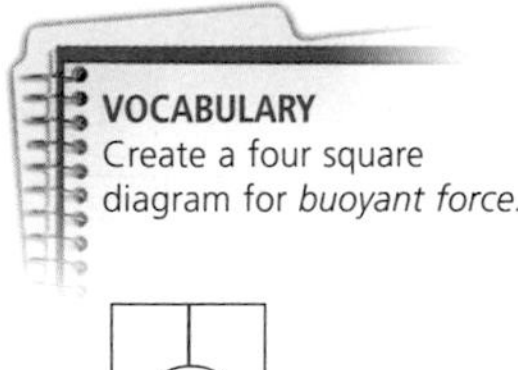

VOCABULARY
Create a four square diagram for *buoyant force.*

RESOURCES FOR DIFFERENTIATED INSTRUCTION

Below Level

UNIT RESOURCE BOOK

- Reading Study Guide A, pp. 183–184
- Decoding Support, p. 196

AUDIO CDS

Additional INVESTIGATION, What Floats Your Boat?, A, B, & C, pp. 210–218; Teacher Instructions, pp. 346–347

Advanced

UNIT RESOURCE BOOK

- Challenge and Extension, p. 189
- Challenge Reading, pp. 192–193

English Learners

UNIT RESOURCE BOOK

Spanish Reading Study Guide, pp. 187–188

AUDIO CDS

- Audio Readings in Spanish
- Audio Readings (English)

Buoyancy

The photograph on the right shows a balloon that has been pushed into a beaker of water. Remember that in a fluid, pressure increases with depth. This means that there is greater pressure acting on the bottom of the balloon than on the top of it. The pressure difference between the top and bottom of the balloon produces a net force that is pushing the balloon upward.

When you push a balloon underwater, the water level rises because the water and the balloon cannot be in the same place at the same time. The volume of the water has not changed, but some of the water has been displaced, or moved, by the balloon. The volume of the displaced water is equal to the volume of the balloon. The buoyant force on the balloon is equal to the weight of the displaced water. A deflated balloon would displace less water and would therefore have a smaller buoyant force on it.

net force

CHECK YOUR READING Why does increasing the volume of an object increase the buoyant force on it when it is in a fluid?

Density and Buoyancy

Whether or not an object floats in a fluid depends on the densities of both the object and the fluid. Density is a measure of the amount of matter packed into a unit volume. The density of an object is equal to its mass divided by its volume, and is commonly measured in grams per cubic centimeter (g/cm^3).

READING TIP Remember that both air and water are fluids, and water has a greater density than air. Therefore, water has a greater buoyant force.

If an object is less dense than the fluid it is in, the fluid the object displaces can weigh more than the object. A wooden ball that is pushed underwater, as in the beaker below and on the left, rises to the top and floats. An object rising in a liquid has a buoyant force acting upon it that is greater than its own weight. If an object is floating in a liquid, the buoyant force is balancing the weight.

If the object is more dense than the fluid it is in, the object weighs more than the fluid it displaces. A glass marble placed in the beaker on the far right sinks to the bottom because glass is denser than water. The weight of the water the marble displaces is less than the weight of the marble. A sinking object has a weight that is greater than the buoyant force on it.

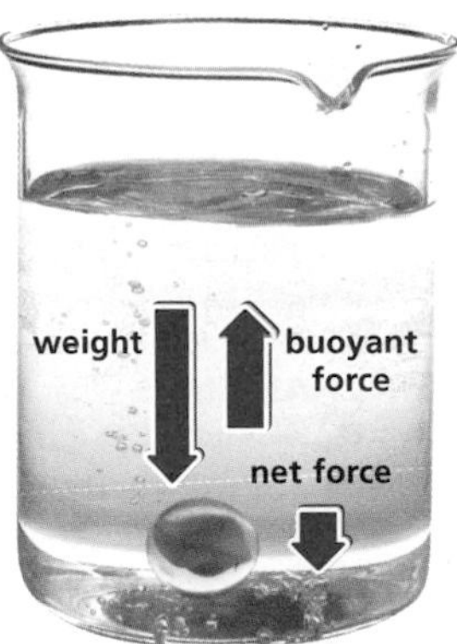

3.4 INSTRUCT

Address Misconceptions

IDENTIFY Ask: Do the materials to build a ship have to be less dense than water? If students answer yes, they hold the misconception that materials that make up boats and ships are less dense than water.

CORRECT Divide some modeling clay in half. Roll half of the clay into a ball, and mold the other half into a bowl. Place the ball of clay into a container of water and observe what happens. Place the clay bowl in the container, open side up, and observe what happens. Have students propose reasons why the bowl floats and the ball sinks. *The air in the bowl produces a form that has less density overall than water. Clay itself is denser than water. The clay ball contains no materials that lessen this density, so it sinks.*

REASSESS Ask: Why would a boat with a hole in the bottom eventually sink? *Water would enter the hole, replacing the air in the boat. At some point, the boat would no longer be less dense than water, and it would sink. One exception to this would be a wooden boat with nothing in it. Most wood is less dense than water, so it probably would float.*

Technology Resources

Visit **ClassZone.com** for background on common student misconceptions.

MISCONCEPTION DATABASE

Ongoing Assessment

Explain how fluids apply forces to objects.

Ask: Does air apply equal amounts of force to a floating balloon? Explain. *No, the force is greater on the part of the balloon that is closer to Earth.*

CHECK YOUR READING *Answer: Increasing the volume of an object increases the volume of the liquid it displaces, which increases the buoyant force.*

DIFFERENTIATE INSTRUCTION

More Reading Support

A How is the density of an object calculated? *Divide its mass by its volume.*

B What is a common unit used for density of objects? *g/cm^3*

Additional Investigation To reinforce Section 3.4 learning goals, use the following full-period investigation:

R **Additional INVESTIGATION,** What Floats Your Boat? A, B, & C, pp. 210–218, 346–347 (Advanced students should complete Levels B and C.)

Below Level Have students place one ice cube in a glass of water and another in a glass of alcohol and explain their observations. *Ice floats in water because ice is less dense than water. It sinks in alcohol because it is denser than alcohol.*

INVESTIGATE
Bernoulli's Principle

PURPOSE Observe how the speed of air affects air pressure

TIPS *15 min.*

- The marks on the straw should be straight and narrow.
- Practice blowing straight across the top of the straw before recording observations. Students should be careful not to blow downward into the marked straw.

WHAT DO YOU THINK? *The water rose in the marked straw. When air was blown softly, water rose in the straw, but not as much as when air was blown with greater intensity.*

CHALLENGE *The air blown across the top of the tube would not change the pressure of the air in the tube.*

Datasheet, Bernoulli's Principle, p. 190

Technology Resources

Customize this student lab as needed or look for an alternative. Print rubrics to assess student lab reports.

Lab Generator CD-ROM

Ongoing Assessment

CHECK YOUR READING *Answer: The faster the fluid moves, the less pressure it exerts on surfaces or openings over which it flows.*

The motion of a fluid affects its pressure.

The motion of a fluid affects the amount of pressure it exerts. A faster-moving fluid exerts less pressure as it flows over the surface of an object than a slower moving fluid. For example, wind blowing over a chimney top decreases the pressure at the top of the chimney. The faster air has less pressure than the slower-moving air in the fireplace. The increased pressure difference more effectively pulls the smoke from a fire out of the fireplace and up the chimney.

Bernoulli's Principle

Bernoulli's principle, named after Daniel Bernoulli (buhr-NOO-lee), a Swiss mathematician who lived in the 1700s, describes the effects of fluid motion on pressure. In general, **Bernoulli's principle** says that an increase in the speed of the motion of a fluid decreases the pressure within the fluid. The faster a fluid moves, the less pressure it exerts on surfaces or openings it flows over.

CHECK YOUR READING What is the relationship between the speed of a fluid and the pressure that the fluid exerts?

INVESTIGATE Bernoulli's Principle

How does the speed of air affect air pressure?

PROCEDURE

1. Use the pen to mark off intervals of 1 cm along the length of one of the straws.
2. Put a drop of food coloring in the cup of water and stir it. Place the marked straw into the cup and hold it upright so that the water level in the straw is at one of the marks. The straw should not touch the bottom of the cup.
3. Position the second straw as shown. Blow across the open end of the marked straw. Observe the level of the water in the marked straw as you blow.
4. Blow harder and then softer. Observe the water level as you change the speed of the air.

WHAT DO YOU THINK?

- What happened to the water in the straw as you blew?
- How did the speed of the air relate to the changes you observed?

CHALLENGE What results would you expect if you blew over the top of a tube with a closed bottom instead of the straw? Explain.

SKILL FOCUS
Observing

MATERIALS
- pen
- ruler
- two clear straws
- clear plastic cup filled with water
- food coloring

TIME
15 minutes

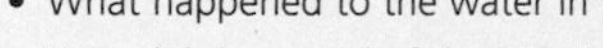

DIFFERENTIATE INSTRUCTION

More Reading Support

C If a moving fluid slows down, what happens to its pressure? *pressure increases*

English Learners English learners may have a difficult time recognizing cause-and-effect relationships when sentences beginning with *If* do not contain clauses in which *then* is stated directly. Copy the following chart on the board and ask students to model the examples provided (from p. 99).

Cause	Effect
If an object is floating in a liquid,	the buoyant force is balancing the weight.
If the object is more dense than the fluid it is in,	the object weighs more than the fluid it displaces.

Applying Bernoulli's Principle

Bernoulli's principle has many applications. One important application is used in airplanes. Airplane wings can be shaped to take advantage of Bernoulli's principle. Certain wing shapes cause the air flowing over the top of the wing to move faster than the air flowing under the wing. Such a design improves the lifting force on a flying airplane.

Many racecars, however, have a device on the rear of the car that has the reverse effect. The device is designed like an upside-down airplane wing. This shape increases the pressure on the top of the car. The car is pressed downward on the road, which increases friction between the tires and the road. With more friction, the car is less likely to skid as it goes around curves at high speeds.

A prairie-dog colony also shows Bernoulli's principle in action. The mounds that prairie dogs build over some entrances to their burrows help to keep the burrows well-ventilated.

1. Air closer to the ground tends to move at slower speeds than air higher up. The air over an entrance at ground level generally moves slower than the air over an entrance in a raised mound.
2. The increased speed of the air over a raised mound entrance decreases the pressure over that opening.
3. The greater air pressure over a ground-level entrance produces an unbalanced force that pushes air through the tunnels and out the higher mound entrance.

? D

Bernoulli's Principle in Nature

Bernoulli's principle explains why having two entrances at different heights helps ventilate a prairie-dog burrow.

1 Air moves more slowly near the ground.

2 The air over the raised entrance moves faster and has less pressure than the slower-moving air near the ground.

3 The pressure difference between the two entrances moves air through the tunnel.

Teach Difficult Concepts

Students may have previously been taught that airplanes fly "because of Bernoulli's principle." Many physicists prefer to explain airplane flight using Newton's third law—the wings push the air down, and the air pushes the wings up. The roles of both Bernoulli's principle and Newton's laws in flight are still being debated among scientists. Encourage interested students to research this debate on the Internet.

It is true that wings can be designed to improve lift by forcing the air flowing over the wing to move faster than the air flowing under the wing. Contrary to what students may have learned, however, the air does not go faster over the longer top side because it is forced to "meet up" at the back of the wing. Wind tunnel experiments, in fact, show that the air does not meet up again.

Ongoing Assessment

Describe how the motion of a fluid affects the pressure it exerts.

Ask: Which exerts more pressure, the water in rapids or water in a slow-moving stream? *water in a slow-moving stream*

DIFFERENTIATE INSTRUCTION

More Reading Support

D Why does air flow from a ground-level entrance to a higher entrance? *The pressure is greater at the lower entrance.*

Alternative Assessment Fluids forced through narrow spaces flow faster. This effect, called the Venturi effect, can be used to demonstrate Bernoulli's principle. Have students design an experiment using a stream table to show that water flows faster through narrow channels.

Advanced Have students who are interested in how pressure affects the motion of fluids read the following article:

R Challenge Reading, pp. 192–193

Real World Example

Squeeze a tube of toothpaste at one end and watch the toothpaste squirt out the other end. Apply pressure to one end of a container of caulk, and the caulk flows out the top. In both instances, pressure applied to one part of the fluid is transmitted through the fluid. Other examples include squeezing a plastic container of ketchup or mustard.

The amount of pressure is reflected in the results of the application. If you apply a lot of pressure to a tube or bottle, then a lot of fluid comes out quickly. A small amount of pressure releases a small amount of fluid.

Mathematics Connection

Although Pascal was a scientist, he is primarily known as a mathematician. A triangle containing a classic numeric pattern is named after him. He worked extensively in the areas of geometry and probability.

Develop Critical Thinking

APPLY Air bags can be used to lift large weights. They can be used when basements are installed under an existing house. Have students use Pascal's principle to devise a plan to move a house using air bags. *Sample answer: Excavate the land around the base of the house, and install beams under the house. Gradually work the air bag under the house, and pump air into it. The house will rise by small amounts as the pressure from the air is transmitted throughout the air bag.*

Ongoing Assessment

Explain how forces are transmitted through fluids.

Ask: Why does toothpaste come out of the tube when you squeeze it? *When you increase pressure on the fluid toothpaste, pressure increases all through the tube, and the fluid finds and takes the path of least resistance, through the opening.*

Forces can be transmitted through fluids.

Imagine you have a bottle full of water. You place the bottle cap on it, but you do not tighten the cap. You give the bottle a hard squeeze and the cap falls off. How was the force you put on the bottle transferred to the bottle cap?

Pascal's Principle

E

In the 1600s Blaise Pascal (pa-SKAL), a French scientist for whom the unit of measure called the pascal was named, experimented with fluids in containers. One of his key discoveries is called Pascal's principle. **Pascal's principle** states that when an outside pressure is applied at any point to a fluid in a container, that pressure is transmitted throughout the fluid with equal strength.

You can use Pascal's principle to transmit a force through a fluid. Some car jacks lift cars using Pascal's principle. These jacks contain liquids that transmit and increase the force that you apply.

1. The part of the jack that moves down and pushes on the liquid is called a piston. As you push down on the piston, you increase the pressure on the liquid.
2. The increase in pressure is equal to your applied force divided by the area of the downward-pushing piston. This increase in pressure is transmitted throughout the liquid.

Pascal's Principle

The pressure from the smaller piston is equal to the pressure pushing up the larger one. The large piston can exert more force because of its greater area.

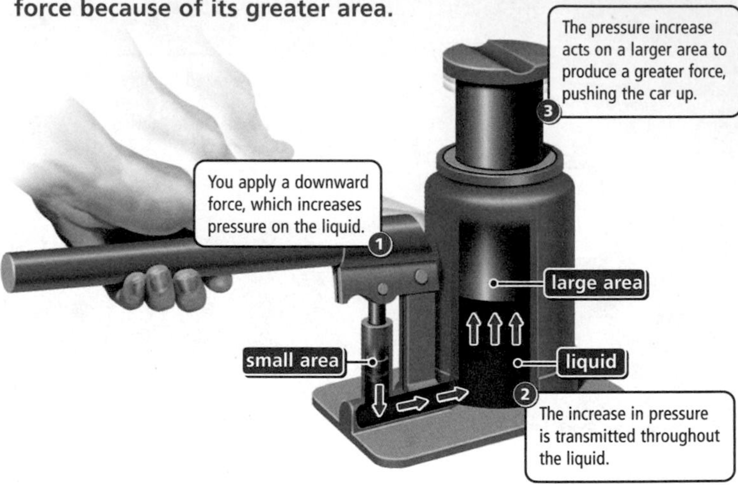
The pressure increase acts on a larger area to produce a greater force, pushing the car up.
3
You apply a downward force, which increases pressure on the liquid.
1
large area
small area
liquid
2
The increase in pressure is transmitted throughout the liquid.

DIFFERENTIATE INSTRUCTION

 **More Reading Support**

E What is Pascal's principle? *If pressure is increased at one point in a fluid, it increases by the same amount everywhere in the fluid.*

Advanced In hydraulics, pressure is applied to a small piston that transfers this pressure to a larger piston, which multiplies the force. Have students use the pressure equation to find the amount of force produced when a 400 N force is applied to a 4 cm^2 piston, which transfers this pressure to a large piston with an area of 40 cm^2. *400 N/4 cm^2 = x/40 cm^2. The resulting force of the large piston is 4000 N.*

 Challenge and Extension, p. 189

❸ The increased pressure pushes upward on another piston, which raises the car. This piston has a large area compared with the first piston, so the upward force is greater than the downward force. A large enough area produces the force needed to lift a car. However, the larger piston does not move upward as far as the smaller one moved downward.

 Describe how pressure is transmitted through a fluid.

Hydraulics

Machines that use liquids to transmit or increase a force are called hydraulic (hy-DRAW-lihk) machines. The advantage to using a liquid instead of a gas is that when you squeeze a liquid, its volume does not change much. The molecules in a liquid are so close together that it is hard to push the molecules any closer. Gas molecules, however, have a lot of space between them. If you apply pressure to a gas, you decrease its volume.

The hydraulic arm on the garbage truck lifts and empties trash cans.

Although hydraulic systems are used in large machines such as garbage trucks, research is being done on using hydraulics on a much smaller scale. Researchers are developing a storage chip similar to a computer chip that uses hydraulics rather than electronics. This chip uses pipes and pumps to move fluid into specific chambers on a rubber chip. Researchers hope that a hydraulic chip system will eventually allow scientists to use a single hand-held device to perform chemical experiments with over a thousand different liquids.

3.4 Review

KEY CONCEPTS

1. Why is there an upward force on objects in water?
2. How does changing the speed of a fluid affect its pressure?
3. If you push a cork into the neck of a bottle filled with air, what happens to the pressure inside the bottle?

CRITICAL THINKING

4. **Infer** Ebony is a dark wood that has a density of 1.2 g/cm^3. Water has a density of 1.0 g/cm^3. Will a block of ebony float in water? Explain.
5. **Analyze** When you use a spray bottle, you force air over a small tube inside the bottle. Explain why the liquid inside the bottle comes out.

CHALLENGE

6. **Synthesize** If you apply a force of 20 N downward on a car jack piston with an area of 2.5 cm^2, what force will be applied to the upward piston if it has an area of 400 cm^2? **Hint:** Remember that pressure equals force divided by area.

ANSWERS

1. The pressure in a fluid increases with depth. The pressure acting on the lower side of an object in a fluid is greater than the pressure on the upper side.

2. A fluid that moves fast exerts less pressure than one that moves more slowly.

3. The pressure increases equally throughout the fluid.

4. No; the density of the ebony wood is greater than the density of water.

5. The moving air over the tube has less pressure than the air in the bottle, so water is pushed up and out of the tube.

6. Find the increase in pressure, then multiply by the area of the large piston.

$$P = \frac{F}{A} = \frac{20\ N}{2.5\ cm^2} = 8\ N/cm^2$$

$$F = PA = 8\ N/cm^2 \cdot 400\ cm^2 = 3200\ N$$

Ongoing Assessment

 *Answer: Pressure is transmitted throughout a fluid with equal strength when an outside pressure is applied at any point.*

Reinforce 

Have students relate the section to the Big Idea.

R Reinforcing Key Concepts, p. 191

3.4 ASSESS & RETEACH

Assess

 Section 3.4 Quiz, p. 45

Reteach

To show an example of Bernoulli's principle, use the following as either a teacher demo or a student activity.

Tape one end of a 6 cm string to each of 2 foam cups. Place the cups about 2 cm apart, with the strings in front of them. Hold the strings on the table. Put your mouth about 5 cm from the cups. Blow hard between the cups and observe that the cups move toward each other. Ask students to explain these results in terms of pressure and velocity of a fluid.

Technology Resources

Have students visit **ClassZone.com** for reteaching of Key Concepts.

 **CONTENT REVIEW**

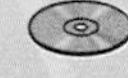 **CONTENT REVIEW CD-ROM**

BACK TO

the BIG idea

Have students use information from this chapter to explain how each of the following terms relates to force: gravity, friction, pressure, fluid. *Sample answer: Gravity is an attractive force of one mass to another mass, and friction is a force that opposes motion. Pressure is not itself a force but describes how force spreads out over an area. A fluid exerts force from all directions upon an object in it.*

KEY CONCEPTS SUMMARY

SECTION 3.1

Ask: In the bottom set of masses, what changes would occur in the arrows if the masses were moved closer together? *They would become larger.*

Ask: If other masses were added to the right of the masses shown, how would the existing arrows change? *Other arrows would be added, but current arrows would not change.*

SECTION 3.2

Ask: How does the force that is needed to start the chair moving compare with the force needed to keep the chair moving? *It is greater.*

Ask: Is air resistance a noticeable force in moving the chair? *no*

SECTION 3.3

Ask: What is the pressure if 50 N of force is applied over 10 m^2? *5 Pa*

Ask: Describe the forces applied by the water on a scuba diver. *Forces are applied by the water in all directions.*

SECTION 3.4

Ask: If a diver displaces water with a weight of 523 N, what is the buoyant force on the diver? *523 N*

Ask: Why might strong winds blowing by a building cause a window to pop out? *The wind creates an area of pressure that is less than the pressure inside the building.*

Review Concepts

- Big Idea Flow Chart, p. T17
- Chapter Outline, pp. T23–T24

Chapter Review

the BIG idea

Newton's laws apply to all forces.

CONTENT REVIEW
CLASSZONE.COM

KEY CONCEPTS SUMMARY

3.1 Gravity is a force exerted by masses.

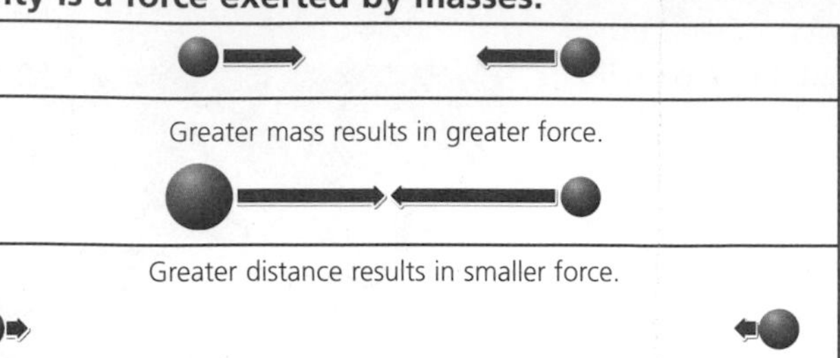

Greater mass results in greater force.
Greater distance results in smaller force.

VOCABULARY
gravity p. 77
weight p. 79
orbit p. 80

3.2 Friction is a force that opposes motion.

Frictional force depends on—
- types of surfaces
- motion of surfaces
- force pressing surfaces together

Air resistance is a type of friction.

friction

VOCABULARY
friction p. 85
fluid p. 88
air resistance p. 89

3.3 Pressure depends on force and area.

$$\text{Pressure} = \frac{\text{Force}}{\text{Area}}$$

Pressure in a fluid acts in all directions.

VOCABULARY
pressure p. 91
pascal p. 92

3.4 Fluids can exert a force on objects.

- Buoyant force is equal to the weight of the displaced fluid.
- A faster-moving fluid produces less pressure than a slower-moving one.
- Pressure is transmitted through fluids.

VOCABULARY
buoyant force p. 98
Bernoulli's principle p. 100
Pascal's principle p. 102

Technology Resources

Have students visit **ClassZone.com** or use the CD-ROM for a cumulative review of concepts.

CONTENT REVIEW

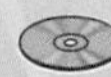
CONTENT REVIEW CD-ROM

Engage students in a whole-class interactive review of Key Concepts. Edit content as you wish.

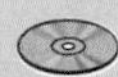
POWER PRESENTATIONS

Reviewing Vocabulary

Write a sentence describing the relationship between each pair of terms.

1. gravity, weight
2. gravity, orbit
3. pressure, pascal
4. fluid, friction
5. density, buoyant force
6. fluid, Bernoulli's principle

Reviewing Key Concepts

Multiple Choice *Choose the letter of the best answer.*

7. Which force keeps Venus in orbit around the Sun?
   - **a.** gravity
   - **b.** friction
   - **c.** hydraulic
   - **d.** buoyancy
8. You and a classmate are one meter apart. If you move farther away, how does the gravitational force between you and your classmate change?
   - **a.** It increases.
   - **b.** It decreases.
   - **c.** It stays the same.
   - **d.** It disappears.
9. You kick a ball on a level sidewalk. It rolls to a stop because
   - **a.** there is no force on the ball
   - **b.** gravity slows the ball down
   - **c.** air pressure is pushing down on the ball
   - **d.** friction slows the ball down
10. You push a chair at a constant velocity using a force of 5 N to overcome friction. You stop to rest, then push again. To start the chair moving again, you must use a force that is
    - **a.** greater than 5 N
    - **b.** equal to 5 N
    - **c.** greater than 0 N but less than 5 N
    - **d.** 0 N
11. How could you place an empty bottle on a table so that it produces the greatest amount of pressure on the table?

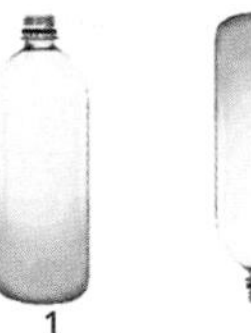

    - **a.** position 1
    - **b.** position 2
    - **c.** position 3
    - **d.** All positions produce the same pressure.
12. As you climb up a mountain, air pressure
    - **a.** increases
    - **b.** decreases
    - **c.** stays the same
    - **d.** changes unpredictably
13. If you squeeze a balloon in the middle, what happens to the air pressure inside the balloon?
    - **a.** It increases only in the middle.
    - **b.** It decreases only in the middle.
    - **c.** It increases throughout.
    - **d.** It decreases throughout.

Short Answer *Write a short answer to each question.*

14. How does the force of attraction between large masses compare with the force of attraction between small masses at the same distance?
15. Explain why a satellite in orbit around Earth does not crash into Earth.
16. You are pushing a dresser with drawers filled with clothing. What could you do to reduce the friction between the dresser and the floor?
17. Why is water pressure greater at a depth of 20 feet than it is at a depth of 10 feet?
18. If you blow over the top of a small strip of paper, the paper bends upward. Why?

ASSESSMENT RESOURCES

UNIT ASSESSMENT BOOK
- Chapter Test A, pp. 46–49
- Chapter Test B, pp. 50–53
- Chapter Test C, pp. 54–57
- Alternative Assessment, pp. 58–59

SPANISH ASSESSMENT BOOK
Spanish Chapter Test, pp. 265–268

Technology Resources

Edit test items and answer choices.

Test Generator CD-ROM

Visit **ClassZone.com** to extend test practice.

Test Practice

Reviewing Vocabulary

1. Weight is the force of gravity acting on a mass.

2. Gravity is the centripetal force that keeps objects in orbit.

3. Pascal is the unit of pressure.

4. Objects moving in a fluid have a frictional force acting upon them.

5. The density of a liquid determines the buoyant force upon it. The greater the density, the greater the force.

6. Bernoulli's principle says that an increase in the speed of the motion of a fluid decreases the pressure within the fluid.

Reviewing Key Concepts

7 a

8. b

9. d

10. a

11. b

12. b

13. c

14. The force between the larger masses is greater.

15. The speed of the satellite is great enough to cause the satellite to orbit around Earth instead of crashing to Earth. Gravity acts as a centripetal force.

16. Reduce weight by removing the drawers and clothes, or place a smooth surface under the dresser to reduce friction.

17. More water is above at 20 feet than at 10 feet, and the increased amount of water applies a greater force.

18. The pressure of the moving air on the top of the paper is less than the pressure of the still air under the paper.

Thinking Critically

19. *The boat is shaped so that it displaces more liquid than the weight of the materials that it is made of. The volume of the boat contains iron and air, and the overall density is less than that of water.*
20. *Solid surface friction does not depend on speed and surface area, while friction between a moving object and fluid does.*
21. *Rubbing produces friction, and friction produces heat.*
22. *The gravity is gradually decreasing because gravity decreases with increased distance.*
23. *You need less initial speed on the Moon because it has less gravity.*
24. *The arrow for air resistance would be greater than that for gravity.*
25. *The buoyant force will increase because the diver is increasing volume but not weight.*
26. *No; without an atmosphere, there would be no air resistance to slow down the skydiver.*
27. *The density of oil is less than the density of water.*
28. *The pressure at the bottom of each flask would be the same because the water levels are the same.*

Using Math Skills in Science

29. $F = mg = 10\ kg \cdot 98\ m/s^2 = 98\ N$
30. $P = FA = 50\ N/0.5\ m^2 = 100\ Pa$

the BIG idea

31. *Sample answer: Gravity, friction with the snowboard, and air resistance act on the snowboarder. The weight of the snowboarder and the snowboard, and friction with the snowboard, act on the ground. Gravity is greater than friction, so it causes the snowboarder to change motion and accelerate.*
32. *Student answers will vary.*

UNIT PROJECTS

Collect schedules, materials lists, and questions. Be sure dates and materials are obtainable, and questions are focused.

 Unit Projects, pp. 5–10

Thinking Critically

19. **APPLY** Explain why an iron boat can float in water, while an iron cube cannot.

20. **COMPARE** How does the friction between solid surfaces compare with the friction between a moving object and a fluid?

21. **APPLY** Explain why a block of wood gets warm when it is rubbed with sandpaper.

22. **PREDICT** The Moon's orbit is gradually increasing. Each year the Moon is about 3.8 cm farther from Earth than the year before. How does this change affect the force of gravity between Earth and the Moon?

23. **APPLY** The Moon has one-sixth the gravity of Earth. Why would it be easier to launch spacecraft into orbit around the Moon than around Earth?

Use the photograph below to answer the next three questions.

24. **APPLY** A skydiver jumps out of a plane. After he reaches terminal velocity, he opens his parachute. Draw a sketch showing the forces of air resistance and gravity on the skydiver after the parachute opens. Use a longer arrow for a greater force.

25. **SYNTHESIZE** Air is a fluid, which produces a small buoyant force on the skydiver. How does this buoyant force change after he opens his parachute? Why?

26. **INFER** The Moon has no atmosphere. Would it be safe to skydive on the Moon? Why or why not?

27. **INFER** When oil and water are mixed together, the two substances separate and the oil floats to the top. How does the density of oil compare with the density of water?

28. **COMPARE** Three flasks are filled with colored water as shown below. How does the water pressure at the bottom of each flask compare with the water pressure at the bottom of the other two?

Using Math Skills in Science

Complete the following calculations.

29. How much force does a 10 kg marble exert on the ground?

30. A force of 50 N is applied on a piece of wood with an area of 0.5 m^2. What is the pressure on the wood?

the BIG idea

31. **ANALYZE** Look again at the picture on pages 74–75. What forces are acting on the snowboarder? on the snow? Use Newton's laws to explain how these forces enable the snowboarder to move down the hill.

32. **SYNTHESIZE** Choose two concepts discussed in this chapter, and describe how Newton's laws relate to those concepts.

UNIT PROJECTS

Check your schedule for your unit project. How are you doing? Be sure that you have placed data or notes from your research into your project folder.

MONITOR AND RETEACH

If students have trouble applying the concepts in items 24–26, watching a videotape of a skydiver can help them. Have them watch the tape without interruption. Then replay the tape, pausing whenever necessary to discuss what forces are acting on the skydiver both before and after the parachute opens. Specifically, discuss gravity, terminal velocity, buoyant force, and air resistance.

Students may benefit from summarizing one or more sections of the chapter.

 Summarizing the Chapter, pp. 219–220

Standardized Test Practice

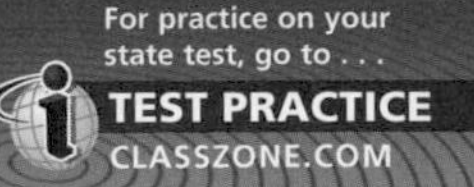

Interpreting Diagrams

Study the diagram and then answer the questions that follow.

Bernoulli's principle states that an increase in the speed of the motion of a fluid decreases the pressure exerted by the fluid. The diagram below relates the movement of a curve ball in baseball to this principle. The ball is shown from above.

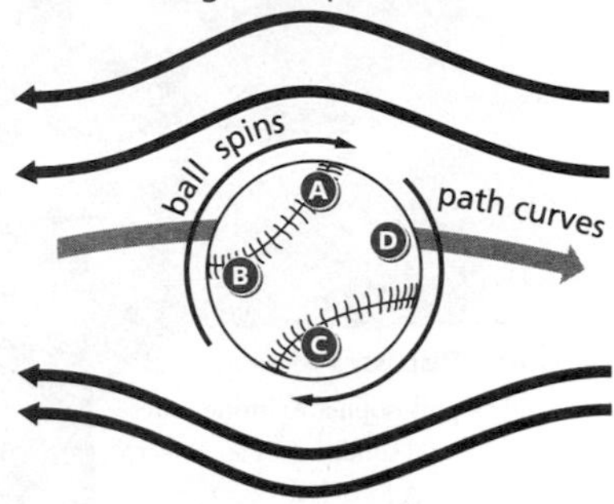

1. To which of these properties does Bernoulli's principle apply?
- **a.** air pressure
- **b.** temperature
- **c.** air resistance
- **d.** density

2. Where is the air moving fastest in the diagram?
- **a.** region A
- **b.** region B
- **c.** region C
- **d.** region D

3. Because the ball is spinning, the air on one side is moving faster than on the other side. This causes the ball to curve due to the
- **a.** air molecules moving slowly and evenly around the ball
- **b.** forward motion of the ball
- **c.** difference in air pressure on the ball
- **d.** changing air temperature around the ball

4. If the baseball were spinning as it moved forward underwater, instead of through the air, how would the pressure of the fluid act on the ball?
- **a.** The water pressure would be the same on all sides.
- **b.** The water pressure would vary as air pressure does.
- **c.** The water pressure would be greatest on the side where air pressure was least.
- **d.** The water pressure would prevent the ball from spinning.

Extended Response

Answer the two questions below in detail. Include some of the terms from the word box. Underline each term you use in your answer.

acceleration	air resistance	density
fluid	friction	gravity
mass	pressure	velocity

5. If a feather and a bowling ball are dropped from the same height, will they fall at the same rate? Explain.

6. A balloon filled with helium or hot air can float in the atmosphere. A balloon filled with air from your lungs falls to the ground when it is released. Why do these balloons behave differently?

METACOGNITIVE ACTIVITY

Have students answer the following questions in their **Science Notebook:**

1. What did you find the most challenging to understand about gravity, friction, and pressure?
2. What questions do you still have about these concepts?
3. How have you solved a problem while working on your Unit Project?

Interpreting Diagrams

1. a | *3. c*

2. c | *4. c*

Extended Response

5. RUBRIC

4 points for a response that correctly answers the question and uses the following terms accurately:

- gravity
- mass
- acceleration

The feather and bowling ball fall at the same rate. The force of gravity acts on both masses equally. Because the bowling ball has more mass, gravity will exert more force on it. However, the greater force does not produce greater acceleration due to inertia. Both fall with the same acceleration.

3 points for a response that correctly answers the question and uses 2 terms accurately

2 points for a response that correctly answers the question and uses 1 term accurately

1 point for a response that correctly answers the question, but doesn't use the terms

6. RUBRIC

4 points for a response that correctly answers the question and uses the following terms accurately:

- fluid
- density

Air and gases, like water, are fluids. Both balloons are in a fluid yet behave differently because of density. The balloon filled with helium or hot air is less dense than the air in the atmosphere. An object that is less dense than the fluid will float. The balloon filled with air from my lungs falls to the ground slowly because the balloon is denser than the air.

3 points for a response that correctly answers the question and uses one term accurately

2 points for a response that correctly answers the question, but doesn't use the terms

FOCUS

Set Learning Goals

Students will

- Compare ancient and modern ideas of force and motion.
- Observe how progress in science aided progress in technology and vice versa.
- Examine the development of technological design and its limits.

National Science Education Standards

A.9.a–g Understandings About Scientific Inquiry

E.6.a–c Understandings About Science and Technology

F.5.a–e, F.5.g Science and Technology In Society

G.1.a–b Science as a Human Endeavor

G.2.a Nature of Science

G.3.a–c History of Science

INSTRUCT

The timeline shows major developments in science and the years in which they occurred. The bottom half addresses developments in technology based on the scientific discoveries in the top half. Gaps in the timeline represent blocks of time that have been omitted.

Social Studies Connection

350 B.C. Ancient philosophers studied a broad range of topics—including the sciences. Philosophers talked and wrote about everything from politics to how objects move. Have students notice how ancient Greeks used forces to achieve speed, move water, and balance weight.

Technology

CATAPULTS were instruments of war, and technology has often been developed for the purpose of improving war weaponry. Point out how catapults, a concept developed by ancient civilizations, are still used today.

TIMELINES in Science

UNDERSTANDING FORCES

In ancient times, people thought that an object would not move unless it was pushed. Scientists came up with ingenious ways to explain how objects like arrows stayed in motion. Over time, they came to understand that all motion could be described by three basic laws. Modern achievements such as suspension bridges and space exploration are possible because of the experiments with motion and forces performed by scientists and philosophers over hundreds of years.

This timeline shows just a few of the many steps on the path toward understanding forces. Notice how scientists used the observations and ideas of previous thinkers as a springboard for developing new theories. The boxes below the timeline show how technology has led to new insights and to applications of those ideas.

350 B.C.

Aristotle Discusses Motion

The Greek philosopher Aristotle states that the natural condition of an object is to be at rest. A force is necessary to keep the object in motion. The greater the force, the faster the object moves.

EVENTS

400 B.C. 350 B.C. 300 B.C.

APPLICATIONS AND TECHNOLOGY

TECHNOLOGY

Catapulting into History

As early as 400 B.C., armies were using objects in motion to do work. Catapults, or machines for hurling stones and spears, were used as military weapons. Five hundred years later, the Roman army used catapults mounted on wheels. In the Middle Ages, young trees were sometimes bent back, loaded with an object, and then released like a large slingshot. Today catapult technology is used to launch airplanes from aircraft carriers. A piston powered by steam propels the plane along the deck of the aircraft carrier until it reaches takeoff speed.

DIFFERENTIATE INSTRUCTION

Below Level To give students a better idea of the amount of time that passed between discoveries, draw a long line across the blackboard. Label the left end 350 B.C., the middle A.D. 750, and the right end 2000. Mark off other year divisions. As you discuss the timeline with students, point to the section of the blackboard timeline in which that event took place.

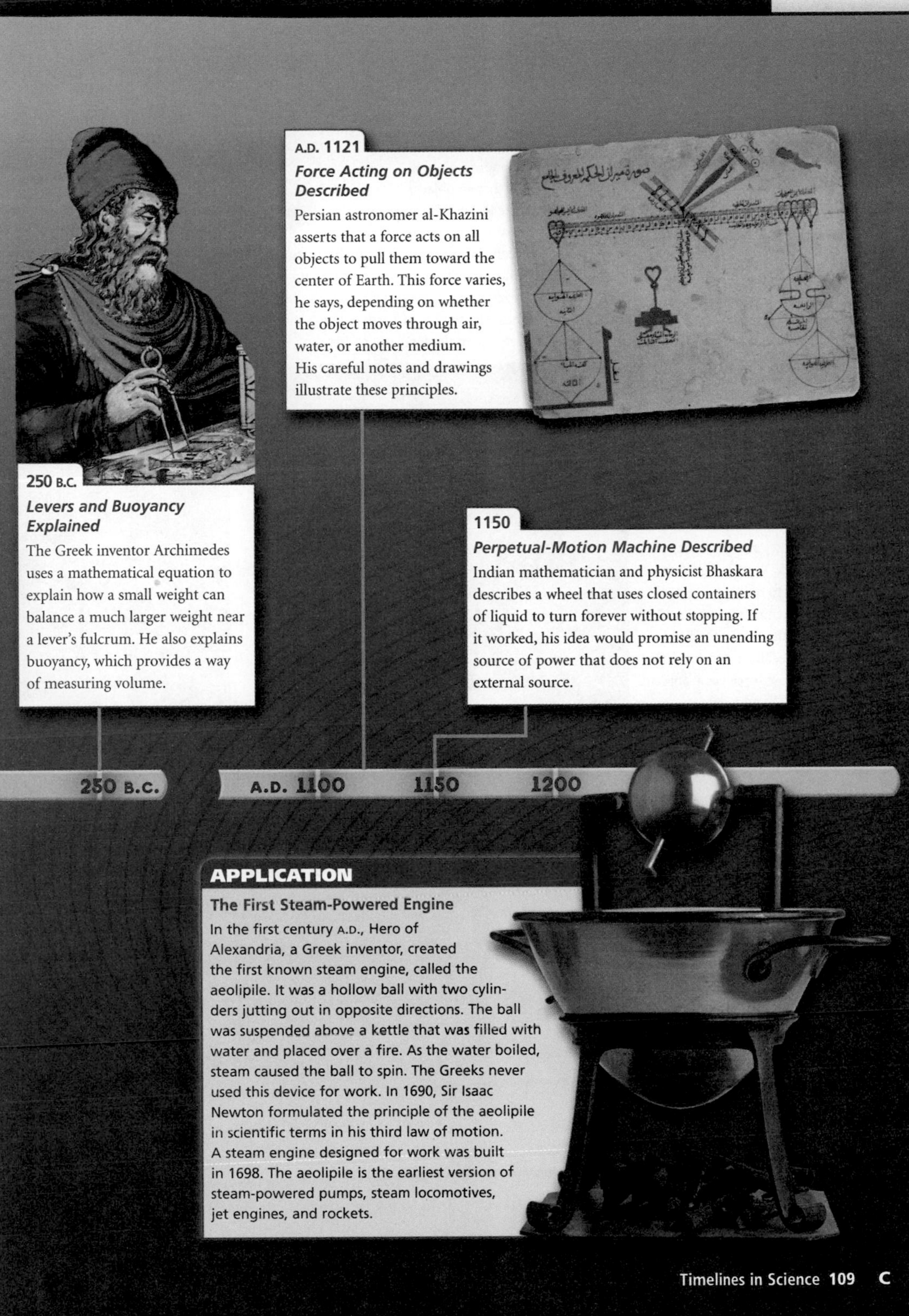

250 B.C.

Levers and Buoyancy Explained

The Greek inventor Archimedes uses a mathematical equation to explain how a small weight can balance a much larger weight near a lever's fulcrum. He also explains buoyancy, which provides a way of measuring volume.

A.D. 1121

Force Acting on Objects Described

Persian astronomer al-Khazini asserts that a force acts on all objects to pull them toward the center of Earth. This force varies, he says, depending on whether the object moves through air, water, or another medium. His careful notes and drawings illustrate these principles.

1150

Perpetual-Motion Machine Described

Indian mathematician and physicist Bhaskara describes a wheel that uses closed containers of liquid to turn forever without stopping. If it worked, his idea would promise an unending source of power that does not rely on an external source.

APPLICATION

The First Steam-Powered Engine

In the first century A.D., Hero of Alexandria, a Greek inventor, created the first known steam engine, called the aeolipile. It was a hollow ball with two cylinders jutting out in opposite directions. The ball was suspended above a kettle that was filled with water and placed over a fire. As the water boiled, steam caused the ball to spin. The Greeks never used this device for work. In 1690, Sir Isaac Newton formulated the principle of the aeolipile in scientific terms in his third law of motion. A steam engine designed for work was built in 1698. The aeolipile is the earliest version of steam-powered pumps, steam locomotives, jet engines, and rockets.

Scientific Process

Stress to students that we should not think of the science of the past as being wrong or of the people of the past as being less intelligent. The ideas they formed were based on the observations available to them at that time. Throughout history, people have tried to explain how the natural world worked; their explanations were simply different from the ones we have today. As important as any discovery was the development of scientific methods of thinking.

Social Studies Connection

A.D. 800 TO 1300 Present-day Iran was once the heart of the Persian Empire. Have students find Iran on a world map. Point out that al-Khazini was an astronomer and physician who lived in the Persian Empire. Emphasize that from about the 800s to the 1300s, some of the world's most innovative science was being done in this region. Ask how geographical features might have helped this civilization advance. *(Ideas from Greece, Egypt, China, and India spread throughout the Persian Empire via sea, river, and overland trade routes.)*

Application

AEOLIPILE Help students make a connection between the simple device shown on page 109 and Watt's steam engine on page 110—and with modern-day steam locomotives and rockets. The aeolipile (EE-uh-lih-PYL) showed the effects produced by steam under pressure. A steam engine converts thermal energy from steam into mechanical energy by allowing the steam to expand and cool. A rocket engine burns fuel, which comes out of the rocket at high speed, creating the thrust needed to propel the rocket.

DIFFERENTIATE INSTRUCTION

Advanced Encourage students to trace the development of ideas from Aristotle to Leonardo to Newton to Einstein. Students might create a visual that represents each new idea as "building on" or "knocking down" the previous idea.

Teach from Visuals

1494 This drawing by Leonardo shows the impossibility of perpetual motion. The wheel spins because gravity pulls down on the weights placed around its rim. While it seems that the wheel should spin constantly, eventually friction will stop it. Have students draw the wheel on a sheet of paper and note the forces acting on it.

Scientific Process

1687 Isaac Newton's theories differed from others. Newton was the first to say that there was an attraction (gravity) between all objects, not just between Earth and other things. Earth attracts a rock, and the rock attracts Earth.

Arts Connection

1600s At about the same time that Galileo and Newton were changing the way we think about motion, Shakespeare wrote the plays *Antony and Cleopatra* (1607) and *The Winter's Tale* (1611). In 1640, Rembrandt painted *The Night Watch.* Stress to students that works of art and literature reflect the scientific beliefs of the time.

Application

STEAM ENGINE James Watt's achievements provide one of the best examples of how advances in scientific knowledge can be applied to improve technology. Watt knew that cooling steam required a great amount of energy. Watt determined that, if he cooled steam outside the cylinder of the engine in a unit called a condenser, he would save a lot of energy. Another important improvement Watt made to the steam engine was finding a new way to change the up-and-down motion of the engine into rotary motion. You may have students suggest ways of changing up-and-down motion to circular motion using gears, levers, and other simple machines.

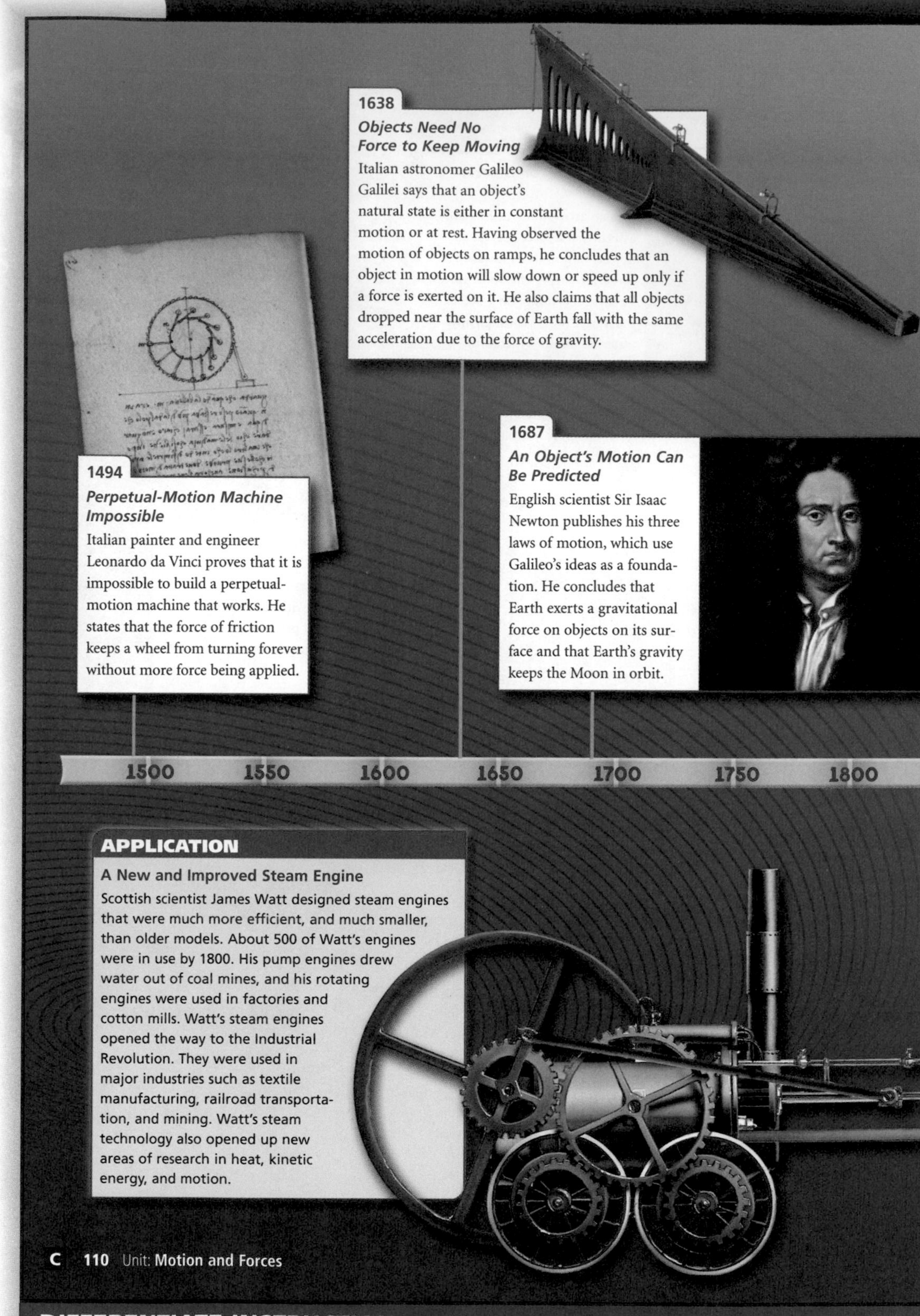

1638
Objects Need No Force to Keep Moving
Italian astronomer Galileo Galilei says that an object's natural state is either in constant motion or at rest. Having observed the motion of objects on ramps, he concludes that an object in motion will slow down or speed up only if a force is exerted on it. He also claims that all objects dropped near the surface of Earth fall with the same acceleration due to the force of gravity.
1494
Perpetual-Motion Machine Impossible
Italian painter and engineer Leonardo da Vinci proves that it is impossible to build a perpetual-motion machine that works. He states that the force of friction keeps a wheel from turning forever without more force being applied.
1687
An Object's Motion Can Be Predicted
English scientist Sir Isaac Newton publishes his three laws of motion, which use Galileo's ideas as a foundation. He concludes that Earth exerts a gravitational force on objects on its surface and that Earth's gravity keeps the Moon in orbit.
1500
1550
1600
1650
1700
1750
1800
APPLICATION
A New and Improved Steam Engine
Scottish scientist James Watt designed steam engines that were much more efficient, and much smaller, than older models. About 500 of Watt's engines were in use by 1800. His pump engines drew water out of coal mines, and his rotating engines were used in factories and cotton mills. Watt's steam engines opened the way to the Industrial Revolution. They were used in major industries such as textile manufacturing, railroad transportation, and mining. Watt's steam technology also opened up new areas of research in heat, kinetic energy, and motion.
C 110 Unit: Motion and Forces

DIFFERENTIATE INSTRUCTION

Below Level Explain to students that a perpetual-motion machine is one that continues working forever without added force. Remind students that they have read about Galileo and Newton in Chapter 2, "Forces."

Advanced Galileo thought about what would happen if he could do his experiments under impossible conditions. Have students drop same-sized solid and hollow metal balls at the same time in a tall container of oil. Observe the difference in falling times. Repeat using water and then air in the container. Ask students to speculate what the time difference would be if there were no fluid at all.

1919

Gravity Bends Light

A solar eclipse confirms German-American physicist Albert Einstein's modification of Newton's laws. Einstein's theory states that the path of a light beam will be affected by nearby massive objects. During the eclipse, the stars appear to shift slightly away from one another because their light has been bent by the Sun's gravity.

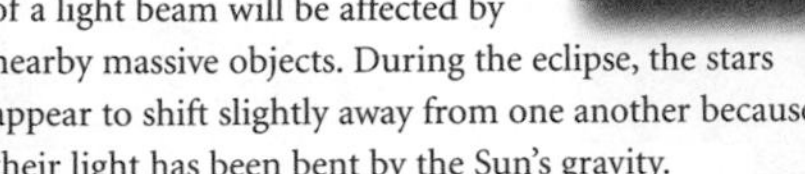

2001

Supercomputers Model Strong Force

Scientists have been using supercomputers to model the force that holds particles in the nucleus of an atom together. This force, called the strong force, cannot be measured directly in the same way that gravity and other forces can. Instead, computer models allow scientists to make predictions that are then compared with experimental results.

RESOURCE CENTER
CLASSZONE.COM
Get current research on force and motion.

1850 1900 1950 2000

INTO THE FUTURE

Since ancient times, scientists and philosophers have tried to explain how forces move objects. We now know that the laws of gravity and motion extend beyond Earth. Engineers have designed powerful spacecraft that can carry robots—and eventually people—to Mars and beyond. Rockets using new technology travel farther on less fuel than liquid-fueled rockets do.

Space travel and related research will continue to unravel the mysteries of forces in the universe. For example, recent observations of outer space provide evidence of an unidentified force causing the universe to expand rapidly. As people venture beyond Earth, we may learn new and unexpected things about the forces we have come to understand so far. The timeline shown here is just the beginning of our knowledge of forces.

TECHNOLOGY

Science Propels Exploration of Outer Space

An increased understanding of forces made space exploration possible. In 1926 American scientist Robert H. Goddard constructed and tested the first liquid-propelled rocket. A replica of Goddard's rocket can be seen at the National Air and Space Museum in Washington, D.C. In 1929 Goddard launched a rocket that carried the first scientific payload, a barometer and a camera.

Many later achievements—including the 1969 walk on the Moon—are a direct result of Goddard's trail-blazing space research.

ACTIVITIES

Reliving History

Bhaskara's design for a perpetual-motion machine involved a wheel with containers of mercury around the rim. As the wheel turned, the mercury would move in such a way that the wheel would always be heavier on one side—and stay in motion. Now we know that this theory goes against the laws of physics. Observe a wheel, a pendulum, or a swing. Think about why it cannot stay in motion forever.

Writing About Science

Suppose you won a trip to outer space. Write a letter accepting or refusing the prize. Give your reasons.

DIFFERENTIATE INSTRUCTION

Below Level Point out that the timeline comes up to the present, including rockets and space travel. The ideas of the Greeks have been added to and perfected by subsequent thinkers. Stress how much time—nearly 2500 years—the timeline covers.

Technology

1960s ONWARD The space program has made countless contributions to scientific knowledge. Lunar rocks gathered during the Apollo missions have provided much information about how the Moon—and even Earth—formed billions of years ago. One astronaut dropped a hammer and a feather on the Moon to prove Galileo's claim that objects always fall at the same rate when there is no air resistance.

INTO THE FUTURE

Have students read the passage. With the class, generate a list of questions on the board that students might have about force. To get students thinking, ask: *Why would a better understanding of force be useful in space travel? What more could we know about the force of gravity? What more might we be able to do with a better understanding of force?*

ACTIVITIES

Reliving History

Using a model would help students understand the principles behind Bhaskara's machine. Encourage students to design a model of such a machine and experiment with keeping it in motion.

Writing Project: The Story Behind the News

Point out that the timeline is written as short newspaper articles, as if the events being reported happened recently. Have students imitate the style: punchy headlines; journalistic reporting of who, what, when, where, and why at the beginning of the article; and details coming later in the report.

Technology Resources

Students can visit **ClassZone.com** for current news about forces.

CHAPTER

Work and Energy

Physical Science UNIFYING PRINCIPLES

PRINCIPLE 1
Matter is made of particles too small to see.

PRINCIPLE 2
Matter changes form and moves from place to place.

PRINCIPLE 3
Energy changes from one form to another, but it cannot be created or destroyed.

PRINCIPLE 4
Physical forces affect the movement of all matter on Earth and throughout the universe.

Unit: Motion and Forces BIG IDEAS

CHAPTER 1
Motion
The motion of an object can be described and predicted.

CHAPTER 2
Forces
Forces change the motion of objects in predictable ways.

CHAPTER 3
Gravity, Friction, and Pressure
Newton's laws apply to all forces.

CHAPTER 4
Work and Energy
Energy is transferred when a force moves an object.

CHAPTER 5
Machines
Machines help people do work by changing the force applied to an object.

CHAPTER 4 KEY CONCEPTS

SECTION 4.1

Work is the use of force to move an object.

1. Force is necessary to do work.
2. Objects that are moving can do work.

SECTION 4.2

Energy is transferred when work is done.

1. Work transfers energy.
2. Work changes potential and kinetic energy.
3. The total amount of energy is constant.

SECTION 4.3

Power is the rate at which work is done.

1. Power can be calculated from work and time.
2. Power can be calculated from energy and time.

The Big Idea Flow Chart is available on p. T25 in the **UNIT TRANSPARENCY BOOK.**

Previewing Content

SECTION

Work is the use of force to move an object. pp. 115–120

1. Force is necessary to do work.

To do **work** on an object, a force must be applied to the object, and the object must move in the direction of the force, as the first diagram below shows. Work is done only by the component of the force that acts in the same direction as the movement of the object, as shown in the second diagram.

Work is done by force that acts in the same direction as the motion of an object.

All of the Applied Force Does Work

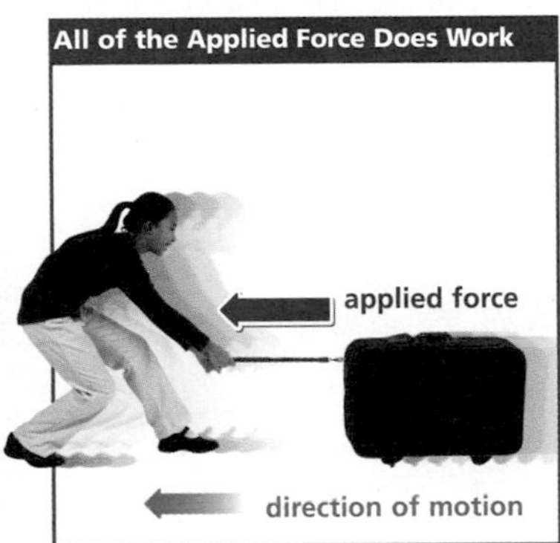

Part of the Applied Force Does Work

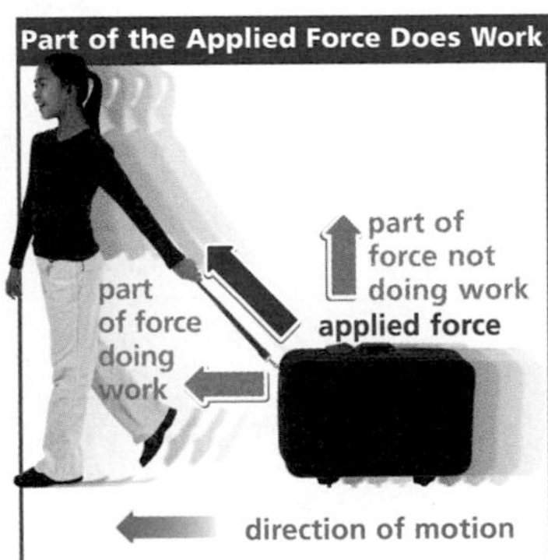

Work can be calculated by multiplying the force applied to an object by the distance the object moves while that force is being applied.

$W = F \cdot d$

The standard unit of measurement of work is the newton-meter, also called a **joule.**

2. Objects that are moving can do work.

Moving objects can do work.

- The gravitational force of Earth does work on water and other natural materials.
- People use moving objects to help them do work.

SECTION

Energy is transferred when work is done. pp. 121–129

1. Work transfers energy.

When work is done on an object, energy is transferred from whatever is exerting the force to the object.

2. Work changes potential and kinetic energy.

All forms of energy can be considered in terms of either potential energy or kinetic energy.

- An object has **potential energy** due to its position or shape. Potential energy due to gravity is called gravitational potential energy (GPE). The GPE of an object can be found by multiplying the object's mass by the acceleration due to Earth's gravity and by the object's height above the ground.

  $GPE = mgh$

- An object has **kinetic energy** when it is moving. Kinetic energy can be calculated by the formula

  $KE = \frac{1}{2}mv^2$

Mechanical energy is an object's combined potential energy and kinetic energy. An object with mechanical energy can do work on another object.

3. The total amount of energy is constant.

The law of **conservation of energy** states that energy is neither created nor destroyed, although it can change into another form. Common forms of energy are mechanical, thermal, chemical, nuclear, and electromagnetic energy.

As a ball rolls down a ramp, the amounts of potential energy and kinetic energy change, but the total energy is the same.

Common Misconceptions

MISCONCEPTION DATABASE
CLASSZONE.COM Background on student misconceptions

WORK REQUIRES A FORCE Students might think that work continues even after a force ceases to be applied. In fact, an object can continue to move after the force is removed, but work is no longer being done.

This misconception is addressed on p. 117.

FORCE AND ENERGY Students may think that force and energy are the same or that energy is a type of force. Actually, force and energy are two different phenomena that are related through the concept of work—the means of transferring energy from one object to another using a force.

This misconception is addressed on p. 124.

SECTION

Power is the rate at which work is done. pp. 130–137

1. Power can be calculated from work and time.
Power is the rate at which work is done. When the power of an object increases, work is done faster. Power can be calculated from work and time.

$P = \frac{W}{t}$ or $P = \frac{F \cdot d}{t}$

- The unit of measurement for power is the **watt,** equal to one joule of work done in one second.
- Another unit of power is **horsepower,** which is based on how much work a horse can do in one minute. It is used primarily to describe engines and motors. One horsepower equals 745 watts.

2. Power can be calculated from energy and time.
Power can be thought of as the rate at which energy is transferred over a certain period of time. Power can be calculated from energy as well as from work. The formula is

$P = \frac{E}{t}$

Common Misconceptions

THE WORD *POWER* The word *power* has many meanings in everyday language. Students often confuse the common meanings of the word for the physical meaning—that of a rate of work or energy transfer.

This misconception is addressed on p. 131.

MISCONCEPTION DATABASE
CLASSZONE.COM Background on student misconceptions

Previewing Labs

Lab Generator CD-ROM
Edit these Pupil Edition labs and generate alternative labs.

EXPLORE the BIG idea

Bouncing Ball, p. 113 Students will observe that mechanical energy is lost when a bouncing ball loses height with each bounce.	**TIME** 10 minutes **MATERIALS** large ball; hard, flat floor
Power Climbing, p. 113 Students will vary their power as they walk and run up a flight of stairs.	**TIME** 10 minutes **MATERIALS** backpack, flight of stairs
Internet Activity: Work, p. 113 Students will manipulate a computer model to determine how varying force and distance affects the amount of work done.	**TIME** 20 minutes **MATERIALS** computer with Internet access

SECTION 

EXPLORE Work, p. 115 Students compare the work done by lifting a book and by holding it without moving it.	**TIME** 5 minutes **MATERIALS** book
INVESTIGATE Work, p. 118 Students determine the amount of work done in lifting a notebook by measuring the force applied and the distance the notebook is moved.	**TIME** 20 minutes **MATERIALS** meter stick, spiral notebook, spring scale

SECTION 

INVESTIGATE Mechanical Energy, p. 125 Students calculate the potential, kinetic, and mechanical energy of a ball as it rolls down a ramp.	**TIME** 20 minutes **MATERIALS** ball, balance, board (60–100 cm long), several books, ruler, 15 cm masking tape, stopwatch, calculator

SECTION 

EXPLORE Power, p. 130 Students explore how time affects work by comparing work done in an initial time with that same work done in half the time.	**TIME** 10 minutes **MATERIALS** 2 plastic cups, 10 marbles, stopwatch
INVESTIGATE Power, p. 133 Students calculate the power it takes to pull an object over a given distance by measuring the force and the time needed.	**TIME** 15 minutes **MATERIALS** meter stick, 30 cm masking tape, 100 g object, spring scale, 60 cm string, stopwatch
CHAPTER INVESTIGATION **Work and Power,** pp. 136–137 Students compare the amount of work and power needed to lift an object straight up with the amount needed to move it up a ramp.	**TIME** 40 minutes **MATERIALS** board (60–100 cm long), chair, meter stick, 60 cm string, small wheeled object, spring scale, stopwatch

 **Additional INVESTIGATION,** Ramp It Up, A, B, & C, pp. 276–284; Teacher Instructions, pp. 346–347

Previewing Chapter Resources

	INTEGRATED TECHNOLOGY	LABS AND ACTIVITIES
CHAPTER 4 **Work and Energy**	 **CLASSZONE.COM** • eEdition Plus • EasyPlanner Plus • Misconception Database • Content Review • Test Practice • Visualization • Resource Centers • Internet Activity: Work • Math Tutorial  **SCILINKS.ORG** SCILINKS  **CD-ROMS** • eEdition • EasyPlanner • Power Presentations • Content Review • Lab Generator • Test Generator 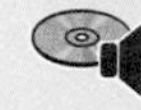 **AUDIO CDS** • Audio Readings • Audio Readings in Spanish	 EXPLORE the Big Idea, p. 113 • Bouncing Ball • Power Climbing • Internet Activity: Work  **UNIT RESOURCE BOOK** Unit Projects, pp. 5–10 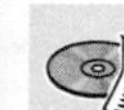 **Lab Generator CD-ROM** Generate customized labs.
 SECTION 4.1 **Work is the use of force to move an object.** pp. 115–120 Time: 2 periods (1 block)  Lesson Plan, pp. 221–222	 • **RESOURCE CENTER,** Work • **MATH TUTORIAL**  **UNIT TRANSPARENCY BOOK** • Big Idea Flow Chart, p. T25 • Daily Vocabulary Scaffolding, p. T26 • Note-Taking Model, p. T27 • 3-Minute Warm-Up, p. T28	 • EXPLORE Work, p. 115 • INVESTIGATE Work, p. 118 • Math in Science, p. 120  **UNIT RESOURCE BOOK** • Datasheet, Work, p. 230 • Additional INVESTIGATION, Ramp It Up, A, B, & C, pp. 276–284 • Math Support, pp. 259, 265 • Math Practice, pp. 260, 266
  SECTION 4.2 **Energy is transferred when work is done.** pp. 121–129 Time: 2 periods (1 block)  Lesson Plan, pp. 232–233	 **VISUALIZATION,** Transfer of Potential and Kinetic Energy  **UNIT TRANSPARENCY BOOK** • Daily Vocabulary Scaffolding, p. T26 • 3-Minute Warm-Up, p. T28 • "Conserving Mechanical Energy" Visual, p. T30	 • INVESTIGATE Mechanical Energy, p. 125 • Think Science, p. 129  **UNIT RESOURCE BOOK** • Datasheet, Mechanical Energy, p. 241 • Math Support, p. 261 • Math Practice, p. 262
 SECTION 4.3 **Power is the rate at which work is done.** pp. 130–137 Time: 4 periods (2 blocks)  Lesson Plan, pp. 243–244	 **RESOURCE CENTER,** Power  **UNIT TRANSPARENCY BOOK** • Big Idea Flow Chart, p. T25 • Daily Vocabulary Scaffolding, p. T26 • 3-Minute Warm-Up, p. T29 • Chapter Outline, pp. T31–T32	 • EXPLORE Power, p. 130 • INVESTIGATE Power, p. 133 • CHAPTER INVESTIGATION, Work and Power, pp. 136–137  **UNIT RESOURCE BOOK** • Datasheet, Power, p. 252 • Math Support, p. 263 • Math Practice, p. 264 • CHAPTER INVESTIGATION, Work and Power, A, B, & C, pp. 267–275

KEY TO ICONS

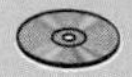 CD/CD-ROM

INTERNET

 Pupil Edition

 Teacher Edition

UNIT RESOURCE BOOK

 UNIT TRANSPARENCY BOOK

UNIT ASSESSMENT BOOK

 SPANISH ASSESSMENT BOOK

SCIENCE TOOLKIT

READING AND REINFORCEMENT

ASSESSMENT

STANDARDS

- Choose Your Own Strategy, B20–25
- Main Idea Web, C38–39
- Daily Vocabulary Scaffolding, H1–8

 **UNIT RESOURCE BOOK**

- Vocabulary Practice, pp. 256–257
- Decoding Support, p. 258
- Summarizing the Chapter, pp. 285–286

Audio Readings CD
Listen to Pupil Edition.

Audio Readings in Spanish CD
Listen to Pupil Edition in Spanish.

- Chapter Review, pp. 139–140
- Standardized Test Practice, p. 141

UNIT ASSESSMENT BOOK

- Diagnostic Test, pp. 60–61
- Chapter Test, A, B, & C, pp. 65–76
- Alternative Assessment, pp. 77–78

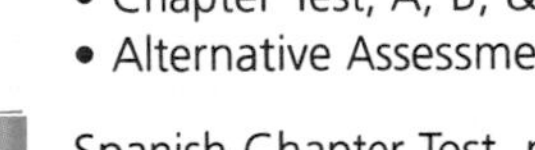

Spanish Chapter Test, pp. 269–272

 **Test Generator CD-ROM**
Generate customized tests.

 **Lab Generator CD-ROM**
Rubrics for Labs

National Standards
A.2-8, A.9.a–c, A.9.e–f, B.2.b, B.3.a, G.1.b

See p. 112 for the standards.

 **UNIT RESOURCE BOOK**

- Reading Study Guide, A & B, pp. 223–226
- Spanish Reading Study Guide, pp. 227–228
- Challenge and Extension, p. 229
- Reinforcing Key Concepts, p. 231

 Ongoing Assessment, pp. 115–117, 119

 Section 4.1 Review, p. 119

 **UNIT ASSESSMENT BOOK**
Section 4.1 Quiz, p. 62

National Standards
A.2–8, A.9.a–c, A.9.e–f, G.1.b

 **UNIT RESOURCE BOOK**

- Reading Study Guide, A & B, pp. 234–237
- Spanish Reading Study Guide, pp. 238–239
- Challenge and Extension, p. 240
- Reinforcing Key Concepts, p. 242
- Challenge Reading, pp. 254–255

 Ongoing Assessment, pp. 121, 123–127

 Section 4.2 Review, p. 128

 **UNIT ASSESSMENT BOOK**
Section 4.2 Quiz, p. 63

National Standards
A.2–8, A.9.a–c, A.9.e–f, B.3.a, G.1.b

 **UNIT RESOURCE BOOK**

- Reading Study Guide, A & B, pp. 245–248
- Spanish Reading Study Guide, pp. 249–250
- Challenge and Extension, p. 251
- Reinforcing Key Concepts, p. 253

 Ongoing Assessment, pp. 131–132, 134–135

 Section 4.3 Review, p. 135

 **UNIT ASSESSMENT BOOK**
Section 4.3 Quiz, p. 64

National Standards
A.2–8, A.9.a–c, A.9.e–f, G.1.b

Previewing Resources for Differentiated Instruction

CHAPTER INVESTIGATION

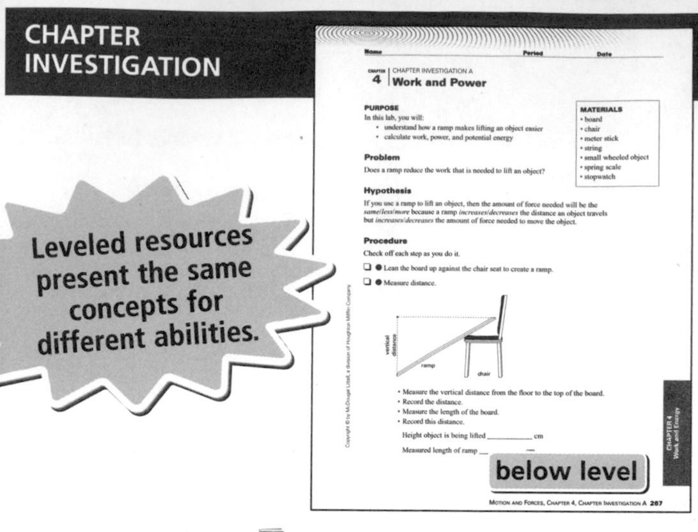

Leveled resources present the same concepts for different abilities.

Name Period Date

CHAPTER 4 | CHAPTER INVESTIGATION A
Work and Power

PURPOSE
In this lab, you will:
- understand how a ramp makes lifting an object easier
- calculate work, power, and potential energy

MATERIALS
- board
- chair
- meter stick
- string
- small wheeled object
- spring scale
- stopwatch

Problem
Does a ramp reduce the work that is needed to lift an object?

Hypothesis
If you use a ramp to lift an object, then the amount of force needed will be the same/less/more because a ramp increases/decreases the distance an object travels but increases/decreases the amount of force needed to move the object.

Procedure
Check off each step as you do it.
- Lean the board up against the chair seat to create a ramp.
- Measure distance.

vertical distance
ramp
chair

- Measure the vertical distance from the floor to the top of the board.
- Record the distance.
- Measure the length of the board.
- Record this distance.

Height object is being lifted ________ cm
Measured length of ramp ________ cm

below level

MOTION AND FORCES, CHAPTER 4, CHAPTER INVESTIGATION A 267

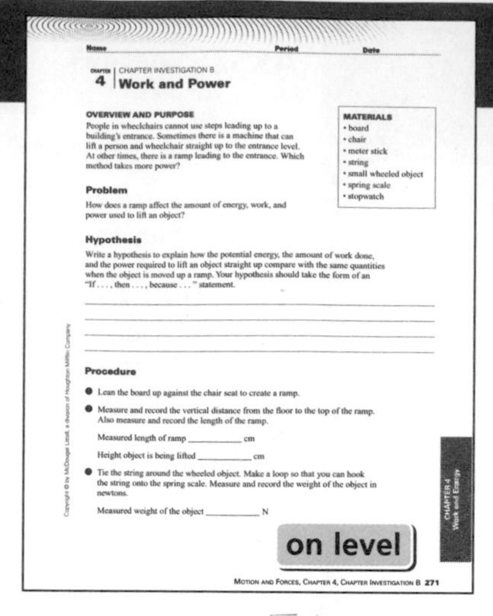

Name Period Date

CHAPTER 4 | CHAPTER INVESTIGATION B
Work and Power

OVERVIEW AND PURPOSE
People in wheelchairs cannot use steps leading up to a building's entrance. Sometimes there is a machine that can lift a person and wheelchair straight up to the entrance level. At other times, there is a ramp leading to the entrance. Which method takes more power?

MATERIALS
- board
- chair
- meter stick
- string
- small wheeled object
- spring scale
- stopwatch

Problem
How does a ramp affect the amount of energy, work, and power used to lift an object?

Hypothesis
Write a hypothesis to explain how the potential energy, the amount of work done, and the power required to lift an object straight up compare with the same quantities when the object is moved up a ramp. Your hypothesis should take the form of an "If . . . , then . . . , because . . ." statement.

Procedure
- Lean the board up against the chair seat to create a ramp.
- Measure and record the vertical distance from the floor to the top of the ramp. Also measure and record the length of the ramp.

Measured length of ramp ________ cm
Height object is being lifted ________ cm

- Tie the string around the wheeled object. Make a loop so that you can hook the string onto the spring scale. Measure and record the weight of the object in newtons.

Measured weight of the object ________ N

on level

MOTION AND FORCES, CHAPTER 4, CHAPTER INVESTIGATION B 271

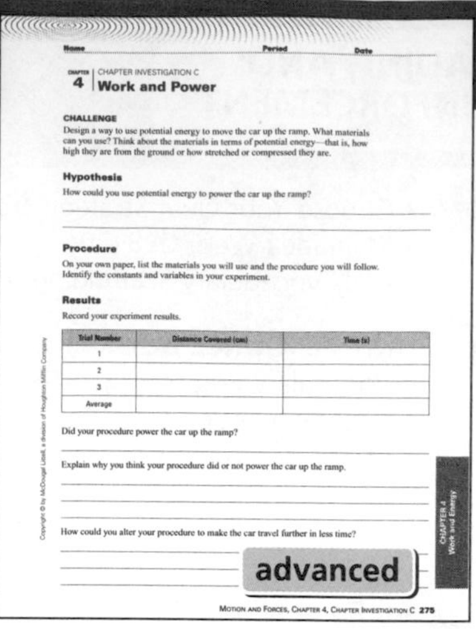

Name Period Date

CHAPTER 4 | CHAPTER INVESTIGATION C
Work and Power

CHALLENGE
Design a way to use potential energy to move the car up the ramp. What materials can you use? Think about the materials in terms of potential energy—that is, how high they are from the ground or how stretched or compressed they are.

Hypothesis
How could you use potential energy to power the car up the ramp?

Procedure
On your own paper, list the materials you will use and the procedure you will follow. Identify the constants and variables in your experiment.

Results
Record your experiment results.

Trial Number	Distance Covered (cm)	Time (s)
1		
2		
3		
Average		

Did your procedure power the car up the ramp?

Explain why you think your procedure did or not power the car up the ramp.

How could you alter your procedure to make the car travel farther in less time?

advanced

MOTION AND FORCES, CHAPTER 4, CHAPTER INVESTIGATION C 275

UNIT RESOURCE BOOK, pp. 267–270 | pp. 271–274 | pp. 271–275

READING STUDY GUIDE

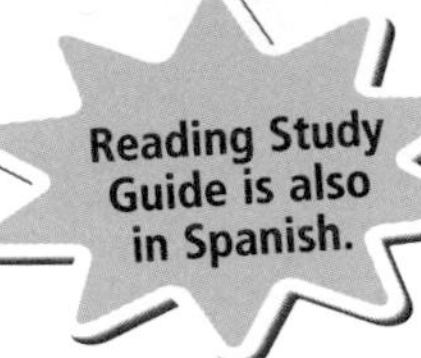

Reading Study Guide is also in Spanish.

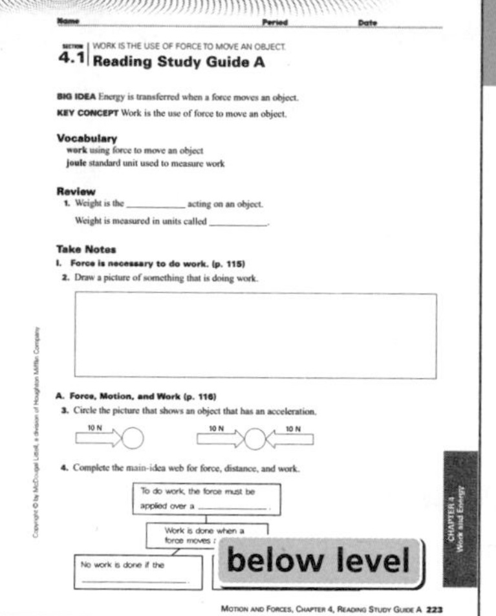

Name Period Date

SECTION 4.1 | WORK IS THE USE OF FORCE TO MOVE AN OBJECT.
Reading Study Guide A

BIG IDEA Energy is transferred when a force moves an object.
KEY CONCEPT Work is the use of force to move an object.

Vocabulary
work using force to move an object
joule standard unit used to measure work

Review
1. Weight is the ________ acting on an object.
Weight is measured in units called ________.

Take Notes
I. Force is necessary to do work. (p. 115)
2. Draw a picture of something that is doing work.

A. Force, Motion, and Work (p. 116)
3. Circle the picture that shows an object that has an acceleration.
4. Complete the main-idea web for force, distance, and work.

below level

MOTION AND FORCES, CHAPTER 4, READING STUDY GUIDE A 223

UNIT RESOURCE BOOK, pp. 223–224

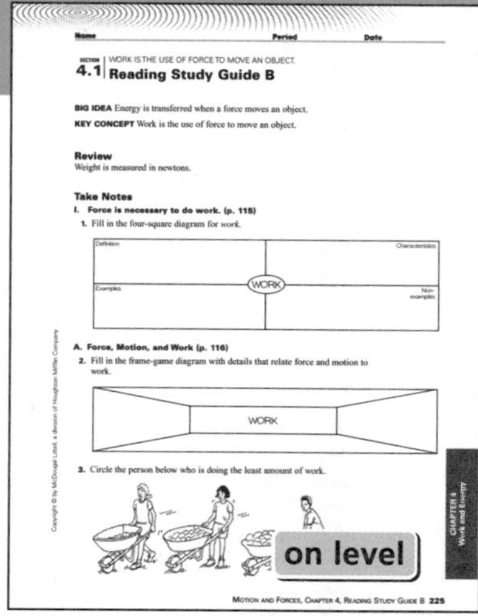

Name Period Date

SECTION 4.1 | WORK IS THE USE OF FORCE TO MOVE AN OBJECT.
Reading Study Guide B

BIG IDEA Energy is transferred when a force moves an object.
KEY CONCEPT Work is the use of force to move an object.

Review
Weight is measured in newtons.

Take Notes
I. Force is necessary to do work. (p. 115)
1. Fill in the four-square diagram for work.

A. Force, Motion, and Work (p. 116)
2. Fill in the frame-game diagram with details that relate force and motion to work.
3. Circle the person below who is doing the least amount of work.

on level

MOTION AND FORCES, CHAPTER 4, READING STUDY GUIDE B 225

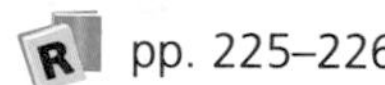

pp. 225–226

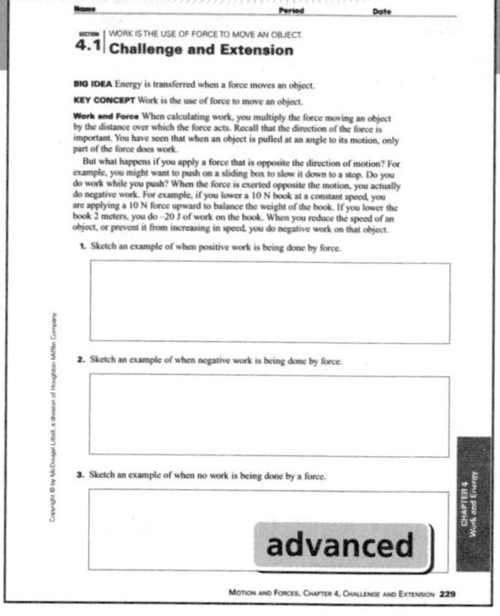

Name Period Date

SECTION 4.1 | WORK IS THE USE OF FORCE TO MOVE AN OBJECT.
Challenge and Extension

BIG IDEA Energy is transferred when a force moves an object.
KEY CONCEPT Work is the use of force to move an object.

1. Sketch an example of when positive work is being done by force.
2. Sketch an example of when negative work is being done by force.
3. Sketch an example of when no work is being done by a force.

advanced

MOTION AND FORCES, CHAPTER 4, CHALLENGE AND EXTENSION 229

p. 229

CHAPTER TEST

Chapter Test is also in Spanish.

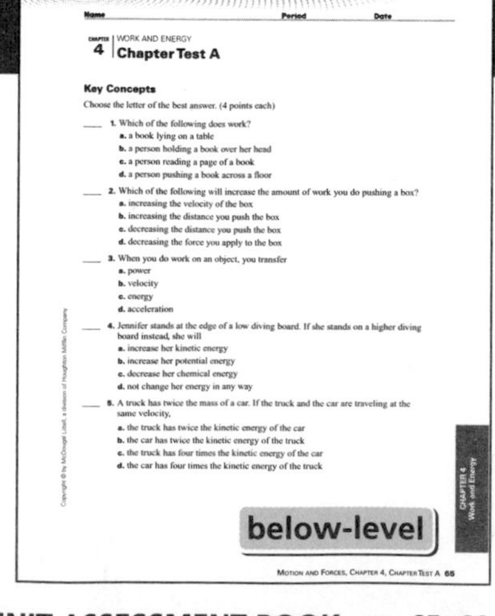

Name Period Date

CHAPTER 4 | WORK AND ENERGY
Chapter Test A

Key Concepts
Choose the letter of the best answer. (4 points each)

1. Which of the following does work?
a. a book lying on a table
b. a person holding a book over her head
c. a person reading a page of a book
d. a person pushing a book across a floor

2. Which of the following will increase the amount of work you do pushing a box?
a. increasing the velocity of the box
b. increasing the distance you push the box
c. decreasing the distance you push the box
d. decreasing the force you apply to the box

3. When you do work on an object, you transfer
a. power
b. velocity
c. energy
d. acceleration

4. Jennifer stands at the edge of a low diving board. If she stands on a higher diving board instead, she will
a. increase her kinetic energy
b. increase her potential energy
c. decrease her chemical energy
d. not change her energy in any way

5. A truck has twice the mass of a car. If the truck and the car are traveling at the same velocity,
a. the truck has twice the kinetic energy of the car
b. the car has twice the kinetic energy of the truck
c. the truck has four times the kinetic energy of the car
d. the car has four times the kinetic energy of the truck

below-level

MOTION AND FORCES, CHAPTER 4, CHAPTER TEST A 65

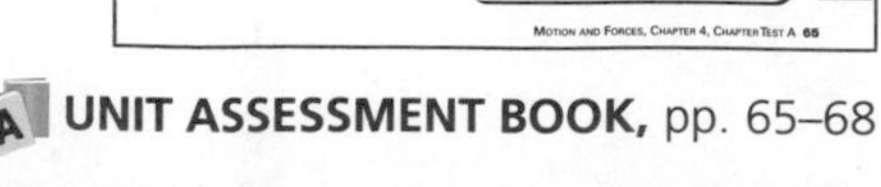

UNIT ASSESSMENT BOOK, pp. 65–68

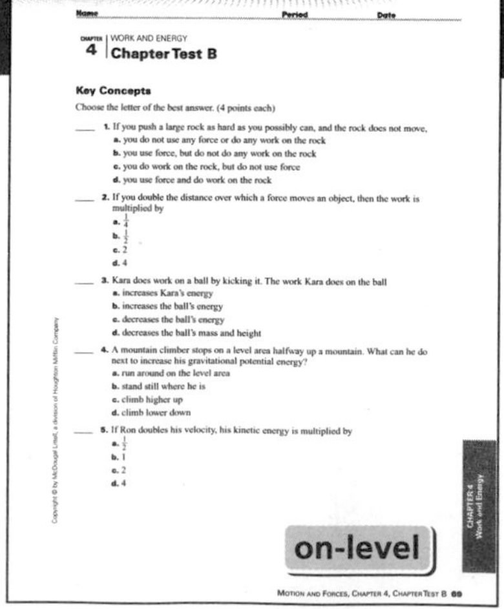

Name Period Date

CHAPTER 4 | WORK AND ENERGY
Chapter Test B

Key Concepts
Choose the letter of the best answer. (4 points each)

1. If you push a large rock as hard as you possibly can, and the rock does not move,
a. you do not use any force or do any work on the rock
b. you use force, but do not do any work on the rock
c. you do work on the rock, but do not use force
d. you use force and do work on the rock

2. If you double the distance over which a force moves an object, then the work is multiplied by

3. Kara does work on a ball by kicking it. The work Kara does on the ball
a. increases Kara's energy
b. increases the ball's energy
c. decreases the ball's energy
d. decreases the ball's mass and height

4. A mountain climber stops on a level area halfway up a mountain. What can he do next to increase his gravitational potential energy?
a. run around on the level area
b. stand still where he is
c. climb higher up
d. climb lower down

5. If Ron doubles his velocity, his kinetic energy is multiplied by

on-level

MOTION AND FORCES, CHAPTER 4, CHAPTER TEST B 69

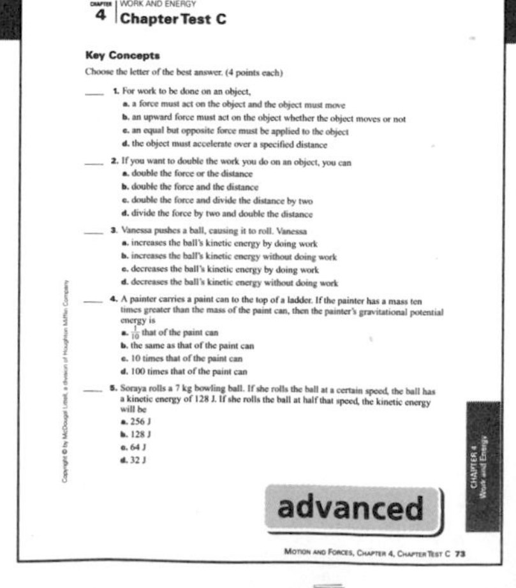

Name Period Date

CHAPTER 4 | WORK AND ENERGY
Chapter Test C

Key Concepts
Choose the letter of the best answer. (4 points each)

1. For work to be done on an object,
a. a force must act on the object and the object must move
b. an upward force must act on the object whether the object moves or not
c. an equal but opposite force must be applied to the object
d. the object must accelerate over a specified distance

2. If you want to double the work you do on an object, you can
a. double the force or the distance
b. double the force and the distance
c. double the force and divide the distance by two
d. divide the force by two and double the distance

3. Vanessa pushes a ball, causing it to roll. Vanessa
a. increases the ball's kinetic energy by doing work
b. increases the ball's kinetic energy without doing work
c. decreases the ball's kinetic energy by doing work
d. decreases the ball's kinetic energy without doing work

4. A painter carries a paint can to the top of a ladder. If the painter has a mass ten times greater than the mass of the paint can, then the painter's gravitational potential energy is
a. $\frac{1}{10}$ that of the paint can
b. the same as that of the paint can
c. 10 times that of the paint can
d. 100 times that of the paint can

5. Soraya rolls a 7 kg bowling ball. If she rolls the ball at a certain speed, the ball has a kinetic energy of 128 J. If she rolls the ball at half that speed, the kinetic energy will be
a. 256 J
b. 128 J
c. 64 J
d. 32 J

advanced

MOTION AND FORCES, CHAPTER 4, CHAPTER TEST C 73

pp. 69–72 | pp. 73–76

TECHNOLOGY

CLASSZONE.COM

CD/CD-ROMS

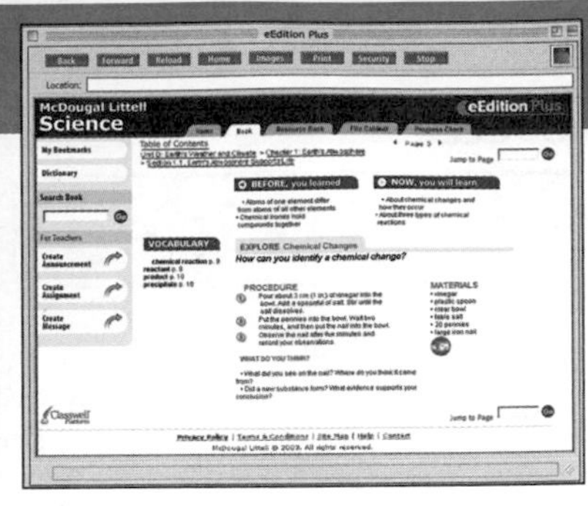

CLASSZONE.COM

VISUAL CONTENT

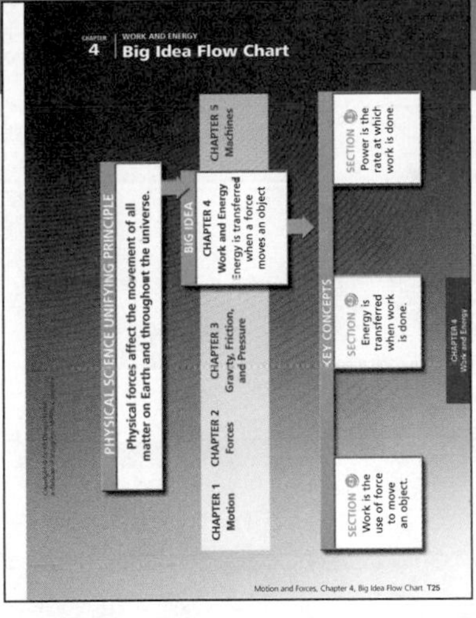

UNIT TRANSPARENCY BOOK, p. T25

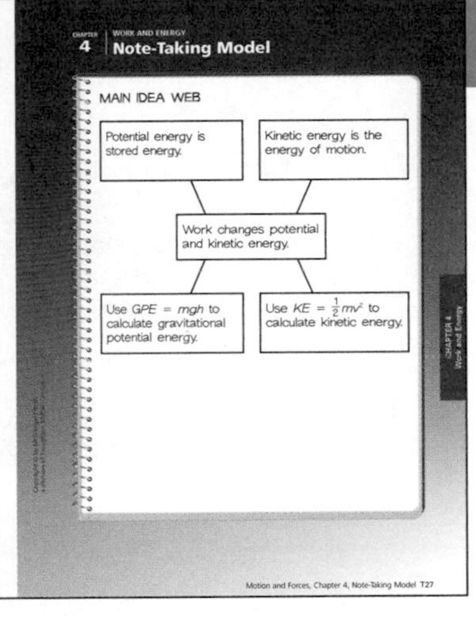

p. T27

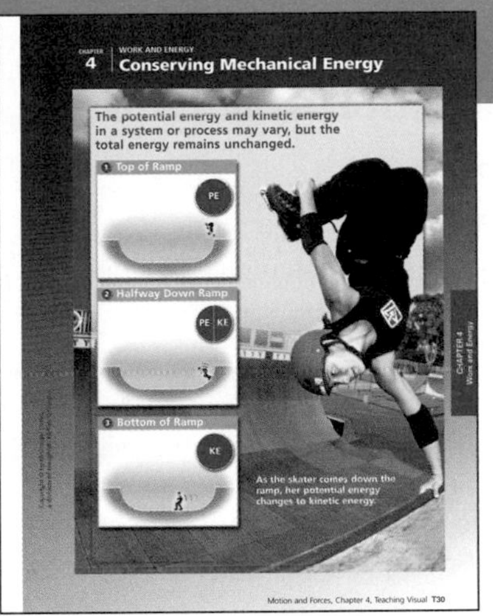

p. T30

MORE SUPPORT

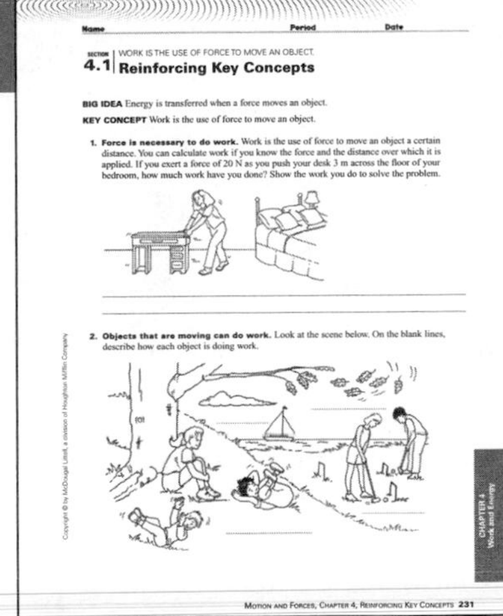

UNIT RESOURCE BOOK, p. 231

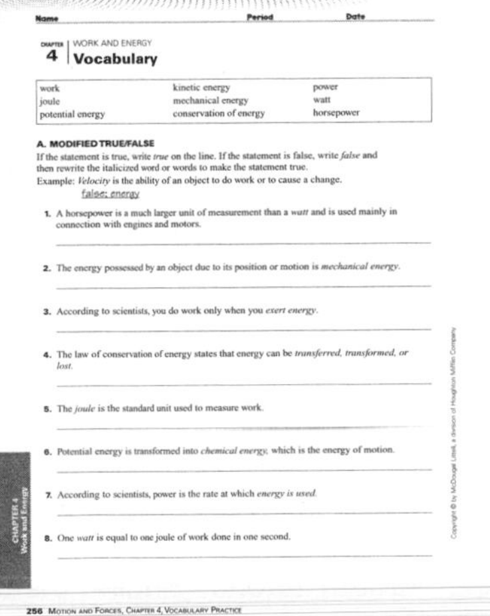

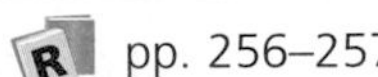
pp. 256–257

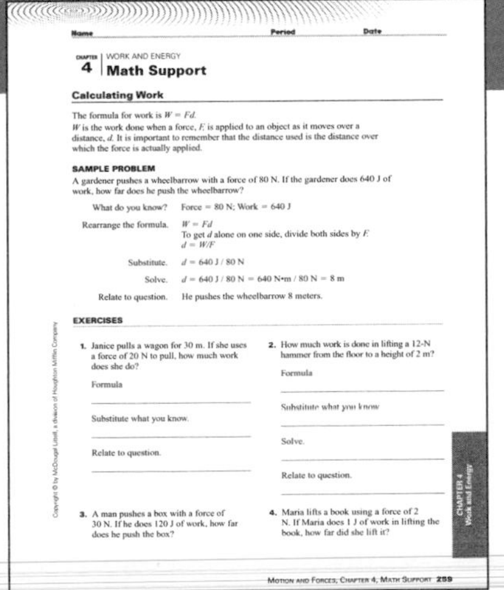

p. 259

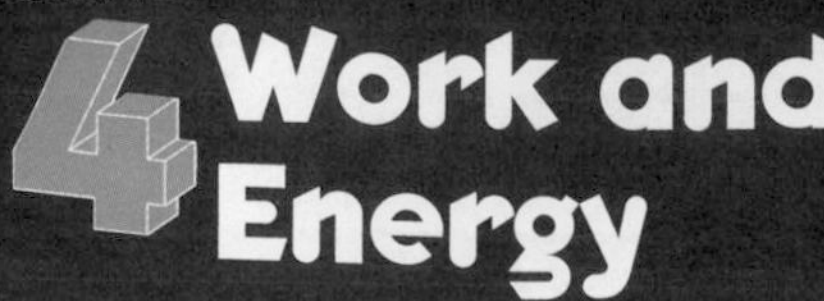

CHAPTER 4 Work and Energy

INTRODUCE

the BIG idea

Have students look at the photograph of the young woman carrying a box and discuss how the question in the blue box links to the Big Idea:

- In what way is the person in the photograph doing work?
- What is a force?
- Did the person in the photograph apply force at any time but not do any work?

National Science Education Standards

Content

B.2.b An object that is not being subjected to a force will continue to move at a constant speed and in a straight line.

B.3.a Energy is a property of many substances and is associated with heat, light, electricity, mechanical motion, sound, nuclei, and the nature of a chemical. Energy is transferred in many ways.

Process

A.2–8 Design and conduct an investigation; use tools to gather and interpret data; use evidence to describe, predict, explain, model; think critically to make relationships between evidence and explanation; recognize different explanations and predictions; communicate scientific procedures and explanations; use mathematics.

A.9.a–c, A.9.e–f Understand scientific inquiry by using different investigations, methods, mathematics, and explanations based on logic, evidence, and skepticism.

G.1.b Science requires different abilities

CHAPTER 4 Work and Energy

the BIG idea

Energy is transferred when a force moves an object.

Key Concepts

SECTION

4.1 Work is the use of force to move an object.
Learn about the relationship between force and work.

4.2 Energy is transferred when work is done.
Learn how energy is related to work.

SECTION

4.3 Power is the rate at which work is done.
Learn to calculate power from work and energy.

Which takes more work, lifting a box or holding a box? Why?

Internet Preview

CLASSZONE.COM

Chapter 4 online resources: Content Review, Simulation, Visualization, two Resource Centers, Math Tutorial, Test Practice

INTERNET PREVIEW

CLASSZONE.COM For student use with the following pages:

Review and Practice
- Content Review, pp. 114, 138
- Math Tutorial: Finding the Mean, p. 120
- Test Practice, p. 141

Activities and Resources
- Internet Activity: Work, p. 113
- Resource Centers: Work, p. 116; Power, p. 132
- Visualization: Transfer of Potential & Kinetic Energy, p. 126

NSTA scilinks.org SCILINKS

Potential and Kinetic Energy
Code: MDL007

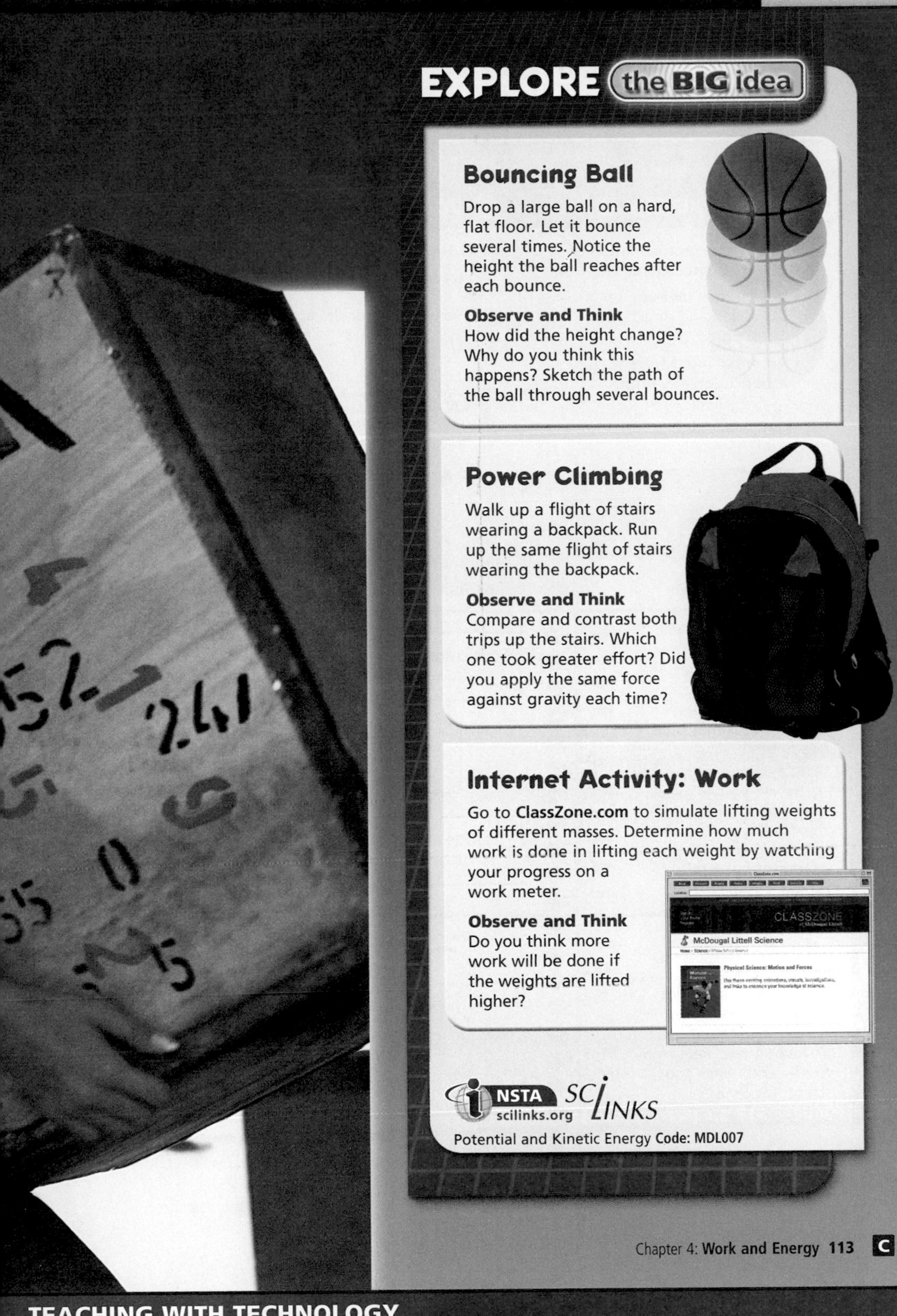

EXPLORE the BIG idea

Bouncing Ball

Drop a large ball on a hard, flat floor. Let it bounce several times. Notice the height the ball reaches after each bounce.

Observe and Think
How did the height change? Why do you think this happens? Sketch the path of the ball through several bounces.

Power Climbing

Walk up a flight of stairs wearing a backpack. Run up the same flight of stairs wearing the backpack.

Observe and Think
Compare and contrast both trips up the stairs. Which one took greater effort? Did you apply the same force against gravity each time?

Internet Activity: Work

Go to **ClassZone.com** to simulate lifting weights of different masses. Determine how much work is done in lifting each weight by watching your progress on a work meter.

Observe and Think
Do you think more work will be done if the weights are lifted higher?

NSTA scilinks.org SCILINKS
Potential and Kinetic Energy **Code: MDL007**

EXPLORE the BIG idea

These inquiry-based activities are appropriate for use at home or as a supplement to classroom instruction.

Bouncing Ball

PURPOSE To observe a decrease in mechanical energy. Students see that a ball loses height as it bounces.

TIP *10 min.* Students should think about what happens to the ball's energy as it bounces.

Answer: The height of the ball decreases with each bounce. Some of the ball's energy of motion is converted to other forms of energy as the ball moves through the air.

REVISIT after p. 126.

Power Climbing

PURPOSE To perform work using different amounts of power

TIP *10 min.* Have students think about how their results would change if they carried a heavier backpack.

Answer: Both trips up the stairs took the same amount of work, but running used more effort because it took less time to climb the stairs. The same force against gravity was applied each time.

REVISIT after p. 131.

Internet Activity: Work

PURPOSE To see how much work is done when applying force to an object

TIP *20 min.* Assign different distances to student groups. Have the class graph distance versus work.

Answer: Yes, more work will be done if the weights are lifted higher. To lift the weights higher, more energy must be used.

REVISIT after p. 118.

TEACHING WITH TECHNOLOGY

Spreadsheet Have students use a spreadsheet program to record their data and do the calculations for "Investigate Mechanical Energy" on p. 125, and the Chapter Investigation, pp. 136–137.

Video Camera Students can use a video camera to tape "Investigate Power" on p. 133. They then can view the tape to make sure they are pulling the object across the floor while using a constant force.

PREPARE

CONCEPT REVIEW

Activate Prior Knowledge

- Set a toy car on a table, then give the car a push. Have students describe what they observe in terms of the force applied, the change in the car's position, its velocity, and its acceleration.
- Ask students to identify correct units for velocity, speed, and acceleration.
- Ask students to define *force* in their own words.

TAKING NOTES

Main Idea Web

Writing notes around each heading is similar to making an outline of the major concepts in the chapter. Summarizing the most important details of a concept in this way will help students learn by organizing new material.

Vocabulary Strategy

Having students choose a note-taking strategy allows them to use the strategy that works best for a particular topic. It is also a good review of the strategies already presented:description wheel, magnet words, and four square.

Vocabulary and Note-Taking Resources

- Vocabulary Practice, pp. 256–257
- Decoding Support, p. 258

- Daily Vocabulary Scaffolding, p. T26
- Note-Taking Model, p. T27

- Choose Your Own Strategy, B20–25
- Main Idea Web, C38–39
- Daily Vocabulary Scaffolding, H1–8

CHAPTER 4

Getting Ready to Learn

CONCEPT REVIEW

- Forces change the motion of objects in predictable ways.
- Velocity is a measure of the speed and direction of an object.
- An unbalanced force produces acceleration.

VOCABULARY REVIEW

velocity p. 22
force p. 41

See Glossary for definitions.

energy, mass

CONTENT REVIEW
CLASSZONE.COM
Review concepts and vocabulary.

TAKING NOTES

MAIN IDEA WEB

Write each new blue heading in a box. Then write notes in boxes around it that give important terms and details about that blue heading.

CHOOSE YOUR OWN STRATEGY

Take notes about new vocabulary terms using one or more of the strategies from earlier chapters—**description wheel, magnet words,** or **four square.** Feel free to mix and match the strategies or use a different strategy.

See the Note-Taking Handbook on pages R45–R51.

SCIENCE NOTEBOOK

Work is the use of force to move an object.

Work = Force · distance

Force is necessary to do work.

Joule is the unit for measuring work.

Work depends on force and distance.

Description Wheel: feature, feature, feature, TERM, feature, feature, feature

Four Square: Definition, Characteristics, TERM, Examples, Nonexamples

Magnet Word: related terms, TERM, related ideas

CHECK READINESS

Administer the Diagnostic Test to determine students' readiness for new science content and their mastery of requisite math skills.

 Diagnostic Test, pp. 60–61

Technology Resources

Students needing content and math skills should visit **ClassZone.com**.

- CONTENT REVIEW
- MATH TUTORIAL

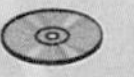 CONTENT REVIEW CD-ROM

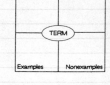

KEY CONCEPT

Work is the use of force to move an object.

 **BEFORE, you learned**

- An unbalanced force produces acceleration
- Weight is measured in newtons

 **NOW, you will learn**

- How force and work are related
- How moving objects do work

VOCABULARY

work p. 115
joule p. 117

EXPLORE Work

How do you work?

PROCEDURE

1. Lift a book from the floor to your desktop. Try to move the book at a constant speed.
2. Now lift the book again, but stop about halfway up and hold the book still for about 30 seconds. Then continue lifting the book to the desktop.

MATERIALS
book

WHAT DO YOU THINK?

- Do you think you did more work the first time you lifted the book or the second time you lifted the book?
- What do you think *work* means?

Force is necessary to do work.

VOCABULARY
You might want to make a description wheel diagram in your notebook for *work*.

What comes to mind when you think of work? Most people say they are working when they do anything that requires a physical or mental effort. But in physical science, **work** is the use of force to move an object some distance. In scientific terms, you do work only when you exert a force on an object and move it. According to this definition of work, reading this page is not doing work. Turning the page, however, would be work because you are lifting the page.

Solving a math problem in your head is not doing work. Writing the answer is work because you are moving the pencil across the paper. If you want to do work, you have to use force to move something.

 How does the scientific definition of work differ from the familiar definition?

4.1 FOCUS

Set Learning Goals

Students will

- Recognize how force and work are related.
- Identify how moving objects do work.
- Determine through an experiment how much work is done when lifting an object.

3-Minute Warm-Up

Display Transparency 28 or copy this exercise on the board:

Decide whether these statements are true. If they are not true, correct them.

1. An object has acceleration if its velocity is changing. *true*
2. Force is measured in joules. *Force is measured in newtons.*
3. Gravity is a force. *true*

 3-Minute Warm-Up, p. T28

4.1 MOTIVATE

EXPLORE Work

PURPOSE To explore the scientific definition of work

TIP *5 min.* Help students understand that work is done only when an object is moving. Ask if work would be done if the book were replaced by a barbell that was too heavy to be lifted. *Work is not done if the barbell does not move.*

WHAT DO YOU THINK? *Many students will say that they did more work the second time. Work is done only when the book is moving.*

Ongoing Assessment

 *Answer: The scientific definition of work involves the use of force to move an object over a distance. The ordinary definition involves making a mental or a physical effort.*

RESOURCES FOR DIFFERENTIATED INSTRUCTION

Below Level

UNIT RESOURCE BOOK
- Reading Study Guide A, pp. 223–224
- Decoding Support, p. 258

AUDIO CDS

 **Additional INVESTIGATION,** Ramp It Up, A, B, & C, pp. 276–284; Teacher Instructions, pp. 346–347

Advanced

UNIT RESOURCE BOOK
Challenge and Extension, p. 229

English Learners

UNIT RESOURCE BOOK
Spanish Reading Study Guide, pp. 227–228

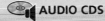

AUDIO CDS
- Audio Readings in Spanish
- Audio Readings (English)

4.1 INSTRUCT

Teach Difficult Concepts

Students may have a difficult time adopting the scientific definition of work. Tell them to imagine a waiter with a tray of food. Ask:

- Does the waiter do work to get the tray up to shoulder height? Why or why not? *Yes; force is applied in the same direction as movement.*
- Is the force the waiter is using to hold up the tray doing work? Why? *No; the force he exerts to hold up the tray does not change its position.*

Teach from Visuals

To help students interpret the "Work" visual, ask:

- Is the same amount of work done when you pull the suitcase with the handle in a horizontal position as when you pull it at an angled position? *No; if you use the same force, less of that force does work when you pull the handle at an angle.*
- If you walk while carrying the suitcase by its small top handle, is the force that you are using to hold up the suitcase doing work? Why? *No; the suitcase is moving forward and you are applying the force upward.*

Ongoing Assessment

Recognize how force and work are related.

Ask: Is work being done if you hold a trumpet? Why? *No; you are not moving it.* Is work being done if you carry a trumpet while marching up the stairs? Why? *Yes; both the force on the trumpet and its motion are in the same direction, so work is being done.*

CHECK YOUR READING *Answer: when the object is not moving and when the force is not acting in the same direction as the motion of the object*

READING VISUALS *Answer: It changes the amount of the applied force that does work.*

Learn more about work.

Force, Motion, and Work

Work is done only when an object that is being pushed or pulled actually moves. If you lift a book, you exert a force and do work. What if you simply hold the book out in front of you? No matter how tired your muscles may become from holding the book still, you are not doing work unless you move the book.

A

The work done by a force is related to the size of the force and the distance over which the force is applied. How much work does it take to push a grocery cart down an aisle? The answer depends on how hard you push the cart and the length of the aisle. If you use the same amount of force, you do more work pushing a cart down a long aisle than a short aisle.

Work is done only by the part of the applied force that acts in the same direction as the motion of an object. Suppose you need to pull a heavy suitcase on wheels. You pull the handle up at an angle as you pull the suitcase forward. Only the part of the force pulling the suitcase forward is doing work. The force with which you pull upward on the handle is not doing work because the suitcase is not moving upward—unless you are going uphill.

Give two examples of when you are applying a force but not doing work.

Work

Work is done by force that acts in the same direction as the motion of an object.

All of the Applied Force Does Work

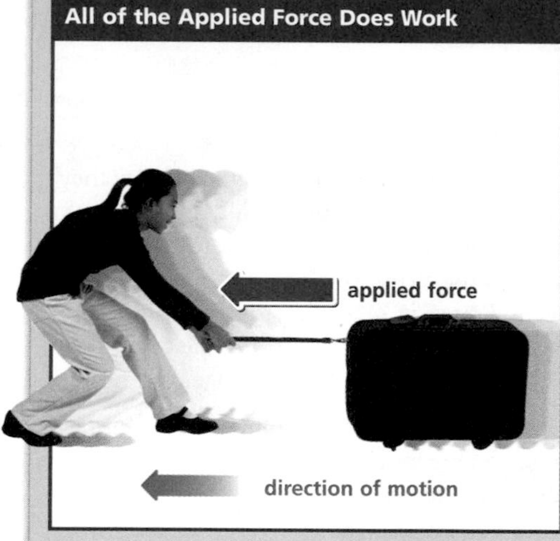

Part of the Applied Force Does Work

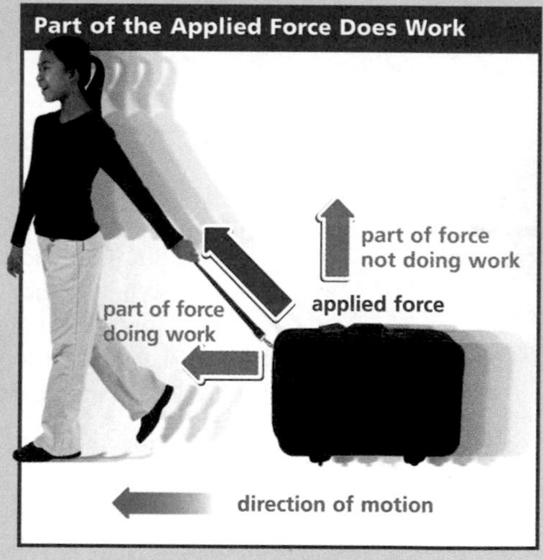

How does changing the direction of the applied force change the amount of the force that is doing work?

DIFFERENTIATE INSTRUCTION

More Reading Support

A Which two things determine how much work is done? *size of the force, distance over which it is applied*

English Learners English learners may have difficulty distinguishing between nouns and verbs when they are used in different contexts. For example, the following sentences use the word *work* in two different ways: *He works all day. He does his work all day.* In the first sentence, *work* is a verb. In the second sentence, *work* is a noun. Throughout this chapter, the word *work* is most often a noun; "Work is the use of force to move an object." Make sure students are reading nouns and verbs correctly.

Calculating Work

This man is doing work when he applies force to lift his body.

Work is a measure of how much force is applied over a certain distance. You can calculate the work a force does if you know the size of the force applied to an object and the distance over which the force acts. The distance involved is the distance the object moved in the direction of that force. The calculation for work is shown in the following formula:

? B

$$\text{Work} = \text{Force} \cdot \text{distance}$$
$$W = Fd$$

You read in previous chapters that you can measure force in newtons. You also know that you can measure distance in meters. When you multiply a force in newtons times a distance in meters, the product is a measurement called the newton-meter (N•m), or the **joule** (jool).

The joule (J) is the standard unit used to measure work. One joule of work is done when a force of one newton moves an object one meter. To get an idea of how much a joule of work is, lift an apple (which weighs about one newton) from your foot to your waist (about one meter).

Use the formula for work to solve the problem below.

Calculating Work

Sample Problem

How much work is done if a person lifts a barbell weighing 450 N to a height of 2 m?

What do you know? force needed to lift = 450 N, distance = 2 m

What do you want to find out? Work

Write the formula: $W = Fd$

Substitute into the formula: $W = 450\text{ N} \cdot 2\text{ m}$

Calculate and simplify: $W = 900\text{ N}\cdot\text{m}$

Check that your units agree: Unit is newton-meter (N•m). Unit of work is joule, which is N•m. Units agree.

Answer: $W = 900\text{ J}$

Practice the Math

1. If you push a cart with a force of 70 N for 2 m, how much work is done?
2. If you did 200 J of work pushing a box with a force of 40 N, how far did you push the box?

REMINDER

You know that $W = Fd$. You can manipulate the formula to find force or distance.

$d = \frac{W}{F}$ and $F = \frac{W}{d}$

DIFFERENTIATE INSTRUCTION

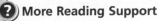

More Reading Support

B How do you calculate work? *Multiply the force on an object by the distance the object moves in the direction of the force.*

Additional Investigation To reinforce Section 4.1 learning goals, use the following full period investigation:

Additional INVESTIGATION, Ramp It Up, A, B, & C, pp. 276–284, 346–347 (Advanced students should complete Levels B and C.)

English Learners English learners may not be familiar with the following words and phrases: *hit a baseball, swim a lap, tap a keyboard, bowling alley* (p. 118), and *escalator* on p. 120. Make sure English learners have sufficient background knowledge.

Teach from Visuals

Remind students that work equals force times distance. Ask: In what two ways could the man in the photograph increase the amount of work he does? *He could increase the force with which he pushes against the floor or increase the distance that he raises his body.*

Address Misconceptions

IDENTIFY Ask: If you shove a ball with a force of 50 newtons and it rolls 3 meters across the floor, how much work have you done? If students answer 150 joules, they may hold the misconception that work is done on the ball even while it rolls.

CORRECT Remind students that work is done only while a force is being applied to an object. For example, when you throw a ball, no work is done after the ball leaves your hand, because the forward force is no longer being applied.

REASSESS Ask students when someone does work while skateboarding. *while the skateboarder is pushing on the ground with his or her foot*

Technology Resources

Visit **ClassZone.com** for background on common student misconceptions.

MISCONCEPTION DATABASE

Develop Mathematics Skills

R
- Math Support, p. 259
- Math Practice, p. 260

History of Science

The joule is named for James Prescott Joule (1818–1889), a British physicist who established the mechanical theory of heat and the law of conservation of energy.

Ongoing Assessment

Practice the Math *Answers:*

1. $W = F \cdot d = 70\text{ N} \cdot 2\text{ m} = 140\text{ N} \cdot \text{m}$
   $W = 140\text{ J}$
2. $d = \frac{W}{F} = \frac{200\text{ J}}{40\text{ N}} = 5\text{ m}$

INVESTIGATE Work

PURPOSE To show the relationship between work, force, and distance by finding the work done in lifting a notebook

TIP *20 min.* Have one student in each group measure the force of all the group members so measurements are consistent.

WHAT DO YOU THINK? *Answers will vary. You will do less work if you are shorter and more work if you are taller. No work is done if you stop moving the book.*

CHALLENGE *The work done would equal students' results times 10, times the number of days in the school year (180 days).*

 Datasheet, Work, p. 230

Technology Resources

Customize this student lab as needed or look for an alternative. Print rubrics to assess student lab reports.

 **Lab Generator CD-ROM**

Metacognitive Strategy

Ask students if they had misconceptions about the concepts that "Investigate Work" illustrated. For example, did they think the weight of the notebook would increase when a taller person lifted it? Have them write about what changed their ideas.

EXPLORE the BIG idea

Revisit "Internet Activity: Work" on p. 113. Have students explain their results.

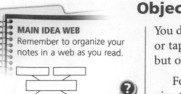

MAIN IDEA WEB
Remember to organize your notes in a web as you read.

Objects that are moving can do work.

You do work when you pick up your books, hit a baseball, swim a lap, or tap a keyboard. These examples show that you do work on objects, but objects can also do work.

For example, in a bowling alley, the bowling balls do work on the pins they hit. Outdoors, the moving air particles in a gust of wind do work that lifts a leaf off the ground. Moving water, such as the water in a river, also does work. If the windblown leaf lands in the water, it might be carried downstream by the current. As the leaf travels downstream, it might go over the edge of a waterfall. In that case, the gravitational force of Earth would pull the leaf and water down.

You can say that an object or person does work on an object, or that the force the object or person is exerting does work. For example, you could say that Earth (an object) does work on the falling water, or that gravity (a force) does work on the water.

INVESTIGATE Work

How much work does it take?

SKILL FOCUS
Measuring

MATERIALS
- meter stick
- spiral notebook
- spring scale

TIME
20 minutes

PROCEDURE

1. Have a partner help you measure how high your shoulders are from the ground. Record the distance in meters. Round to the nearest tenth of a meter.
2. Attach the notebook to the spring scale. Then slowly lift the notebook to your shoulder to see how much force you are exerting. Record the amount in newtons.
3. Calculate the work you did while lifting one notebook. Use this information to estimate how much work you do every day when you pick up all your notebooks to take them to school. (**Hint:** Work equals force times distance.)

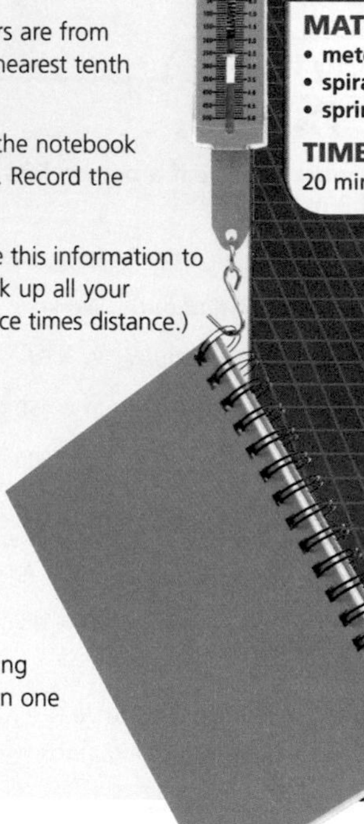

WHAT DO YOU THINK?

- Approximately how much work does it take to pick up your notebook?
- How would the amount of work you do change if you were shorter? taller?
- How much work are you doing on the notebook if you have stopped to talk to a friend?

CHALLENGE If you pick up a notebook 10 times a day during the school year, how much work do you do on the notebook in one year? (Assume that there are 180 school days in a year.)

DIFFERENTIATE INSTRUCTION

More Reading Support

C How can a bowling ball do work? *by exerting a force on bowling pins when it hits them*

Advanced Challenge students to compare the amount of work done while lifting a notebook on Earth and when standing on the Moon. Students should conduct research about the Moon to find that the force of gravity is one-sixth of that on Earth. *Since the weight of the notebook on the Moon is only one-sixth of its weight on Earth, only one-sixth of the amount of work would be done when lifting the notebook on the Moon.*

 Challenge and Extension, p. 229

APPLY How could you increase the work done by this water wheel?

Throughout history, people have taken advantage of the capability of objects in motion to do work. Many early cultures built machines such as water wheels to use the force exerted by falling water, and windmills to use the force exerted by moving air. In a water wheel like the one in the photograph, gravity does work on the water. As the water falls, it also can do work on any object that is put in its path. Falling water can turn a water wheel or the turbine of an electric generator.

The water wheel shown above uses the work done by water to turn gears that run a mill and grind grain. In the same way, windmills take advantage of the force of moving air particles. The wind causes the sails of a windmill to turn. The turning sails do work to run machinery or an irrigation system.

CHECK YOUR READING Describe how a water wheel does work.

4.1 Review

KEY CONCEPTS

1. If you push very hard on an object but it does not move, have you done work? Explain.
2. What two factors do you need to know to calculate how much work was done in any situation?
3. Was work done on a book that fell from a desk to the floor? If so, what force was involved?

CRITICAL THINKING

4. **Synthesize** Work is done on a ball when a soccer player kicks it. Is the player still doing work on the ball as it rolls across the ground? Explain.
5. **Calculate** Tina lifted a box 0.5 m. The box weighed 25 N. How much work did Tina do on the box?

CHALLENGE

6. **Analyze** Ben and Andy each pushed an empty grocery cart. Ben used twice the force, but they both did the same amount of work. Explain.

ANSWERS

1. No; the object must move for work to be done.

2. force and distance

3. yes; the force of gravity

4. No; the player is no longer exerting a force on the ball.

5. $W = F \cdot d$
$= 25\ N \cdot 0.5\ m$
$W = 12.5\ J$

6. Ben used twice the force, but his cart moved only half as far as Andy's cart.

Ongoing Assessment

Identify how moving objects do work.

Ask: How can water falling over a dam do work? *It can turn turbines and run machines.*

CHECK YOUR READING *Answer: A water wheel uses work done by water to turn gears that run a mill and grind grain.*

PHOTO CAPTION Answer: increase the amount of water that is flowing onto the wheel

Reinforce the BIG idea

Have students relate the section to the Big Idea.

 Reinforcing Key Concepts, p. 231

4.1 ASSESS & RETEACH

Assess

 Section 4.1 Quiz, p. 62

Reteach

Have students make a table in which they list three activities that are considered work according to the customary definition but not the scientific definition, three activities that are considered work by the scientific definition but not the ordinary definition, and three activities that fit both definitions.

Sample answers: customary but not scientific—reading, thinking, dreaming; scientific but not customary—blinking, breathing, turning a page; both customary and scientific—raking leaves, moving furniture, climbing a hill

Technology Resources

Have students visit **ClassZone.com** for reteaching of Key Concepts.

 **CONTENT REVIEW**

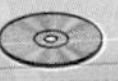 **CONTENT REVIEW CD-ROM**

MATH IN SCIENCE
Math Skills Practice for Science

Set Learning Goal

To learn how extreme values (outliers) in a data set can affect the mean of the data set

Present the Science

Ask students if they have ridden on an escalator (a moving staircase between floors of a building). Tell them that an escalator is like a large conveyor belt set on an incline. The steps are pulled along tracks by chains that move around gears at the top and bottom of the staircase. A typical escalator uses a 100 horsepower motor—about half the power of a large automobile engine—to turn the gears. Discuss an escalator in terms of *Work = Force times distance.*

Develop Algebra Skills

- Remind students to divide the total by the correct number of values after they drop outliers.
- Ask: What is another term for *mean? average*

DIFFERENTIATION TIP Below level: If students have trouble understanding the concept of outliers, have them show the eight data entries on a number line. The high outlier, 4850 J, will be obvious.

Close

Ask students to give an example of how their school lives are affected by means. *Their course grades are means of test scores, homework scores, and other scores.*

- Math Support, p. 265
- Math Practice, p. 266

Technology Resources

Students can visit **ClassZone.com** for practice in finding the mean of a data set.

MATH TUTORIAL

MATH in SCIENCE

MATH TUTORIAL
CLASSZONE.COM
Click on Math Tutorial for more help with finding the mean.

SKILL: WORKING WITH AVERAGES

Eliminating Extreme Values

A value that is far from most others in a set of data is called an outlier. Outliers make it difficult to find a value that might be considered average. Extremely high or extremely low values can throw off the mean. That is why the highest and lowest figures are ignored in some situations.

Example

The data set below shows the work an escalator does to move 8 people of different weights 5 meters. The work was calculated by multiplying the force needed to move each person by a distance of 5 meters.

4850 J 1600 J 3400 J 2750 J
2950 J 1750 J 3350 J 3800 J

The mean amount of work done is 3056 J.

(1) To calculate an adjusted mean, begin by identifying a high outlier in the data set.

High outlier: 4850

(2) Discard this value and find the new mean.

1600 J + 3400 J + 2750 J + 2950 J + 1750 J + 3350 J + 3800 J = 19,600 J

$$\text{Mean} = \frac{19{,}600\text{ J}}{7} = 2800\text{ J}$$

ANSWER The mean amount of work done for this new data set is 2800 J.

Answer the following questions.

1. After ignoring the high outlier in the data set, does this new mean show a more typical level of work for the data set? Why or why not?
2. Do you think the lowest value in the data set is an outlier? Remove it and calculate the new average. How did this affect the results?
3. Suppose the heaviest person in the original data set were replaced by a person weighing the same as the lightest person. What would be the new mean for the data set?

CHALLENGE The median of a data set is the middle value when the values are written in numerical order. Find the median of the adjusted data set (without the high outlier). Compare it with the original and adjusted means. Why do you think it is closer to one than the other?

C 120 Unit: Motion and Forces

ANSWERS

1. Yes; the new mean is not distorted by a very large or a very small value in the data set.

2. The new average to the nearest whole number is 3000 J. The mean is less than the mean in the original set of data.

3. After replacing 4850 J with 1600 J, the new mean for the data set would be 2650 J.

CHALLENGE *Median: 1600 1750 2750 (2950) 3350 3400 3800*
The median is closer to the adjusted mean because both calculations have been done without the high outlier.

KEY CONCEPT

Energy is transferred when work is done.

BEFORE, you learned
- Work is the use of force to move an object
- Work can be calculated

NOW, you will learn
- How work and energy are related
- How to calculate mechanical, kinetic, and potential energy
- What the conservation of energy means

VOCABULARY

potential energy p. 122
kinetic energy p. 122
mechanical energy p. 125
conservation of energy p. 126

THINK ABOUT

How is energy transferred?

School carnivals sometimes include dunk tanks. The goal is to hit a target with a ball, causing a person sitting over a tank of water to fall into the water. You do work on the ball as you throw with your arm. If your aim is good, the ball does work on the target. How do you transfer your energy to the ball?

Work transfers energy.

MAIN IDEA WEB
Remember to add boxes to your main idea web as you read.

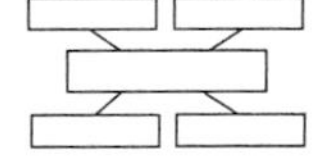

When you change the position and speed of the ball in the carnival game, you transfer energy to the ball. Energy is the ability of a person or an object to do work or to cause a change. When you do work on an object, some of your energy is transferred to the object. You can think of work as the transfer of energy. In fact, both work and energy are measured in the same unit, the joule.

The man in the photograph above converts one form of energy into another form when he uses his muscles to toss the ball. You can think of the man and the ball as a system, or a group of objects that affect one another. Energy can be transferred from the man to the ball, but the total amount of energy in the system does not change.

CHECK YOUR READING How are work and energy related?

RESOURCES FOR DIFFERENTIATED INSTRUCTION

Below Level
UNIT RESOURCE BOOK
- Reading Study Guide A, pp. 234–235
- Decoding Support, p. 258

AUDIO CDS

Advanced
UNIT RESOURCE BOOK
- Challenge and Extension, p. 240
- Challenge Reading, pp. 254–255

English Learners
UNIT RESOURCE BOOK
Spanish Reading Study Guide, pp. 238–239

AUDIO CDS
- Audio Readings in Spanish
- Audio Readings (English)

4.2 FOCUS

Set Learning Goals
Students will
- Recognize how work and energy are related.
- Demonstrate how to calculate mechanical, kinetic, and potential energy.
- Explain the law of conservation of energy.
- Analyze data through experimentation to determine energy changes in a rolling ball.

3-Minute Warm-Up
Display Transparency 28 or copy this exercise on the board:

Draw a picture of yourself pushing or pulling an object. Add arrows to show the direction of the motion and the applied force. *Refer students to the photographs on p. 116, if needed.* Exchange your drawing with a partner and indicate whether all of the applied force in your partner's picture is doing work.

 3-Minute Warm-Up, p. T28

4.2 MOTIVATE

THINK ABOUT
PURPOSE To have students think about how energy is transferred

DISCUSS Have students look at the photograph. Ask:
- What is the source of the energy that allows the thrower to fling the ball toward the target? *the muscles in the thrower's arm*
- What change of energy occurs when the ball is thrown? *Energy is transferred from the man to the ball.*

Answer: You do work to change the ball's speed and position.

Ongoing Assessment
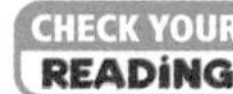
CHECK YOUR READING *Answer: When work is done, energy is transferred.*

4.2 INSTRUCT

Develop Critical Thinking

APPLY Have students apply their knowledge of potential energy to answer the questions below about a person using a bow and arrow.

- When the person pulls back on the bowstring, what happens to the energy of the bow? *Its potential energy increases.*
- What kind of potential energy does the bow have? *elastic potential energy*
- What happens to the bow's potential energy when the arrow is released? *It is transferred to the arrow as kinetic energy.*
- What determines how far the arrow will travel? *the amount of work that was done on the bow*

Teach from Visuals

To help students interpret the visuals of potential and kinetic energy, ask:

- In which picture is the boy's potential energy at a maximum? *the first picture*
- In which picture does the boy have the most kinetic energy? *the second picture*
- What kind of energy does the trampoline have when the boy lands on it? *elastic potential energy*

Work changes potential and kinetic energy.

READING TiP

The word *potential* comes from the Latin word *potentia*, which means "power." The word *kinetic* comes from the Greek word *kinetos,* which means "moving."

When you throw a ball, you transfer energy to it and it moves. By doing work on the ball, you can give it **kinetic energy** (kuh-NEHT-ihk), which is the energy of motion. Any moving object has some kinetic energy. The faster an object moves, the more kinetic energy it has.

When you do work to lift a ball from the ground, you give the ball a different type of energy, called potential energy. **Potential energy** is stored energy, or the energy an object has due to its position or its shape. The ball's position in your hand above the ground means that it has the potential to fall to the ground. The higher you lift the ball, the more work you do, and the more potential energy the ball has.

You can also give some objects potential energy by changing their shape. For example, if you are holding a spring, you can do work on the spring by squeezing it. After you do the work, the spring has potential energy because it is compressed. This type of potential energy is called elastic potential energy. Just as position gives the spring the potential to fall, compression gives the spring the potential to expand.

Potential and Kinetic Energy

Potential Energy

The boy has potential energy based on his position because gravity will pull him back down.

Kinetic Energy

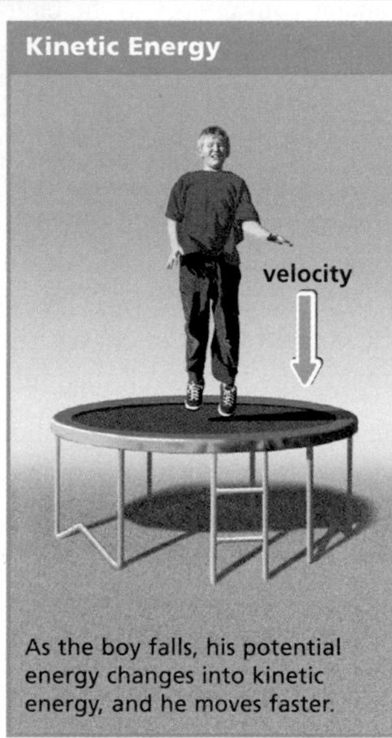

As the boy falls, his potential energy changes into kinetic energy, and he moves faster.

Potential Energy

The trampoline has potential energy because it is stretched.

DIFFERENTIATE INSTRUCTION

More Reading Support

A What is potential energy? *energy stored in an object because of its position or shape*

B What is kinetic energy? *energy that an object has due to its motion*

English Learners English learners may not have prior knowledge of *carnivals* on p. 121, *divers* on p. 123, *in-line skating* on p. 127, and *treadmills* on p. 129. You may also want to call their attention to the phrase "due to" on pp. 122, 123, 125, and explain that "due to" often means "because of."

Calculating Gravitational Potential Energy

Potential energy caused by gravity is called gravitational potential energy. Scientists must take gravitational potential energy into account when launching a spacecraft. Designers of roller coasters must make sure that roller-coaster cars have enough potential energy at the top of a hill to reach the top of the next hill. You can use the following formula to calculate the gravitational potential energy of an object:

Gravitational Potential Energy = mass · gravitational acceleration · height

$$GPE = mgh$$

Recall that g is the acceleration due to Earth's gravity. It is equal to 9.8 m/s^2 at Earth's surface.

The diver in the photograph below has given herself gravitational potential energy by climbing to the diving board. If you know her mass and the height of the board, you can calculate her potential energy.

Calculating Potential Energy

Sample Problem

What is the gravitational potential energy of a girl who has a mass of 40 kg and is standing on the edge of a diving board that is 5 m above the water?

What do you know? mass = 40 kg, gravitational acceleration = 9.8 m/s^2, height = 5 m

What do you want to find out? Gravitational Potential Energy

Write the formula: $GPE = mgh$

Substitute into the formula: $GPE = 40\ kg \cdot 9.8\ m/s^2 \cdot 5\ m$

Calculate and simplify: $GPE = 1960\ kg\ m^2/s^2$

Check that your units agree: $kg\ m^2/s^2 = kg \cdot m/s^2 \cdot m = N{\cdot}m = J$

Unit of energy is J. Units agree.

Answer: $GPE = 1960\ J$

Practice the Math

1. An apple with a mass of 0.1 kg is attached to a branch of an apple tree 4 m from the ground. How much gravitational potential energy does the apple have?
2. If you lift a 2 kg box of toys to the top shelf of a closet, which is 3 m high, how much gravitational potential energy will the box of toys have?

REMINDER

A newton (N) is a kg · m/s^2, and a joule (J) is a N·m.

The formula for gravitational potential energy is similar to the formula for work ($W = Fd$). The formula for GPE also has a force (mg) multiplied by a distance (h). To understand why mg is a force, remember two things: force equals mass times acceleration, and g is the acceleration due to Earth's gravity.

Develop Algebra Skills

Students must be careful to check whether a problem gives mass or weight. In the formula $GPE = mgh$, mass is multiplied by gravity and height. Because weight is defined as mass times gravity, g drops out when the problem gives weight.

Real World Example

Juggling is a good example of the conversion of gravitational potential energy and kinetic energy. The juggler's hand does work when it is tossing the ball upward to give the ball its initial kinetic energy. After the ball leaves the juggler's hand, the force of gravity acts on it. Because gravity acts to pull the ball downward, the ball loses kinetic energy and slows to a stop at its highest point. When the ball begins to move downward, gravity increases the ball's kinetic energy.

Develop Mathematics Skills

R
- Math Support, p. 261
- Math Practice, p. 262

Ongoing Assessment

Practice the Math *Answers:*

1. $GPE = mgh$
   $= 0.1\ kg \cdot 9.8\ m/s^2 \cdot 4\ m$
   $GPE = 3.92\ J$
2. $GPE = mgh$
   $= 2\ kg \cdot 9.8\ m/s^2 \cdot 3\ m$
   $GPE = 58.8\ J$

DIFFERENTIATE INSTRUCTION

More Reading Support

C What is gravitational potential energy? *potential energy that is caused by gravity*

Below Level If students are having difficulty with calculations, the value for g in the formula for gravitational potential energy can be rounded to 10 m/s^2. This will make calculations easier.

Address Misconceptions

IDENTIFY Ask: If a space shuttle is moving in a straight line through space at a constant speed, what can you say about its energy and any force exerted on it? If students respond that the shuttle has both force and energy, they may not fully understand that force and energy are two different phenomena.

CORRECT Ask if the space shuttle has energy. *Yes, it has kinetic energy due to its motion through space.* Ask if force is being exerted on the space shuttle. *No force is being exerted on the shuttle once it has been set in motion.*

REASSESS Ask students to explain the relationship between force and energy in the example of a book being pushed over the floor. *The force of the push, exerted over a distance, provides the book with kinetic energy.*

Technology Resources

Visit **ClassZone.com** for background on common student misconceptions.

 **MISCONCEPTION DATABASE**

Integrate the Sciences

All molecules have thermal energy because their atoms are in constant motion. The higher the temperature, the faster the molecules move, and the more thermal energy they have. This concept is crucial in chemical and biochemical reactions because molecules must have a certain amount of energy to react with each other. In industry, a reaction often can be made to occur by heating the reactants. In biological systems, high temperatures can damage the cells of a living organism, but enzymes can lower the amount of energy the reactants need to react with each other.

Ongoing Assessment

 *Answers:*

1. $KE = \frac{1}{2}mv^2$

$= \frac{1}{2} \cdot 0.002 \text{ kg} \cdot (15 \text{ m/s})^2$

$= \frac{1}{2} \cdot 0.002 \text{ kg} \cdot 225 \text{ m}^2/\text{s}^2$

$KE = 0.225 \text{ J}$

2. *the car*

Calculating Kinetic Energy

The girl on the swing at left has kinetic energy. To find out how much kinetic energy she has at the bottom of the swing's arc, you must know her mass and her velocity. Kinetic energy can be calculated using the following formula:

$$\text{Kinetic Energy} = \frac{\text{mass} \cdot \text{velocity}^2}{2}$$

$$KE = \frac{1}{2}mv^2$$

Notice that velocity is squared while mass is not. Increasing the velocity of an object has a greater effect on the object's kinetic energy than increasing the mass of the object. If you double the mass of an object, you double its kinetic energy. Because velocity is squared, if you double the object's velocity, its kinetic energy is four times greater.

Calculating Kinetic Energy

Sample Problem

What is the kinetic energy of a girl who has a mass of 40 kg and a velocity of 3 m/s?

What do you know? mass = 40 kg, velocity = 3 m/s

What do you want to find out? Kinetic Energy

Write the formula: $KE = \frac{1}{2}mv^2$

Substitute into the formula: $KE = \frac{1}{2} \cdot 40 \text{ kg} \cdot (3 \text{ m/s})^2$

Calculate and simplify: $KE = \frac{1}{2} \cdot 40 \text{ kg} \cdot \frac{9 \text{ m}^2}{\text{s}^2}$

$= \frac{360 \text{ kg} \cdot \text{m}^2}{2 \text{ s}^2}$

$= 180 \text{ kg} \cdot \text{m}^2/\text{s}^2$

Check that your units agree: $\frac{\text{kg} \cdot \text{m}^2}{\text{s}^2} = \frac{\text{kg} \cdot \text{m}}{\text{s}^2} \cdot \text{m} = \text{N} \cdot \text{m} = \text{J}$

Unit of energy is J. Units agree.

Answer: $KE = 180$ J

Practice the Math

1. A grasshopper with a mass of 0.002 kg jumps up at a speed of 15 m/s. What is the kinetic energy of the grasshopper?
2. A truck with a mass of 6000 kg is traveling north on a highway at a speed of 17 m/s. A car with a mass of 2000 kg is traveling south on the same highway at a speed of 30 m/s. Which vehicle has more kinetic energy?

DIFFERENTIATE INSTRUCTION

 **More Reading Support**

D How do you calculate kinetic energy? *multiply an object's mass by the square of its velocity and divide by two*

Advanced Have students graph a linear relationship (kinetic energy versus mass) and a nonlinear relationship (kinetic energy versus the velocity). The first graph will be a straight line, and the second graph will be a steep curve. The differences between these two graphs will show students that the kinetic energy of an object is affected much more by a given percentage change in velocity than by the same percentage change in mass.

 Challenge and Extension, p. 240

Calculating Mechanical Energy

Mechanical energy is the energy possessed by an object due to its motion or position—in other words, it is the object's combined potential energy and kinetic energy. A thrown baseball has mechanical energy as a result of both its motion (kinetic energy) and its position above the ground (gravitational potential energy). Any object that has mechanical energy can do work on another object.

VOCABULARY
Use a vocabulary strategy to help you remember *mechanical energy.*

Once you calculate an object's kinetic and potential energy, you can add the two values together to find the object's mechanical energy.

Mechanical Energy = Potential Energy + Kinetic Energy

$$ME = PE + KE$$

For example, a skateboarder has a potential energy of 200 joules due to his position at the top of a hill and a kinetic energy of 100 joules due to his motion. His total mechanical energy is 300 joules.

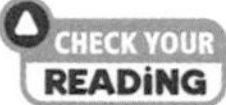
How is mechanical energy related to kinetic and potential energy?

INVESTIGATE Mechanical Energy

How does mechanical energy change?

PROCEDURE

1. Find and record the mass of the ball.
2. Build a ramp with the board and books. Measure and record the height of the ramp. You will place the ball at the top of the ramp, so calculate the ball's potential energy at the top of the ramp using mass and height.
3. Mark a line on the floor with tape 30 cm from the bottom of the ramp.
4. Place the ball at the top of the ramp and release it without pushing. Time how long the ball takes to travel from the end of the ramp to the tape.
5. Calculate the ball's speed using the time you measured in step 4. Use this speed to calculate the ball's kinetic energy after it rolled down the ramp.

WHAT DO YOU THINK?

- At the top of the ramp, how much potential energy did the ball have? kinetic energy? mechanical energy?
- Compare the ball's mechanical energy at the top of the ramp with its mechanical energy at the bottom of the ramp. Are they the same? Why or why not?

CHALLENGE Other than gravity, what forces could have affected the movement of the ball?

SKILL FOCUS
Analyzing data

MATERIALS
- ball
- balance
- board
- books
- ruler
- tape
- stopwatch
- calculator

TIME
20 minutes

DIFFERENTIATE INSTRUCTION

 **More Reading Support**

E What is mechanical energy? *the energy an object has due to its motion and its position*

Inclusion Visual learners and students with cognitive disabilities will benefit if they see the energy exchange of the ball graphically or in a picture. Give them energy diagrams that answer the investigate questions. Have students sketch the investigate setup and label the kinetic energy and potential energy at the top and the bottom of the ramp and at the finish line.

INVESTIGATE
Mechanical Energy

PURPOSE To analyze the relationship among kinetic, potential, and mechanical energy

TIP *20 min.* To prevent giving the ball extra energy by pushing it as you release it, place a pencil in front of the ball so that it does not roll. Lift the pencil to start the ball rolling.

WHAT DO YOU THINK? *top—all of the energy is PE; bottom—all of the energy is KE. At the bottom, the ME should be the same or less than it was at the top. Some potential energy will be converted into energy that rotates the ball as it rolls; this is measured as a loss in mechanical energy.*

CHALLENGE *friction*

 Datasheet, Mechanical Energy, p. 241

Technology Resources

Customize this student lab as needed or look for an alternative. Print rubrics to assess student lab reports.

 **Lab Generator CD-ROM**

Teaching with Technology

Students can use a spreadsheet to record data and make calculations.

Ongoing Assessment

Demonstrate how to calculate mechanical, kinetic, and potential energy.

Ask: What is the total mechanical energy of a rolling ball that has a potential energy of 479 J and a kinetic energy of 247 J? ME = PE + KE = *479 J + 247 J = 726 J*

CHECK YOUR READING *Answer: Mechanical energy is the total kinetic and potential energy that an object has.*

Real World Example

The first hill of a roller coaster is always the highest. A roller-coaster car has maximum potential energy and minimum kinetic energy at the top of the first hill. This potential energy changes to kinetic energy when the car begins to move downhill. Because some energy is lost to friction, the car could not climb the second hill if it were as high as the first hill. Each hill must be lower than the previous hill.

EXPLORE the BIG idea

Revisit "Bouncing Ball" on p. 113. Have students explain their results.

Ongoing Assessment

Explain the law of conservation of energy.

Ask: How can energy be added to an object and still follow the law of conservation of energy? *Another object in the system must lose energy.*

PHOTO CAPTION Answer: Tell students that an anchor attached to the pendulum nudges the pendulum in the right direction each time it swings. This is the energy the pendulum needs to overcome friction and keep swinging. Some clocks have a key-wound spring inside the clock.

Observe how potential and kinetic energy are transferred on an amusement park ride.

The total amount of energy is constant.

F

You know that energy is transferred when work is done. No matter how energy is transferred or transformed, all of the energy is still present somewhere in one form or another. This is known as the **law of conservation of energy.** As long as you account for all the different forms of energy involved in any process, you will find that the total amount of energy never changes.

Conserving Mechanical Energy

Look at the photograph of the in-line skater on page 127. As she rolls down the ramp, the amounts of kinetic energy and potential energy change. However, the total—or the mechanical energy—stays the same. In this example, energy lost to friction is ignored.

1. At the top of the ramp, the skater has potential energy because gravity can pull her downward. She has no velocity; therefore, she has no kinetic energy.
2. As the skater rolls down the ramp, her potential energy decreases because the elevation decreases. Her kinetic energy increases because her velocity increases. The potential energy lost as the skater gets closer to the ground is converted into kinetic energy. Halfway down the ramp, half of her potential energy has been converted to kinetic energy.
3. At the bottom of the ramp, all of the skater's energy is kinetic. Gravity cannot pull her down any farther, so she has no more gravitational potential energy. Her mechanical energy—the total of her potential and kinetic energy—stays the same throughout.

Losing Mechanical Energy

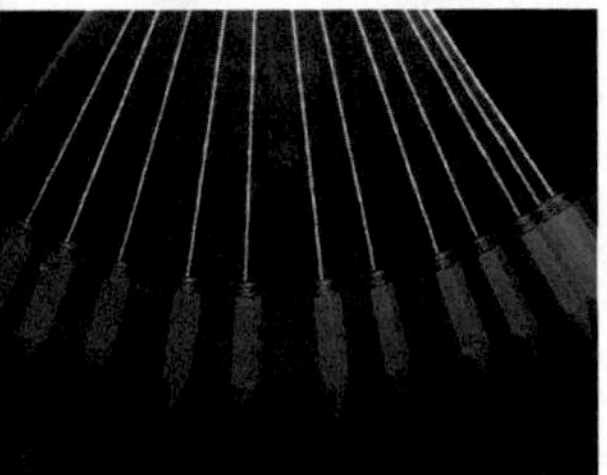

APPLY Energy must occasionally be added to a pendulum to keep it swinging. What keeps a grandfather clock's pendulum swinging regularly?

A pendulum is an object that is suspended from a fixed support so that it swings freely back and forth under the influence of gravity. As a pendulum swings, its potential energy is converted into kinetic energy and then back to potential energy in a continuous cycle. Ideally, the potential energy at the top of each swing would be the same as it was the previous time. However, the height of the pendulum's swing actually decreases slightly each time, until finally the pendulum stops altogether.

In most energy transformations, some of the energy is transformed into heat. In the case of the pendulum, there is friction between the string and the support, as well as air resistance from the air around the pendulum. The mechanical energy is used to do work against friction and air resistance. This process transforms the mechanical energy into heat. The mechanical energy has not been destroyed; it has simply changed form and been transferred from the pendulum.

DIFFERENTIATE INSTRUCTION

More Reading Support

F What is the law of conservation of energy? *Energy cannot be created or destroyed.*

English Learners English learners may be unfamiliar with some informal writing styles. Writers often use sentence fragments and leave readers to substitute the remainder of the sentence themselves. For example, in "Investigate Mechanical Energy" on p. 125, " . . . how much potential energy did the ball have? Kinetic energy? Mechanical energy?" readers are expected to substitute the terms into the initial question:" . . . how much kinetic energy did the ball have?" Be aware of informal writing styles and assist English learners in understanding them.

Conserving Mechanical Energy

The potential energy and kinetic energy in a system or process may vary, but the total energy remains unchanged.

① Top of Ramp

100% PE

At the top of the ramp, the skater's mechanical energy is equal to her potential energy because she has no velocity.

② Halfway Down Ramp

50% PE | 50% KE

As the skater goes down the ramp, she loses height but gains speed. The potential energy she loses is equal to the kinetic energy she gains.

③ Bottom of Ramp

100% KE

As the skater speeds along the bottom of the ramp, all of the potential energy has changed to kinetic energy. Her mechanical energy remains unchanged.

Fabiola da Silva is a professional in-line skater who was born in Brazil but now lives in California.

READING VISUALS How do the skater's kinetic and potential energy change as she skates up and down the ramp? (Assume she won't lose any energy to friction.)

DIFFERENTIATE INSTRUCTION

Below Level Have students give an oral summary of the energy changes that take place in the visual. The changes can be displayed graphically as well.

Advanced Have students interested in finding out about how pole vaulting works read the following article:

R Challenge Reading, pp. 254–255

Teach from Visuals

To help students interpret the "Conserving Mechanical Energy" visual, ask:

- Why does the skater have no kinetic energy at the top of the ramp? *She is momentarily at rest.*
- Why does the skater have no potential energy at the bottom of the ramp? *The bottom of the ramp is the lowest height.*

This visual is also available as T30 in the Unit Transparency Book.

Teach Difficult Concepts

Students may have a difficult time understanding that work must be done for an object's total energy to increase. Explain that an object's energy cannot increase unless additional energy comes from somewhere. To help students understand, you might try the following demonstration.

Teacher Demo

With a sharp knife, carefully cut a racquetball in half. Trim one of the halves so that it is slightly smaller than a hemisphere. Turn it inside out and drop it, bulge side up, on a hard surface. The ball will snap and rebound to a height much greater than that from which it was dropped. Ask students how the ball can bounce higher than its original height. *The work required to turn the hemisphere inside out is stored as potential energy. As the dropped ball hits the hard surface, the potential energy is released and converted to kinetic energy, which allows the ball to rebound to a greater height.*

Ongoing Assessment

READING VISUALS *Answer: As the skater goes down the ramp, kinetic and potential energy become equal. At the bottom of the ramp, all of the potential energy has changed to kinetic energy. As the skater goes up the ramp, the opposite happens. At the top of the ramp, all the kinetic energy has changed to potential energy.*

Develop Critical Thinking

PROVIDE EXAMPLES Have students make a set of four flash cards, one for each form of energy mentioned in the text. Students should write the energy form on one side of a card and list examples on the other side. Encourage students to apply their knowledge of the forms of energy by providing examples that are not in the text.

Integrate the Sciences

Organisms are continually transforming energy from one type to another. A plant transforms electromagnetic energy (light) to chemical energy (sugar). Chemical energy fuels muscles, which can transform it to mechanical energy (climbing and sliding down a slide). Shivering converts chemical energy to thermal energy.

Reinforce (the BIG idea)

Have students relate the section to the Big Idea.

 Reinforcing Key Concepts, p. 242

4.2 ASSESS & RETEACH

Assess

 Section 4.2 Quiz, p. 63

Reteach

Have students draw a picture and label the energy exchanges that take place when a skier rides a ski lift to the top of a slope and then skis down the slope. Students should realize that work is done on the skier by the lift and that the energy needed to do that work comes from the fuel used to run the lift.

Technology Resources

Have students visit **ClassZone.com** for reteaching of Key Concepts.

 CONTENT REVIEW

 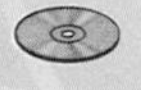 CONTENT REVIEW CD-ROM

MAIN IDEA WEB
Include common forms of energy in your web.

Forms of Energy

As you have seen, mechanical energy is a combination of kinetic energy and potential energy. Other common forms of energy are discussed below. Each of these forms of energy is also a combination of kinetic energy and potential energy. Chemical energy, for example, is potential energy when it is stored in bonds.

Thermal energy is the energy an object has due to the motion of its molecules. The faster the molecules in an object move, the more thermal energy the object has.

Chemical energy is the energy stored in chemical bonds that hold chemical compounds together. If a molecule's bonds are broken or rearranged, energy is released or absorbed. Chemical energy is used to light up fireworks displays. It is also stored in food and in matches.

Nuclear energy is the potential energy stored in the nucleus of an atom. In a nuclear reaction, a tiny portion of an atom's mass is turned into energy. The source of the Sun's energy is nuclear energy. Nuclear energy can be used to run power plants that provide electricity.

Electromagnetic energy is the energy associated with electrical and magnetic interactions. Energy that is transferred by electric charges or current is often called electrical energy. Another type of electromagnetic energy is radiant energy, the energy carried by light, infrared waves, and x-rays.

It is possible to transfer, or convert, one energy form into one or more other forms. For example, when you rub your hands together on a cold day, you convert mechanical energy to thermal energy. Your body converts chemical energy stored in food to thermal and mechanical energy (muscle movement).

4.2 Review

KEY CONCEPTS

1. Explain the relationship between work and energy.
2. How are potential energy and kinetic energy related to mechanical energy?
3. When one form of energy changes into one or more other forms of energy, what happens to the total amount of energy?

CRITICAL THINKING

4. **Infer** Debra used 250 J of energy to roll a bowling ball. When the ball arrived at the end of the lane, it had only 200 J of energy. What happened to the other 50 J?
5. **Calculate** A satellite falling to Earth has a kinetic energy of 182.2 billion J and a potential energy of 1.6 billion J. What is its mechanical energy?

CHALLENGE

6. **Apply** At what point in its motion is the kinetic energy of the end of a pendulum greatest? At what point is its potential energy greatest? When its kinetic energy is half its greatest value, how much potential energy did it gain?

C 128 Unit: **Motion and Forces**

ANSWERS

1. When work is done, energy is transferred.

2. Mechanical energy is the total potential and kinetic energy that an object has.

3. The total amount of energy stays the same.

4. The rest of the energy was lost to friction.

5. ME = PE + KE
= *182.2 billion J + 1.6 billion J*
ME = *183.8 billion J*

6. A pendulum's kinetic energy is greatest at the bottom of its swing. Its potential energy is greatest at the top of its swing. When its kinetic energy is half its greatest value, the potential energy gained is equal to half the kinetic energy.

Think SCIENCE

SKILL: ISOLATING VARIABLES

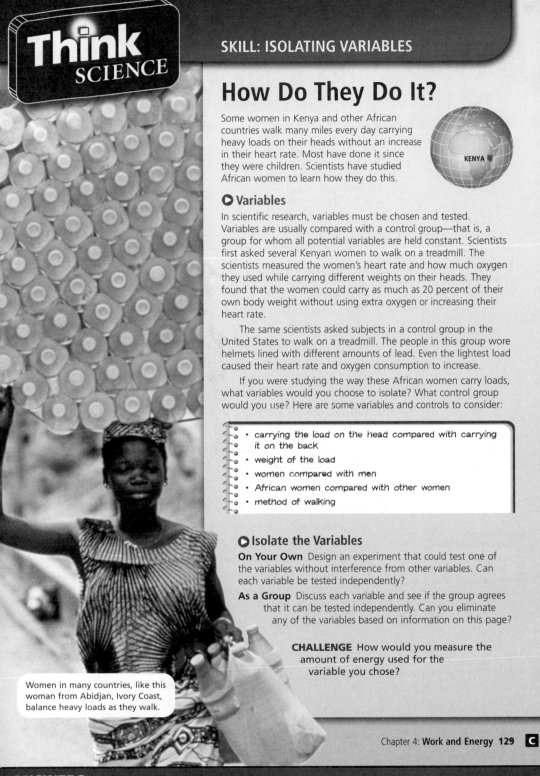

How Do They Do It?

Some women in Kenya and other African countries walk many miles every day carrying heavy loads on their heads without an increase in their heart rate. Most have done it since they were children. Scientists have studied African women to learn how they do this.

Variables

In scientific research, variables must be chosen and tested. Variables are usually compared with a control group—that is, a group for whom all potential variables are held constant. Scientists first asked several Kenyan women to walk on a treadmill. The scientists measured the women's heart rate and how much oxygen they used while carrying different weights on their heads. They found that the women could carry as much as 20 percent of their own body weight without using extra oxygen or increasing their heart rate.

The same scientists asked subjects in a control group in the United States to walk on a treadmill. The people in this group wore helmets lined with different amounts of lead. Even the lightest load caused their heart rate and oxygen consumption to increase.

If you were studying the way these African women carry loads, what variables would you choose to isolate? What control group would you use? Here are some variables and controls to consider:

- carrying the load on the head compared with carrying it on the back
- weight of the load
- women compared with men
- African women compared with other women
- method of walking

Isolate the Variables

On Your Own Design an experiment that could test one of the variables without interference from other variables. Can each variable be tested independently?

As a Group Discuss each variable and see if the group agrees that it can be tested independently. Can you eliminate any of the variables based on information on this page?

CHALLENGE How would you measure the amount of energy used for the variable you chose?

Women in many countries, like this woman from Abidjan, Ivory Coast, balance heavy loads as they walk.

ANSWERS

See students' plans. The researchers did not test the women by having them carry the loads on their backs, nor did they compare women with men. These variables can be eliminated based on information on the page.

CHALLENGE *Measure test subjects' heart rates and oxygen consumption.*

THINK SCIENCE
Scientific Methods of Thinking

Set Learning Goal

To isolate variables when designing an experiment to learn why some African women can carry large loads without increasing their heart rate

Present the Science

In a recent study, two European researchers learned that some African women unconsciously modify their gait when carrying large loads on their heads and thereby use less energy. The energy they save is applied to carrying the weight.

Guide the Activity

- Students should understand how a treadmill works before they begin to consider variables.
- Ask students if they can think of other variables to test.
- Have students compare a walking human to an upside-down pendulum in terms of potential and kinetic energy.

COOPERATIVE LEARNING STRATEGY

Have groups discuss why they think the women can carry so much weight. The group should brainstorm the variables they might test, choose one, and design an experiment to test it.

If you want students to design an experiment that tests a variable, display the following resources to help them along.

- Identifying Variables and Constants, A25–26
- Setting Up Experimental Groups and Controls, A27–28

Close

Ask: Humans transfer potential and kinetic energy when they walk, but only about 65 percent of the energy is actually transferred. Where does the remaining 35 percent of the energy for walking come from? *the muscles*

4.3 FOCUS

Set Learning Goals

Students will

- Explain how power relates to work and time.
- Explain how power relates to energy and time.
- Describe some common uses of power.
- Measure the power needed to pull an object.

3-Minute Warm-Up

Display Transparency 29 or copy this exercise on the board:

Match the correct term to each definition.

Definitions

1. stored energy *b*
2. the use of force to move an object a certain distance *d*
3. the energy of motion *c*

Terms

a. mechanical energy c. kinetic energy
b. potential energy d. work

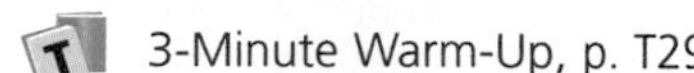

4.3 MOTIVATE

EXPLORE Power

PURPOSE To investigate whether time affects the amount of work done

TIP *10 min.* Have a partner set the timer and record the times.

WHAT DO YOU THINK? *The same amount of work was done both times. The amount of work done is independent of the time needed to do the work. Answers will vary.*

KEY CONCEPT

Power is the rate at which work is done.

BEFORE, you learned

- Mechanical energy is a combination of kinetic energy and potential energy
- Mechanical energy can be calculated
- Work transfers energy

NOW, you will learn

- How power is related to work and time
- How power is related to energy and time
- About common uses of power

VOCABULARY

power p. 130
watt p. 131
horsepower p. 132

EXPLORE Power

How does time affect work?

PROCEDURE

1. Place the cups side by side. Put all of the marbles in one cup.
2. Place each marble, one by one, into the other cup. Time how long it takes to do this.
3. Set the timer for half that amount of time. Then repeat step 2 in that time.

WHAT DO YOU THINK?

- Did you do more work the first time or the second time? Why?
- What differences did you notice between the two tries?

MATERIALS

- 2 plastic cups
- 10 marbles
- stopwatch

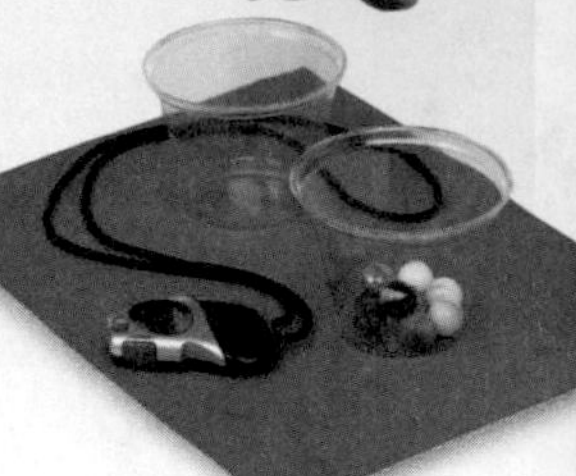

Power can be calculated from work and time.

VOCABULARY Use a vocabulary strategy to help you remember the meaning of *power*.

If you lift a book one meter, you do the same amount of work whether you lift the book quickly or slowly. However, when you lift the book quickly, you increase your **power**—the rate at which you do work. A cook increases his power when he beats eggs rapidly instead of stirring them slowly. A runner increases her power when she breaks into a sprint to reach the finish line.

The word *power* has different common meanings. It is used to mean a source of energy, as in a power plant, or strength, as in a powerful engine. When you talk about a powerful swimmer, for example, you would probably say that the swimmer is very strong or very fast. If you use the scientific definition of power, you would instead say that a powerful swimmer is one who does the work of moving herself through the water in a short time.

RESOURCES FOR DIFFERENTIATED INSTRUCTION

Below Level

UNIT RESOURCE BOOK

- Reading Study Guide A, pp. 245–246
- Decoding Support, p. 258

AUDIO CDS

Advanced

UNIT RESOURCE BOOK

Challenge and Extension, p. 251

English Learners

UNIT RESOURCE BOOK

Spanish Reading Study Guide, pp. 249–250

AUDIO CDS

- Audio Readings in Spanish
- Audio Readings (English)

Each of the swimmers shown in the photograph above is doing work—that is, she is using a certain force to move a certain distance. It takes time to cover that distance. The power a swimmer uses depends on the force, the distance, and the time it takes to cover that distance. The more force the swimmer uses, the more power she has. Also, the faster she goes, the more power she has because she is covering the same distance in a shorter time. Swimmers often increase their speed toward the end of a race, which increases their power, making it possible for them to reach the end of the pool in less time.

CHECK YOUR READING Summarize in your own words the difference between work and power.

Calculating Power from Work

You know that a given amount of work can be done by a slow-moving swimmer over a long period of time or by a fast-moving swimmer in a short time. Likewise, a given amount of work can be done by a low-powered motor over a long period of time or by a high-powered motor in a short time.

Because power is a measurement of how much work is done in a given time, power can be calculated based on work and time. To find power, divide the amount of work by the time it takes to do the work.

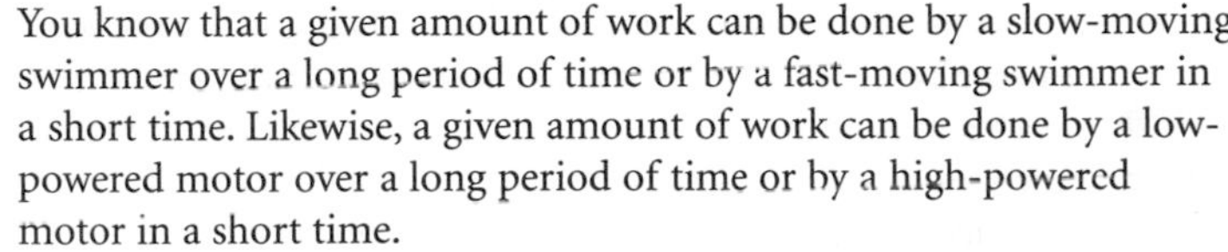

$$\text{Power} = \frac{\text{Work}}{\text{time}} \qquad P = \frac{W}{t}$$

READING TIP W (in italicized type) is the letter that represents the variable *Work*. W, not italicized, is the abbreviation for watt.

Remember that work is measured in joules. Power is often measured in joules of work per second. The unit of measurement for power is the **watt** (W). One watt is equal to one joule of work done in one second. If an object does a large amount of work, its power is usually measured in units of 1000 watts, or kilowatts.

4.3 INSTRUCT

Address Misconceptions

IDENTIFY The word *power* has multiple uses in daily language. Ask students to write a few sentences using the word *power.* Students will probably identify *power* with *strength,* rather than using the word in the physical sense—that is, as a rate.

CORRECT To emphasize the meaning of power as a rate, discuss the following. Suppose students have a lawn-mowing business in which they work six hours each day and get paid per lawn. If they take two hours to mow each lawn, they can mow only three lawns in one day. If they take only one hour to mow each lawn, they can mow six lawns per day. Ask students how they can maximize their earnings. *Students should realize that the faster they mow (increasing their power), the more money they can earn.*

REASSESS It is commonly said that electric meters measure power. Ask students what electric meters actually measure. *Electric meters measure a unit of kilowatt-hours (kWh). Meters measure the total amount of energy that is transferred. Electric meters do not measure power.*

Technology Resources

Visit **ClassZone.com** for background on common student misconceptions.

MISCONCEPTION DATABASE

EXPLORE the BIG idea

Revisit "Power Climbing" on p. 113. Have students explain their results.

Ongoing Assessment

Explain how power relates to work and time.

Ask: When runners speed up at the end of a race, what are they doing? *They are increasing their power to reach the finish line sooner.*

CHECK YOUR READING *Answer: Power is the rate at which work is done.*

DIFFERENTIATE INSTRUCTION

More Reading Support

A How do you calculate power if you know the time and the amount of work done? *Power is work divided by time.*

English Learners Place the words *power, watt,* and *horsepower* on the classroom's Science Word Wall with abbreviated definitions for quick reference. English learners may not have prior knowledge of *conveyor belt* on p. 132 (under "Practice the Math"), and may be confused by the phrase "beats eggs rapidly" on p. 130. Explain that a conveyer belt is a mechanical piece of equipment that moves items on an assembly line and that beating eggs means stirring them quickly.

History of Science

James Watt did not invent the steam engine. However, the improvements he made to an engine invented by Thomas Newcomen made Watt's engine the first economically feasible engine, which contributed to the Industrial Revolution. Watt defined 1 horsepower as 1 horse lifting 33,000 pounds (14,968 kg) a distance of 1 foot in 1 minute. One horsepower is actually about 50 percent more than the rate at which an average horse can work.

Develop Algebra Skills

In question *2* of Practice the Math, students should note that the elevator takes 8 seconds to go up 2 floors, but the height of only a single floor is given.

- Math Support, p. 263
- Math Practice, p. 264

Ongoing Assessment

Practice the Math *Answer:*

1. $P = \frac{W}{t}$

$\frac{10\ J}{20\ s} = 0.5\ J/s$

$P = 0.5\ W$

2. $P = \frac{W}{t}$, $W = F \cdot d$

$P = \frac{F \cdot d}{t} = \frac{1710\ N \cdot 8\ m}{8\ s}$

$P = 1710\ N \cdot m/s = 1710\ J/s$
$= 1710\ W$

Find out more about power.

Calculating Power from Work

Sample Problem

An Antarctic explorer uses 6000 J of work to pull his sled for 60 s. What power does he need?

What do you know? Work = 6000 J, time = 60 s

What do you want to find out? Power

Write the formula: $P = \frac{W}{t}$

Substitute into the formula: $P = \frac{6000\ J}{60\ s}$

Calculate and simplify: $P = 100\ J/s = 100\ W$

Check that your units agree: $\frac{J}{s} = W$

Unit of power is W. Units agree.

Answer: $P = 100\ W$

Practice the Math

1. If a conveyor belt uses 10 J to move a piece of candy a distance of 3 m in 20 s, what is the conveyor belt's power?
2. An elevator uses a force of 1710 N to lift 3 people up 1 floor. Each floor is 4 m high. The elevator takes 8 s to lift the 3 people up 2 floors. What is the elevator's power?

Horsepower

James Watt, the Scottish engineer for whom the watt is named, improved the power of the steam engine in the mid-1700s. Watt also developed a unit of measurement for power called the horsepower. **Horsepower** is based on what it sounds like—the amount of work a horse can do in a minute. In Watt's time, people used horses to do many different types of work. For example, horses were used on farms to pull plows and wagons. ? B

Both the horse and the tractor use power to pull objects around a farm.

Watt wanted to explain to people how powerful his steam engine was compared with horses. After observing several horses doing work, Watt concluded that an average horse could move 150 pounds a distance of 220 feet in 1 minute. Watt called this amount of power 1 horsepower. A single horsepower is equal to 745 watts. Therefore, a horsepower is a much larger unit of measurement than a watt.

Today horsepower is used primarily in connection with engines and motors. For example, you may see a car advertised as having a 150-horsepower engine. The power of a motorboat, lawn mower, tractor, or motorcycle engine is also referred to as horsepower.

DIFFERENTIATE INSTRUCTION

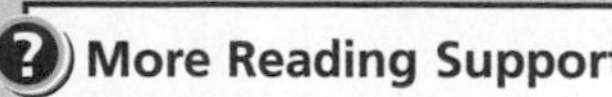

More Reading Support

B What is horsepower based on? *the amount of work a horse can do in one minute*

Advanced To give students practice in converting units, ask them to estimate how many 75-watt engines it would take to equal the power of a 100-horsepower engine. *Since 1 horsepower is equal to 745 W, 100 hp = 74,500 W. Approximately 1000 75-watt engines equal a 100-horsepower engine.*

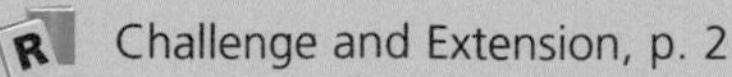

Challenge and Extension, p. 251

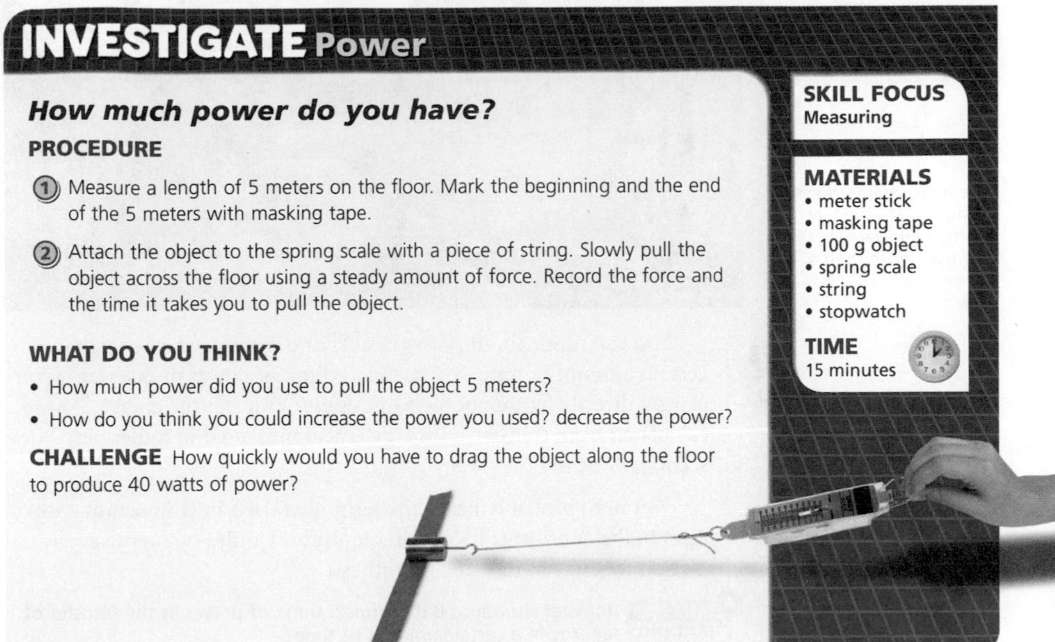

How much power do you have?

PROCEDURE

1. Measure a length of 5 meters on the floor. Mark the beginning and the end of the 5 meters with masking tape.
2. Attach the object to the spring scale with a piece of string. Slowly pull the object across the floor using a steady amount of force. Record the force and the time it takes you to pull the object.

WHAT DO YOU THINK?

- How much power did you use to pull the object 5 meters?
- How do you think you could increase the power you used? decrease the power?

CHALLENGE How quickly would you have to drag the object along the floor to produce 40 watts of power?

MATERIALS
- meter stick
- masking tape
- 100 g object
- spring scale
- string
- stopwatch

TIME
15 minutes

Power can be calculated from energy and time.

Sometimes you may know that energy is being transferred, but you cannot directly measure the work done by the forces involved. For example, you know that a television uses power. But there is no way to measure all the work every part of the television does in terms of forces and distance. Because work measures the transfer of energy, you can also think of power as the amount of energy transferred over a period of time.

Calculating Power from Energy

C

When you turn on a television, it starts using energy. Each second the television is on, a certain amount of electrical energy is transferred from a local power plant to your television. If you measure how much energy your television uses during a given time period, you can find out how much power it needs by using the following formula:

$$\text{Power} = \frac{\text{Energy}}{\text{time}} \qquad P = \frac{E}{t}$$

This formula should look familiar to you because it is very similar to the formula used to calculate power from work.

DIFFERENTIATE INSTRUCTION

 **More Reading Support**

C How do you calculate power if you know the amount of energy and the time? *Power is energy divided by time.*

Inclusion A spring scale with large numerals will be helpful to students with visual impairments while they are performing "Investigate Power." Students with physical ailments can do the timing.

Alternative Assessment Have students write a brochure that a tractor manufacturer could give to customers. It should explain the concept of power and describe the power of the engine in terms of horsepower.

INVESTIGATE Power

PURPOSE To measure the work used to pull an object and to calculate power

TIPS *15 min.* Allow students a few minutes to explore, and then suggest the following:

- If students need help in reading spring scales, refer them to p. R16 in the Lab Handbook at the end of their textbook.
- Any object can be used as long as the force needed to pull it is easy to measure. A small sack of beans is a good choice.
- A 100 g object weighs about one-fourth of a pound.

WHAT DO YOU THINK? *Students who pull faster should record more power used. Increase the power by pulling the object faster; decrease the power by pulling the object more slowly.*

CHALLENGE *Answers will depend on the friction between the object and the floor.*

 Datasheet, Power, p. 252

Technology Resources

Customize this student lab as needed or look for an alternative. Print rubrics to assess student lab reports.

 **Lab Generator CD-ROM**

Metacognitive Strategy

Have students identify content that challenges their prior understandings about power. Ask: How is the scientific definition of power different from the way you use the word in everyday conversation? What surprised you about the scientific meaning of power?

Teaching with Technology

If you have a video camera, have students videotape when they pull the object along the floor so that they can see if they are using a steady amount of force.

Ongoing Assessment

Explain how power relates to energy and time.

Ask: What is the relationship between power, energy, and time? *Power is directly related to energy and inversely related to time. It is the rate at which energy is transferred.*

CHECK YOUR READING *Answer: Calculate power from energy and time when you cannot measure the work used to transfer the energy, as a television set.*

Practice the Math *Answers:*

1. $P = \frac{E}{t}$

$= \frac{100\ J}{2\ s}$

$P = 50\ J/s = 50\ W$

2. $E = P \cdot t$

$= 1.1\ W \cdot 10\ s$

$E = 11\ W \cdot s = 11\ J$

The photograph shows Hong Kong, China, at night. Every second, the city uses more than 4 billion joules of electrical energy!

D You can think about power as any kind of transfer of energy in a certain amount of time. It is useful to think of power in this way if you cannot directly figure out the work used to transfer the energy. Power calculated from transferred energy is also measured in joules per second, or watts.

You have probably heard the term *watt* used in connection with light bulbs. A 60-watt light bulb requires 60 joules of energy every second to shine at its rated brightness.

CHECK YOUR READING In what situations is it useful to think of power as the transfer of energy in a certain amount of time?

REMINDER Remember that energy and work are both measured in joules.

Calculating Power from Energy

Sample Problem

A light bulb used 600 J of energy in 6 s. What is the power of the light bulb?

What do you know?	Energy = 600 J, time = 6 s
What do you want to find out?	Power
Write the formula:	$P = \frac{E}{t}$
Substitute into the formula:	$P = \frac{600\ J}{6\ s}$
Calculate and simplify:	$P = 100\ J/s$
Check that your units agree:	Unit is J/s. Unit for power is W, which is also J/s. Units agree.
Answer:	$P = 100\ W$

Practice the Math

1. A laptop computer uses 100 J every 2 seconds. How much power is needed to run the computer?
2. The power needed to pump blood through your body is about 1.1 W. How much energy does your body use when pumping blood for 10 seconds?

DIFFERENTIATE INSTRUCTION

More Reading Support

D When is it useful to calculate power from energy rather than from work? *when you cannot figure out the amount of work used to transfer the energy*

Below Level Have students write the formulas for calculating power from work and power from energy. If they see the formulas side by side, they will realize how similar the formulas are.

Everyday Power

Many appliances in your home rely on electricity for energy. Each appliance requires a certain number of joules per second, the power it needs to run properly. An electric hair dryer uses energy. For example, a 600-watt hair dryer needs 600 joules per second. The wattage of the hair dryer indicates how much energy per second it needs to operate.

The dryer works by speeding up the evaporation of water on the surface of hair. It needs only two main parts to do this: a heating coil and a fan turned by a motor.

1. When the hair dryer is plugged into an outlet and the switch is turned on, electrical energy moves electrons in the wires, creating a current.
2. This current runs an electric motor that turns the fan blades. Air is drawn into the hair dryer through small holes in the casing. The turning fan blades push the air over the coil.
3. The current also makes the heating coil become hot.
4. The fan pushes heated air out of the dryer.

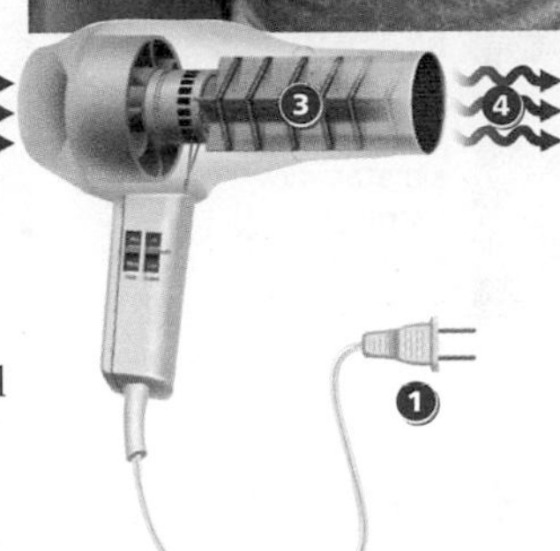

Most hair dryers have high and low settings. At the high power setting, the temperature is increased, more air is pushed through the dryer, and the dryer does its work faster. Some dryers have safety switches that shut off the motor when the temperature rises to a level that could burn your scalp. Insulation keeps the outside of the dryer from becoming hot to the touch.

Many other appliances, from air conditioners to washing machines to blenders, need electrical energy to do their work. Take a look around you at all the appliances that help you during a typical day.

4.3 Review

KEY CONCEPTS

1. How is power related to work?
2. Name two units used for power, and give examples of when each unit might be used.
3. What do you need to know to calculate how much energy a light bulb uses?

CRITICAL THINKING

4. **Apply** Discuss different ways in which a swimmer can increase her power.
5. **Calculate** Which takes more power: using 15 N to lift a ball 2 m in 5 seconds or using 100 N to push a box 2 m in 1 minute?

CHALLENGE

6. **Analyze** A friend tells you that you can calculate power by using a different formula from the one given in this book. The formula your friend gives you is as follows:

   Power = force • speed

   Do you think this is a valid formula for power? Explain.

ANSWERS

1. Power is the rate at which work is done. The faster you do work, the greater your power.

2. Power can be measured in joules per second (watts) or in horsepower. Examples will vary but might include light bulbs for watts and cars for horsepower.

3. how many watts it needs and how long it is on

4. by adding force and increasing speed

5. using 15 N to lift a ball 2 m in 5 seconds

6. Yes; speed is distance divided by time, so power equals force times distance divided by time. Since work is force times distance, the formula can be simplified to work divided by time.

Teach from Visuals

As they read the numbered text, advise students to follow the numbers under the photograph of the young woman drying her hair.

Ongoing Assessment

Describe some sources that provide power for electricity.

Ask: What are some sources of energy that provide power for electricity? *Chemical energy from fossil fuels, nuclear energy from nuclear reactions, energy from wind and water, and solar energy from the Sun are all common sources of energy for electrical power.*

Reinforce the BIG idea

Have students relate the section to the Big Idea.

 Reinforcing Key Concepts, p. 253

4.3 ASSESS & RETEACH

Assess

 Section 4.3 Quiz, p. 64

Reteach

Have students explain in their own words the relationships among power, work, energy, and time. *Energy and work vary directly with power and inversely with time. As the time it takes to do a certain amount of work increases, power decreases. As work or energy increases, power increases.*

Technology Resources

Have students visit **ClassZone.com** for reteaching of Key Concepts.

 **CONTENT REVIEW**

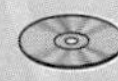 **CONTENT REVIEW CD-ROM**

CHAPTER INVESTIGATION

Focus

PURPOSE Students will relate work done to the potential energy gained, and power to the work and the time.

OVERVIEW Students will calculate the potential energy, the work done, and the power needed to lift a wheeled object straight up and to raise it to the same height using a ramp. Students should find that

- the work done equals or exceeds the gain in potential energy as the car moves up the ramp.
- power depends on the time it takes the student to move the car. The longer the time taken to move the car, the less power the car has.

Lab Preparation

- If appropriate, have students bring in small wheeled objects from home. A toy car or truck is a good choice.
- Review how to use and read a spring scale. Remind students that a spring scale measures force in newtons.
- Prior to the investigation, have students read through the investigation, write their hypothesis, and prepare their data tables. Or you may wish to copy and distribute datasheets and rubrics.

 UNIT RESOURCE BOOK, pp. 267–275

 SCIENCE TOOLKIT, F14

Lab Management

If necessary, substitute a wristwatch for a stopwatch.

INCLUSION Spring scales with large numbers are useful if you have students with visual impairments. Spring scales that are easy to hold are useful if you have physically challenged students.

Teaching with Technology

Students can use spreadsheet software to record their data and calculate their results. Spreadsheets will simplify number manipulation.

CHAPTER INVESTIGATION

Work and Power

OVERVIEW AND PURPOSE People in wheelchairs cannot use steps leading up to a building's entrance. Sometimes there is a machine that can lift a person and wheelchair straight up to the entrance level. At other times, there is a ramp leading to the entrance. Which method takes more power?

Problem

How does a ramp affect the amount of energy, work, and power used to lift an object?

Hypothesize

Write a hypothesis to explain how the potential energy, the amount of work done, and the power required to lift an object straight up compare with the same quantities when the object is moved up a ramp. Your hypothesis should take the form of an "If . . . , then . . . , because . . ." statement.

MATERIALS
- board
- chair
- meter stick
- string
- small wheeled object
- spring scale
- stopwatch

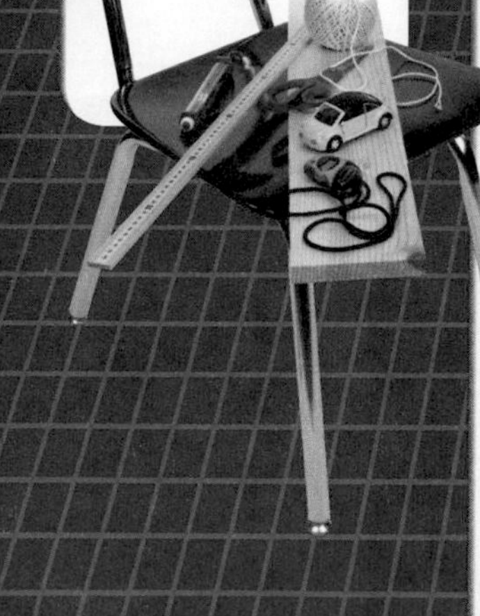

Procedure

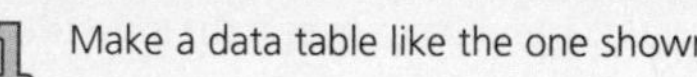
1. Make a data table like the one shown.

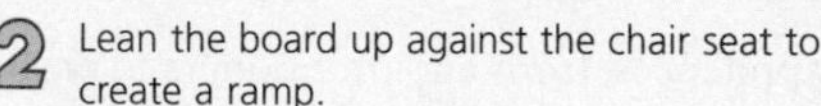
2. Lean the board up against the chair seat to create a ramp.

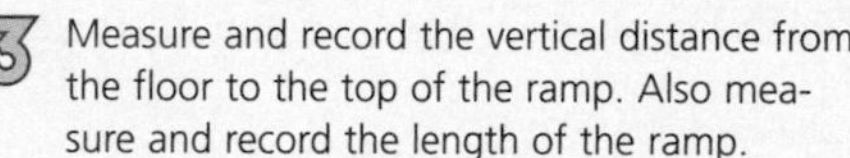
3. Measure and record the vertical distance from the floor to the top of the ramp. Also measure and record the length of the ramp.

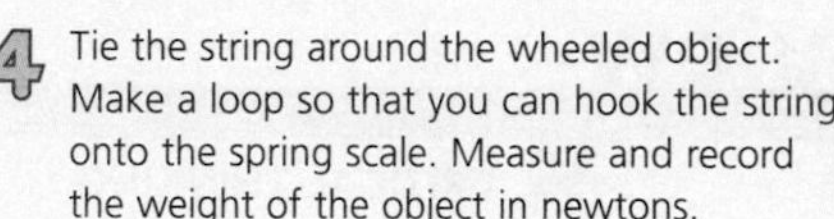
4. Tie the string around the wheeled object. Make a loop so that you can hook the string onto the spring scale. Measure and record the weight of the object in newtons.

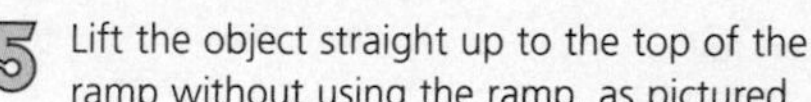
5. Lift the object straight up to the top of the ramp without using the ramp, as pictured.

C 136 Unit: Motion and Forces

INVESTIGATION RESOURCES

 CHAPTER INVESTIGATION, Work and Power
- Level A, pp. 267–270
- Level B, pp. 271–274
- Level C, p. 275

Advanced students should complete Levels B & C.

 Writing a Lab Report, D12–13

Technology Resources

Customize this student lab as needed or look for an alternative. Print rubrics to assess student lab reports.

 **Lab Generator CD-ROM**

6 On the spring scale, read and record the newtons of force needed to lift the object. Time how long it takes to lift the object from the floor to the top of the ramp. Conduct three trials and average your results. Record your measurements in the data table.

7 Drag the object from the bottom of the ramp to the top of the ramp with the spring scale, and record the newtons of force that were needed to move the object and the time it took. Conduct three trials and average your results.

Observe and Analyze

1. **RECORD OBSERVATIONS** Draw the setup of the procedure. Be sure your data table is complete.
2. **IDENTIFY VARIABLES AND CONSTANTS** List the variables and constants in your notebook.
3. **CALCULATE**

   Potential Energy Convert centimeters to meters. Then calculate the gravitational potential energy (GPE) of the object at the top of the ramp. (Recall that weight equals mass times gravitational acceleration.)

   $$\text{Gravitational Potential Energy} = \text{weight} \cdot \text{height}$$

   Work Calculate the work done, first when the object was lifted and then when it was pulled. Use the appropriate distance.

   $$\text{Work} = \text{Force} \cdot \text{distance}$$

   Power Calculate the power involved in both situations.

   $$\text{Power} = \frac{\text{Work}}{\text{time}}$$

Conclude

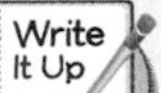

1. **COMPARE** How did the distance through which the object moved when it was pulled up the ramp differ from the distance when it was lifted straight up? How did the amount of force required differ in the two situations?
2. **COMPARE** How does your calculated value for potential energy compare with the values you obtained for work done?
3. **INTERPRET** Answer the question posed in the problem.
4. **ANALYZE** Compare your results with your hypothesis. Did your results support your hypothesis?
5. **IDENTIFY LIMITS** What possible limitations or sources of error could you have experienced?
6. **APPLY** A road going up a hill usually winds back and forth instead of heading straight to the top. How does this affect the work a car does to get to the top? How does it affect the power involved?

INVESTIGATE Further

CHALLENGE Design a way to use potential energy to move the car up the ramp. What materials can you use? Think about the materials in terms of potential energy—that is, how high they are from the ground or how stretched or compressed they are.

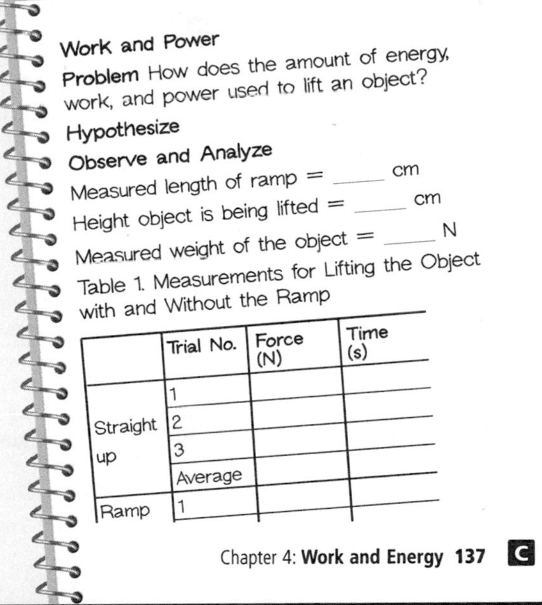

Work and Power

Problem How does the amount of energy, work, and power used to lift an object?

Hypothesize

Observe and Analyze

Measured length of ramp = _____ cm

Height object is being lifted = _____ cm

Measured weight of the object = _____ N

Table 1. Measurements for Lifting the Object with and Without the Ramp

	Trial No.	Force (N)	Time (s)
Straight up	1		
	2		
	3		
	Average		
Ramp	1		

Observe and Analyze

SAMPLE DATA Weight of object: 0.75 N; Height of object being lifted, 48 cm; Length of ramp, 68 cm; Force (3 trials), 0.7 N, 0.75 N, 0.75 N; Time, 1 s, 0.5 s, 0.5 s; Force (3 trials), 0.55 N, 0.5 N, 0.5 N; Time, 6 s, 5 s, 4 s.

1. *See students' diagrams.*
2. *Variables include distance, force, and time. Constants include mass of the object, force of gravity, and vertical height of the chair seat.*
3. *See students' data tables.*

Conclude

1. *The ramp distance was greater than the distance straight up. The amount of force required to move the object up the ramp was less than the amount of force required to lift it straight up. If more work was done in moving the car up the ramp than in moving it straight up, it was due to friction on the ramp.*
2. *Answers will vary.*
3. *A ramp can decrease the amount of force needed to move an object, but the amount of work is the same. For a given speed, power will be less when using a ramp.*
4. *Students' answers will vary.*
5. *Sources of error might include not pulling the object at a constant force, not measuring to the nearest centimeter, and not reading the scale correctly.*
6. *The work that a car does is the same as one heading straight to the top. The power may be less if the car takes longer to get there.*

INVESTIGATE Further

CHALLENGE Sample answer: Attach the car to a heavy object with a string. Place the object on the chair, then drop it to the floor. As it falls, the car will be pulled up the ramp.

Post-Lab Discussion

- Discuss the variables and constants in the lab. Students should understand that the gravitational potential energy is the same when lifting something straight up and when using a ramp because mass, gravitational acceleration, and height are constants. Only distance, force, and time—the factors used when calculating work and power—vary.
- Movers' trucks usually have ramps that workers use to load and unload heavy objects such as furniture. Ask students what advantages a ramp provides.

BACK TO

the BIG idea

This demonstration illustrates that energy is transferred when a force moves an object. You may want to use a Newton's cradle if one is available. Tie two steel balls to an overhead bar. Pull one ball to the side, then release it. It will hit the second ball and move it. Ask students to describe the force and the energy transfer and explain why the second ball moves. *A force is applied to the first ball when it is pulled. When the ball is released, its gravitational potential energy changes to kinetic energy, which is transferred to the second ball during the collision.*

KEY CONCEPTS SUMMARY

SECTION 4.1

Ask: In which diagram is more work done if the distance is the same? *the left-hand diagram*

Ask: Why is less work done in the right-hand picture? *Part of the force does no work because it is not applied in the same direction as the motion.*

SECTION 4.2

Ask: Which form of energy does the skater have at the top of the ramp? *gravitational potential energy*

Ask: What happens to this potential energy as the skater travels down the ramp? *It changes to kinetic energy.*

SECTION 4.3

Ask: What can you infer about the power of the leading swimmer? *She has more power than the other swimmers.*

Ask: What can you infer about the leading swimmer's time? *She is covering the distance in less time than the other swimmers.*

Review Concepts

- Big Idea Flow Chart, p. T25
- Chapter Outline, pp. T31–T32

4 Chapter Review

the BIG idea

Energy is transferred when a force moves an object.

KEY CONCEPTS SUMMARY

4.1 Work is the use of force to move an object.

Work is done by a force that acts in the same direction as the motion of an object.

Work = Force • distance

VOCABULARY
work p. 115
joule p. 117

4.2 Energy is transferred when work is done.

The amounts of potential energy and kinetic energy in a system or process may vary, but the total amount of energy remains unchanged.

$GPE = mgh$

$KE = \frac{1}{2}mv^2$

$ME = PE + KE$

VOCABULARY
potential energy p. 122
kinetic energy p. 122
mechanical energy p. 125
conservation of energy p. 126

4.3 Power is the rate at which work is done.

Power can be calculated from work and time.

$$\text{Power} = \frac{\text{Work}}{\text{time}}$$

Power can be calculated from energy and time.

$$\text{Power} = \frac{\text{Energy}}{\text{time}}$$

Power is measured in watts (W) and sometimes horsepower (hp).

VOCABULARY
power p. 130
watt p. 131
horsepower p. 132

Technology Resources

Have students visit **ClassZone.com** or use the CD-ROM for a cumulative review of concepts.

 CONTENT REVIEW

 CONTENT REVIEW CD-ROM

Engage students in a whole-class interactive review of Key Concepts. Edit content as you wish.

 POWER PRESENTATIONS

Reviewing Vocabulary

Make a four square diagram for each of the terms listed below. Write the term in the center. Define it in one square. Write characteristics, examples, and formulas (if appropriate) in the other squares. A sample is shown below.

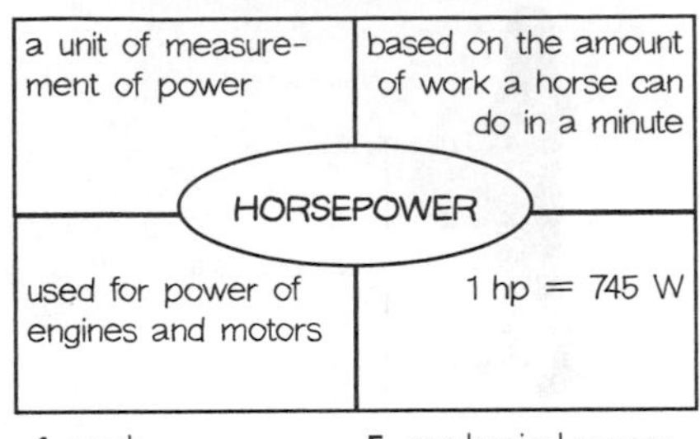

a unit of measurement of power
based on the amount of work a horse can do in a minute
HORSEPOWER
used for power of engines and motors
1 hp = 745 W

1. work
2. joule
3. potential energy
4. kinetic energy
5. mechanical energy
6. power
7. watt

Reviewing Key Concepts

Multiple Choice *Choose the letter of the best answer.*

8. Work can be calculated from
   a. force and speed
   b. force and distance
   c. energy and time
   d. energy and distance
9. If you balance a book on your head, you are not doing work on the book because
   a. doing work requires moving an object
   b. you are not applying any force to the book
   c. the book is doing work on you
   d. the book has potential energy
10. Energy that an object has because of its position or shape is called
    a. potential energy
    b. kinetic energy
    c. thermal energy
    d. chemical energy
11. Suppose you are pushing a child on a swing. During what space of time are you doing work on the swing?
    a. while you hold it back before letting go
    b. while your hands are in contact with the swing and pushing forward
    c. after you let go of the swing and it continues to move forward
    d. all the time the swing is in motion
12. A falling ball has a potential energy of 5 J and a kinetic energy of 10 J. What is the ball's mechanical energy?
    a. 5 J
    b. 10 J
    c. 15 J
    d. 50 J
13. The unit that measures one joule of work done in one second is called a
    a. meter
    b. watt
    c. newton-meter
    d. newton
14. By increasing the speed at which you do work, you increase your
    a. force
    b. work
    c. energy
    d. power
15. A ball kicked into the air will have the greatest gravitational potential energy
    a. as it is being kicked
    b. as it starts rising
    c. at its highest point
    d. as it hits the ground

Short Answer *Answer each of the following questions in a sentence or two.*

16. How can you tell if a force you exert is doing work?
17. How does a water wheel do work?
18. State the law of conservation of energy. How does it affect the total amount of energy in any process?
19. Explain why a swing will not stay in motion forever after you have given it a push. What happens to its mechanical energy?
20. What are two ways to calculate power?
21. Why did James Watt invent a unit of measurement based on the work of horses?

ASSESSMENT RESOURCES

UNIT ASSESSMENT BOOK
- Chapter Test A, pp. 65–68
- Chapter Test B, pp. 69–72
- Chapter Test C, pp. 73–76
- Alternative Assessment, pp. 77–78

SPANISH ASSESSMENT BOOK
Spanish Chapter Test, pp. 269–272

Technology Resources

Edit test items and answer choices.

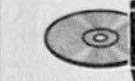

Test Generator CD-ROM

Visit **ClassZone.com** to extend test practice.

Test Practice

Reviewing Vocabulary

Sample answers:

1. *work: the use of force to move an object a certain distance;* W = F · d*; example: turning a page; nonexample: pushing an immovable object*
2. *joule: unit for measuring work; same as newton-meter; a force in newtons times a distance in meters; 1 J = force of 1 N moving an object 1 m*
3. *potential energy: energy due to an object's position or shape; stored energy; example: a person at the top of a diving board;* GPE = mgh
4. *kinetic energy: energy due to an object's motion; most kinetic energy when object moves fastest; example: moving car;* $KE = \frac{1}{2}mv^2$
5. *mechanical energy: energy due to an object's position or motion; any object with mechanical energy can do work on another object; example: thrown ball;* ME = PE + KE
6. *power: rate at which work is done; to increase power, do activity faster;* P = W/t; P = E/t
7. *watt: unit of measurement for power; 1 W = 1 J of work done in 1 s; kilowatt = 1000 W*

Reviewing Key Concepts

8. b
9. a
10. a
11. b
12. c
13. b
14. d
15. c

16. *It moves an object.*
17. *Falling water works on gears moving the wheel. The wheel grinds grain.*
18. *No matter how energy is transferred or transformed, all of it is still present in one form or another. The total amount of energy in a process never changes.*
19. *The ME goes into heating the chain through friction, moving air molecules the swing hits, and creating sound. The swing eventually converts all of its ME and stops.*
20. *by dividing the amount of work done by the time it takes to do it; by dividing energy used in a given period by time*
21. *to show people how powerful his engine was by comparing it with the more familiar work of horses*

Thinking Critically

22. *The barbell has potential energy when it is resting on the ground. The weightlifter gives the barbell mechanical energy by doing work on it. As it goes up, it has both kinetic energy and gravitational potential energy.*
23. *Winding up a car gives it potential energy. When released, it becomes kinetic energy.*
24. *No; motion in the direction of the force is necessary to do work.*
25. *Power increases as the chair moves faster because the time is less. Work increases with distance.*
26. *It became thermal energy due to friction between the chair and the floor.*
27. *As the ball falls, gravitational potential energy changes to kinetic energy. Some energy changes to thermal energy as the ball hits the floor.*
28. *from chemical energy in the muscles, which transfer it to an object*

Using Math Skills in Science

29. 225 J	*32. 120 J*
30. 9 m	*33. 25 W*
31. 4.9 J	*34. 2400 J*

the BIG idea

35. *The young woman lifting the box does work because she is using force to hold the box. The box has potential energy, and the woman has kinetic energy if she is walking. Power is the rate at which the work of lifting and carrying the box is done.*
36. *Answers should include force applied over a distance (for example, work) and should trace the energy transfers.*

UNIT PROJECTS

Students should have begun designing their models or presentations by this time. Remind them to continue researching as needed. Encourage them to try different solutions to problems.

 Unit Projects, pp. 5–10

Thinking Critically

22. **SYNTHESIZE** A weightlifter holds a barbell above his head. How do the barbell's potential energy, kinetic energy, and mechanical energy change as it is lifted and then lowered to the ground?

23. **SYNTHESIZE** What happens when you wind up a toy car and release it? Describe the events in terms of energy.

Use the photograph below to answer the next three questions.

24. **APPLY** When the boy first pushes on the chair, the chair does not move due to friction. Is the boy doing work? Why or why not?

25. **ANALYZE** For the first two seconds, the boy pushes the chair slowly at a steady speed. After that, he pushes the chair at a faster speed. How does his power change if he is using the same force at both speeds? How does his work change?

26. **SYNTHESIZE** As the boy pushes the chair, he does work. However, when he stops pushing, the chair stops moving and does not have any additional kinetic or potential energy. What happened to the energy he transferred by doing work on the chair?

27. **APPLY** A bouncing ball has mechanical energy. Each bounce, however, reaches a lower height than the last. Describe what happens to the mechanical, potential, and kinetic energy of the ball as it bounces several times.

28. **CONNECT** When you do work, you transfer energy. Where does the energy you transfer come from?

Using Math Skills in Science

Complete the following calculations.

29. Use the information in the photograph below to calculate the work the person does in lifting the box.

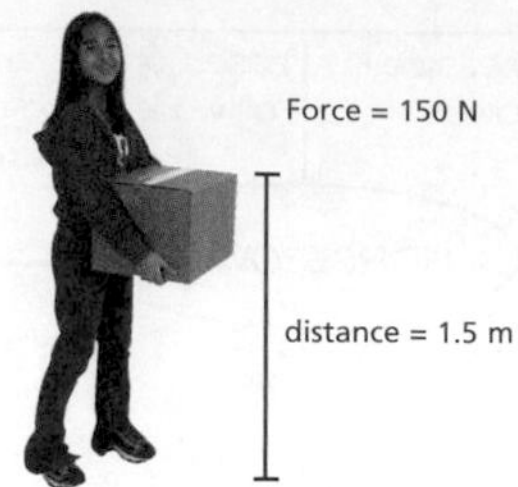

30. If you did 225 J of work to pull a wagon with a force of 25 N, how far did you pull it?

31. A kite with a mass of 0.05 kg is caught on the roof of a house. The house is 10 m high. What is the kite's gravitational potential energy? (Recall that $g = 9.8\ m/s^2$.)

32. A baseball with a mass of 0.15 kg leaves a pitcher's hand traveling 40 m/s toward the batter. What is the baseball's kinetic energy?

33. Suppose it takes 150 J of force to push a cart 10 m in 60 s. Calculate the power.

34. If an electric hair dryer uses 1200 W, how much energy does it need to run for 2 s?

the BIG idea

35. **SYNTHESIZE** Look back at the photograph of the person lifting a box on pages 112–113. Describe the picture in terms of work, potential energy, kinetic energy, and power.

36. **WRITE** Think of an activity that involves work. Write a paragraph explaining how the work is transferring energy and where the transferred energy goes.

UNIT PROJECTS

If you need to create graphs or other visuals for your project, be sure you have grid paper, poster board, markers, or other supplies.

MONITOR AND RETEACH

If students have trouble applying the concepts in questions 24–26, have volunteers take turns reenacting the scenario in the photograph. As they push the chair, they can describe for the class what is happening in terms of force, work, speed, friction, and energy.

Students may benefit from summarizing one or more sections of the chapter.

 Summarizing the Chapter, pp. 285–286

Standardized Test Practice

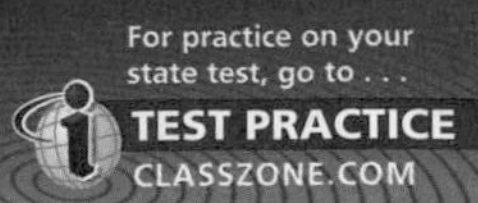
For practice on your state test, go to . . .
TEST PRACTICE
CLASSZONE.COM

Understanding Experiments

Read the following description of an experiment. Then answer the questions that follow.

James Prescott Joule is well known for a paddle-wheel experiment he conducted in the mid-1800s. He placed a paddle wheel in a bucket of water. Then he set up two weights on either side of the bucket. As the weights fell, they turned the paddle wheel. Joule recorded the temperature of the water before and after the paddle wheel began turning. He found that the water temperature increased as the paddle wheel turned.

Based on this experiment, Joule concluded that the falling weights released mechanical energy, which was converted into heat by the turning wheel. He was convinced that whenever mechanical force is exerted, heat is produced.

1. Which principle did Joule demonstrate with this experiment?
- **a.** When energy is converted from one form to another, some energy is lost.
- **b.** The amount of momentum in a system does not change as long as there are no outside forces acting on the system.
- **c.** One form of energy can be converted into another form of energy.
- **d.** When one object exerts a force on another object, the second object exerts an equal and opposite force on the first object.

2. Which form of energy was released by the weights in Joule's experiment?
- **a.** electrical
- **b.** mechanical
- **c.** nuclear
- **d.** heat

3. Which form of energy was produced in the water?
- **a.** chemical
- **b.** electrical
- **c.** nuclear
- **d.** heat

4. Based on Joule's finding that movement causes temperature changes in water, which of the following would be a logical prediction?
- **a.** Water held in a container should increase in temperature.
- **b.** Water at the base of a waterfall should be warmer than water at the top.
- **c.** Water with strong waves should be colder than calm water.
- **d.** Water should increase in temperature with depth.

Extended Response

Answer the two questions below in detail. Include some of the terms from the word box. Underline each term you use in your answer.

potential energy	conservation of energy	force
kinetic energy	power	work

5. A sledder has the greatest potential energy at the top of a hill. She has the least amount of potential energy at the bottom of a hill. She has the greatest kinetic energy when she moves the fastest. Where on the hill does the sledder move the fastest? State the relationship between kinetic energy and potential energy in this situation.

6. Andre and Jon are moving boxes of books from the floor to a shelf in the school library. Each box weighs 15 lb. Andre lifts 5 boxes in one minute. Jon lifts 5 boxes in 30 seconds. Which person does more work? Which person applies more force? Which person has the greater power? Explain your answers.

METACOGNITIVE ACTIVITY

Have students answer the following questions in their **Science Notebook:**

1. Which concepts about work and energy did you find to be the most challenging?
2. Predict the content of the next chapter.
3. What information are you having a difficult time finding that is needed to complete your Unit Project?

Understanding Experiments

1. c *2. b* *3. d* *4. b*

Extended Response

5. RUBRIC

4 points for a response that correctly answers the question and uses the following terms accurately:

- potential energy
- kinetic energy

At the top of the hill, the sledder has potential energy but no kinetic energy. As the sledder moves down the hill, her potential energy decreases because the elevation decreases. Her kinetic energy increases because her velocity increases. Halfway down the hill, the sledder's potential energy is equal to her kinetic energy. Therefore, the sledder moves the fastest at the bottom of the hill.

3 points for a response that correctly answers the question and uses one term accurately
2 points for a response that correctly answers the question but does not use the terms accurately

6. RUBRIC

4 points for a response that correctly answers all the questions and uses the following terms accurately:

- energy
- work
- power

Jon and Andre both did the same amount of work lifting the boxes, since the potential energy they gave each box was the same. Power is the rate at which work is done. Jon has the greater power because he lifted the same weight of boxes the same distance in half the time.

3 points for a response that correctly answers two questions and uses two terms accurately
2 points for a response that correctly answers one question and uses one term accurately
1 point for a response that correctly answers the questions, but does not use the terms

CHAPTER

Machines

Physical Science

UNIFYING PRINCIPLES

PRINCIPLE 1

Matter is made of particles too small to see.

PRINCIPLE 2

Matter changes form and moves from place to place.

PRINCIPLE 3

Energy changes from one form to another, but it cannot be created or destroyed.

PRINCIPLE 4

Physical forces affect the movement of all matter on Earth and throughout the universe.

Unit: Motion and Forces

BIG IDEAS

CHAPTER 1
Motion

The motion of an object can be described and predicted.

CHAPTER 2
Forces

Forces change the motion of objects in predictable ways.

CHAPTER 3
Gravity, Friction, and Pressure

Newton's laws apply to all forces.

CHAPTER 4
Work and Energy

Energy is transferred when a force moves an object.

CHAPTER 5
Machines

Machines help people do work by changing the force applied to an object.

CHAPTER 5

KEY CONCEPTS

SECTION 5.1

Machines help people do work.

1. Machines change the way force is applied.
2. Work transfers energy.
3. Output work is always less than input work.

SECTION 5.2

Six simple machines have many uses.

1. There are six simple machines.
2. The mechanical advantage of a machine can be calculated.

SECTION 5.3

Modern technology uses compound machines.

1. Compound machines are combinations of simple machines.
2. Modern technology creates new uses for machines.

T The Big Idea Flow Chart is available on p. T33 in the **UNIT TRANSPARENCY BOOK.**

Previewing Content

SECTION

Machines help people do work.
pp. 145–153

1. Machines change the way force is applied.
A **machine** is a device that helps people do work. It does not change the amount of work done.

- If a machine decreases the amount of force needed to do the work, the distance over which that force must be applied increases.
- A machine can change the direction of an applied force.

Input force is the force exerted on a machine. Output force is the force exerted on an object by a machine. The number of times a machine multiplies the input force is the machine's **mechanical advantage.**

2. Work transfers energy.
A machine increases the potential or kinetic energy of an object by doing work on it. For a certain amount of work, if distance increases, force decreases, as shown in the figure below.

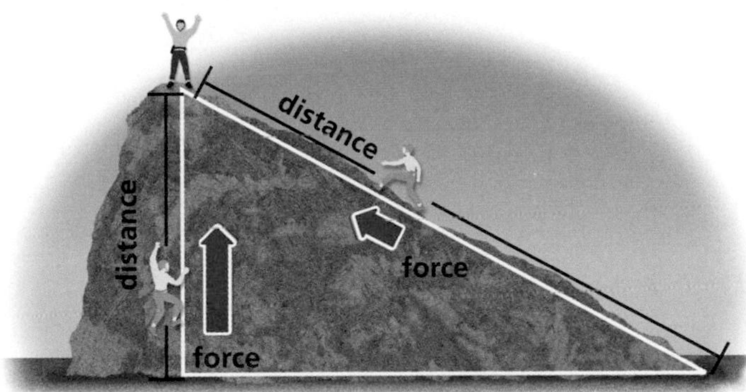

3. Output work is always less than input work.
Efficiency is the ratio of a machine's output work to the input work.

Efficiency (%) = Output work/Input work · 100

No real machine is 100 percent efficient. Machines lose energy to friction, which is why we lubricate moving parts. Another source of loss of efficiency is air resistance, which is the reason for streamlined designs for vehicles and cyclists' helmets.

SECTION

Six simple machines have many uses. pp. 154–163

1. There are six simple machines.
The lever and the inclined plane are the two main types of **simple machines.** Other simple machines are based on these two.

- A **lever** is a solid bar that rotates on a fixed point called a **fulcrum.** There are three classes of levers based on the relative locations of the input force, the output force, and the fulcrum.
- A **wheel and axle** is a wheel attached to a shaft. It acts like a rotating collection of levers. The input force can be applied to either part, which transfers force to the other part.
- A **pulley** is a wheel with an axle and a grooved rim. A rope or a cable moves in the groove. Pulleys can be either fixed or movable. A combination of the two types is called a block and tackle.
- An **inclined plane** is a sloping surface that supports the weight of an object while the object moves from one level to another.
- A **wedge** has a thick end and a thin end. A wedge can be used to cut, split, pierce objects or to hold objects together.
- A **screw** is an inclined plane wrapped around a cylinder or a cone to form a spiral. Screws can be used to hold things together or to raise and lower objects.

2. The mechanical advantage of a machine can be calculated.
If a machine were 100 percent efficient, its ideal mechanical advantage would be output force/input force, or

$$MA = \frac{F_{out}}{F_{in}}$$

- For an inclined plane, divide the length of the incline by the height of the incline.
- For a wheel and axle, divide the radius where the input force is applied by the radius where the output force is applied.
- For a lever, calculate this by dividing distance from input force to fulcrum by the distance from output force to fulcrum.
- For pulleys, the mechanical advantage is equal to the number of ropes that support the weight.

Common Misconceptions

MISCONCEPTION DATABASE
CLASSZONE.COM Background on student misconceptions

WORK Students may think that machines reduce work and energy. In fact, machines reduce the amount of effort required to do work by changing the size or the direction of the force. Because the machine must transfer a certain amount of energy, it must do an equivalent amount of work.

 This misconception is addressed on p. 148.

EFFICIENCY Students might think that machines can be 100 percent efficient. In reality, any machine that has a moving part loses some energy, and thus efficiency, to friction. Only ideal machines have 100 percent efficiency.

 This misconception is addressed on p. 150.

Previewing Content

SECTION

5.3 Modern technology uses compound machines. pp. 164–171

1. Compound machines are combinations of simple machines.

A machine that is made of two or more simple machines is a **compound machine.** Compound machines often have many moving parts and must overcome more friction than simple machines.

The mechanical advantage of a compound machine equals the product of the mechanical advantages of all the simple machines that make up the compound machine.
For example, if a lever with a *MA* of 2 acts in series with a lever with a *MA* of 3, the mechanical advantage of the lever combination will be $2 \cdot 3$, or 6.

- The mechanical advantage of a gear system comprising two wheel-and-axle systems is found by dividing the number of teeth on the output wheel by the number of teeth on the input wheel. For example, a wheel with 16 teeth turns another wheel with 24 teeth. The wheel with 16 teeth is the input wheel and the wheel with 24 teeth is the output wheel.

  $$MA = \frac{24}{16} = 1.5$$

  Thus, 1.5 is the mechanical advantage.
- For these two gears, the smaller input gear has half as many teeth as the output gear. The mechanical advantage is 2.

2. Modern technology creates new uses for machines.

Sophisticated machines are often based on or contain several simple machines.

- Machines built from individual atoms and molecules of material are the result of **nanotechnology.** Most nanomachines are still in the experimental stage.
- **Robots** are machines that work automatically or by remote control. They do jobs in places where it is difficult or dangerous for people to work.

Previewing Labs

Lab Generator CD-ROM
Edit these Pupil Edition labs and generate alternative labs.

EXPLORE the BIG idea

Changing Direction, p. 143 Students examine a window blind to learn that machines can change the direction of a force.	**TIME** 10 minutes **MATERIALS** window blind
Shut the Door! p. 143 Students experiment with a door to test the relationship between force and the length of a lever arm.	**TIME** 10 minutes **MATERIALS** a door on hinges
Internet Activity: Machines, p. 143 Students analyze tools to learn about different machines.	**TIME** 20 minutes **MATERIALS** computer with Internet access

SECTION

EXPLORE Machines, p. 145 Students examine small machines to learn how they help people work.	**TIME** 15 minutes **MATERIALS** various small machines
INVESTIGATE Efficiency, p. 151 Students collect and analyze data for a block moving on a ramp and use the results to calculate efficiency.	**TIME** 20 minutes **MATERIALS** board, books, meter stick, wooden block with eyehook, spring scale, sandpaper

SECTION

EXPLORE Changing Forces, p. 154 Students adjust the fulcrum of a lever to determine how the input force changes.	**TIME** 15 minutes **MATERIALS** 2 pencils, small book
INVESTIGATE Pulleys, p. 157 Students create a block-and-tackle pulley system, calculate mechanical advantage, and infer how the input and output forces vary.	**TIME** 20 minutes **MATERIALS** 100 g mass, spring scale, 2 pulleys with rope, ring stand with ring or clamp

SECTION

CHAPTER INVESTIGATION **Design a Machine,** pp. 170–171 Students design a machine, build it, test it, and calculate its mechanical advantage and its efficiency.	**TIME** 40 minutes **MATERIALS** 500 g object; 100 g object; meter stick; spring scale; single, double, or triple pulleys with rope; board; stick or pole

R **Additional INVESTIGATION,** Levers and Fulcrums, A, B, & C, pp. 335–343; Teacher Instructions, pp. 346–347

Previewing Chapter Resources

Chapter 5 Machines

INTEGRATED TECHNOLOGY

CLASSZONE.COM
- eEdition Plus
- EasyPlanner Plus
- Misconception Database
- Content Review
- Test Practice
- Resource Centers
- Simulation
- Internet Activity: Machines
- Math Tutorial

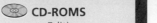

SCILINKS.ORG

SCILINKS

CD-ROMS
- eEdition
- EasyPlanner
- Power Presentations
- Content Review
- Lab Generator
- Test Generator

AUDIO CDS
- Audio Readings
- Audio Readings in Spanish

LABS AND ACTIVITIES

EXPLORE the Big Idea, p. 143
- Changing Direction
- Shut the Door!
- Internet Activity: Machines

UNIT RESOURCE BOOK
Unit Projects, pp. 5–10

Lab Generator CD-ROM
Generate customized labs.

SECTION

Machines help people do work.
pp. 145–153

Time: 2 periods (1 block)

 Lesson Plan, pp. 287–288

MATH TUTORIAL

UNIT TRANSPARENCY BOOK
- Big Idea Flow Chart, p. T33
- Daily Vocabulary Scaffolding, p. T34
- Note-Taking Model, p. T35
- 3-Minute Warm-Up, p. T36

- EXPLORE Machines, p. 145
- INVESTIGATE Efficiency, p. 151
- Math in Science, p. 153

UNIT RESOURCE BOOK
- Datasheet, Efficiency, p. 296
- Math Support, p. 324
- Math Practice, p. 325

Six simple machines have many uses.
pp. 154–163

Time: 2 periods (1 block)

Lesson Plan, pp. 298–299

- **SIMULATION,** Mechanical Advantage
- **RESOURCE CENTER,** Artificial Limbs

UNIT TRANSPARENCY BOOK
- Daily Vocabulary Scaffolding, p. T34
- 3-Minute Warm-Up, p. T36

- EXPLORE Changing Forces, p. 154
- INVESTIGATE Pulleys, p. 157
- Connecting Sciences, p. 163

UNIT RESOURCE BOOK
- Datasheet, Pulleys, p. 307
- Additional INVESTIGATION, Levers and Fulcrums, A, B, & C, pp. 335–343

SECTION

Modern technology uses compound machines.
pp. 164–171

Time: 4 periods (2 blocks)

 Lesson Plan, pp. 309–310

RESOURCE CENTERS, Nanomachines, Robots

UNIT TRANSPARENCY BOOK
- Big Idea Flow Chart, p. T33
- Daily Vocabulary Scaffolding, p. T34
- 3-Minute Warm-Up, p. T37
- "A Robot at Work" Visual, p. T38
- Chapter Outline, pp. T39–T40

CHAPTER INVESTIGATION, Design a Machine, pp. 170–171

UNIT RESOURCE BOOK
CHAPTER INVESTIGATION, Design a Machine, A, B, & C, pp. 326–334

KEY TO ICONS 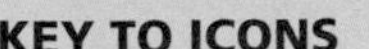 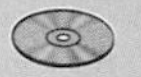 CD/CD-ROM 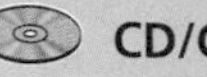 INTERNET  Pupil Edition · Teacher Edition · UNIT RESOURCE BOOK  UNIT TRANSPARENCY BOOK · UNIT ASSESSMENT BOOK  SPANISH ASSESSMENT BOOK · SCIENCE TOOLKIT

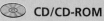

READING AND REINFORCEMENT	ASSESSMENT	STANDARDS
 • Word Triangle, B18–19 • Choose Your Own Strategy, C36, C38–39, C42, C43 • Daily Vocabulary Scaffolding, H1–8  **UNIT RESOURCE BOOK** • Vocabulary Practice, pp. 321–322 • Decoding Support, p. 323 • Summarizing the Chapter, pp. 344–345  **Audio Readings CD** Listen to Pupil Edition.  **Audio Readings in Spanish CD** Listen to Pupil Edition in Spanish.	 • Chapter Review, pp. 173–174 • Standardized Test Practice, p. 175  **UNIT ASSESSMENT BOOK** • Diagnostic Test, pp. 79–80 • Chapter Test, A, B, & C, pp. 84–95 • Alternative Assessment, pp. 96–97 • Unit Test, A, B, & C, pp. 98–109  • Spanish Chapter Test, pp. 225–228 • Spanish Unit Test, pp. 277–280  **Test Generator CD-ROM** Generate customized tests.  **Lab Generator CD-ROM** Rubrics for Labs	National Standards A.1–8, A.9.a–g, E.2–5, E.6.c–e, F.5.a–c, F.5.e, G.1.b See p. 142 for the standards.
 **UNIT RESOURCE BOOK** • Reading Study Guide, A & B, pp. 289–292 • Spanish Reading Study Guide, pp. 293–294 • Challenge and Extension, p. 295 • Reinforcing Key Concepts, p. 297	 Ongoing Assessment, pp. 146–147, 149–152  Section 5.1 Review, p. 152  **UNIT ASSESSMENT BOOK** Section 5.1 Quiz, p. 81	National Standards A.2–8, A.9.a–c, A.9.e–f, G.1.b
 **UNIT RESOURCE BOOK** • Reading Study Guide, A & B, pp. 300–303 • Spanish Reading Study Guide, pp. 304–305 • Challenge and Extension, p. 306 • Reinforcing Key Concepts, p. 308	 Ongoing Assessment, pp. 155–159, 161  Section 5.2 Review, p. 162  **UNIT ASSESSMENT BOOK** Section 5.2 Quiz, p. 82	National Standards A.2–8, A.9.a–c, A.9.e–f, G.1.b
UNIT RESOURCE BOOK • Reading Study Guide, A & B, pp. 311–314 • Spanish Reading Study Guide, pp. 315–316 • Challenge and Extension, p. 317 • Reinforcing Key Concepts, p. 318 • Challenge Reading, pp. 319–320	 Ongoing Assessment, pp. 164–165, 167–169  Section 5.3 Review, p. 169 **UNIT ASSESSMENT BOOK** Section 5.3 Quiz, p. 83	National Standards A.1–8, A.9.a–g, E.2–5, E.6.c–e, F.5.a–c, F.5.e, G.1.b

Previewing Resources for Differentiated Instruction

CHAPTER INVESTIGATION

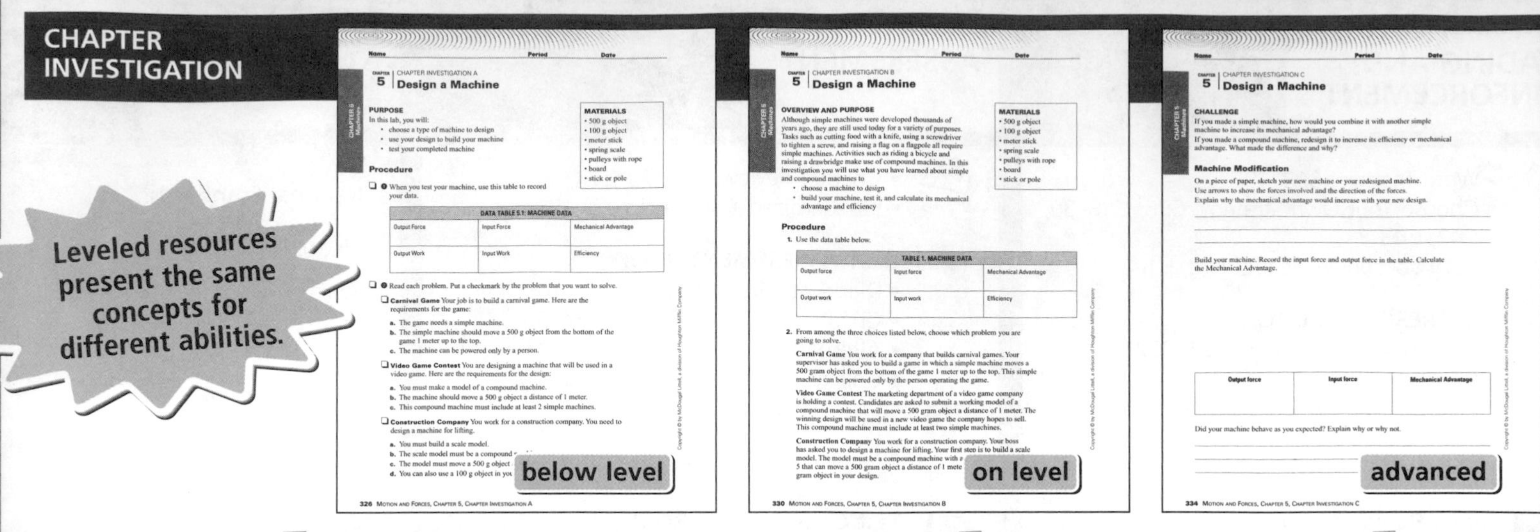
Leveled resources present the same concepts for different abilities.

Name Period Date

CHAPTER 5 | CHAPTER INVESTIGATION A
Design a Machine

PURPOSE
In this lab, you will:
- choose a type of machine to design
- use your design to build your machine
- test your completed machine

MATERIALS
- 500 g object
- 100 g object
- meter stick
- spring scale
- pulleys with rope
- board
- stick or pole

Procedure

When you test your machine, use this table to record your data.

DATA TABLE 5.1: MACHINE DATA

Output Force	Input Force	Mechanical Advantage
Output Work	Input Work	Efficiency

Read each problem. Put a checkmark by the problem that you want to solve.

Carnival Game Your job is to build a carnival game. Here are the requirements for the game:
a. The game needs a simple machine.
b. The simple machine should move a 500 g object from the bottom of the game 1 meter up to the top.
c. The machine can be powered only by a person.

Video Game Contest You are designing a machine that will be used in a video game. Here are the requirements for the design:
a. You must make a model of a compound machine.
b. The machine should move a 500 g object a distance of 1 meter.
c. This compound machine must include at least 2 simple machines.

Construction Company You work for a construction company. You need to design a machine for lifting.
a. You must build a scale model.
b. The scale model must be a compound ...
c. The model must move a 500 g object ...
d. You can also use a 100 g object in you...

below level

326 Motion and Forces, Chapter 5, Chapter Investigation A

Name Period Date

CHAPTER 5 | CHAPTER INVESTIGATION B
Design a Machine

OVERVIEW AND PURPOSE
Although simple machines were developed thousands of years ago, they are still used today for a variety of purposes. Tasks such as cutting food with a knife, using a screwdriver to tighten a screw, and raising a flag on a flagpole all require simple machines. Activities such as riding a bicycle and raising a drawbridge make use of compound machines. In this investigation you will use what you have learned about simple and compound machines to
- choose a machine to design
- build your machine, test it, and calculate its mechanical advantage and efficiency

MATERIALS
- 500 g object
- 100 g object
- meter stick
- spring scale
- pulleys with rope
- board
- stick or pole

Procedure
1. Use the data table below.

TABLE 1. MACHINE DATA

Output force	Input force	Mechanical Advantage
Output work	Input work	Efficiency

2. From among the three choices listed below, choose which problem you are going to solve.

Carnival Game You work for a company that builds carnival games. Your supervisor has asked you to build a game in which a simple machine moves a 500 gram object from the bottom of the game 1 meter up to the top. This simple machine can be powered only by the person operating the game.

Video Game Contest The marketing department of a video game company is holding a contest. Candidates are asked to submit a working model of a compound machine that will move a 500 gram object a distance of 1 meter. The winning design will be used in a new video game the company hopes to sell. This compound machine must include at least two simple machines.

Construction Company You work for a construction company. Your boss has asked you to design a machine for lifting. Your first step is to build a scale model. The model must be a compound machine with ... 5 that can move a 500 gram object a distance of 1 mete... gram object in your design.

on level

330 Motion and Forces, Chapter 5, Chapter Investigation B

Name Period Date

CHAPTER 5 | CHAPTER INVESTIGATION C
Design a Machine

CHALLENGE
If you made a simple machine, how would you combine it with another simple machine to increase its mechanical advantage?
If you made a compound machine, redesign it to increase its efficiency or mechanical advantage. What made the difference and why?

Machine Modification
On a piece of paper, sketch your new machine or your redesigned machine. Use arrows to show the forces involved and the direction of the forces. Explain why the mechanical advantage would increase with your new design.

Build your machine. Record the input force and output force in the table. Calculate the Mechanical Advantage.

Output force	Input force	Mechanical Advantage

Did your machine behave as you expected? Explain why or why not.

advanced

334 Motion and Forces, Chapter 5, Chapter Investigation C

R UNIT RESOURCE BOOK, pp. 326–329 | R pp. 330–333 | R pp. 330–334

READING STUDY GUIDE

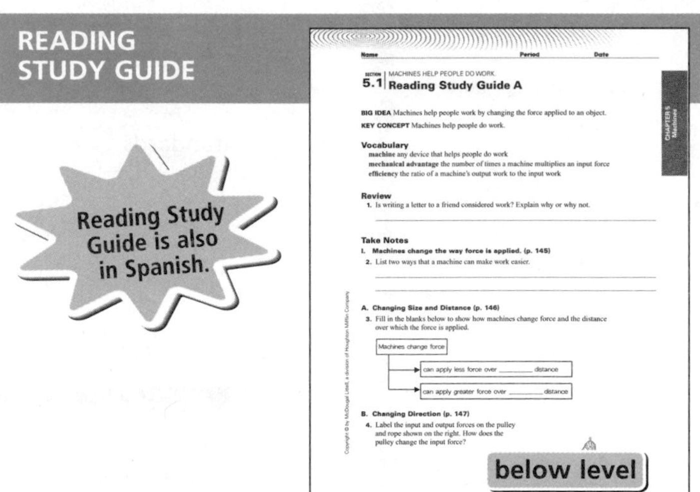
Reading Study Guide is also in Spanish.

Name Period Date

SECTION 5.1 | MACHINES HELP PEOPLE DO WORK.
Reading Study Guide A

BIG IDEA Machines help people work by changing the force applied to an object.
KEY CONCEPT Machines help people do work.

Vocabulary
machine any device that helps people do work
mechanical advantage the number of times a machine multiplies an input force
efficiency the ratio of a machine's output work to the input work

Review
1. Is writing a letter to a friend considered work? Explain why or why not.

Take Notes
I. Machines change the way force is applied. (p. 145)
2. List two ways that a machine can make work easier.

A. Changing Size and Distance (p. 146)
3. Fill in the blanks below to show how machines change force and the distance over which the force is applied.

Machines change force
→ can apply less force over ________ distance
→ can apply greater force over ________ distance

B. Changing Direction (p. 147)
4. Label the input and output forces on the pulley and rope shown on the right. How does the pulley change the input force?

below level

Motion and Forces, Chapter 5, Reading Study Guide A 289

R UNIT RESOURCE BOOK, pp. 289–290

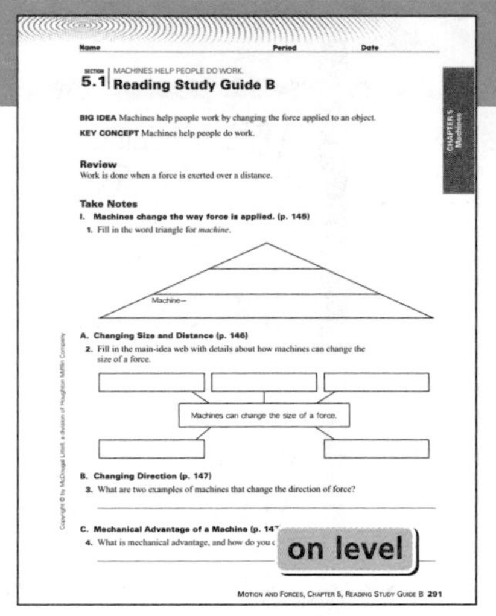
Name Period Date

SECTION 5.1 | MACHINES HELP PEOPLE DO WORK.
Reading Study Guide B

BIG IDEA Machines help people work by changing the force applied to an object.
KEY CONCEPT Machines help people do work.

Review
Work is done when a force is exerted over a distance.

Take Notes
I. Machines change the way force is applied. (p. 145)
1. Fill in the word triangle for *machine*.

Machine—

A. Changing Size and Distance (p. 146)
2. Fill in the main-idea web with details about how machines can change the size of a force.

Machines can change the size of a force.

B. Changing Direction (p. 147)
3. What are two examples of machines that change the direction of force?

C. Mechanical Advantage of a Machine (p. 14...
4. What is mechanical advantage, and how do you ...

on level

Motion and Forces, Chapter 5, Reading Study Guide B 291

R pp. 291–292

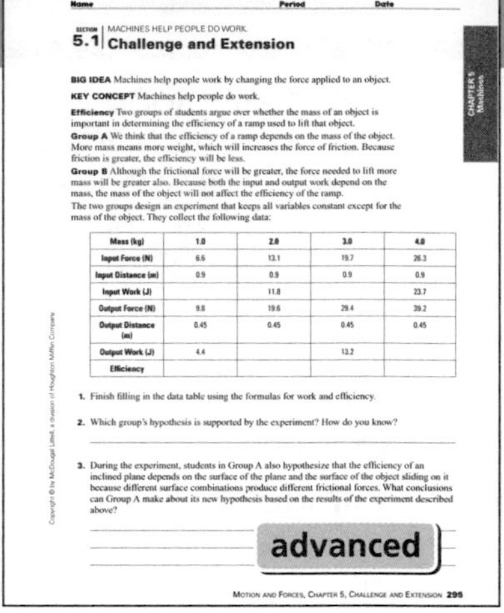
Name Period Date

SECTION 5.1 | MACHINES HELP PEOPLE DO WORK.
Challenge and Extension

BIG IDEA Machines help people work by changing the force applied to an object.
KEY CONCEPT Machines help people do work.

Efficiency Two groups of students argue over whether the mass of an object is important in determining the efficiency of a ramp used to lift that object.
Group A We think that the efficiency of a ramp depends on the mass of the object. More mass means more weight, which will increase the force of friction. Because friction is greater, the efficiency will be less.
Group B Although the frictional force will be greater, the force needed to lift more mass will be greater also. Because both the input and output work depend on the mass, the mass of the object will not affect the efficiency of the ramp.
The two groups design an experiment that keeps all variables constant except for the mass of the object. They collect the following data:

Mass (kg)	1.0	2.0	3.0	4.0
Input Force (N)	6.6	13.1	19.7	26.2
Input Distance (m)	0.9	0.9	0.9	0.9
Input Work (J)		11.8		23.7
Output Force (N)	9.8	19.6	29.4	39.2
Output Distance (m)	0.45	0.45	0.45	0.45
Output Work (J)	4.4		13.2	
Efficiency				

1. Finish filling in the data table using the formulas for work and efficiency.
2. Which group's hypothesis is supported by the experiment? How do you know?
3. During the experiment, students in Group A also hypothesize that the efficiency of an inclined plane depends on the surface of the plane and the surface of the object sliding on it because different surface combinations produce different frictional forces. What conclusions can Group A make about its new hypothesis based on the results of the experiment described above?

advanced

Motion and Forces, Chapter 5, Challenge and Extension 295

R p. 295

CHAPTER TEST

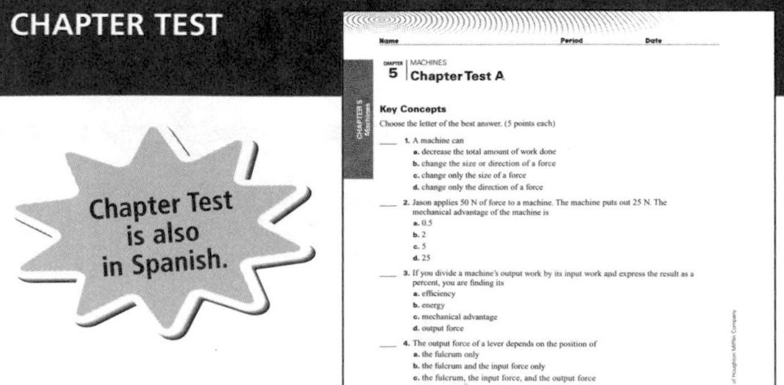
Chapter Test is also in Spanish.

Name Period Date

CHAPTER 5 | MACHINES
Chapter Test A

Key Concepts
Choose the letter of the best answer. (5 points each)

___ 1. A machine can
a. decrease the total amount of work done
b. change the size or direction of a force
c. change only the size of a force
d. change only the direction of a force

___ 2. Jason applies 50 N of force to a machine. The machine puts out 25 N. The mechanical advantage of the machine is
a. 0.5
b. 2
c. 5
d. 25

___ 3. If you divide a machine's output work by its input work and express the result as a percent, you are finding its
a. efficiency
b. energy
c. mechanical advantage
d. output force

___ 4. The output force of a lever depends on the position of
a. the fulcrum only
b. the fulcrum and the input force only
c. the fulcrum, the input force, and the output force
d. the input force only

___ 5. Suki turns the wheel of a wheel and axle. The force applied by the axle
a. is less than the force Suki exerts
b. acts over less distance than the force Suki exerts
c. decreases the total work Suki does
d. decreases as the distance from the wheel increases

below level

84 Motion and Forces, Chapter 5, Chapter Test A

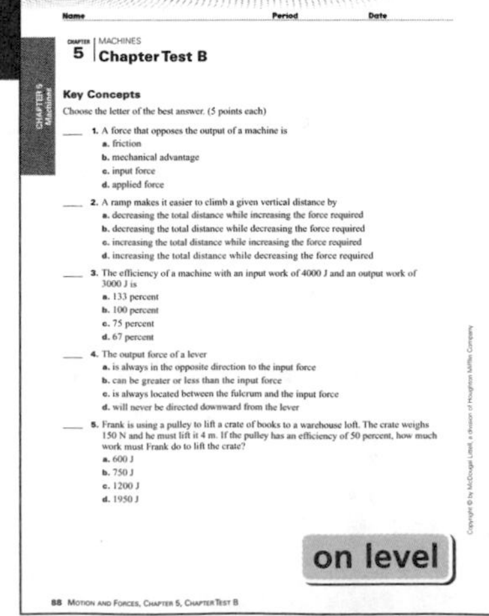
Name Period Date

CHAPTER 5 | MACHINES
Chapter Test B

Key Concepts
Choose the letter of the best answer. (5 points each)

___ 1. A force that opposes the output of a machine is
a. friction
b. mechanical advantage
c. input force
d. applied force

___ 2. A ramp makes it easier to climb a given vertical distance by
a. decreasing the total distance while increasing the force required
b. decreasing the total distance while decreasing the force required
c. increasing the total distance while increasing the force required
d. increasing the total distance while decreasing the force required

___ 3. The efficiency of a machine with an input work of 4000 J and an output work of 3000 J is
a. 133 percent
b. 100 percent
c. 75 percent
d. 67 percent

___ 4. The output force of a lever
a. is always in the opposite direction to the input force
b. can be greater or less than the input force
c. is always located between the fulcrum and the input force
d. will never be directed downward from the lever

___ 5. Frank is using a pulley to lift a crate of books to a warehouse loft. The crate weighs 150 N and he must lift it 4 m. If the pulley has an efficiency of 50 percent, how much work must Frank do to lift the crate?
a. 600 J
b. 750 J
c. 1200 J
d. 1950 J

on level

88 Motion and Forces, Chapter 5, Chapter Test B

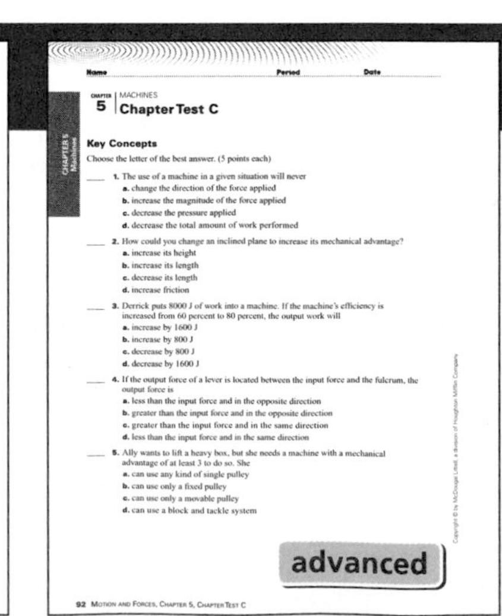
Name Period Date

CHAPTER 5 | MACHINES
Chapter Test C

Key Concepts
Choose the letter of the best answer. (5 points each)

___ 1. The use of a machine in a given situation will never
a. change the direction of the force applied
b. increase the magnitude of the force applied
c. decrease the pressure applied
d. decrease the total amount of work performed

___ 2. How could you change an inclined plane to increase its mechanical advantage?
a. increase its height
b. increase its length
c. decrease its length
d. increase friction

___ 3. Derrick puts 8000 J of work into a machine. If the machine's efficiency is increased from 60 percent to 80 percent, the output work will
a. increase by 1600 J
b. increase by 800 J
c. decrease by 800 J
d. decrease by 1600 J

___ 4. If the output force of a lever is located between the input force and the fulcrum, the output force is
a. less than the input force and in the opposite direction
b. greater than the input force and in the opposite direction
c. greater than the input force and in the same direction
d. less than the input force and in the same direction

___ 5. Ally wants to lift a heavy box, but she needs a machine with a mechanical advantage of at least 3 to do so. She
a. can use any kind of single pulley
b. can use only a fixed pulley
c. can use only a movable pulley
d. can use a block and tackle system

advanced

92 Motion and Forces, Chapter 5, Chapter Test C

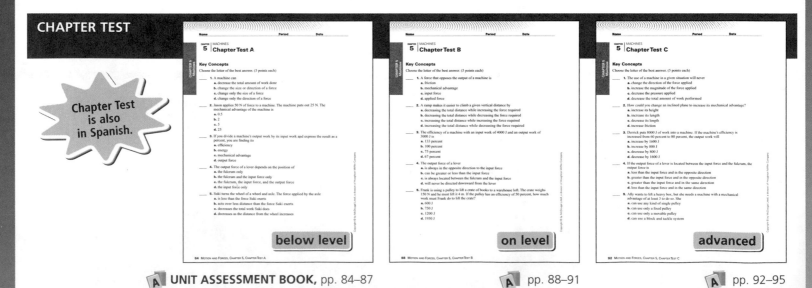

A UNIT ASSESSMENT BOOK, pp. 84–87 | A pp. 88–91 | A pp. 92–95

TECHNOLOGY

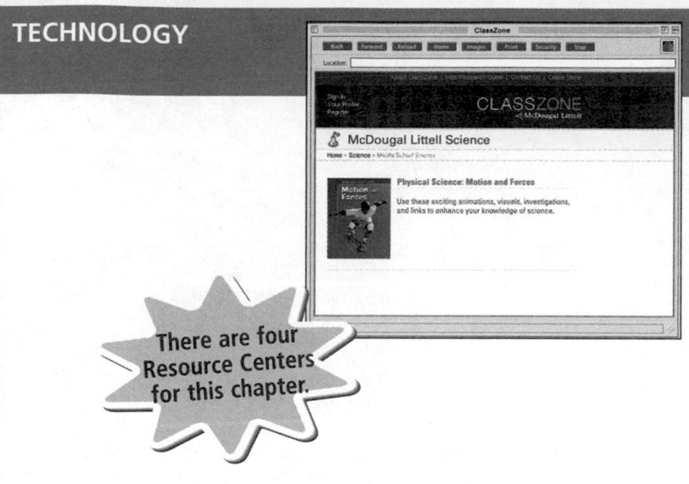

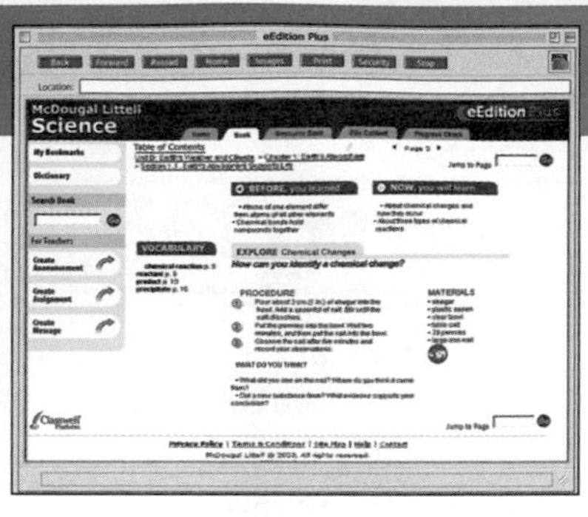

CLASSZONE.COM

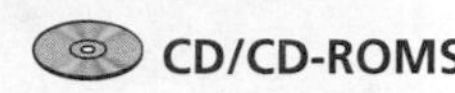
CD/CD-ROMS

CLASSZONE.COM

VISUAL CONTENT

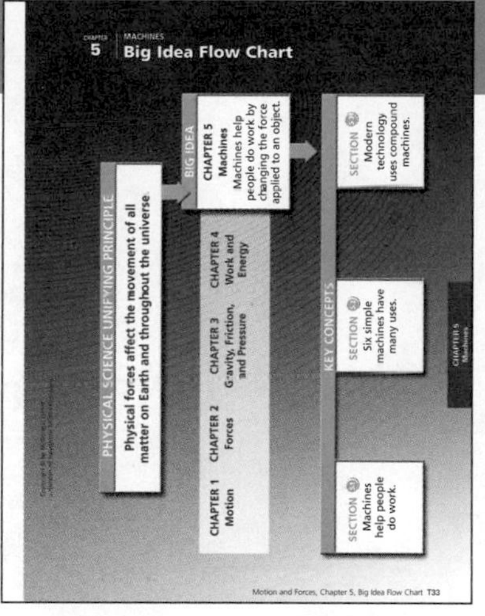

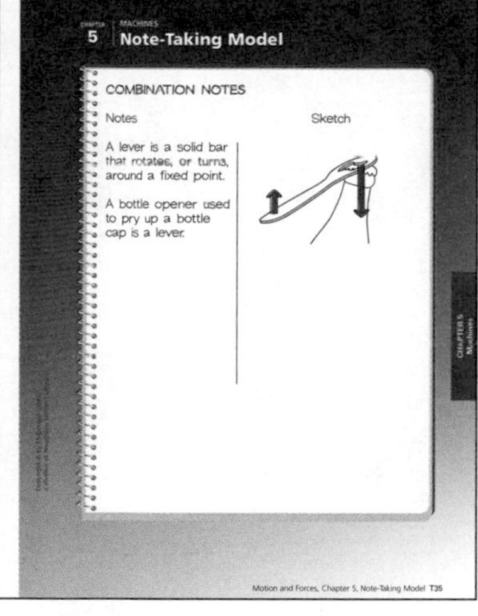

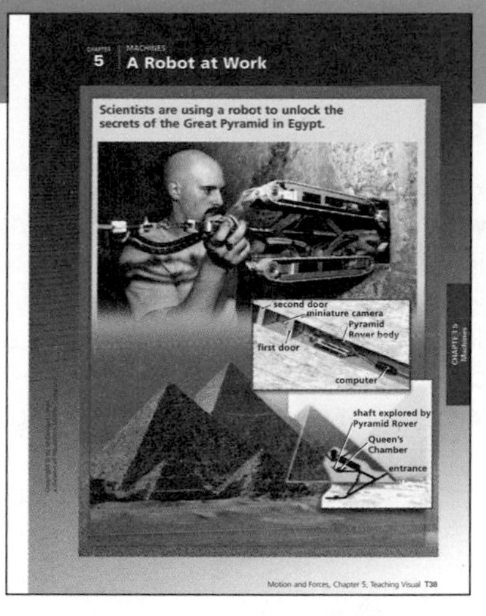

UNIT TRANSPARENCY BOOK, p. T33

p. T35

p. T38

MORE SUPPORT

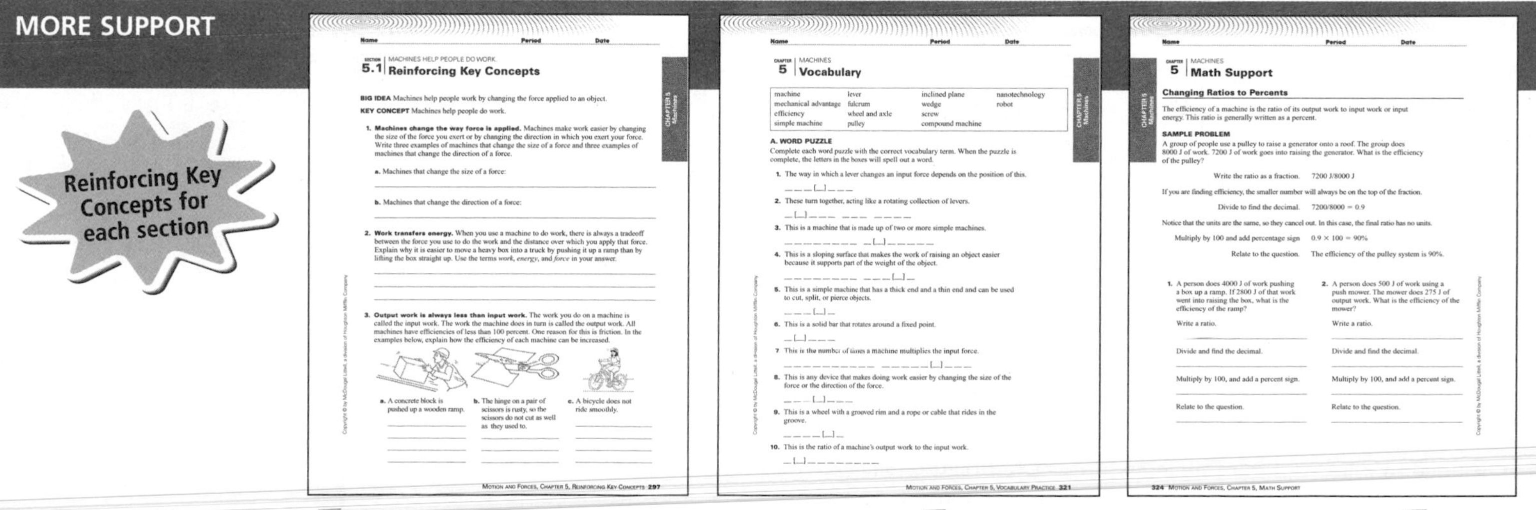

SECTION 5.1 MACHINES HELP PEOPLE DO WORK
Reinforcing Key Concepts

CHAPTER 5 MACHINES
Vocabulary

CHAPTER 5 MACHINES
Math Support
Changing Ratios to Percents

UNIT RESOURCE BOOK, p. 297

pp. 321–322

p. 324

CHAPTER 5 Machines

INTRODUCE

the BIG idea

Have students look at the photograph of the sculpture and discuss how the question in the box links to the Big Idea:

- What work is this machine doing?
- Can you find places where something changes the direction of a force?
- What types of machines do you see in the sculpture?

National Science Education Standards

Process

A.1–8 Identify questions that can be answered through scientific investigations; design and conduct an investigation; use tools to gather and interpret data; use evidence to describe, predict, explain, model; think critically to make relationships between evidence and explanation; recognize different explanations and predictions; communicate scientific procedures and explanations; use mathematics.

A.9.a–g Understand scientific inquiry by using different investigations, methods, mathematics, technology, explanations based on logic, evidence, and skepticism. Data often results in new investigations.

E.2–5 Design, implement, and evaluate a solution or product; communicate technological design.

E.6.c–e Science drives technology and technology drives science; perfectly designed solutions don't exist, all technology has tradeoffs; all designs have limits.

F.5.a–c, F.5.e Science influences society; societal challenges inspire scientific research; technology influences society; scientists work in different settings.

G.1.b Science requires different abilities.

CHAPTER

Machines

the BIG idea

Machines help people do work by changing the force applied to an object.

Key Concepts

SECTION

Machines help people do work.
Learn about machines and how they are used to do work.

SECTION

Six simple machines have many uses.
Learn about levers and inclined planes and the other simple machines that are related to them.

SECTION

Modern technology uses compound machines.
Learn how scientists are using nanotechnology and robots to create new ways for machines to do work.

Internet Preview

CLASSZONE.COM

Chapter 5 online resources: Content Review, Simulation, four Resource Centers, Math Tutorial, Test Practice

Balls move through this sculpture. What do you think keeps the balls in motion?

INTERNET PREVIEW

CLASSZONE.COM For student use with the following pages:

Review and Practice

- Content Review, pp. 144, 172
- Math Tutorial: Percents and Fractions, p. 153
- Test Practice, p. 175

Activities and Resources

- Internet Activity: Machines, p. 143
- Simulation: Mechanical Advantage, p. 161
- Resource Centers: Artificial Limbs, p. 163; Nanomachines, p. 167; Robots, p. 169

Simple Machines
Code: MDL008

EXPLORE the BIG idea

Changing Direction
Observe how a window blind works. Notice how you use a downward force to pull the blind up. Look around you for other examples.
Observe and Think
Why does changing the direction of a force make work easier?

Shut the Door!
Find a door that swings freely on its hinges. Stand on the side where you can push the door to close it. Open the door. Push the door closed several times, placing your hand closer to or farther from the hinge each time.
Observe and Think Which hand placement made it easiest to shut the door? Why do you think that is so?

Internet Activity: Machines
Go to ClassZone.com to learn more about the simple machines in everyday objects. Select an item and think about how it moves and does its job. Then test your knowledge of simple machines.
Observe and Think
What other objects contain simple machines?
CLASSZONE
McDougal Littell Science
NSTA
scilinks.org
SCILINKS
Simple Machines Code: MDL008
Chapter 5: Machines 143
C

TEACHING WITH TECHNOLOGY

Video Camera You might want to tape the movement of parts of simple and compound machines pictured in "Explore Machines" on p. 145. Clips could be shown at regular speed or in slow motion to help students examine how the machines work.

Digital Camera Have students photograph their examples of compound machines from the Chapter Investigation on pp. 170–171. They can classify the different machines according to the simple machines they contain.

EXPLORE the BIG idea

These inquiry-based activities are appropriate for use at home or as a supplement to classroom instruction.

Changing Direction

PURPOSE To show that machines can change the direction of force. Students observe that a machine can change not only the size but also the direction of a force.

TIP *10 min.* Be sure the blind is firmly attached so that it does not fall when force is applied. Instruct students in how to use the blind before beginning the activity.

Answer: It is easier for people to pull down rather than up.

REVISIT after p. 147.

Shut the Door!

PURPOSE To learn about the relationship between force and the length of a lever. Students will observe that the length of the lever arm affects the amount of effort required.

TIP *10 min.* Students also could use a book cover to do this activity.

Answer: It is easiest to close when you push the edge farthest from the hinge.

REVISIT after p. 161.

Internet Activity: Machines

PURPOSE To introduce students to different machines. Students will examine different tools and relate them to simple machines.

TIP *20 min.* Students might want to have sketches of each simple machine available when they examine the tools.

Sample answers: doorknobs, cars, elevators

REVISIT after p. 162.

PREPARE

CONCEPT REVIEW

Activate Prior Knowledge

- Have a student lift a stack of books straight up from a desk.
- Ask students to list all the energy transformations that take place when this work is done. ***chemical energy to kinetic energy in the arm; potential energy to kinetic energy of the books***

TAKING NOTES

Choose Your Own Strategy

Having tried different note-taking techniques, students can choose the strategies that best fit their individual learning styles.

Vocabulary Strategy

The word triangle helps tie concepts to concrete objects. Drawings in word triangles are especially useful because they present information visually. Tell students to make drawings that help them remember what the terms mean. Often, the drawings will not look like physical objects. They may wish to use symbols in their drawings, such as arrows, to represent motion and direction.

Vocabulary and Note-Taking Resources

- Vocabulary Practice, pp. 321–322
- Decoding Support, p. 323

- Daily Vocabulary Scaffolding, p. T34
- Note-Taking Model, p. T35

- Word Triangle, B18–19
- Combination Notes, C36
- Main Idea Web, C38–39
- Supporting Main Ideas, C42
- Outline, C43
- Daily Vocabulary Scaffolding, H1–8

CHAPTER 5

Getting Ready to Learn

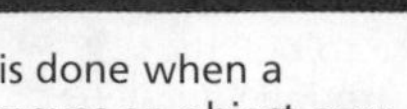

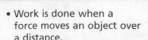

CONCEPT REVIEW

- Work is done when a force moves an object over a distance.
- Energy can be converted from one form to another.
- Energy is transferred when work is done.

VOCABULARY REVIEW

work p. 115
mechanical energy p. 125
power p. 130

See Glossary for definitions.

energy, technology

CONTENT REVIEW
CLASSZONE.COM
Review concepts and vocabulary.

TAKING NOTES

CHOOSE YOUR OWN STRATEGY

Take notes using one or more of the strategies from earlier chapters—**outline, combination notes, supporting main ideas,** and **main idea web.** Feel free to mix and match the strategies, or use an entirely different note-taking strategy.

VOCABULARY STRATEGY

Draw a **word triangle** diagram for each new vocabulary term. On the bottom line, write and define the term. Above that, write a sentence that uses the term correctly. At the top, draw a small picture to show what the term looks like.

SCIENCE NOTEBOOK

Outline
I. Main idea
A. Supporting idea
1. Detail
2. Detail
B. Supporting idea

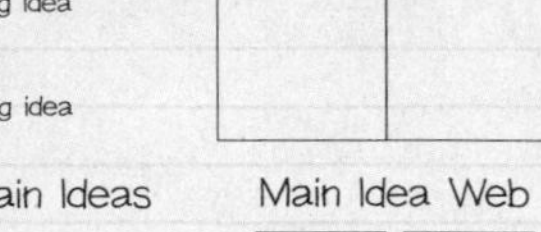

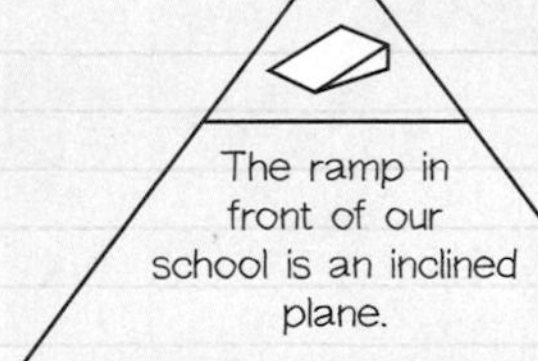

See the Note-Taking Handbook on pages R45–R51.

CHECK READINESS

Administer the Diagnostic Test to determine students' readiness for new science content and their mastery of requisite math skills.

 Diagnostic Test, pp. 79–80

Technology Resources

Students needing content and math skills should visit **ClassZone.com**.

- CONTENT REVIEW
- MATH TUTORIAL

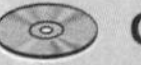 CONTENT REVIEW CD-ROM

KEY CONCEPT

Machines help people do work.

BEFORE, you learned
- Work is done when a force is exerted over a distance
- Some work can be converted to heat or sound energy

NOW, you will learn
- How machines help you do work
- How to calculate a machine's efficiency

VOCABULARY
machine p. 145
mechanical advantage p. 147
efficiency p. 150

EXPLORE Machines

How do machines help you work?

PROCEDURE
1. Look at one of the machines closely. Carefully operate the machine and notice how each part moves.
2. Sketch a diagram of the machine. Try to show all of the working parts. Add arrows and labels to show the direction of motion for each part.

MATERIALS
various small machines

WHAT DO YOU THINK?
- What is the function of the machine?
- How many moving parts does it have?
- How do the parts work together?
- How does this machine make work easier?

Machines change the way force is applied.

For thousands of years, humans have been improving their lives with technology. Technology is the use of knowledge to create products or tools that make life easier. The simplest machine is an example of technology.

A **machine** is any device that helps people do work. A machine does not decrease the amount of work that is done. Instead, a machine changes the way in which work is done. Recall that work is the use of force to move an object. If, for example, you have to lift a heavy box, you can use a ramp to make the work easier. Moving the box up a ramp—which is a machine—helps you do the work by reducing the force you need to lift the box.

VOCABULARY
Make a word triangle diagram in your notebook for *machine*.

RESOURCES FOR DIFFERENTIATED INSTRUCTION

Below Level
UNIT RESOURCE BOOK
- Reading Study Guide A, pp. 289–290
- Decoding Support, p. 323

AUDIO CDS

Advanced
UNIT RESOURCE BOOK
Challenge and Extension, p. 295

English Learners
UNIT RESOURCE BOOK
Spanish Reading Study Guide, pp. 293–294

AUDIO CDS
- Audio Readings in Spanish
- Audio Readings (English)

5.1 FOCUS

Set Learning Goals
Students will
- Explain how machines help people do work.
- Calculate a machine's efficiency.
- Analyze data from an experiment investigating the efficiency of a ramp.

3-Minute Warm-Up
Display Transparency 36 or copy this exercise on the board:

Write a short paragraph describing a situation in which you are doing work on an object. What type of energy does the object gain? How do you know?

Examples should include some use of force to move an object, such as kicking a ball or lifting a box. If the object's velocity increases, the object gains kinetic energy. If the object's height increases, the object gains potential energy.

T 3-Minute Warm-Up, p. T36

5.1 MOTIVATE

EXPLORE Machines
PURPOSE To learn how machines make work easier

TIP *15 min.* The machines could include a nutcracker, an eggbeater, a bottle opener, a screwdriver, pliers, and scissors.

WHAT DO YOU THINK? *Sample answer: Pliers grip something tightly. They have two moving parts screwed together that come together as the handles are squeezed. Work is easier because the pliers apply lots of pressure to a small area more easily than someone could without them.*

Teaching with Technology
Videotape some of the machines from "Explore Machines" while they are in use. Show the tape in class—first at regular speed and then in slow motion—so students can see the different parts involved.

 INSTRUCT

Teach from Visuals

To help students interpret the visual of the boy raking leaves, ask:

- What do the arrows tell you about the size of the force the boy's hand applies to the rake and the size of the force the rake applies to the leaves? *The hand applies more force to the rake than the rake applies to the leaves.*
- If the boy pulls the rake until it is straight up and down, how does the distance the bottom of the rake moves compare with the distance his lower hand moves? *The rake bottom moves a larger distance than the hand.*

Ongoing Assessment

CHECK YOUR READING *Answer: The rake exerts an output force on the leaves.*

If machines do not reduce the amount of work required, how do they help people do work? Machines make work easier by changing

- the size of the force needed to do the work and the distance over which the force is applied
- the direction in which the force is exerted

Machines can be powered by different types of energy. Electronic machines, such as computers, use electrical energy. Mechanical machines, such as a rake, use mechanical energy. Often this mechanical energy is supplied by the person who is using the machine.

Changing Size and Distance

Some machines help you do work by changing the size of the force needed. Have you ever tried to open a door by turning the doorknob's shaft instead of the handle? This is not easy to do. It takes less force to turn the handle of the doorknob than it does to turn the shaft. Turning the handle makes opening the door easier, even though you must turn it through a greater distance.

A rake is a machine that changes a large force over a short distance to a smaller force over a larger distance.

If a machine—such as a doorknob attached to a shaft—allows you to exert less force, you must apply that force over a greater distance. The total amount of work remains the same whether it is done with a machine or not. You can think of this in terms of the formula for calculating work—work is force times distance. Because a machine does not decrease the amount of work to be done, less force must mean greater distance. ? A

A doorknob allows you to apply a smaller force over a greater distance. Some machines allow you to apply a greater input force over a shorter distance. Look at the boy using a rake, which is a machine. The boy moves his hands a short distance to move the end of the rake a large distance, allowing him to rake up more leaves.

Input force is the force exerted on a machine. Output force is the force that a machine exerts on an object. The boy in the photograph is exerting an input force on the rake. As a result, the rake exerts an output force on the leaves. The work the boy puts into the rake is the same as the work he gets out of the rake. However, the force he applies is greater than the force the rake can apply to the leaves. The output force is less than the input force, but it acts over a longer distance.

CHECK YOUR READING How can a rake help you do work? Use the word *force* in your answer.

DIFFERENTIATE INSTRUCTION

More Reading Support

A What is work? *force times distance*

English Learners English learners may be unfamiliar with the use of dashes in writing. A dash can be used to show a sudden break in a sentence, such as a change in thought or an interjection. Use the following example from the middle of this page: You can think of this in terms of the formula for calculating work—work is force time distance.

Changing Direction

Machines also can help you work by changing the direction of a force. Think of raising a flag on a flagpole. You pull down on the rope, and the flag moves up. The rope system is a machine that changes the direction in which you exert your force. The rope system does not change the size of the force, however. The force pulling the flag upward is equal to your downward pull.

A shovel is a machine that can help you dig a hole. Once you have the shovel in the ground, you push down on the handle to lift the dirt up. You can use some of the weight of your body as part of your input force. That would not be possible if you were lifting the dirt by using only your hands. A shovel also changes the size of the force you apply, so you need less force to lift the dirt.

APPLY How does the rope system help the man raise the flag?

Mechanical Advantage of a Machine

When machines help you work, there is an advantage—or benefit—to using them. The number of times a machine multiplies the input force is called the machine's **mechanical advantage** (MA). To find a machine's mechanical advantage, divide the output force by the input force.

$$\text{Mechanical Advantage} = \frac{\text{Output Force}}{\text{Input Force}}$$

For machines that allow you to apply less force over a greater distance—such as a doorknob—the output force is greater than the input force. Therefore, the mechanical advantage of this type of machine is greater than 1. For example, if the input force is 10 newtons and the output force is 40 newtons, the mechanical advantage is 40 N divided by 10 N, or 4.

For machines that allow you to apply greater force over a shorter distance—such as a rake—the output force is less than the input force. In this case, the mechanical advantage is less than 1. If the input force is 10 newtons and the output force is 5 newtons, the mechanical advantage is 0.5. However, such a machine allows you to move an object a greater distance.

Sometimes changing the direction of the force is more useful than decreasing the force or the distance. For machines that change only the direction of a force—such as the rope system on a flagpole—the input force and output force are the same. Therefore, the mechanical advantage of the machine is 1.

EXPLORE the BIG idea

Revisit "Changing Direction" on p. 143. Have students explain their results.

Teach Difficult Concepts

Provide some simple mechanical-advantage problems to help students learn how to solve them. Review how to rearrange the variables in an equation to isolate the one you are trying to find. Use these sample problems:

- The output force of a machine is 600 N, and the input force is 200 N. What is the mechanical advantage of the machine?
  $MA = \frac{F_{out}}{F_{in}} = \frac{600\ N}{200\ N} = 3$
- A machine has an input force of 150 N and a mechanical advantage of 0.5. What is the output force?
  $F_{out} = MA \cdot F_{in} = 0.5 \cdot 150\ N = 75\ N$
- The output force of a machine is 135 N, and the mechanical advantage is 2.5. What is the input force?
  $F_{in} = \frac{F_{out}}{MA} = \frac{135\ N}{2.5} = 54\ N$

Teach from Visuals

Have students look at the visual on p. 146. Ask: Is the mechanical advantage of the rake greater than 1 or less than 1? How do you know? *The mechanical advantage is less than 1 because the input force is greater than the output force.*

Ongoing Assessment

Explain how machines make work easier.

Ask: Is it easier to lift a heavy box or to push it up a ramp? Why? *easier to push it up a ramp; the ramp changes the size of the force needed to move the box.*

PHOTO CAPTION Answer: The rope system changes the direction of the force that he must exert. The input force and output force are the same, however.

DIFFERENTIATE INSTRUCTION

More Reading Support

B How can using a rope system to open or close curtains make the work easier? *It changes the direction of the applied force.*

Below Level Have students describe what is meant by the word *advantage*. Ask questions that lead to concrete examples, such as, What advantage does a track team with faster runners have over another team? Responses should include the concept that if you have a greater advantage, the results are more favorable. Compare the responses with the concept that increased mechanical advantage results in a more favorable output force compared with input force.

Address Misconceptions

IDENTIFY Ask: How does a machine affect the amount of work done, and how does this affect energy? If students say that machines reduce the amount of work and less energy is expended, they may hold the misconception that machines reduce the amount of work and energy needed.

CORRECT Place a pencil under a meter stick at the 20 cm mark. Place or tape a 100 g mass at the 0 cm mark. Place a 200 g mass (weighing 1.9 N) on the meter stick so that it just pushes up the 100 g mass. Remove the 200 g mass so the meter stick returns to its original position, and measure the distance the 200 g mass moved through (the height). Repeat with a 50 g (0.49 N) mass. Using the weights as the force, have students calculate and compare the work done in both cases.

REASSESS How did the forces compare in the two examples? How did the work compare? *The second force was less, the work was the same.*

Technology Resources

Visit **ClassZone.com** for background on common student misconceptions.

 **MISCONCEPTION DATABASE**

Teach Difficult Concepts

Refer students to the formula for gravitational potential energy, $GPE = mgh$, where g, the gravitational acceleration, is 9.8 m/s^2. Ask the following questions:

- A girl with a mass of 50 kg starts at the bottom of a cliff that is 30 m high. After she has climbed to the top, what is her gravitational potential energy? GPE = mgh = *50 kg · 9.8 m/s^2 · 30 m = 14,700 J*
- Does her gravitational potential energy at the top depend on her path or the way she climbed up the cliff? *no*
- Where does this potential energy come from? *the work the girl does climbing*
- How much work must the girl do to get to the top of the cliff? W = F · d = ma · d = *50 kg · 9.8 m/s^2 · 30 m = 14,700 J*

NOTE-TAKING STRATEGY
Remember to organize your notes in a chart or web as you read.

Work transfers energy.

Machines transfer energy to objects on which they do work. Every time you open a door, the doorknob is transferring mechanical energy to the shaft. A machine that lifts an object gives it potential energy. A machine that causes an object to start moving, such as a baseball bat hitting a ball, gives the object kinetic energy.

Energy

When you lift an object, you transfer energy to it in the form of gravitational potential energy—that is, potential energy caused by gravity. The higher you lift an object, the more work you must do and the more energy you give to the object. This is also true if a machine lifts an object. The gravitational potential energy of an object depends on its height above Earth's surface, and it equals the work required to lift the object to that height.

Recall that gravitational potential energy is the product of an object's mass, gravitational acceleration, and height *(GPE = mgh)*. In the diagram on page 149, the climber wants to reach the top of the hill. The higher she climbs, the greater her potential energy. This energy comes from the work the climber does. The potential energy she gains equals the amount of work she does.

Work

As you have seen, when you use a machine to do work, there is always an exchange, or tradeoff, between the force you use to do the work and the distance over which you apply that force. You apply less force over a longer distance or greater force over a shorter distance.

To reach the top of the hill, the climber must do work. Because she needs to increase her potential energy by a certain amount, she must do the same amount of work to reach the top of the hill whether she climbs a steep slope or a gentle slope.

The sloping surface of the hill acts like a ramp, which is a simple machine called an inclined plane. You know that machines make work easier by changing the size or direction of a force. How does this machine make the climber's work easier?

As the climber goes up the hill, she is doing work against gravity.

1. One side of the hill is a very steep slope—almost straight up. If the climber takes the steep slope, she climbs a shorter distance, but she must use more force.
2. Another side of the hill is a long, gentle slope. Here the climber travels a greater distance but uses much less effort.

DIFFERENTIATE INSTRUCTION

More Reading Support

C How can a machine increase the amount of energy an object has? *by doing work on the object*

Advanced Have students design a physical model that shows how, for a given amount of work, there must be a trade-off between force and distance. Have students present their models to the class. *Example: Students can use blocks to represent a unit of work. The student can arrange a set number of blocks into different-shaped rectangles, with one side representing force and the other side representing distance.* This is also a good activity for students who have physical disabilities, particularly vision impairments.

R Challenge and Extension, p. 295

If the climber uses the steep slope, she must lift almost her entire weight. The inclined plane allows her to exert her input force over a longer distance; therefore, she can use just enough force to overcome the net force pulling her down the inclined plane. This force is less than her weight. In many cases, it is easier for people to use less force over a longer distance than it is for them to use more force over a shorter distance.

? D

Energy and Work

To reach the top of the hill, the climber must do at least as much work as the amount of potential energy she needs to gain.

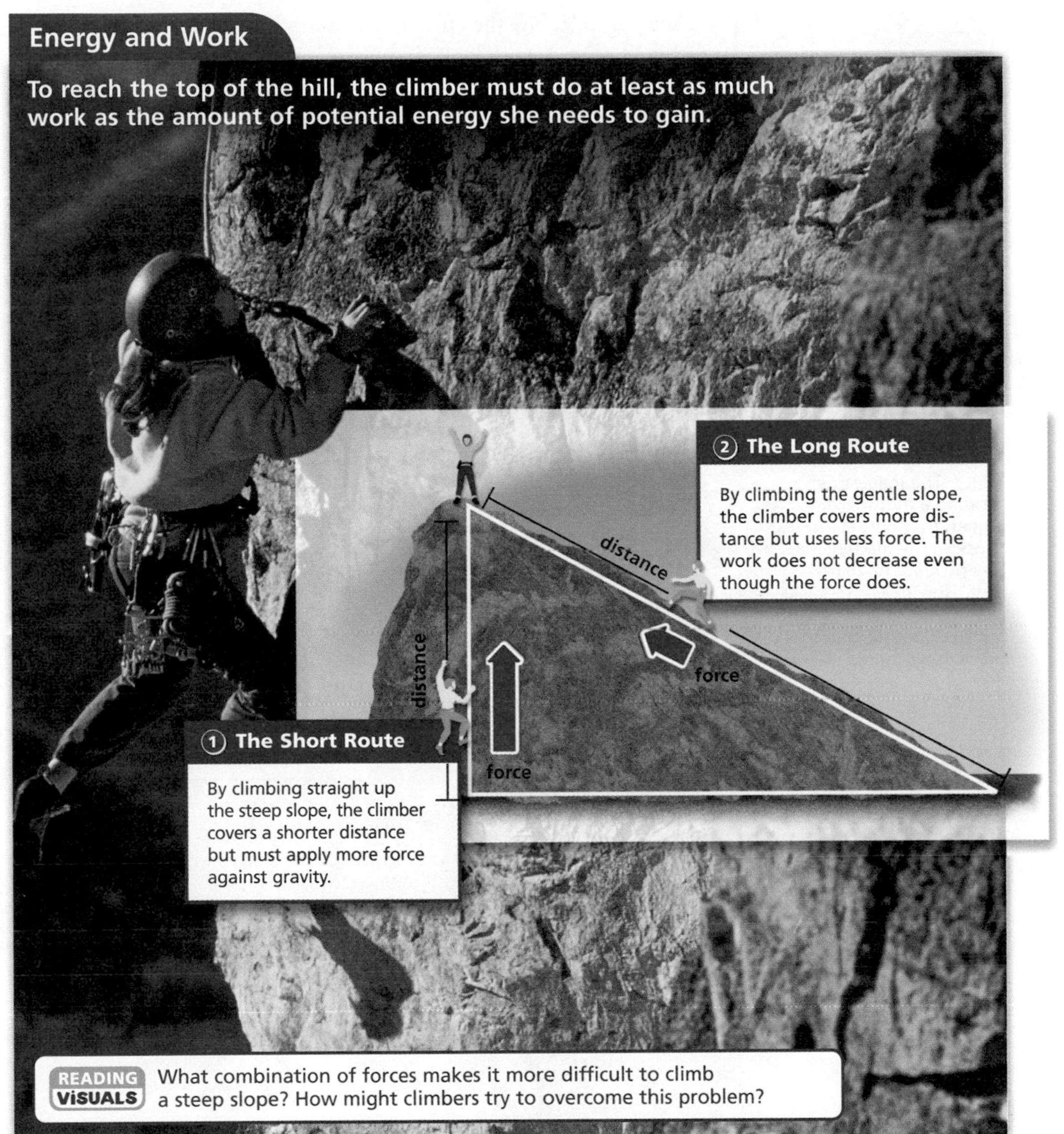

① The Short Route
By climbing straight up the steep slope, the climber covers a shorter distance but must apply more force against gravity.
② The Long Route
By climbing the gentle slope, the climber covers more distance but uses less force. The work does not decrease even though the force does.
distance
force
distance
force

READING VISUALS What combination of forces makes it more difficult to climb a steep slope? How might climbers try to overcome this problem?

Teach from Visuals

To help students interpret the visual of the rock climber, ask:

- Will more work be done in climbing the face of the steep slope or in going up the gentle slope? *The work will be the same.*
- Will the distance the person has to walk be greater or less if the gentle slope is used? *The distance will be greater.*
- Will the force applied be greater or less if the gentle slope is used? *The force will be less.*

Develop Number Sense

Remind students that units that look different are not necessarily unrelated. Show them that gravitational potential energy (discussed on p. 148) equals the work done in the visual.

$F = m \cdot a$ and $W = F \cdot d$, so $W = m \cdot a \cdot d$ (mass times acceleration times distance)

Gravitational potential energy equals *mgh* (mass times gravitational acceleration times height [distance]). So when you find gravitational potential energy, you also have found the amount of work it takes to climb to the given height.

Ongoing Assessment

READING VISUALS *Answer: Gravity makes it more difficult to climb a steep slope. Climbers must lift almost their entire weight and use enough force to overcome the net force pulling them down the steep slope. They could overcome the problem by using a harness and other climbing equipment.*

DIFFERENTIATE INSTRUCTION

More Reading Support

D What happens when someone is climbing straight up a steep slope? *They cover a shorter distance, but use more force*

Below Level Ask students to quickly answer the question, Which would be easier to climb: a set of steps or a telephone pole of the same height? *The steps would be easier to climb.* Have them relate their conclusions to the amount of work done. *The amount of work done in climbing the stairs is the same as the amount required to climb a pole, though less force is required.*

Address Misconceptions

IDENTIFY Ask: What is the maximum efficiency of a real machine? If students answer "100 percent," they hold the misconception that machines can be 100 percent efficient.

CORRECT Have students conduct the Investigation on p. 151 while using materials that will minimize friction. They could cover the ramp with waxed paper or oil. No matter how little friction is present, the ramp will not be 100 percent efficient.

REASSESS Ask: Electric motors are quite efficient. How do you know that they are not 100 percent efficient? *Machines containing electric motors become warm, showing a transformation of input energy to heat.*

Technology Resources

Visit **ClassZone.com** for background on common student misconceptions.

MISCONCEPTION DATABASE

Develop Number Sense

Students might have a difficult time remembering whether to divide output work by input work or vice versa to calculate efficiency.

- Remind students that some friction is present in all real machines.
- Because friction is present, efficiency is always less than 100 percent. Since output work is always less than input work, output work must be divided by input work in order for efficiency to be less than 100%.
- If students still have difficulty after they have heard this reasoning, tell them that they can find efficiency by remembering the simple mnemonic **DOBI**, which stands for **D**ivide **O**utput work **B**y **I**nput work.

Ongoing Assessment

Answer: A machine's efficiency is the ratio of its output work to the input work.

PHOTO CAPTION Answer: 30 percent

Output work is always less than input work.

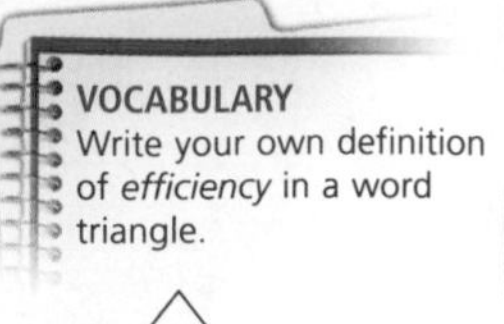

VOCABULARY
Write your own definition of *efficiency* in a word triangle.

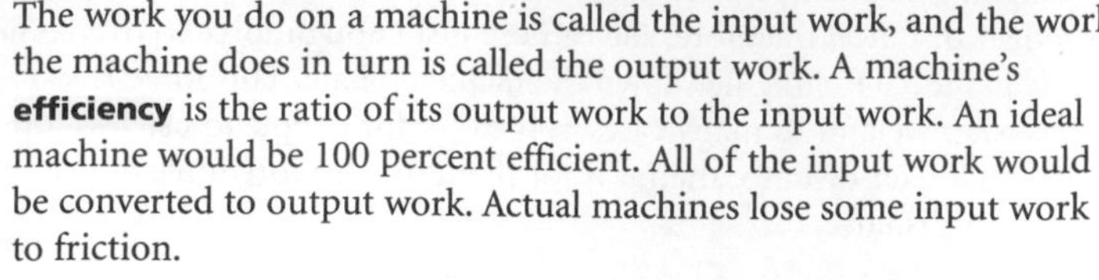

E

The work you do on a machine is called the input work, and the work the machine does in turn is called the output work. A machine's **efficiency** is the ratio of its output work to the input work. An ideal machine would be 100 percent efficient. All of the input work would be converted to output work. Actual machines lose some input work to friction.

You can calculate the efficiency of a machine by dividing the machine's output work by its input work and multiplying that number by 100.

$$\text{Efficiency (\%)} = \frac{\text{Output work}}{\text{Input work}} \cdot 100$$

Recall that work is measured in joules. Suppose you do 600 J of work in using a rope system to lift a box. The work done on the box is 540 J. You would calculate the efficiency of the rope system as follows:

F

$$\text{Efficiency} = \frac{540\ \text{J}}{600\ \text{J}} \cdot 100 = 90\%$$

CHECK YOUR READING What is a machine's efficiency? How does it affect the amount of work a machine can do?

APPLY The mail carrier is riding a motorized human transport machine. Suppose the machine has an efficiency of 70 percent. How much work is lost in overcoming friction on the sidewalk and in the motor?

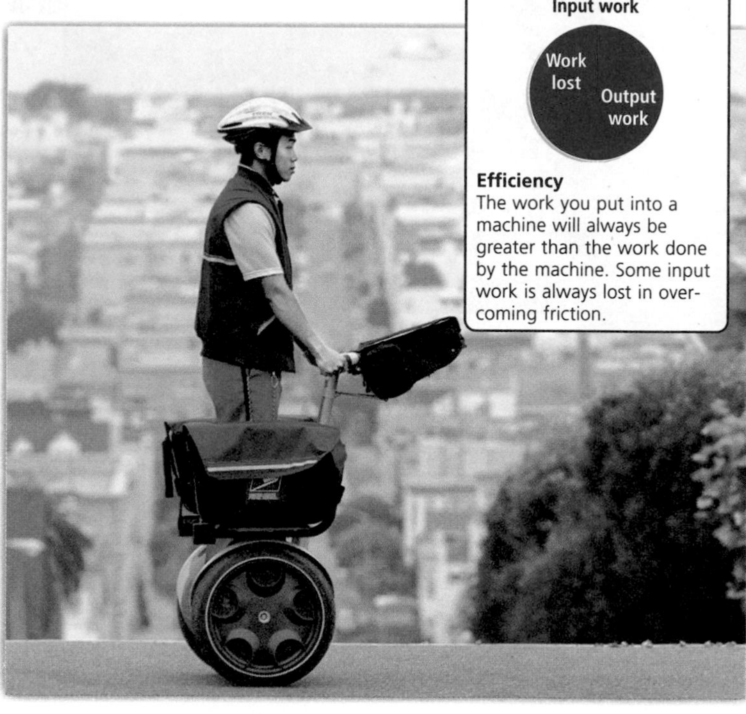

Efficiency
The work you put into a machine will always be greater than the work done by the machine. Some input work is always lost in overcoming friction.

DIFFERENTIATE INSTRUCTION

More Reading Support

E Which is greater—a machine's output work or its input work? *input*

F If a machine loses 10 percent of the input work to friction, what is its efficiency? *90 percent*

Efficiency and Energy

You know that work transfers energy and that machines make work easier. The more mechanical energy is lost in the transfer to other forms of energy, the less efficient the machine. Machines lose some energy in the form of heat due to friction. The more moving parts a machine has, the more energy it loses to friction because the parts rub together. Machines can lose energy to other processes as well.

G For example, a car engine has an efficiency of only about 25 percent. It loses much of the energy supplied by its fuel to heat from combustion. By comparison, a typical electric motor has more than an 80 percent efficiency. That means the motor converts more than 80 percent of the input energy into mechanical energy, or motion.

H Many appliances come with energy guides that can help a buyer compare the energy efficiency of different models. A washing machine with the highest energy rating may not always save the most energy, however, because users may have to run those machines more often.

INVESTIGATE Efficiency

What is the efficiency of a ramp?

PROCEDURE

1. Build a ramp as shown. Measure the vertical height of the ramp and the length of the ramp in centimeters. Convert these distances to meters and record.
2. Attach the block to the spring scale and measure the force in newtons needed to lift the block straight up. Record this force as the output force. Multiply the output force by the height of the ramp in meters to get the output work. Record the output work.
3. Use the spring scale to pull the block up the ramp with a constant force. Record the force measured on the spring scale as the input force. Multiply the input force by the length of the ramp in meters to get the input work. Record the input work.
4. Use the input work and output work from steps 2 and 3 to calculate the efficiency of the ramp. Record your results.

WHAT DO YOU THINK?

- How did your input work compare with your output work?
- What could you do to increase the efficiency of the ramp?

CHALLENGE Would adding sandpaper on the surface of the ramp increase or decrease the efficiency of the ramp? Why? Test your hypothesis.

SKILL FOCUS
Analyzing data

MATERIALS
- board
- books
- meter stick
- wooden block with eye hook
- spring scale

for Challenge:
- sandpaper

TIME
20 minutes

INVESTIGATE Efficiency

PURPOSE To determine the efficiency of a ramp

TIP *20 min.* If necessary, point out that the length of the ramp is not equal to the horizontal distance along the desktop.

WHAT DO YOU THINK? *Output work should be less than input work. One way to increase efficiency is to reduce the friction between the board and the block.*

CHALLENGE *Adding sandpaper would decrease efficiency because it would increase friction.*

Datasheet, Efficiency, p. 296

Technology Resources

Customize this student lab as needed or look for an alternative. Print rubrics to assess student lab reports.

Lab Generator CD-ROM

Ongoing Assessment

Calculate a machine's efficiency.

Ask: If someone does 500 joules of work on a pair of pliers and the pliers do 300 joules of work on a wire, what is the efficiency of the pliers?

Input work = 500 J

Output work = 300 J

$$E = \frac{300\ J}{500\ J} \cdot 100 = 60\%$$

DIFFERENTIATE INSTRUCTION

More Reading Support

G Which is more efficient—a car engine or an electric motor? *electric motor*

H What does an appliance's energy rating reflect? *the energy efficiency of the machine*

Inclusion For students who are visual and kinesthetic learners, supply mathematical models that they can use to determine the relationship between efficiency and the energy lost to friction. A ten-by-ten grid (or graph paper) can be a model. Students can shade in the efficiency percentage and find the amount of energy lost to friction in the unshaded part.

Ongoing Assessment

 *Answer: reduce friction by using oil or grease*

Reinforce the BIG idea

Have students relate the section to the Big Idea.

 Reinforcing Key Concepts, p. 297

5.1 ASSESS & RETEACH

Assess

 Section 5.1 Quiz, p. 81

Reteach

No real machine is 100 percent efficient. The law of conservation of energy states that, under normal circumstances, energy is conserved. Ask students to explain how both of these statements can be true. *Some energy that is put into the machine is transformed into other forms of energy, such as heat from friction. The total amount of energy in the system is conserved.*

Technology Resources

Have students visit **ClassZone.com** for reteaching of Key Concepts.

 CONTENT REVIEW

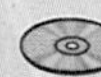 CONTENT REVIEW CD-ROM

Proper maintenance can help keep a bicycle running as efficiently as possible.

Increasing Efficiency

Because all machines lose input work to friction, one way to improve the efficiency of a machine is by reducing friction. Oil is used to reduce friction between the moving parts of car engines. The use of oil makes engines more efficient.

Another machine that loses input work is a bicycle. Bicycles lose energy to friction and to air resistance. Friction losses result from the meeting of the gears, from the action of the chain on the sprocket, and from the tires changing shape against the pavement. A bicycle with poorly greased parts or other signs of poor maintenance requires more force to move. For a mountain bike that has had little maintenance, as much as 15 percent of the total work may be lost to friction. A well-maintained Olympic track bike, on the other hand, might lose only 0.5 percent.

 What is a common way to increase a machine's efficiency?

5.1 Review

KEY CONCEPTS

1. In what ways can a machine change a force?
2. How is a machine's efficiency calculated?
3. Why is a machine's actual output work always less than its input work?

CRITICAL THINKING

4. **Apply** How would the input force needed to push a wheelchair up a ramp change if you increased the height of the ramp but not its length?
5. **Compare** What is the difference between mechanical advantage and efficiency?

CHALLENGE

6. **Apply** Draw and label a diagram to show how to pull down on a rope to raise a load of construction materials.

ANSWERS

1. *by changing the size of a force, changing its direction, or both*

2. *Divide the output work by the input work, and multiply by 100.*

3. *Some input work is always lost to friction.*

4. *More input force would be needed.*

5. *Mechanical advantage is the number of times a machine multiplies the input force. Efficiency is the ratio of a machine's output work to input work.*

6. *Students' diagrams should show that pulling down on a rope helps to raise the load by changing the direction of the force.*

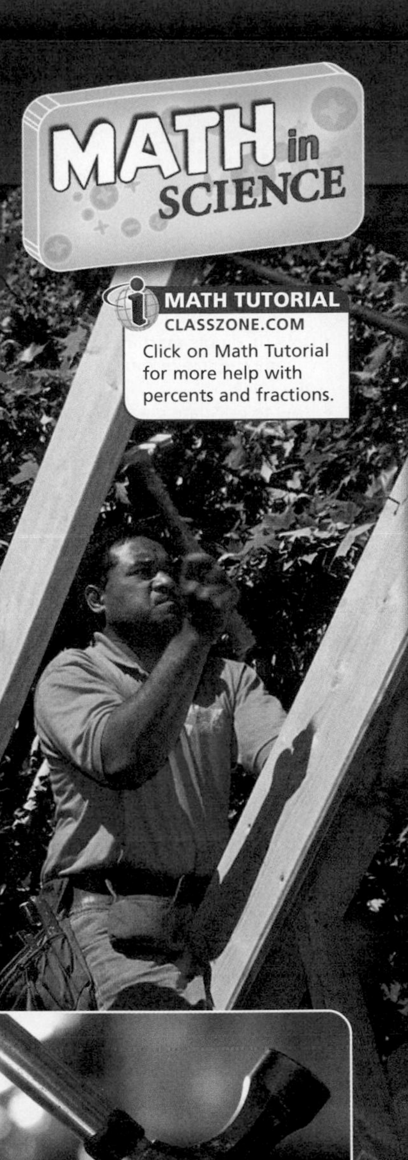

MATH TUTORIAL
CLASSZONE.COM
Click on Math Tutorial for more help with percents and fractions.

No machine, no matter how large or small, is 100 percent efficient. Some of the input energy is lost to sound, heat, or other products.

SKILL: CHANGING RATIOS TO PERCENTS

How Efficient Are Machines?

A hammer is used to pound in nails. It can also be used to pry nails out of wood. When used to pry nails, a hammer is a machine called a lever. Like all machines, the hammer is not 100 percent efficient.

Efficiency is the amount of work a machine does divided by the amount of work that is done on the machine. To calculate efficiency, you must first find the ratio of the machine's output work to the input work done on the machine. A ratio is the comparison of two numbers by means of division. You convert the ratio to a decimal by dividing. Then convert the decimal to a percent.

Example

A person is doing 1000 joules of work on a hammer to pry up a nail. The hammer does 925 joules of work on the nail to pull it out of the wood.

(1) Find the ratio of output work to input work.

$$\frac{\text{Output work}}{\text{Input work}} = \frac{925\ \text{J}}{1000\ \text{J}} = 0.925$$

(2) To convert the decimal to a percent, multiply 0.925 by 100 and add a percent sign.

$$0.925 \cdot 100 = 92.5\%$$

ANSWER The efficiency of the hammer is 92.5 percent. This means that the hammer loses 7.5 percent of the input work to friction and other products.

Answer the following questions.

1. A construction worker does 1000 J of work in pulling down on a rope to lift a weight tied to the other end. If the output work of the rope system is 550 J, what is the ratio of output work to input work? What is the efficiency of the rope system?
2. If a machine takes in 20,000 J and puts out 5000 J, what is its efficiency?
3. You do 6000 J of work to pull a sled up a ramp. After you reach the top, you discover that the sled had 3600 J of work done on it. What is the efficiency of the ramp?

CHALLENGE If you put 7000 J of work into a machine with an efficiency of 50 percent, how much work will you get out?

ANSWERS

1. $550\ J : 1000\ J;\ E = \frac{550\ J}{1000\ J} \cdot 100 = 55\%$

2. $E = \frac{5000\ J}{20,000\ J} \cdot 100 = 25\%$

3. $E = \frac{3600\ J}{6000\ J} \cdot 100 = 60\%$

CHALLENGE $W_{out} = E \cdot W_{in} = 0.5 \cdot 7000\ J = 3500\ J$

MATH IN SCIENCE

Math Skills Practice for Science

Set Learning Goal

To calculate the efficiency of a machine by changing ratios to percents

Present the Science

The efficiency of a machine is the ratio of its output work to the input work. Output work is defined as work done by the machine on something else. Input work is work done on the machine. Some input work is lost to undesired work, such as work done against friction. Such work is not considered output work.

Develop Number Sense

Students have to know how to write a ratio that compares output work to input work and be able to change this ratio to a percent.

- If students have difficulty identifying work as either output or input, they should remember that output work is always smaller than input work.
- Remind students that energy is conserved. If they find that the efficiency of a machine is 74 percent, for example, they should keep in mind that 26 percent of the energy input is lost to friction or other unwanted products, such as sound.

Close

Ask: What is the input work if the output work of a machine is 500 joules and the efficiency is 49 percent?

$$W_{in} = \frac{W_{out}}{E} = \frac{500\ J}{.49} = 1020\ J$$

- Math Support, p. 324
- Math Practice, p. 325

Technology Resources

Students can visit **ClassZone.com** for practice working with ratios and percents.

MATH TUTORIAL

5.2 FOCUS

Set Learning Goals

Students will

- Describe how six simple machines change the size or direction of a force.
- Calculate mechanical advantage.
- Infer through experimentation the mechanical advantage of a pulley system.

3-Minute Warm-Up

Display Transparency 36 or copy this exercise on the board:

Think about all the chores a landscaping crew must do to get a yard or park in shape in the spring. They must mow, weed, and trim. They might have to plant flowers and grass seed and to water the plants. List the machines needed for landscaping. Explain how each machine helps the workers do work. *Sample answer: A hoe helps workers by cutting into the ground and pulling apart compacted soil.*

 3-Minute Warm-Up, p. T36

5.2 MOTIVATE

EXPLORE Changing Forces

PURPOSE To change a force applied to an object

TIP *15 min.* A hexagonal pencil will stay in place better than a round pencil.

WHAT DO YOU THINK? *Moving the bottom pencil closer to the book requires someone to use less force to lift the book; moving the pencil farther from the book requires more force. It is easier to lift the book when the bottom pencil is closer to the book and harder when it is farther away.*

KEY CONCEPT

5.2 Six simple machines have many uses.

BEFORE, you learned

- Machines help you work by changing the size or direction of a force
- The number of times a machine multiplies the input force is the machine's mechanical advantage

NOW, you will learn

- How six simple machines change the size or direction of a force
- How to calculate mechanical advantage

VOCABULARY

simple machine p. 154
lever p. 155
fulcrum p. 155
wheel and axle p. 156
pulley p. 156
inclined plane p. 158
wedge p. 158
screw p. 159

EXPLORE Changing Forces

How can you change a force?

PROCEDURE

1. Lay one pencil on a flat surface. Place the other pencil on top of the first pencil and perpendicular to it, as shown. Place the book on one end of the top pencil.
2. Push down on the free end of the top pencil to raise the book.
3. Change the position of the bottom pencil so that it is closer to the book and repeat step 2. Then move the bottom pencil closer to the end of the pencil you are pushing on and repeat step 2.

MATERIALS

- 2 pencils
- small book

WHAT DO YOU THINK?

- How did changing the position of the bottom pencil affect how much force you needed to lift the book?
- At which position is it easiest to lift the book? most difficult?

NOTE-TAKING STRATEGY
As you read, remember to take notes about the main ideas and supporting details.

There are six simple machines.

You have read about how a ramp and a shovel can help you do work. A ramp is a type of inclined plane, and a shovel is a type of lever. An inclined plane and a lever are both simple machines. **Simple machines** are the six machines on which all other mechanical machines are based. In addition to the inclined plane and the lever, simple machines include the wheel and axle, pulley, wedge, and screw. As you will see, the wheel and axle and pulley are related to the lever, and the wedge and screw are related to the inclined plane. You will read about each of the six simple machines in detail in this section.

RESOURCES FOR DIFFERENTIATED INSTRUCTION

Below Level

UNIT RESOURCE BOOK

- Reading Study Guide A, pp. 300–301
- Decoding Support, p. 323

AUDIO CDS

 **Additional INVESTIGATION,** Levers and Fulcrums, A, B, & C, pp. 335–343; Teacher Instructions, pp. 346–347

Advanced

UNIT RESOURCE BOOK

Challenge and Extension, p. 306

English Learners

UNIT RESOURCE BOOK

Spanish Reading Study Guide, pp. 304–305

 **AUDIO CDS**

- Audio Readings in Spanish
- Audio Readings (English)

Lever

A **lever** is a solid bar that rotates, or turns, around a fixed point. The bar can be straight or curved. The fixed point is called the **fulcrum.** A lever can multiply the input force. It can also change the direction of the input force. If you apply a force downward on one end of a lever, the other end can lift a load.

The way in which a lever changes an input force depends on the positions of the fulcrum, the input force, and the output force in relation to one another. Levers with different arrangements have different uses. Sometimes a greater output force is needed, such as when you want to pry up a bottle cap. At other times you use a greater input force on one end to get a higher speed at the other end, such as when you swing a baseball bat. The three different arrangements, sometimes called the three classes of levers, are shown in the diagram below.

CHECK YOUR READING What two parts are needed to make a lever?

Levers

Levers can be classified according to where the fulcrum is.

First-Class Lever

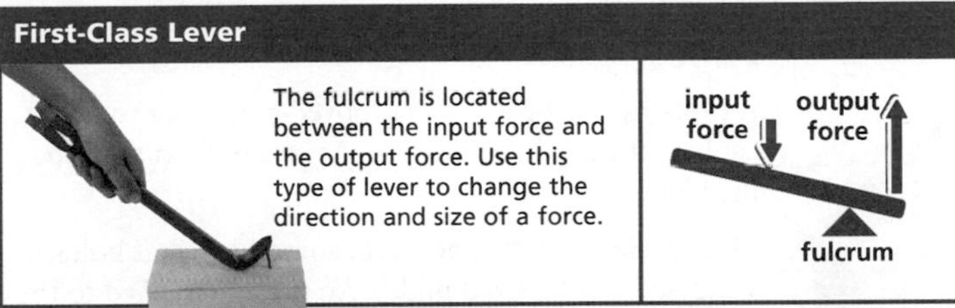

The fulcrum is located between the input force and the output force. Use this type of lever to change the direction and size of a force.

Second-Class Lever

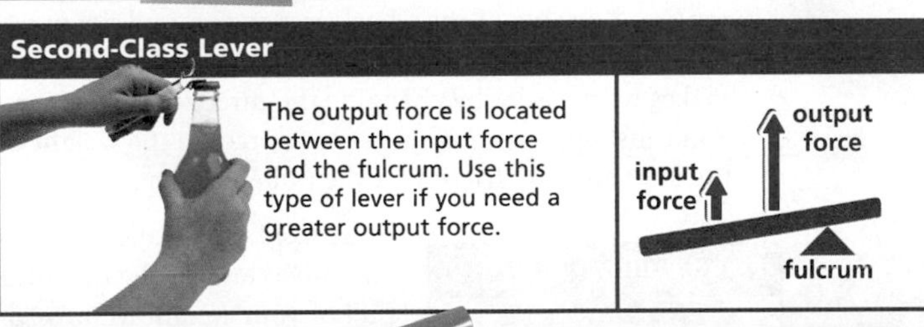

The output force is located between the input force and the fulcrum. Use this type of lever if you need a greater output force.

Third-Class Lever

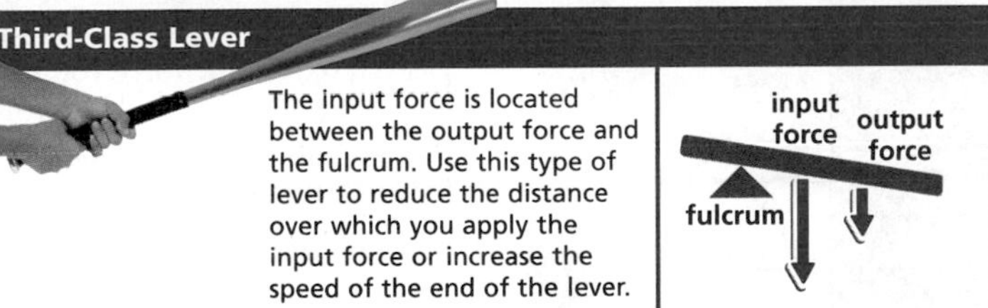

The input force is located between the output force and the fulcrum. Use this type of lever to reduce the distance over which you apply the input force or increase the speed of the end of the lever.

READING TIP The lengths of the arrows in the diagram represent the size of the force.

5.2 INSTRUCT

Develop Critical Thinking

CLASSIFY Ask students how they might measure mass by using a lever. Have students apply their knowledge of levers to identify the lever used in a triple-beam balance and to classify it. *The input force is supplied by an object on the pan. The output force is supplied by the masses on the beams. The fulcrum is between these two forces, so it is a first-class lever.* Forces are equal when the mass on the pan equals the mass on the beam.

Ongoing Assessment

CHECK YOUR READING *Answer: a solid bar and a fulcrum*

DIFFERENTIATE INSTRUCTION

More Reading Support

A What is a lever? *a solid bar that rotates around a fixed point*

B Levers are classified according to what? *where the fulcrum is*

Additional Investigation To reinforce Section 5.2 learning goals, use the following full-period investigation:

R **Additional INVESTIGATION,** Levers and Fulcrums, A, B, & C, pp. 335–343, 346–347

English Learners English learners may not be familiar with verbs such as *would, could, should,* and *might* that indicate a hypothetical situation. Help students by substituting more familiar verbs in these sentences. For example, "suppose you could" can be read as "suppose you can."

Develop Critical Thinking

APPLY A screwdriver is a type of wheel and axle. Have students apply their knowledge of a wheel and axle to describe what type of screwdriver they would use to apply a large force, such as one required to put a screw into a hard piece of wood. *one with a large-diameter handle*

Teacher Demo

Provide groups of students with old doorknobs that have been taken apart so students can view their operation. While students are examining their doorknobs, point out the various parts of the wheel and axle that they contain. Elicit from students that the knob is the wheel, and the shaft is the axle. Have them describe the output force of the axle compared with the input force of the wheel in a doorknob. *The output force acts over a shorter distance than the input force.*

Ongoing Assessment

CHECK YOUR READING *Answer: Force on the axle increased distance and decreased force from the wheel. Force on the wheel results in decreased distance and increased force from the axle.*

Wheel and Axle

Wheel and Axle

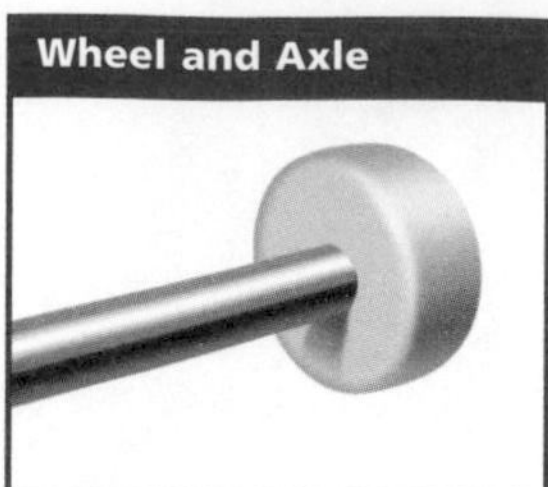

A **wheel and axle** is a simple machine made of a wheel attached to a shaft, or axle. The wheels of most means of transportation—such as a bicycle and a car—are attached to an axle. The wheel and axle act like a rotating collection of levers. The axle at the wheel's center is like a fulcrum. Other examples of wheels and axles are screwdrivers, steering wheels, doorknobs, and electric fans. C

Depending on your purpose for using a wheel and axle, you might apply a force to turn the wheel or the axle. If you turn the wheel, your input force is transferred to the axle. Because the axle is smaller than the wheel, the output force acts over a shorter distance than the input force. A driver applies less force to a steering wheel to get a greater turning force from the axle, or steering column. This makes it easier to steer the car.

If, instead, you turn the axle, your force is transferred to the wheel. Because the wheel is larger than the axle, the force acts over a longer distance. A car also contains this use of a wheel and axle. The engine turns the drive axles, which turn the wheels.

CHECK YOUR READING Compare the results of putting force on the axle with putting force on the wheel.

Pulley

A **pulley** is a wheel with a grooved rim and a rope or cable that rides in the groove. As you pull on the rope, the wheel turns.

A pulley that is attached to something that holds it steady is called a fixed pulley. An object attached to the rope on one side of the wheel rises as you pull down on the rope on the other side of the wheel. The fixed pulley makes work easier by changing the direction of the force. You must apply enough force to overcome the weight of the load and any friction in the pulley system. D

Fixed Pulley

A fixed pulley allows you to take advantage of the downward pull of your weight to move a load upward. It does not, however, reduce the force you need to lift the load. Also, the distance you pull the rope through is the same distance that the object is lifted. To lift a load two meters using a fixed pulley, you must pull down two meters of rope.

C 156

DIFFERENTIATE INSTRUCTION

More Reading Support

C A wheel and axle is a form of what other type of simple machine? *lever*

D What is a pulley that is attached to something that holds it steady called? *a fixed pulley*

Advanced

R Challenge and Extension, p. 306

In a movable pulley setup, one end of the rope is fixed, but the wheel can move. The load is attached to the wheel. The person pulling the rope provides the output force that lifts the load. A single movable pulley does not change the direction of the force. Instead, it multiplies the force. Because the load is supported by two sections of rope, you need only half the force you would use with a fixed pulley to lift it. However, you must pull the rope through twice the distance.

CHECK YOUR READING How does a single fixed pulley differ from a single movable pulley?

Movable Pulley

A combination of fixed and movable pulleys is a pulley system called a block and tackle. A block and tackle is used to haul and lift very heavy objects. By combining fixed and movable pulleys, you can use more rope sections to support the weight of an object. This reduces the force you need to lift the object. The mechanical advantage of a single pulley can never be greater than 2. If engineers need a pulley system with a mechanical advantage greater than 2, they often use a block-and-tackle system.

INVESTIGATE Pulleys

What is the mechanical advantage of a pulley system?

PROCEDURE

1. Hang the mass on the spring scale to find its weight in newtons. Record this weight as your output force.
2. Tie the top of one pulley to the ring stand.
3. Attach the mass to the second pulley.
4. Attach one end of the second pulley's rope to the bottom of the first pulley. Then thread the free end of the rope through the second pulley. Loop the rope up and over the first pulley, as shown.
5. Attach the spring scale to the free end of the rope. Pull down to lift the mass. Record the force you used as your input force. Calculate the mechanical advantage of this pulley system. **Hint:** The mechanical advantage can be calculated by dividing the output force by the input force.

WHAT DO YOU THINK?

- How did your input force compare with your output force?
- What caused the results you observed?

CHALLENGE Explain what the mechanical advantage would be for a pulley system that includes another movable pulley.

SKILL FOCUS
Inferring

MATERIALS
- 100 g mass
- spring scale
- 2 pulleys with rope
- ring stand

TIME
20 minutes

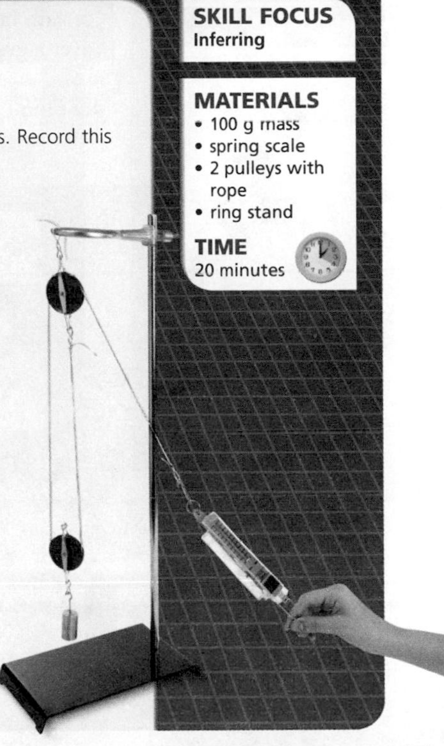

INVESTIGATE Pulleys

PURPOSE To infer the mechanical advantage of a pulley system

TIPS *20 min.* Use a clamp or a ring on the ring stand to hold the pulleys.

WHAT DO YOU THINK? *Students should observe that the input force is about half of the output force. The force must be applied over a greater distance (more rope is pulled).*

CHALLENGE *Because the mechanical advantage of a single movable pulley is 2, the mechanical advantage of the system with two such pulleys would be 4.*

Datasheet, Pulleys, p. 307

Technology Resources

Customize this student lab as needed or look for an alternative. Print rubrics to assess student lab reports.

Lab Generator CD-ROM

Metacognitive Strategy

Ask students to write a paragraph identifying specific steps in the procedure that confused them. Ask them to describe how the picture in the investigation helped them perform any troublesome steps of the procedure.

Ongoing Assessment

CHECK YOUR READING *Answer: A fixed pulley changes the direction of force; its mechanical advantage is 1. A movable pulley decreases the size of the force and increases its distance; its mechanical advantage is 2.*

DIFFERENTIATE INSTRUCTION

More Reading Support

E What is in a fixed position in a movable pulley? *one end of the rope*

F Where is the load attached on a movable pulley? *to the wheel*

Alternative Assessment Have interested students work in groups to brainstorm limitations to pulley systems. Would there be a situation in which a pulley system could not work to lift an object? *Answers might include a weight so heavy that the output force would break the rope or an object placed in a location where no pulley could be set up.*

Real World Example

Anyone who has driven over mountains has observed that the roads curve and do not go straight up and down. Because a mountain road is an inclined plane, increasing its length by zigzagging decreases the steepness of the incline. A less steep slope decreases the amount of force needed to get to the top of the mountain. Thus, a heavy truck can drive up a winding road to the top of a mountain because the engine has to exert less force. That same truck has less trouble going down a winding road because the upward force from the incline reduces the net downward force and makes it easier to control its speed.

Integrate the Sciences

The teeth of different animals indicate the differences in their diets.

- Carnivores, such as wolves, have sharp, wedgelike teeth with small angles. These teeth apply great force to tear apart meat.
- Some herbivores, such as many types of rodents, have wedgelike teeth in front, which are useful for gnawing.

Ongoing Assessment

CHECK YOUR READING *Answer: Inclined planes allow you to exert less force over a greater distance. They help support the weight of the object being raised.*

Inclined Plane

Inclined Plane

Recall that it is difficult to lift a heavy object straight up because you must apply a force great enough to overcome the downward pull of the force of gravity. For this reason people often use ramps. A ramp is an **inclined plane,** a simple machine that is a sloping surface. The photograph at the left shows the interior of the Guggenheim Museum in New York City. The levels of the art museum are actually one continuous inclined plane.

Inclined planes make the work of raising an object easier because they support part of the weight of the object while it is being moved from one level to another. The surface of an inclined plane applies a reaction force on the object resting on it. This extra force on the object helps to act against gravity. If you are pushing an object up a ramp, you have to push with only enough force to overcome the smaller net force that pulls the object down parallel to the incline.

The less steep an inclined plane is, the less force you need to push or pull an object on the plane. This is because a less steep plane supports more of an object's weight than a steeper plane. However, the less steep an inclined plane is, the farther you must go to reach a certain height. While you use less force, you must apply that force over a greater distance.

CHECK YOUR READING How do inclined planes help people do work? Your answer should mention force.

Wedge

Wedge

A **wedge** is a simple machine that has a thick end and a thin end. Wedges are used to cut, split, or pierce objects—or to hold objects together. A wedge is a type of inclined plane, but inclined planes are stationary, while wedges often move to do work.

Some wedges are single, movable inclined planes, such as a doorstop, a chisel, or an ice scraper. Another kind of wedge is made of two back-to-back inclined planes. Examples include the blade of an axe or a knife. In the photograph at the left, a sculptor is using a chisel to shape stone. The sculptor applies an input force on the chisel by tapping its thicker end with a mallet. That force pushes the thinner end of the chisel into the stone. As a result, the sides of the thinner end exert an output force that separates the stone.

DIFFERENTIATE INSTRUCTION

More Reading Support

G What is an inclined plane? *a simple machine that is a sloping surface*

C 158 Unit: Motion and Forces

The angle of the cutting edge determines how easily a wedge can cut through an object. Thin wedges have small angles and need less input force to cut than do thick wedges with large angles. That is why a sharp knife blade cuts more easily than a dull one.

H You also can think of a wedge that cuts objects in terms of how it changes the pressure on a surface. The thin edges of a wedge provide a smaller surface area for the input force to act on. This greater pressure makes it easier to break through the surface of an object. A sharp knife can cut through an apple skin, and a sharp chisel can apply enough pressure to chip stone.

A doorstop is a wedge that is used to hold objects together. To do its job, a doorstop is pressed tip-first under a door. As the doorstop is moved into position, it lifts the door slightly and applies a force to the bottom of the door. In return, the door applies pressure to the doorstop and causes the doorstop to press against the floor with enough force to keep the doorstop—and the door—from moving.

Screw

I A **screw** is an inclined plane wrapped around a cylinder or cone to form a spiral. A screw is a simple machine that can be used to raise and lower weights as well as to fasten objects. Examples of screws include drills, jar lids, screw clamps, and nuts and bolts. The spiraling inclined plane that sticks out from the body of the screw forms the threads of the screw.

In the photograph at right, a person is using a screwdriver, which is a wheel and axle, to drive a screw into a piece of wood. Each turn of the screwdriver pushes the screw farther into the wood. As the screw is turned, the threads act like wedges, exerting an output force on the wood. If the threads are very close together, the force must be applied over a greater distance—that is, the screw must be turned many times—but less force is needed.

Screw

The advantage of using a screw instead of a nail to hold things together is the large amount of friction that keeps the screw from turning and becoming loose. Think of pulling a nail out of a piece of wood compared with pulling a screw from the same piece of wood. The nail can be pulled straight out. The screw must be turned through a greater distance to remove it from the wood.

Notice that the interior of the Guggenheim Museum shown on page 158 is not only an inclined plane. It is also an example of a screw. The inclined plane is wrapped around the museum's atrium, which is an open area in the center.

CHECK YOUR READING Explain how a screw moves deeper into the wood as it is turned.

Teacher Demo

This demo may help students understand the relationship between a wedge and a screw. Cut a piece of paper into a right triangle. Darken the long edge (the hypotenuse) with a marker or pen. Show students how this resembles an inclined plane. Now place one of the short edges of the triangle against a pencil so it runs along and parallel to the pencil. Wrap the triangular piece of paper tightly around the pencil. Point out how the darkened edge now resembles the spiral of a screw, which can be seen as one inclined ramp wrapped around a wedge.

Ongoing Assessment

Describe how six simple machines change the size or direction of a force.

Ask: Explain how a pulley changes the direction or size of a force. *A fixed pulley changes the direction of a force by using the downward pull of your weight to move a load upward; a movable pulley multiplies the force but does not change the direction of force.*

CHECK YOUR READING *Answer: As the screw moves deeper into the wood, the threads act like wedges, exerting an output force on the wood. The force is applied over a greater distance if the threads are very close together, so less force is needed.*

DIFFERENTIATE INSTRUCTION

More Reading Support

H What does the tip of a wedge apply to a surface? *greater pressure*

I What is the inclined plane on a screw called? *the thread*

Below Level Have students use a table to summarize the effect on force of the six simple machines. Have them create a table with three columns and label them "Simple Machine," "Changes Direction of Force," and "Changes Size of Force." They should draw diagrams of all six simple machines and decide how they change force.

Real World Example

The force required for everyday tasks can sometimes be reduced by using simple machines. For a person in a wheelchair, going up and over a curb or up stairs requires a great amount of force. Ramps allow people in wheelchairs to apply less force over a greater distance. Pulleys allow people who are weak or have limited use of their arms or legs to be more mobile. Suspended pulleys might help these people get in and out of bed, a chair, a bathtub, or a vehicle.

The mechanical advantage of a machine can be calculated.

Recall that the number of times a machine multiplies the input force is the machine's mechanical advantage. You can calculate a machine's mechanical advantage using this formula:

$$\text{Mechanical Advantage} = \frac{\text{Output Force}}{\text{Input Force}}$$

$$MA = \frac{F_{out}}{F_{in}}$$

This formula works for all machines, regardless of whether they are simple machines or more complicated machines.

If a machine decreases the force you use to do work, the distance over which you have to apply that force increases. It is possible to use this idea to calculate the mechanical advantage of a simple machine without knowing what the input and output forces are. To make this calculation, however, you must assume that your machine is not losing any work to friction. In other words, you must assume that your machine is 100 percent efficient. The mechanical advantage that you calculate when making this assumption is called the ideal mechanical advantage.

J

READING TIP

Scientists often consider the way in which an object will behave under ideal conditions, such as when there is no friction.

Inclined Plane You can calculate the ideal mechanical advantage of an inclined plane by dividing its length by its height.

$$\text{Ideal Mechanical Advantage} = \frac{\text{length of incline}}{\text{height of incline}}$$

$$IMA = \frac{l}{h}$$

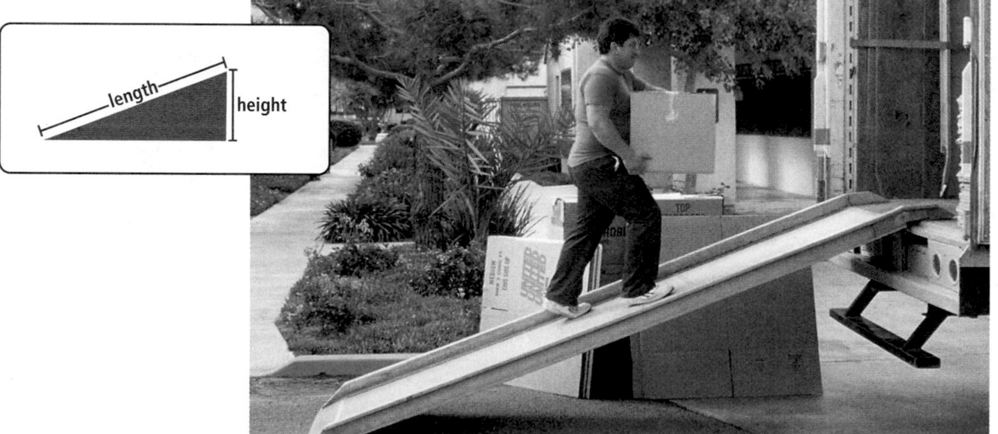

DIFFERENTIATE INSTRUCTION

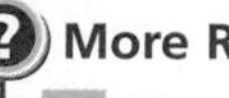

More Reading Support

J How does ideal mechanical advantage differ from actual mechanical advantage? *Ideal mechanical advantage assumes no loss of work to friction.*

Inclusion Allow students with learning disabilities to use calculators to help with mechanical-advantage calculations. Remind them that although efficiency is always less than 1 (100 percent), mechanical advantage can be greater than, less than, or equal to 1.

C 160 Unit: Motion and Forces

Be sure to use the length of the incline in your calculation, as shown in the diagram, and not the length of the base. If the mover in the photograph on page 160 increased the length of the ramp, he would increase the ramp's mechanical advantage. However, he would also increase the distance over which he had to carry the box.

SIMULATION
CLASSZONE.COM

Explore the mechanical advantage of an inclined plane.

Wheel and Axle To calculate the ideal mechanical advantage of a wheel and axle, use the following formula:

$$\text{Ideal Mechanical Advantage} = \frac{\text{Radius of input}}{\text{Radius of output}}$$

$$IMA = \frac{R_{in}}{R_{out}}$$

REMINDER

The radius is the distance from the center of the wheel or axle to any point on its circumference.

The Ferris wheel below is a giant wheel and axle. A motor applies an input force to the Ferris wheel's axle, which turns the wheel. In this example, the input force is applied to the axle, so the radius of the axle is the input radius in the formula above. The output force is applied by the wheel, so the radius of the wheel is the output radius.

For a Ferris wheel, the input force is greater than the output force. The axle turns through a shorter distance than the wheel does. The ideal mechanical advantage of this type of wheel and axle is less than 1.

Sometimes, as with a steering wheel, the input force is applied to turn the wheel instead of the axle. Then the input radius is the wheel's radius, and the output radius is the axle's radius. In this case, the input force on the wheel is less than the output force applied by the axle. The ideal mechanical advantage of this type of wheel and axle is greater than 1.

Teacher Demo

Demonstrate that beams of different length on a lever need different amounts of force. Turn a broom upside down. Have two students tightly hold the broom handle just below the brush. The handle should be off the floor. Have another student sit on the floor near the broom handle. Have the student on the floor touch the handle near its end with one finger. Ask students to explain how the seated student so easily made the broom handle move. ***The students' hands acted as a fulcrum. The long lever beam of the handle multiplied the small force applied at its end.***

EXPLORE the BIG idea

Revisit "Shut the Door!" on p. 143. Have students explain their results.

Ongoing Assessment

Calculate mechanical advantage.

Ask: What is the mechanical advantage if the output force is 60 N and the input force is 15 N?

$MA = \frac{F_{out}}{F_{in}} = 60\text{ N}/15\text{ N} = 4$

DIFFERENTIATE INSTRUCTION

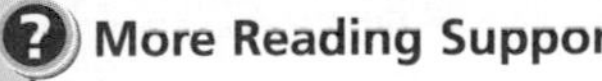

More Reading Support

K What is the ideal mechanical advantage of a wheel and axle in which the radius of input is 12 centimeters and the radius of output is 2 centimeters? *6*

Advanced Have students pick two simple machines and show mathematically the relationship between the ideal mechanical advantage and the shape of the machine. Remind students that for an ideal machine, output work equals input work.

Example: For an ideal inclined plane, input work equals input force times the length of the plane. Output work equals output force times height of plane. For an ideal machine, output work equals input work, so $F_{out} \cdot$ height. = $F_{in} \cdot$ length Rearranging gives F_{out}/F_{in} = length/height.

EXPLORE (the BIG idea)

Revisit "Internet Activity: Machines" on p. 143. Ask students to name more machines that contain simple machines.

Reinforce (the BIG idea)

Have students relate the section to the Big Idea.

 Reinforcing Key Concepts, p. 308

5.2 ASSESS & RETEACH

Assess

 Section 5.2 Quiz, p. 82

Reteach

Have pairs of students make up questions about force and simple machines based on the photograph on p. 146. Have them trade questions and answer them for review. Questions might include the following:

- How do you know that the efficiency of the rake is less than 100 percent? *It is a real machine.*
- Which simple machine is used in the photograph? *a lever*
- How would you calculate the mechanical advantage of the rake? *by dividing the distance from the input force to the point where the rake prongs connect to the handle (fulcrum) by the distance from the output force to the fulcrum.*

Technology Resources

Have students visit **ClassZone.com** for reteaching of Key Concepts.

 CONTENT REVIEW

 CONTENT REVIEW CD-ROM

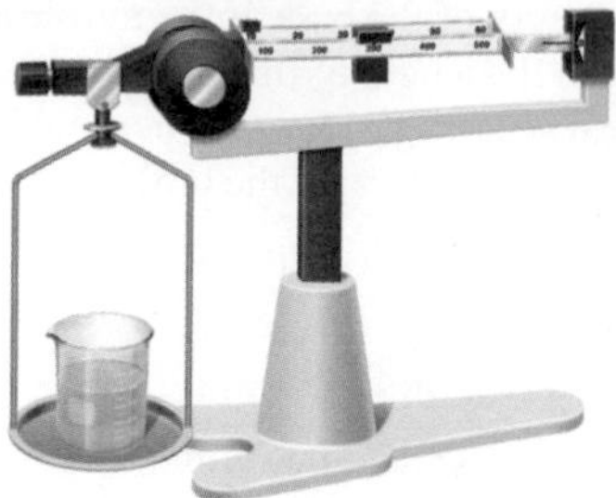

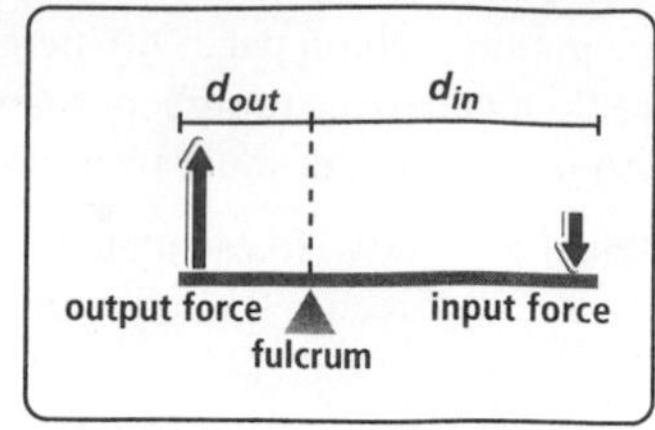

Lever The beam balance above is a lever. The beam is the solid bar that turns on a fixed point, or fulcrum. The fulcrum is the beam's balance point. When you slide the weight across the beam, you are changing the distance between the input force and the fulcrum. The mechanical advantage depends on the distances of the input force and output force from the fulcrum. The output force is applied to balance the beaker.

To calculate the ideal mechanical advantage of a lever, use the following formula:

$$\textbf{Ideal Mechanical Advantage} = \frac{\textbf{distance from input force to fulcrum}}{\textbf{distance from output force to fulcrum}}$$

$$IMA = \frac{d_{in}}{d_{out}}$$

This formula applies to all three arrangements of levers. If the distance from the input force to the fulcrum is greater than the distance from the output force to the fulcrum, the ideal mechanical advantage is greater than 1. The beam balance is an example of this type of lever.

5.2 Review

KEY CONCEPTS

1. Name the six simple machines and give an example of each.
2. Explain how a screw changes the size of the force needed to push it into wood.
3. To calculate mechanical advantage, what two things do you need to know?

CRITICAL THINKING

4. **Synthesize** How is a pulley similar to a wheel and axle?
5. **Calculate** What is the ideal mechanical advantage of a wheel with a diameter of 30 cm fixed to an axle with a diameter of 4 cm if the axle is turned?

CHALLENGE

6. **Infer** How can you increase a wedge's mechanical advantage? Draw a diagram to show your idea.

ANSWERS

1. Sample answer: lever, bottle opener; wheel and axle, doorknob; pulley, flagpole system; inclined plane, ramp; wedge, chisel; screw, jar lid

2. A small input force is applied over a long distance.

3. output force and input force

4. A pulley contains a wheel that turns on an axle.

5. $MA = \frac{4\ cm}{30\ cm} = 0.13$

6. Make the wedge longer and thinner.

Other parts of the human body can act like simple machines. For example, teeth work like wedges.

PHYSICAL SCIENCE AND LIFE SCIENCE

A Running Machine

Marlon Shirley, who lives in Colorado, lost his lower left leg due to an accident at the age of five. He is a champion sprinter who achieved his running records while using a prosthesis (prahs-THEE-sihs), or a device used to replace a body part. Like his right leg, his prosthetic leg is a combination of simple machines that convert the energy from muscles in his body to move him forward. The mechanical system is designed to match the forces of his right leg.

Legs as Levers

Compare Marlon Shirley's artificial leg with his right leg. Both legs have long rods—one made of bone and the other of metal—that provide a strong frame. These rods act as levers. At the knee and ankle, movable joints act as fulcrums for these levers to transfer energy between the runner's body and the ground.

How Does It Work?

1. As the foot—real or artificial—strikes the ground, the leg stops moving forward and downward and absorbs the energy of the change in motion. The joints in the ankle and knee act as fulcrums as the levers transfer the energy to the muscle in the upper leg. This muscle acts like a spring to store the energy.
2. When the runner begins the next step, the energy is transferred back into the leg from the upper leg muscle. The levers in the leg convert the energy into forward motion of the runner's body.

The people who design prosthetic legs study the natural motion of a runner to learn exactly how energy is distributed and converted to motion so that they can build an artificial leg that works well with the real leg.

EXPLORE

1. **VISUALIZE** Run across a room, paying close attention to the position of one of your ankles and knees as you move. Determine where the input force, output force, and fulcrum are in the lever formed by your lower leg.
2. **CHALLENGE** Use the library or the Internet to learn more about mechanical legs used in building robots that walk. How do the leg motions of these robots resemble your walking motions? How are they different?

CLASSZONE.COM
Find out more about artificial limbs.

EXPLORE

1. ***VISUALIZE*** *Input force comes from the knee area, and the output force comes from the foot. The ankle serves as the fulcrum.*
2. ***CHALLENGE*** *Answers will vary, but should reflect that different machines in the mechanical legs correspond to human body parts (such as movable joints that function as fulcrums) and that these machines work together to closely resemble a person's walking motion.*

CONNECTING SCIENCES
Integration of Sciences

Set Learning Goal

To understand how parts of a human leg act as levers

Integrate the Sciences

Several types of levers are present in the human body. For example, when you stand on your toes, your foot is a lever. The fulcrum is at the ankle joint. The output force is at the arch area, and the input force is the tendon at the back of the heel.

A lever system is formed by the head and the neck when the head is lowered toward the chest. The neck muscles provide the input force, and the weight of the head supplies the output force. The fulcrum is between these two forces.

Discussion Questions

Ask: Name all the levers in a leg. *thigh, lower leg, foot, toes*

Ask: Where are the fulcrums in the system of levers that is a leg? *hip, knee, ankle, toe joints*

Ask: What part of the human leg would be hardest to simulate? Why? *Sample answer: the muscles, because of their elasticity and self-contained energy*

Close

Ask: Which joint in the arm acts as a fulcrum for the lever formed by the bones in the lower arm? *the elbow*

Technology Resources

Have students visit **ClassZone.com** to find out more about artificial limbs

RESOURCE CENTER, Artificial Limbs

5.3 FOCUS

Set Learning Goals

Students will

- Recognize how simple machines can be combined.
- Describe how scientists have developed extremely small machines.
- Explain how robots are used.

3-Minute Warm-Up

Display Transparency 37 or copy this exercise on the board:

All machines are based on six simple machines. Draw a picture of each of the following machines.

lever

inclined plane

wedge

screw

pulley

wheel and axle

3-Minute Warm-Up, p. T37

5.3 MOTIVATE

THINK ABOUT

PURPOSE To introduce the concept that compound machines are made up of simple machines

DISCUSS Discuss with students how tow trucks for cars and light trucks would differ from tow trucks used to pull heavy trucks, such as tractor-trailers. *In the picture, students might recognize a pulley, a wheel and axle, a wedge, and an inclined plane.*

Ongoing Assessment

CHECK YOUR READING *Answer: Compound machines are made of two or more simple machines.*

KEY CONCEPT

5.3 Modern technology uses compound machines.

BEFORE, you learned

- Simple machines change the size or direction of a force
- All machines have an ideal and an actual mechanical advantage

NOW, you will learn

- How simple machines can be combined
- How scientists have developed extremely small machines
- How robots are used

VOCABULARY

compound machine p. 164
nanotechnology p. 167
robot p. 169

THINK ABOUT

How does a tow truck do work?

When a car is wrecked or disabled, the owner might call a towing service. The service sends a tow truck to take the car to be repaired. Tow trucks usually are equipped with a mechanism for freeing stuck vehicles and towing, or pulling, them. Look at the tow truck in the photograph at the right. What simple machines do you recognize?

Compound machines are combinations of simple machines.

Like the tow truck pictured above, many of the more complex devices that you see or use every day are combinations of simple machines. For example, a pair of scissors is a combination of two levers. The cutting edges of those levers are wedges. A fishing rod is a lever with the fishing line wound around a wheel and axle, the reel. A machine that is made of two or more simple machines is called a **compound machine.**

VOCABULARY Remember to write a definition for *compound machine* in a word triangle.

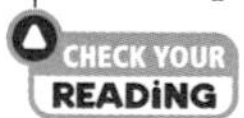

In a very complex compound machine, such as a car, the simple machines may not be obvious at first. However, if you look carefully at a compound machine, you should be able to identify forms of levers, pulleys, and wheels and axles.

CHECK YOUR READING How are simple machines related to compound machines?

RESOURCES FOR DIFFERENTIATED INSTRUCTION

Below Level

UNIT RESOURCE BOOK

- Reading Study Guide A, pp. 311–312
- Decoding Support, p. 323

Advanced

UNIT RESOURCE BOOK

- Challenge and Extension, p. 317
- Challenge Reading, pp. 319–320

English Learners

UNIT RESOURCE BOOK

Spanish Reading Study Guide, pp. 315–316

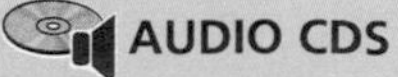

- Audio Readings in Spanish
- Audio Readings (English)

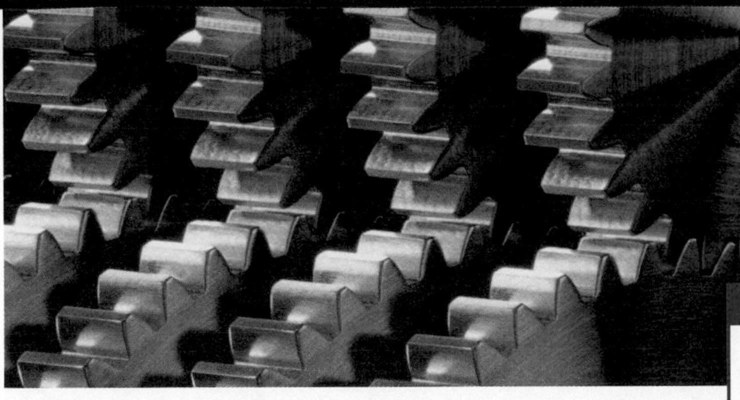

The gears in the photograph and diagram are spur gears, the most common type of gear.

Gears

Gears

Gears are based on the wheel and axle. Gears have teeth on the edge of the wheel that allow one gear to turn another. A set of gears forms a compound machine in which one wheel and axle is linked to another.

A

Two linked gears that are the same size and have the same number of teeth will turn at the same speed. They will move in opposite directions. In order to make them move in the same direction, a third gear must be added between them. The gear that turns another gear applies the input force; the gear that is turned exerts the output force. A difference in speed between two gears—caused by a difference in size and the distance each turns through—produces a change in force.

CHECK YOUR READING How do gears form a compound machine?

Mechanical Advantage of Compound Machines

The mechanical advantage of any compound machine is equal to the product of the mechanical advantages of all the simple machines that make up the compound machine. For example, the ideal mechanical advantage of a pair of scissors would be the product of the ideal mechanical advantages of its two levers and two wedges.

The mechanical advantage of a pair of gears with different diameters can be found by counting the teeth on the gears. The mechanical advantage is the ratio of the number of teeth on the output gear to the number of teeth on the input gear. If there are more than two gears, count only the number of teeth on the first and last gears in the system. This ratio is the mechanical advantage of the whole gear system.

Compound machines typically must overcome more friction than simple machines because they tend to have many moving parts. Scissors, for example, have a lower efficiency than one lever because there is friction at the point where the two levers are connected. There is also friction between the blades of the scissors as they close.

5.3 INSTRUCT

Develop Algebra Skills

To clarify the procedure for calculating mechanical advantage, provide students with practice problems. Sample problems:

- A compound machine contains two simple machines, one with a mechanical advantage of 3 and another with a mechanical advantage of 2. What is the mechanical advantage of the compound machine? $MA = 2 \cdot 3 = 6$
- A gear with 15 teeth turns another gear with 20 teeth. What is the mechanical advantage of the gear system?

$$MA = \frac{20}{15} = 1.3$$

Ongoing Assessment

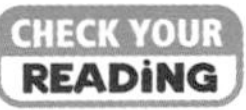

One wheel and axle turns another wheel and axle.

DIFFERENTIATE INSTRUCTION

More Reading Support

A How does a gear differ from a simple wheel and axle? *It has teeth on the edge of the wheel.*

English Learners Be aware that English learners do not always have the same background knowledge as the rest of the class. Certain technical terms mentioned in this chapter may be new to English Learners, such as *prosthesis* (p. 163), *artery* (p. 166), *Jaws of Life* (p. 166), *remote control* (p. 168), and *fiber optic cable* (p. 168).

Teach from Visuals

Remind students that hydraulics can produce great force. In a hydraulic system, fluid is contained in two connected chambers, each of which contains a free-moving piston. The force applied to the smaller piston causes the larger piston to exert a greater force.

Integrate the Sciences

Some scientists working with nanotechnology base their work on cells, proteins, and molecules in the human body. For example, one group of researchers manipulated strands of DNA molecules to produce nano-sized tweezers. Two strands of DNA form the tweezers, while a third strand provides the fuel to close or open the arms. Such tweezers could be used to build other nanomachines or to manipulate the materials needed to build nanocircuits.

Ongoing Assessment

PHOTO CAPTION Answer: levers and wedges are some simple machines in the Jaws of Life tool.

APPLY What simple machines do you see in this Jaws of Life cutting tool?

Modern technology creates new uses for machines.

Sophisticated modern machinery is often based on or contains simple machines. Consider Jaws of Life tools, which are used to help rescue people who have been in accidents. These cutters, spreaders, and rams are powered by hydraulics, the use of fluids to transmit force. When every second counts, these powerful machines can pry open metal vehicles or collapsed concrete structures quickly and safely. The cutters are a compound machine made up of two levers—much like a pair of scissors. Their edges are wedges.

Contrast this equipment with a drill-like machine so small that it can be pushed easily through human arteries. Physicians attach the tiny drill to a thin, flexible rod and push the rod through a patient's artery to an area that is blocked. The tip rotates at extremely high speeds to break down the blockage. The tiny drill is a type of wheel and axle.

Microtechnology and Nanotechnology

Manufacturers make machines of all sizes by shaping and arranging pieces of metal, plastic, and other materials. Scientists have used technology to create very small machines through miniaturization—the making of smaller and smaller, or miniature, parts. Micromachines are too small to be seen by the naked eye but are visible under a microscope. There is a limit, however, to how far micromachines can be shrunk. **? B**

READING TIP

Micro-means "one-millionth." For example, a microsecond is one-millionth of a second. *Nano*-means "one-billionth." A nanosecond is one-billionth of a second.

To develop even tinier machines, scientists needed a new approach. Scientists have used processes within the human body as their model. For example, inside the body a protein molecule carries materials back and forth within a cell on regular paths that are similar to little train tracks. The natural machines in the human body inspired scientists to develop machines that could be 1000 times smaller than the diameter of a human hair.

DIFFERENTIATE INSTRUCTION

More Reading Support

B How is technology involved in miniaturization? *It allows scientists to create machines with miniature parts.*

Alternative Assessment American cartoonist Rube Goldberg (1883–1970) was famous for his cartoons of humorous and complicated machines. Today any device that is more complicated than practical is called a "Rube Goldberg machine." Show students a few examples of a Rube Goldberg machine. Ask students to design their own Rube Goldberg machine that uses the principles they have learned from the chapter.

C 166 Unit: Motion and Forces

These extremely tiny machines are products of **nanotechnology,** the science and technology of building electronic circuits and devices from single atoms and molecules. Scientists say that they create these machines, called nanomachines, from the bottom up. Instead of shaping already formed material—such as metal and plastic—they guide individual atoms of material to arrange themselves into the shapes needed for the machine parts.

Tools enable scientists to see and manipulate single molecules and atoms. The scanning tunneling microscope can create pictures of individual atoms. To manipulate atoms, special tools are needed to guide them into place. Moving and shaping such small units presents problems, however. Atoms tend to attach themselves to other atoms, and the tools themselves are also made of atoms. Thus it is difficult to pick up an atom and place it in another position using a tool because the atom might attach itself to the tool.

CHECK YOUR READING Compare the way in which nanomachines are constructed with the way in which larger machines are built.

RESOURCE CENTER CLASSZONE.COM

Learn more about nanomachines.

Nanomachines are still mostly in the experimental stage. Scientists have many plans for nanotechnology, including protecting computers from hackers and performing operations inside the body. For example, a nanomachine could be injected into a person's bloodstream, where it could patrol and search out infections before they become serious problems. When the machine had completed its work, it could switch itself off and be passed out of the body. Similar nanomachines could carry anti-cancer drugs to specific cells in the body.

Nanotechnology could also be used to develop materials that repel water and dirt and make cleaning jobs easy. Nanoscale biosensors could be used to detect harmful substances in the environment. Another possible use for nanotechnology is in military uniforms that can change color—the perfect camouflage.

In the future, nanotechnology may change the way almost everything is designed and constructed. As with any new technology, it will be important to weigh both the potential risks and benefits.

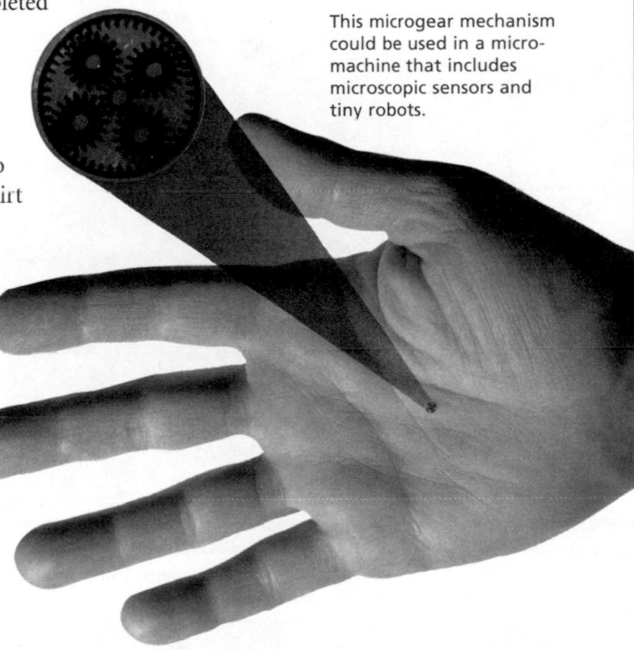

This microgear mechanism could be used in a micromachine that includes microscopic sensors and tiny robots.

Real World Example

Ask students: When machines perform tasks, what happens to the people who formerly were doing those tasks? New processes have always made old processes obsolete. In some instances, new jobs are created by new technology. After all, someone needs to build and maintain robots, for example. However, these jobs often require skills that the workers might not have.

Ongoing Assessment

Describe how modern technology has enabled scientists to create extremely small machines.

Ask: How did scientists learn how to build a nanomachine? *They studied processes that build natural machines in the human body.*

Point out to students that these processes are common to nearly all life forms on Earth. Much nano research is based on bacteria.

CHECK YOUR READING *Answer: Larger machines are constructed by shaping and arranging already formed pieces of material. Nanomachines are constructed from individual atoms of material that arrange themselves into the shape that is needed.*

DIFFERENTIATE INSTRUCTION

More Reading Support

C What is the design and building of very tiny machines called? *nanotechnology*

Alternative Assessment Have students write a short story or create a comic book about a way in which they think microtechnology or nanotechnology could be used in the future. Encourage creativity based on accurate science. Allow class time for students to share their products.

Teach from Visuals

To help students interpret the visual "A Robot at Work," ask:

- Which tasks did the Pyramid Rover perform? *climbing a ramp, drilling, collecting and sending information, sensing conditions*
- Where else would a similar robot be useful? *Sample answers: cave exploration, looking for people in a collapsed mine or building*

This visual is also available as T38 in the Unit Transparency Book.

Social Studies Connection

The Great Pyramid of Egypt is one of dozens of large pyramids built by ancient Egyptians to serve as burial places for their kings and queens. It is unknown exactly how workers lifted 1 million stones with an average weight of 2.5 tons to form this pyramid. One theory is that the Egyptians used ramps lubricated with mud. Another is that they used levers to raise each block a bit at a time.

Real World Example

A robot helped astronauts build the International Space Station. The space station includes a robotic arm, a 58-foot crane known as Canadarm 2. This robot has more dexterity than a human arm and will eventually have its own robot "hand."

Ongoing Assessment

READING VISUALS *Sample answer: wheels and axles to turn the treads, a screw for the drill, and levers to help it expand in size*

A Robot at Work

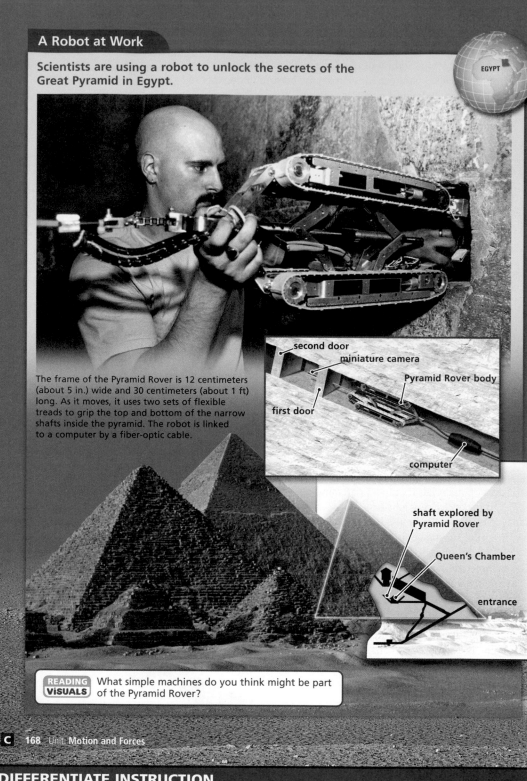

Scientists are using a robot to unlock the secrets of the Great Pyramid in Egypt.

The frame of the Pyramid Rover is 12 centimeters (about 5 in.) wide and 30 centimeters (about 1 ft) long. As it moves, it uses two sets of flexible treads to grip the top and bottom of the narrow shafts inside the pyramid. The robot is linked to a computer by a fiber-optic cable.

READING VISUALS What simple machines do you think might be part of the Pyramid Rover?

C 168 Unit: Motion and Forces

DIFFERENTIATE INSTRUCTION

Advanced Have students use the Internet to find information about the Pyramid Rover or other robots. Designate a classroom bulletin board to which students can add information as it is acquired. Ask them to explain how other robots they learn about are used.

Challenge and Extension, p. 317

Have students who are interested in learning more about how modern technology creates new uses for machines read the following article:

Challenge Reading, pp. 319–320

Robots

Humans have always taken risks to do jobs in places that are dangerous or difficult to get to. More and more often, robots can be used to do these jobs. A **robot** is a machine that works automatically or by remote control. When many people hear the word *robot,* they think of a machine that looks or moves like a person. However, most robots do not resemble humans at all. That is because they are built to do things humans cannot do or to go places where it is difficult for humans to go.

Find out more about the Pyramid Rover and other robots.

The Pyramid Rover, shown on page 168, is an example of a robot developed to go where people cannot. After a camera revealed a door at the end of an eight-inch-square shaft inside the Great Pyramid, the Pyramid Rover was sent through the shaft to explore the area. While researchers remained in the Queen's Chamber in the center of the pyramid, the robot climbed the shaft until it came to a door. Using ultrasound equipment mounted on the robot, researchers determined that the door was three inches thick. The robot drilled a hole in the door for a tiny camera and a light to pass through. The camera then revealed another sealed door!

Many companies use robots to manufacture goods quickly and efficiently. Robots are widely used for jobs such as welding, painting, and assembling products. Robots do some repetitive work better than humans, because robots do not get tired or bored. Also, they do the task in exactly the same way each time. Robots are very important to the automobile and computer industries.

How are robots better than humans at some jobs?

5.3 Review

KEY CONCEPTS

1. How do you estimate the mechanical advantage of a compound machine?
2. What are some uses of nanotechnology? Can you think of other possible uses for nanomachines?
3. What are three types of jobs that robots can do?

CRITICAL THINKING

4. **Synthesize** What factors might limit how large or how small a machine can be?
5. **Infer** How do you think the size of a gear compared with other gears in the same system affects the speed of its rotation?

CHALLENGE

6. **Apply** Robots might be put to use replacing humans in firefighting and other dangerous jobs. Describe a job that is dangerous. Tell what a robot must be able to do and what dangers it must be able to withstand to accomplish the required tasks.

ANSWERS

1. You multiply the mechanical advantages of all the simple machines that make up the compound machine.

2. protecting computers from hackers, environmental monitoring, medicine

3. Sample answers: exploring dangerous areas, exploring distant places, and repetitive work in manufacturing

4. the space in which it works; it cannot be smaller than atoms

5. the larger the gear, the slower its rotation

6. Sample answer: Demolition: a robot could place explosives, set detonators, and relay information in buildings that are unsafe; it must be able to travel over unstable surfaces.

Ongoing Assessment

Explain how robots are used to go where humans cannot.

Ask: Why might a robot be sent into space to study another planet? *The trip is too dangerous and too long for a human to attempt. Also, a robot does not have to return to Earth.*

CHECK YOUR READING *Answer: Robots can be designed to go into places too small or too dangerous for humans. They do repetitive work in the same way every time and do not become tired or bored.*

Reinforce the BIG idea

Have students relate the section to the Big Idea.

 Reinforcing Key Concepts, p. 318

Assess

A Section 5.3 Quiz, p. 83

Reteach

Have students examine a fishing rod, line, and reel. Have them describe the simple machines used in this equipment. *The rod itself is a lever, the hand of the person using it is the fulcrum. The reel is a wheel and axle. The eyelet at the end of the pole, through which the line passes, forms a pulley.*

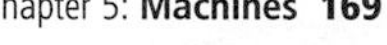

Technology Resources

Have students visit **ClassZone.com** for reteaching of Key Concepts.

 **CONTENT REVIEW**

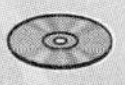 **CONTENT REVIEW CD-ROM**

CHAPTER INVESTIGATION

Focus

PURPOSE To build and use a compound machine made from simple machines

OVERVIEW Students will design one of three types of machine. They will

- build the machine
- test it to make sure it will move a 500-gram object 1 meter
- calculate its mechanical advantage
- calculate its efficiency

Lab Preparation

- Have students read the lab for homework and choose a machine to build.
- Tell students that they can bring in additional materials but that you have to approve them beforehand.
- Have students read through the investigation and prepare their data tables. Or you may wish to copy and distribute datasheets and rubrics.

 UNIT RESOURCE BOOK, pp. 326–334

 SCIENCE TOOLKIT, F13

Lab Management

- Have students individually decide what type of machine they want to build. Then group students according to what machine they chose.
- If enough equipment is available, students can work in groups of two or three.
- Show students how to thread the pulley.

SAFETY You should preapprove materials that students want to bring from home.

INCLUSION Some students may have difficulty calculating the mechanical advantage and efficiency of the machine. Review how to make the calculations before beginning the investigation.

Teaching with Technology

Use a digital camera to photograph students' machines. Display images and have students identify the simple machines they contain.

CHAPTER INVESTIGATION

Design a Machine

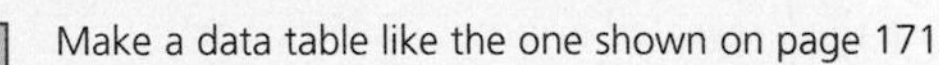

DESIGN — YOUR OWN —

OVERVIEW AND PURPOSE

Although simple machines were developed thousands of years ago, they are still used today for a variety of purposes. Tasks such as cutting food with a knife, using a screwdriver to tighten a screw, and raising a flag on a flagpole all require simple machines. Activities such as riding a bicycle and raising a drawbridge make use of compound machines. In this investigation you will use what you have learned about simple and compound machines to

- choose a machine to design
- build your machine, test it, and calculate its mechanical advantage and efficiency

MATERIALS

- 500 g object
- 100 g object
- meter stick
- spring scale
- pulleys with rope
- board
- stick or pole

Procedure

1. Make a data table like the one shown on page 171.
2. From among the three choices listed below, choose which problem you are going to solve.

Carnival Game You work for a company that builds carnival games. Your supervisor has asked you to build a game in which a simple machine moves a 500-gram object from the bottom of the game 1 meter up to the top. This simple machine can be powered only by the person operating the game.

Video Game Contest The marketing department of a video game company is holding a contest. Candidates are asked to submit a working model of a compound machine that will move a 500-gram object a distance of 1 meter. The winning design will be used in a new video game the company hopes to sell. This compound machine must include at least 2 simple machines.

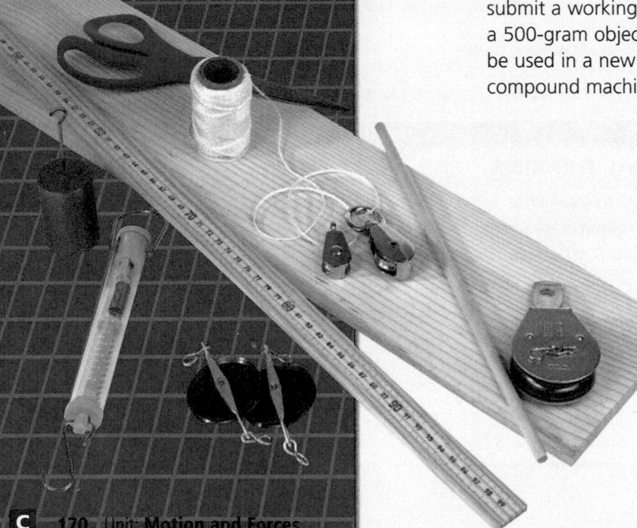

Construction Company You work for a construction company. Your boss has asked you to design a machine for lifting. Your first step is to build a scale model. The model must be a compound machine with a mechanical advantage of 5 that can move a 500-gram object a distance of 1 meter. You also can use a 100-gram object in your design.

C 170 Unit: Motion and Forces

INVESTIGATION RESOURCES

 CHAPTER INVESTIGATION, Design a Machine
- Level A, pp. 326–329
- Level B, pp. 330–333
- Level C, p. 334

Advanced students should complete Levels B & C.

 Writing a Lab Report, D12–13

Technology Resources

Customize this student lab as needed or look for an alternative. Print rubrics to assess student lab reports.

 **Lab Generator CD-ROM**

3 Brainstorm design ideas on paper. Think of different types of machines you might want to build. Choose one machine to build.

4 Build your machine. Use your machine to perform the task of moving a 500-gram object a distance of 1 meter.

If you chose the third problem, test your compound machine to determine if it has a mechanical advantage of 5. If not, modify your machine and retest it.

5 Record all measurements in your data table.

Observe and Analyze

1. **RECORD OBSERVATIONS** Make a sketch of your machine.
2. **CALCULATE** Use your data to calculate the mechanical advantage and efficiency of your machine. Use the formulas below.

$$\text{Mechanical Advantage} = \frac{\text{Output Force}}{\text{Input Force}}$$

$$\text{Efficiency (\%)} = \frac{\text{Output work}}{\text{Input work}} \cdot 100$$

3. **ANALYZE**

Carnival Game Add arrows to the drawing of your machine to show the forces involved and the direction of those forces. If your goal was to move the ball from the top of the game to the bottom at a constant speed, how would your machine and diagram have to be changed?

Video Game Contest Does your machine change the size of the force, the direction of the force, or both? If you used a pulley system (two or more pulleys working together), describe the advantages of using such a system.

Construction Company Determine whether force or distance is changed by each simple machine in your compound machine. In what ways might you improve your machine to increase its efficiency?

Conclude

1. **INFER** How might changing the arrangement of the parts in your machine affect the machine's mechanical advantage?
2. **IDENTIFY LIMITS** What was the hardest part about designing and constructing your machine?
3. **APPLY** If you needed to lift a large rock from a hole at a construction site, which type of simple machine would you use and why? Which type of compound machine would be useful?

INVESTIGATE Further

CHALLENGE If you made a simple machine, how would you combine it with another simple machine to increase its mechanical advantage?

If you made a compound machine, redesign it to increase its efficiency or mechanical advantage. What made the difference and why?

Draw a plan for the new machine. Circle the parts that were changed. If you have time, build your new machine.

Design a Machine

Observe and Analyze

Table 1. Machine Data

Output force	Input force	Mechanical Advantage
Output work	Input work	Efficiency

Sketch

Observe and Analyze

1. *See students' drawings.*
2. *Mechanical advantage should use force as measured by the spring scale. Work done can be determined by force and distance.*
3. *Sample answers: Carnival game: The input and output would be reversed. Video game contest: A pulley system multiplies force. Construction company: Increase the efficiency by reducing friction.*

Conclude

1. *Students should specify whether they think the mechanical advantage will increase, decrease, or stay the same. Students should explain their answers.*
2. *Have students identify specific problems and difficulties in designing and constructing their designs.*
3. *Sample answer: A lever could be placed under the edge of the rock, and you could increase the force by controlling the location of the fulcrum. A compound machine could use a lever to pry the rock out of the ground and a pulley to raise it.*

INVESTIGATE Further

CHALLENGE Combine the first simple machine with another simple machine that has a mechanical advantage greater than 1. For a compound machine, decrease friction, perhaps by oiling the machine, to increase efficiency. Increase its mechanical advantage by increasing the mechanical advantage of the simple machines in it. Check students' plans for new machines.

Post-Lab Discussion

- Have each lab group demonstrate their machine for the rest of the class. For each machine, identify appropriate problems for technological design, design solutions, and implement a proposed design to improve the machine.
- Make a table to tally the number of times each type of simple machine was used in the compound machines.

BACK TO

the BIG idea

Have each student list five machines they have used today.

- For each, have them describe whether it is a simple or a compound machine.
- For compound machines, have students name the simple machines that are included in it.
- Invite students to share their lists with the rest of the class.

KEY CONCEPTS SUMMARY

SECTION 5.1
Ask: If the slope of the inclined plane were made steeper, what would happen to the size of the force arrow? *It would increase.*

Ask: If the person climbed straight up to the top position, would she do more or less work than if she walked up the slope? *The amount of work would be the same.*

SECTION 5.2
Have students sketch the first two simple machines shown with force arrows. Label arrows as input force or output force. Check students' drawings. *Pulley—input force: person pulling rope; output force: lifts the load. Wheel and axle—input force: turning the wheel; output force: turning force of the axle.*

SECTION 5.3
Have students describe the flow of force through each simple machine shown. *The lever is turned and that applies a force to the wheel and axle.*

Review Concepts

- Big Idea Flow Chart, p. T33
- Chapter Outline, pp. T39–T40

5 Chapter Review

the BIG idea

Machines help people do work by changing the force applied to an object.

KEY CONCEPTS SUMMARY

Machines help people do work.

When you use a machine to do work, there is always an exchange, or tradeoff, between the force you use and the distance over which you apply that force. You can use less force over a greater distance or a greater force over a shorter distance to do the same amount of work.

VOCABULARY
machine p. 145
mechanical advantage p. 147
efficiency p. 150

5.2 Six simple machines have many uses.

Simple machines change the size and/or direction of a force.

changes direction

changes size

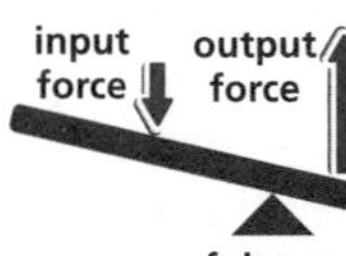

changes both

VOCABULARY
simple machine p. 154
lever p. 155
fulcrum p. 155
wheel and axle p. 156
pulley p. 156
inclined plane p. 158
wedge p. 158
screw p. 159

Modern technology uses compound machines.

- Compound machines are combinations of simple machines.

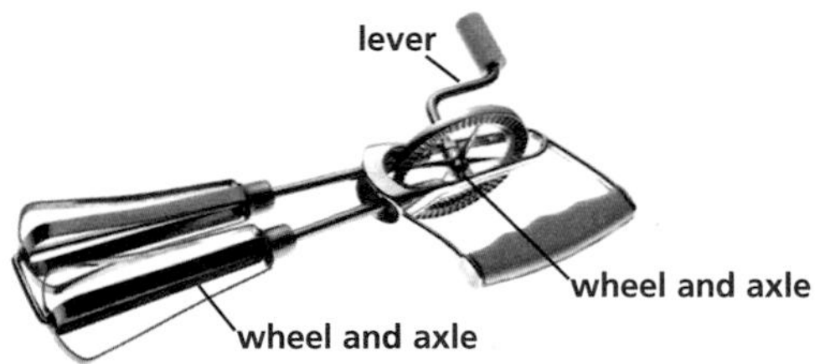

- Modern technology creates new uses for machines.
  —Microtechnology and nanotechnology
  —Robots

VOCABULARY
compound machine p. 164
nanotechnology p. 167
robot p. 169

C 172 Unit: Motion and Forces

Technology Resources

Have students visit **ClassZone.com** or use the CD-ROM for a cumulative review of concepts.

 **CONTENT REVIEW**

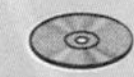 **CONTENT REVIEW CD-ROM**

Engage students in a whole-class interactive review of Key Concepts. Edit content as you wish.

 **POWER PRESENTATIONS**

Reviewing Vocabulary

Write the name of the simple machine shown in each illustration. Give an example from real life for each one.

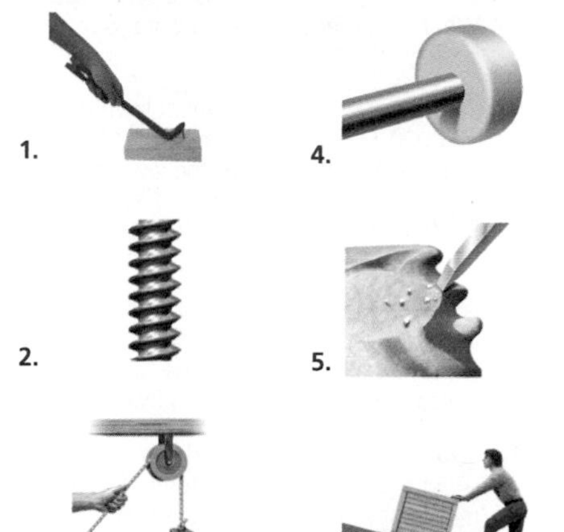

Copy the chart below, and write the definition for each term in your own words. Use the meaning of the term's root to help you.

Term	Root Meaning	Definition
7. machine	having power	
8. nanotechnology	one-billionth	
9. simple machine	basic	
10. efficiency	to accomplish	
11. compound machine	put together	
12. robot	work	
13. fulcrum	to support	

Reviewing Key Concepts

Multiple Choice *Choose the letter of the best answer.*

14. Machines help you work by
- **a.** decreasing the amount of work that must be done
- **b.** changing the size and/or direction of a force
- **c.** decreasing friction
- **d.** conserving energy

15. To calculate mechanical advantage, you need to know
- **a.** time and energy
- **b.** input force and output force
- **c.** distance and work
- **d.** size and direction of a force

16. A machine in which the input force is equal to the output force has a mechanical advantage of
- **a.** 0
- **b.** between 0 and 1
- **c.** 1
- **d.** more than 1

17. You can increase a machine's efficiency by
- **a.** increasing force
- **b.** reducing work
- **c.** increasing distance
- **d.** reducing friction

18. Levers turn around a
- **a.** fixed point called a fulcrum
- **b.** solid bar that rotates
- **c.** wheel attached to an axle
- **d.** sloping surface called an inclined plane

19. When you bite into an apple, your teeth act as what kind of simple machine?
- **a.** lever
- **b.** pulley
- **c.** wedge
- **d.** screw

Short Answer *Answer each of the following questions in a sentence or two.*

20. Describe the simple machines that make up scissors.

21. How do you calculate the mechanical advantage of a compound machine?

22. How did scientists use processes inside the human body as a model for making nanomachines?

Reviewing Vocabulary

Sample answers

1. *lever, crowbar*
2. *screw, jar lid*
3. *pulley, system for raising a flag*
4. *wheel and axle, doorknob*
5. *wedge, doorstop*
6. *inclined plane, ramp*
7. *any device that helps you do work*
8. *the science and technology of building tiny devices from single atoms and molecules*
9. *a basic machine on which all other machines are based*
10. *the percentage of the input work that a machine can return in output work*
11. *a machine that is made up of two or more simple machines*
12. *a machine that works automatically or by remote control*
13. *a fixed point on which a lever rotates*

Reviewing Key Concepts

14. *b*
15. *b*
16. *c*
17. *d*
18. *a*
19. *c*
20. *A pair of scissors is made up of two levers with cutting edges that are wedges.*
21. *Multiply the mechanical advantages of all the simple machines in it.*
22. *Scientists noticed that molecules in the body act as natural machines and learned how they are built.*

ASSESSMENT RESOURCES

UNIT ASSESSMENT BOOK
- Chapter Test A, pp. 84–87
- Chapter Test B, pp. 88–91
- Chapter Test C, pp. 92–95
- Alternative Assessment, pp. 96–97
- Unit Test, A, B, & C, pp. 98–109

SPANISH ASSESSMENT BOOK
- Spanish Chapter Test, pp. 273–276
- Spanish Unit Test, pp. 277–280

Technology Resources

Edit test items and answer choices.

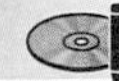 **Test Generator CD-ROM**

Visit **ClassZone.com** to extend test practice.

 **Test Practice**

Thinking Critically

23. *A screw is an inclined plane wrapped around a cylinder.*
24. *Sample answer: Use a block and tackle. The fixed and movable pulleys increase the mechanical advantage of the machine.*
25. *The amount of work would be the same because each person adds the same amount of potential energy. The person on the shorter path must use more force.*
26. *The MA of the board is less, but the amount of work will not change.*
27. *fixed pulley changes the direction of force and its MA is 1; movable pulley decreases the size of the force and increases its distance; its MA is 2*
28. *2.67; 2.67*
29. *The mechanical advantage would increase and this would require the man to exert less input force.*
30. *The input force would increase.*

Using Math Skills in Science

31. *MA = 5 N/10 N = 0.5*
32. *E = 90,000 J/125,000 J · 100 = 72%*
33. *80%*
34. *MA = 21 cm/3 cm = 7*
35. *MA = 24/24 = 1*

the BIG idea

36. *wheel and axle, pulley, inclined plane; the balls move down the inclined plane and continue down the tracks, which are moved by the wheels and the pulley system to help the balls complete their paths.*
37. *Answers will vary depending on the machine students choose.*
38. *Answers should include the concept that work will be done on a smaller scale.*

UNIT PROJECTS

Have students present their projects. Use the appropriate rubrics from the URB to evaluate their work.

 Unit Projects, pp. 5–10

Thinking Critically

23. SYNTHESIZE How is a screw related to an inclined plane?

24. INFER Which simple machine would you use to raise a very heavy load to the top of a building? Why?

25. APPLY If you reached the top of a hill by using a path that wound around the hill, would you do more work than someone who climbed a shorter path? Why or why not? Who would use more force?

26. APPLY You are using a board to pry a large rock out of the ground when the board suddenly breaks apart in the middle. You pick up half of the board and use it to continue prying up the rock. The fulcrum stays in the same position. How has the mechanical advantage of the board changed? How does it change your work?

27. SYNTHESIZE What is the difference between a single fixed pulley and a single movable pulley? Draw a diagram to illustrate the difference.

Use the information in the diagram below to answer the next three questions.

28. SYNTHESIZE What is the mechanical advantage of the ramp? By how many times does the ramp multiply the man's input force?

29. SYNTHESIZE If the ramp's length were longer, what effect would this have on its mechanical advantage? Would this require the man to exert more or less input force?

30. INFER If the ramp's length stayed the same but the height was raised, how would this change the input force required?

Using Math Skills in Science

Complete the following calculations.

31. You swing a hockey stick with a force of 10 N. The stick applies 5 N of force on the puck. What is the mechanical advantage of the hockey stick?

32. Your input work on a manual lawn mower is 125,000 J. The output work is 90,000 J. What is the efficiency of the lawn mower?

33. If a car engine has a 20 percent efficiency, what percentage of the input work is lost?

34. A steering wheel has a radius of 21 cm. The steering column on which it turns has a radius of 3 cm. What is the mechanical advantage of this wheel and axle?

35. Two gears with the same diameter form a gear system. Each gear has 24 teeth. What is the mechanical advantage of this gear system?

the BIG idea

36. DRAW CONCLUSIONS Look back at the photograph on pages 142–143. Name the simple machines you see in the photograph. How do you think they work together to move balls through the sculpture? How has your understanding changed as to the way in which machines help people work?

37. SYNTHESIZE Think of a compound machine you have used recently. Explain which simple machines it includes and how they helped you do work.

38. PREDICT How do you think nanotechnology will be useful in the future? Give several examples.

UNIT PROJECTS

Evaluate all of the data, results, and information from your project folder. Prepare to present your project to the class. Be ready to answer questions posed by your classmates about your results.

C 174 Unit: **Motion and Forces**

MONITOR AND RETEACH

If students have trouble applying the concepts in items 28–30, have them model the scenario using a ramp, a mass, and a spring scale. Ask: **1.** What distances do you measure to determine mechanical advantage of an inclined plane? *ramp length and height* **2.** How does increasing the length of a ramp affect the amount of force used? *It decreases it.* **3.** Can you use this model to answer items 28–30? Explain. *In theory, yes, by using similar reasoning; but the values would be different.* Students may benefit from summarizing sections of the chapter.

 Summarizing the Chapter, pp. 344–345

Standardized Test Practice

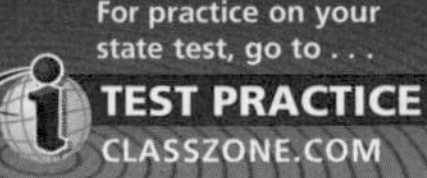
For practice on your state test, go to . . .
TEST PRACTICE
CLASSZONE.COM

Analyzing Graphics

The Archimedean screw is a mechanical device first used more than 2000 years ago. It consists of a screw inside a cylinder. One end of the device is placed in water. As the screw is turned with a handle, its threads carry water upward. The Archimedean screw is still used in some parts of the world to pump water for irrigating fields. It can also be used to move grain in mills.

Study the illustration of an Archimedean screw. Then answer the questions that follow.

1. Which type of simple machine moves water in the cylinder?
a. block and tackle
b. pulley
c. screw
d. wedge

2. Which type of simple machine is the handle?
a. wheel and axle
b. inclined plane
c. pulley
d. wedge

3. What is the energy source for the Archimedean screw?
a. the water pressure inside the screw
b. the person who is turning the handle
c. falling water that is turning the screw
d. electrical energy

4. How is the Archimedean screw helping the person in the illustration do work?
a. by decreasing the input force needed to lift the water
b. by decreasing the work needed to lift the water
c. by decreasing the distance over which the input force is applied
d. by keeping the water from overflowing its banks

5. If the threads on the Archimedean screw are closer together, the input force must be applied over a greater distance. This means that the person using it must turn the handle
a. with more force
b. fewer times but faster
c. in the opposite direction
d. more times with less effort

Extended Response

Answer the two questions below in detail.

6. A playground seesaw is an example of a lever. The fulcrum is located at the center of the board. People seated at either end take turns applying the force needed to move the other person. If one person weighs more than the other, how can they operate the seesaw? Consider several possibilities in your answer.

7. Picture two gears of different sizes turning together. Suppose you can apply a force to turn the larger gear or the smaller gear, and it will turn the other. Discuss what difference it would make whether you turned the larger or smaller gear. Describe the input work you would do on the gear you are turning and the output work that gear would do on the other gear.

METACOGNITIVE ACTIVITY

Have students answer the following questions in their **Science Notebook:**

1. How can you tell when you have made a mistake in doing calculations?
2. What constraint did you experience when building your machine as part of the Chapter Investigation?
3. Now that you have completed the chapters on forces, motion, and machines, what would you have done differently with your Unit Project? Refer back to the formula for efficiency on p. 150.

Analyzing Graphics

1. c
2. a
3. b
4. a
5. d

Extended Response

6. RUBRIC

4 points for a response that correctly answers the question and gives 3 or more correct and varied examples.

Sample answer: Assuming the seesaw (the lever) cannot be moved, there are several possibilities: 1. The heavier person can sit closer to the fulcrum. 2. The heavier person can stay seated and the lighter person can be positioned farther away from the fulcrum. 3. Both people move at the same time. In all cases, they try to balance the board first.

3 points correctly answers the question and gives 2 or more correct and varied examples
2 points correctly answers the question and gives 1 or more correct and varied examples
1 point correctly answers the question

7. RUBRIC

4 points for a response that correctly describes the situation and covers the following key concepts accurately:

- compound machine
- input force
- output force
- mechanical advantage

Sample answer: A set of gears forms a compound machine. The gear that turns another gear applies the input force; the gear that is turned exerts the output force. One turn of the larger gear will turn the smaller gear several times. You will need to turn the smaller gear several times in order to make the larger gear turn once. The mechanical advantage is greater when the smaller gear turns the larger gear.

3 points for a response that covers 3 key concepts accurately
2 points for a response that covers 2 key concepts accurately
1 point for a response that covers 1 key concept accurately

Student Resource Handbooks

Scientific Thinking Handbook R2

Making Observations R2
Predicting and Hypothesizing R3
Inferring R4
Identifying Cause and Effect R5
Recognizing Bias R6
Identifying Faulty Reasoning R7
Analyzing Statements R8

Lab Handbook R10

Safety Rules R10
Using Lab Equipment R12
The Metric System and SI Units R20
Precision and Accuracy R22
Making Data Tables and Graphs R23
Designing an Experiment R28

Math Handbook R36

Describing a Set of Data R36
Using Ratios, Rates, and Proportions R38
Using Decimals, Fractions, and Percents R39
Using Formulas R42
Finding Areas R43
Finding Volumes R43
Using Significant Figures R44
Using Scientific Notation R44

Note-Taking Handbook R45

Note-Taking Strategies R45
Vocabulary Strategies R50

Scientific Thinking Handbook

SCIENTIFIC THINKING HANDBOOK

Making Observations

An **observation** is an act of noting and recording an event, characteristic, behavior, or anything else detected with an instrument or with the senses.

Observations allow you to make informed hypotheses and to gather data for experiments. Careful observations often lead to ideas for new experiments. There are two categories of observations:

- **Quantitative observations** can be expressed in numbers and include records of time, temperature, mass, distance, and volume.
- **Qualitative observations** include descriptions of sights, sounds, smells, and textures.

EXAMPLE

A student dissolved 30 grams of Epsom salts in water, poured the solution into a dish, and let the dish sit out uncovered overnight. The next day, she made the following observations of the Epsom salt crystals that grew in the dish.

To determine the mass, the student found the mass of the dish before and after growing the crystals and then used subtraction to find the difference.

The student measured several crystals and calculated the mean length. (To learn how to calculate the mean of a data set, see page R36.)

Table 1. Observations of Epsom Salt Crystals

Quantitative Observations	Qualitative Observations
• mass = 30 g • mean crystal length = 0.5 cm • longest crystal length = 2 cm	• Crystals are clear. • Crystals are long, thin, and rectangular. • White crust has formed around edge of dish. Epsom salt crystals

Photographs or sketches are useful for recording qualitative observations.

MORE ABOUT OBSERVING

- Make quantitative observations whenever possible. That way, others will know exactly what you observed and be able to compare their results with yours.
- It is always a good idea to make qualitative observations too. You never know when you might observe something unexpected.

Predicting and Hypothesizing

A **prediction** is an expectation of what will be observed or what will happen. A **hypothesis** is a tentative explanation for an observation or scientific problem that can be tested by further investigation.

SCIENTIFIC THINKING HANDBOOK

EXAMPLE

Suppose you have made two paper airplanes and you wonder why one of them tends to glide farther than the other one.

1. Start by asking a question.
2. Make an educated guess. After examination, you notice that the wings of the airplane that flies farther are slightly larger than the wings of the other airplane.
3. Write a prediction based upon your educated guess, in the form of an "If . . . , then . . ." statement. Write the independent variable after the word *if*, and the dependent variable after the word *then.*
4. To make a hypothesis, explain why you think what you predicted will occur. Write the explanation after the word *because*.

1. Why does one of the paper airplanes glide farther than the other?

2. The size of an airplane's wings may affect how far the airplane will glide.

3. Prediction: If I make a paper airplane with larger wings, then the airplane will glide farther.

To read about independent and dependent variables, see page R30.

4. Hypothesis: If I make a paper airplane with larger wings, then the airplane will glide farther, because the additional surface area of the wing will produce more lift.

Notice that the part of the hypothesis after *because* adds an explanation of why the airplane will glide farther.

MORE ABOUT HYPOTHESES

- The results of an experiment cannot prove that a hypothesis is correct. Rather, the results either support or do not support the hypothesis.
- Valuable information is gained even when your hypothesis is not supported by your results. For example, it would be an important discovery to find that wing size is not related to how far an airplane glides.
- In science, a hypothesis is supported only after many scientists have conducted many experiments and produced consistent results.

SCIENTIFIC THINKING HANDBOOK

Inferring

An **inference** is a logical conclusion drawn from the available evidence and prior knowledge. Inferences are often made from observations.

EXAMPLE

A student observing a set of acorns noticed something unexpected about one of them. He noticed a white, soft-bodied insect eating its way out of the acorn.

The student recorded these observations.

Observations

- There is a hole in the acorn, about 0.5 cm in diameter, where the insect crawled out.
- There is a second hole, which is about the size of a pinhole, on the other side of the acorn.
- The inside of the acorn is hollow.

Here are some inferences that can be made on the basis of the observations.

Inferences

- The insect formed from the material inside the acorn, grew to its present size, and ate its way out of the acorn.
- The insect crawled through the smaller hole, ate the inside of the acorn, grew to its present size, and ate its way out of the acorn.
- An egg was laid in the acorn through the smaller hole. The egg hatched into a larva that ate the inside of the acorn, grew to its present size, and ate its way out of the acorn.

When you make inferences, be sure to look at all of the evidence available and combine it with what you already know.

MORE ABOUT INFERENCES

Inferences depend both on observations and on the knowledge of the people making the inferences. Ancient people who did not know that organisms are produced only by similar organisms might have made an inference like the first one. A student today might look at the same observations and make the second inference. A third student might have knowledge about this particular insect and know that it is never small enough to fit through the smaller hole, leading her to the third inference.

Identifying Cause and Effect

In a **cause-and-effect relationship,** one event or characteristic is the result of another. Usually an effect follows its cause in time.

There are many examples of cause-and-effect relationships in everyday life.

Cause	Effect
Turn off a light.	Room gets dark.
Drop a glass.	Glass breaks.
Blow a whistle.	Sound is heard.

Scientists must be careful not to infer a cause-and-effect relationship just because one event happens after another event. When one event occurs after another, you cannot infer a cause-and-effect relationship on the basis of that information alone. You also cannot conclude that one event caused another if there are alternative ways to explain the second event. A scientist must demonstrate through experimentation or continued observation that an event was truly caused by another event.

EXAMPLE

Make an Observation

Suppose you have a few plants growing outside. When the weather starts getting colder, you bring one of the plants indoors. You notice that the plant you brought indoors is growing faster than the others are growing. You cannot conclude from your observation that the change in temperature was the cause of the increased plant growth, because there are alternative explanations for the observation. Some possible explanations are given below.

- The humidity indoors caused the plant to grow faster.
- The level of sunlight indoors caused the plant to grow faster.
- The indoor plant's being noticed more often and watered more often than the outdoor plants caused it to grow faster.
- The plant that was brought indoors was healthier than the other plants to begin with.

To determine which of these factors, if any, caused the indoor plant to grow faster than the outdoor plants, you would need to design and conduct an experiment.

See pages R28–R35 for information about designing experiments.

SCIENTIFIC THINKING HANDBOOK

Recognizing Bias

Television, newspapers, and the Internet are full of experts claiming to have scientific evidence to back up their claims. How do you know whether the claims are really backed up by good science?

SCIENTIFIC THINKING HANDBOOK

Bias is a slanted point of view, or personal prejudice. The goal of scientists is to be as objective as possible and to base their findings on facts instead of opinions. However, bias often affects the conclusions of researchers, and it is important to learn to recognize bias.

When scientific results are reported, you should consider the source of the information as well as the information itself. It is important to critically analyze the information that you see and read.

SOURCES OF BIAS

There are several ways in which a report of scientific information may be biased. Here are some questions that you can ask yourself:

1. Who is sponsoring the research?

   Sometimes, the results of an investigation are biased because an organization paying for the research is looking for a specific answer. This type of bias can affect how data are gathered and interpreted.

2. Is the research sample large enough?

   Sometimes research does not include enough data. The larger the sample size, the more likely that the results are accurate, assuming a truly random sample.

3. In a survey, who is answering the questions?

   The results of a survey or poll can be biased. The people taking part in the survey may have been specifically chosen because of how they would answer. They may have the same ideas or lifestyles. A survey or poll should make use of a random sample of people.

4. Are the people who take part in a survey biased?

   People who take part in surveys sometimes try to answer the questions the way they think the researcher wants them to answer. Also, in surveys or polls that ask for personal information, people may be unwilling to answer questions truthfully.

SCIENTIFIC BIAS

It is also important to realize that scientists have their own biases because of the types of research they do and because of their scientific viewpoints. Two scientists may look at the same set of data and come to completely different conclusions because of these biases. However, such disagreements are not necessarily bad. In fact, a critical analysis of disagreements is often responsible for moving science forward.

R6

Identifying Faulty Reasoning

Faulty reasoning is wrong or incorrect thinking. It leads to mistakes and to wrong conclusions. Scientists are careful not to draw unreasonable conclusions from experimental data. Without such caution, the results of scientific investigations may be misleading.

EXAMPLE

Scientists try to make generalizations based on their data to explain as much about nature as possible. If only a small sample of data is looked at, however, a conclusion may be faulty. Suppose a scientist has studied the effects of the El Niño and La Niña weather patterns on flood damage in California from 1989 to 1995. The scientist organized the data in the bar graph below.

Flood and Storm Damage in California

Estimated damage (millions of dollars)
2000
1500
1000
500
0
1989 1992 1995
Starting year of season (July 1–June 30)
Weak–moderate El Niño
Strong El Niño
SOURCE: *Governor's Office of Emergency Services, California*

The scientist drew the following conclusions:

1. The La Niña weather pattern has no effect on flooding in California.
2. When neither weather pattern occurs, there is almost no flood damage.
3. A weak or moderate El Niño produces a small or moderate amount of flooding.
4. A strong El Niño produces a lot of flooding.

For the six-year period of the scientist's investigation, these conclusions may seem to be reasonable. However, a six-year study of weather patterns may be too small of a sample for the conclusions to be supported. Consider the following graph, which shows information that was gathered from 1949 to 1997.

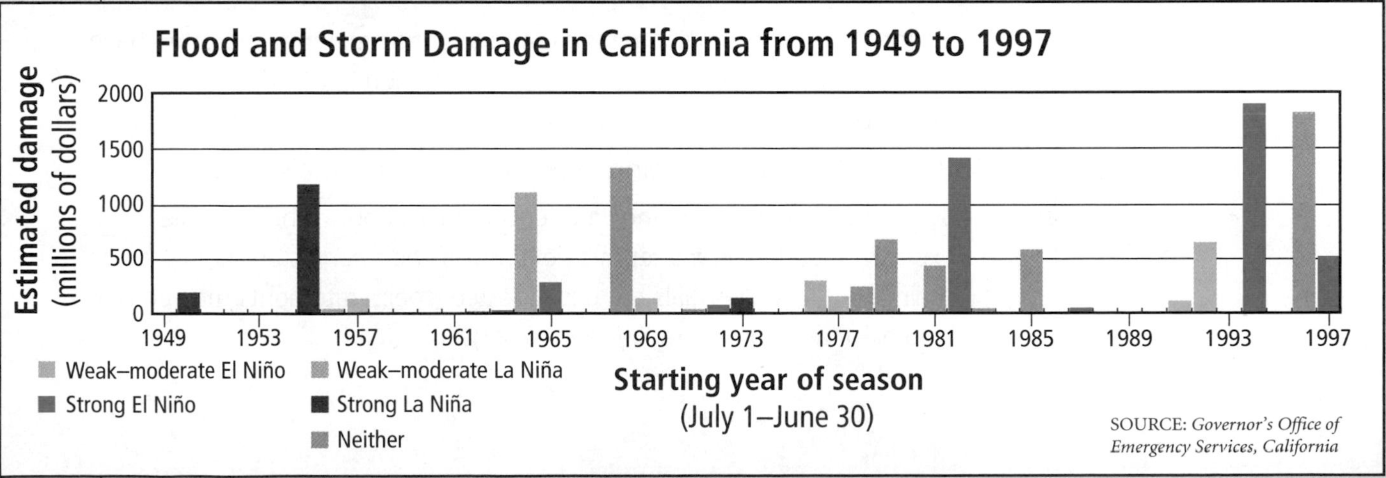

The only one of the conclusions that all of this information supports is number 3: a weak or moderate El Niño produces a small or moderate amount of flooding. By collecting more data, scientists can be more certain of their conclusions and can avoid faulty reasoning.

Analyzing Statements

To **analyze** a statement is to examine its parts carefully. Scientific findings are often reported through media such as television or the Internet. A report that is made public often focuses on only a small part of research. As a result, it is important to question the sources of information.

Evaluate Media Claims

To **evaluate** a statement is to judge it on the basis of criteria you've established. Sometimes evaluating means deciding whether a statement is true.

Reports of scientific research and findings in the media may be misleading or incomplete. When you are exposed to this information, you should ask yourself some questions so that you can make informed judgments about the information.

1. **Does the information come from a credible source?**

   Suppose you learn about a new product and it is stated that scientific evidence proves that the product works. A report from a respected news source may be more believable than an advertisement paid for by the product's manufacturer.

2. **How much evidence supports the claim?**

   Often, it may seem that there is new evidence every day of something in the world that either causes or cures an illness. However, information that is the result of several years of work by several different scientists is more credible than an advertisement that does not even cite the subjects of the experiment.

3. **How much information is being presented?**

   Science cannot solve all questions, and scientific experiments often have flaws. A report that discusses problems in a scientific study may be more believable than a report that addresses only positive experimental findings.

4. **Is scientific evidence being presented by a specific source?**

   Sometimes scientific findings are reported by people who are called experts or leaders in a scientific field. But if their names are not given or their scientific credentials are not reported, their statements may be less credible than those of recognized experts.

R8

Differentiate Between Fact and Opinion

Sometimes information is presented as a fact when it may be an opinion. When scientific conclusions are reported, it is important to recognize whether they are based on solid evidence. Again, you may find it helpful to ask yourself some questions.

1. **What is the difference between a fact and an opinion?**

   A **fact** is a piece of information that can be strictly defined and proved true. An **opinion** is a statement that expresses a belief, value, or feeling. An opinion cannot be proved true or false. For example, a person's age is a fact, but if someone is asked how old they feel, it is impossible to prove the person's answer to be true or false.

2. **Can opinions be measured?**

   Yes, opinions can be measured. In fact, surveys often ask for people's opinions on a topic. But there is no way to know whether or not an opinion is the truth.

HOW TO DIFFERENTIATE FACT FROM OPINION

Human Activities and the Environment

Unfortunately, human use of fossil fuels is one of the most significant developments of the past few centuries. Humans rely on fossil fuels, a non-renewable energy resource, for more than 90 percent of their energy needs.

This careless misuse of our planet's resources has resulted in pollution, global warming, and the destruction of fragile ecosystems. For example, oil pipelines carry more than one million barrels of oil each day across tundra regions. Transporting oil across such areas can only result in oil spills that poison the land for decades.

Opinions
Notice words or phrases that express beliefs or feelings. The words *unfortunately* and *careless* show that opinions are being expressed.

Opinion
Look for statements that speculate about events. These statements are opinions, because they cannot be proved.

Facts
Statements that contain statistics tend to be facts. Writers often use facts to support their opinions.

Lab Handbook

Safety Rules

Before you work in the laboratory, read these safety rules twice. Ask your teacher to explain any rules that you do not completely understand. Refer to these rules later on if you have questions about safety in the science classroom.

Directions

- Read all directions and make sure that you understand them before starting an investigation or lab activity. If you do not understand how to do a procedure or how to use a piece of equipment, ask your teacher.
- Do not begin any investigation or touch any equipment until your teacher has told you to start.
- Never experiment on your own. If you want to try a procedure that the directions do not call for, ask your teacher for permission first.
- If you are hurt or injured in any way, tell your teacher immediately.

Dress Code

goggles

apron

gloves

- Wear goggles when
  — using glassware, sharp objects, or chemicals
  — heating an object
  — working with anything that can easily fly up into the air and hurt someone's eye
- Tie back long hair or hair that hangs in front of your eyes.
- Remove any article of clothing—such as a loose sweater or a scarf—that hangs down and may touch a flame, chemical, or piece of equipment.
- Observe all safety icons calling for the wearing of eye protection, gloves, and aprons.

Heating and Fire Safety

fire safety

heating safety

- Keep your work area neat, clean, and free of extra materials.
- Never reach over a flame or heat source.
- Point objects being heated away from you and others.
- Never heat a substance or an object in a closed container.
- Never touch an object that has been heated. If you are unsure whether something is hot, treat it as though it is. Use oven mitts, clamps, tongs, or a test-tube holder.
- Know where the fire extinguisher and fire blanket are kept in your classroom.
- Do not throw hot substances into the trash. Wait for them to cool or use the container your teacher puts out for disposal.

Electrical Safety

electrical safety

- Never use lamps or other electrical equipment with frayed cords.
- Make sure no cord is lying on the floor where someone can trip over it.
- Do not let a cord hang over the side of a counter or table so that the equipment can easily be pulled or knocked to the floor.
- Never let cords hang into sinks or other places where water can be found.
- Never try to fix electrical problems. Inform your teacher of any problems immediately.
- Unplug an electrical cord by pulling on the plug, not the cord.

Chemical Safety

chemical safety

poison

fumes

- If you spill a chemical or get one on your skin or in your eyes, tell your teacher right away.
- Never touch, taste, or sniff any chemicals in the lab. If you need to determine odor, waft. Wafting consists of holding the chemical in its container 15 centimeters (6 in.) away from your nose, and using your fingers to bring fumes from the container to your nose.
- Keep lids on all chemicals you are not using.
- Never put unused chemicals back into the original containers. Throw away extra chemicals where your teacher tells you to.
- Pour chemicals over a sink or your work area, not over the floor.
- If you get a chemical in your eye, use the eyewash right away.
- Always wash your hands after handling chemicals, plants, or soil.

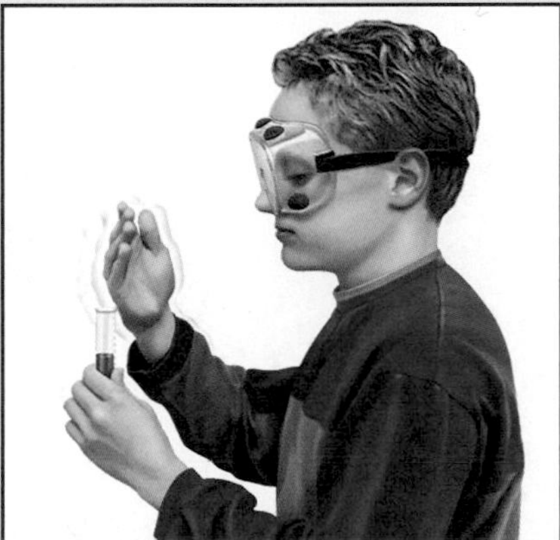
Wafting

Glassware and Sharp-Object Safety

sharp objects

- If you break glassware, tell your teacher right away.
- Do not use broken or chipped glassware. Give these to your teacher.
- Use knives and other cutting instruments carefully. Always wear eye protection and cut away from you.

Animal Safety

- Never hurt an animal.
- Touch animals only when necessary. Follow your teacher's instructions for handling animals.
- Always wash your hands after working with animals.

Cleanup

disposal

- Follow your teacher's instructions for throwing away or putting away supplies.
- Clean your work area and pick up anything that has dropped to the floor.
- Wash your hands.

LAB HANDBOOK

Using Lab Equipment

Different experiments require different types of equipment. But even though experiments differ, the ways in which the equipment is used are the same.

Beakers

- Use beakers for holding and pouring liquids.
- Do not use a beaker to measure the volume of a liquid. Use a graduated cylinder instead. (See page R16.)
- Use a beaker that holds about twice as much liquid as you need. For example, if you need 100 milliliters of water, you should use a 200- or 250-milliliter beaker.

Test Tubes

- Use test tubes to hold small amounts of substances.
- Do not use a test tube to measure the volume of a liquid.
- Use a test tube when heating a substance over a flame. Aim the mouth of the tube away from yourself and other people.
- Liquids easily spill or splash from test tubes, so it is important to use only small amounts of liquids.

Test-Tube Holder

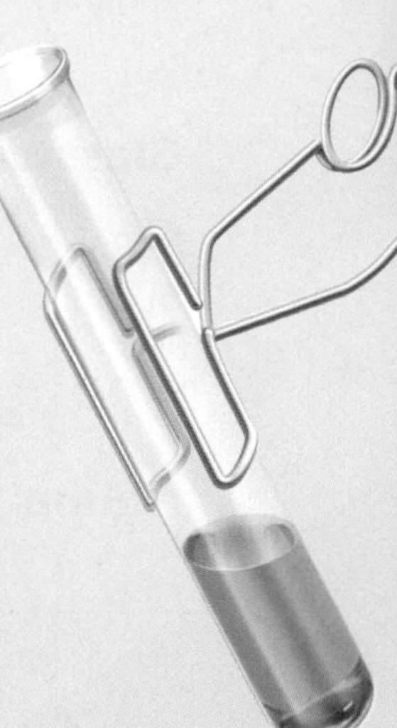

- Use a test-tube holder when heating a substance in a test tube.
- Use a test-tube holder if the substance in a test tube is dangerous to touch.
- Make sure the test-tube holder tightly grips the test tube so that the test tube will not slide out of the holder.
- Make sure that the test-tube holder is above the surface of the substance in the test tube so that you can observe the substance.

LAB HANDBOOK

Test-Tube Rack

- Use a test-tube rack to organize test tubes before, during, and after an experiment.
- Use a test-tube rack to keep test tubes upright so that they do not fall over and spill their contents.
- Use a test-tube rack that is the correct size for the test tubes that you are using. If the rack is too small, a test tube may become stuck. If the rack is too large, a test tube may lean over, and some of its contents may spill or splash.

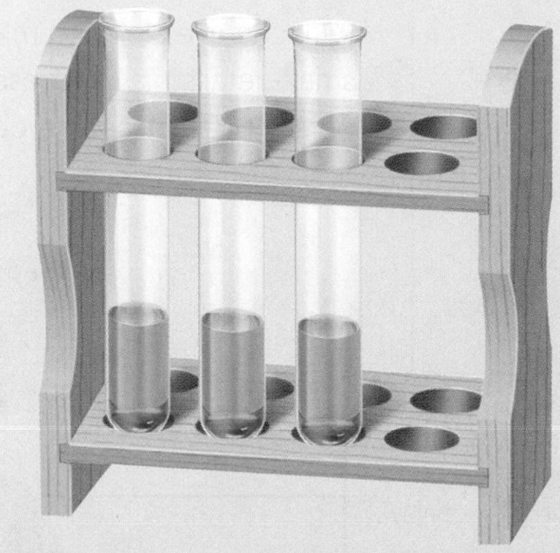

Forceps

- Use forceps when you need to pick up or hold a very small object that should not be touched with your hands.
- Do not use forceps to hold anything over a flame, because forceps are not long enough to keep your hand safely away from the flame. Plastic forceps will melt, and metal forceps will conduct heat and burn your hand.

Hot Plate

- Use a hot plate when a substance needs to be kept warmer than room temperature for a long period of time.
- Use a hot plate instead of a Bunsen burner or a candle when you need to carefully control temperature.
- Do not use a hot plate when a substance needs to be burned in an experiment.
- Always use "hot hands" safety mitts or oven mitts when handling anything that has been heated on a hot plate.

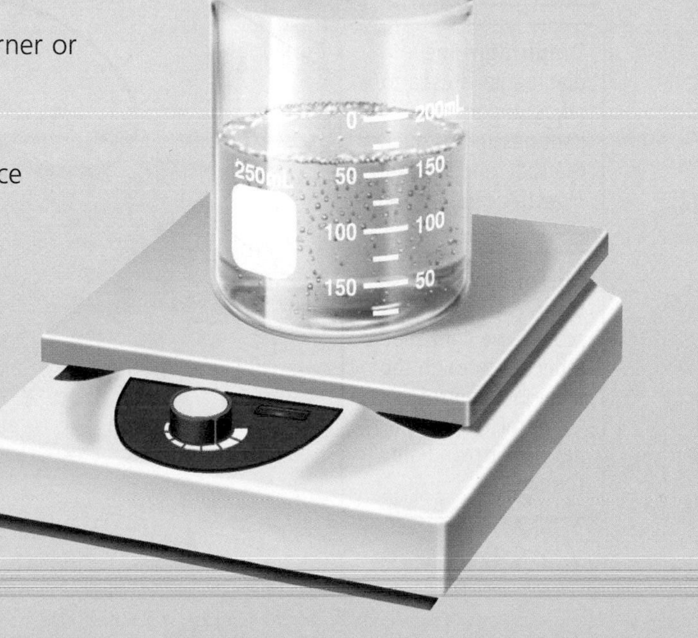

Microscope

Scientists use microscopes to see very small objects that cannot easily be seen with the eye alone. A microscope magnifies the image of an object so that small details may be observed. A microscope that you may use can magnify an object 400 times—the object will appear 400 times larger than its actual size.

Body The body separates the lens in the eyepiece from the objective lenses below.

Nosepiece The nosepiece holds the objective lenses above the stage and rotates so that all lenses may be used.

High-Power Objective Lens This is the largest lens on the nosepiece. It magnifies an image approximately 40 times.

Stage The stage supports the object being viewed.

Diaphragm The diaphragm is used to adjust the amount of light passing through the slide and into an objective lens.

Mirror or Light Source Some microscopes use light that is reflected through the stage by a mirror. Other microscopes have their own light sources.

Eyepiece Objects are viewed through the eyepiece. The eyepiece contains a lens that commonly magnifies an image 10 times.

Coarse Adjustment This knob is used to focus the image of an object when it is viewed through the low-power lens.

Fine Adjustment This knob is used to focus the image of an object when it is viewed through the high-power lens.

Low-Power Objective Lens This is the smallest lens on the nosepiece. It magnifies an image approximately 10 times.

Arm The arm supports the body above the stage. Always carry a microscope by the arm and base.

Stage Clip The stage clip holds a slide in place on the stage.

Base The base supports the microscope.

VIEWING AN OBJECT

1. Use the coarse adjustment knob to raise the body tube.
2. Adjust the diaphragm so that you can see a bright circle of light through the eyepiece.
3. Place the object or slide on the stage. Be sure that it is centered over the hole in the stage.
4. Turn the nosepiece to click the low-power lens into place.
5. Using the coarse adjustment knob, slowly lower the lens and focus on the specimen being viewed. Be sure not to touch the slide or object with the lens.
6. When switching from the low-power lens to the high-power lens, first raise the body tube with the coarse adjustment knob so that the high-power lens will not hit the slide.
7. Turn the nosepiece to click the high-power lens into place.
8. Use the fine adjustment knob to focus on the specimen being viewed. Again, be sure not to touch the slide or object with the lens.

MAKING A SLIDE, OR WET MOUNT

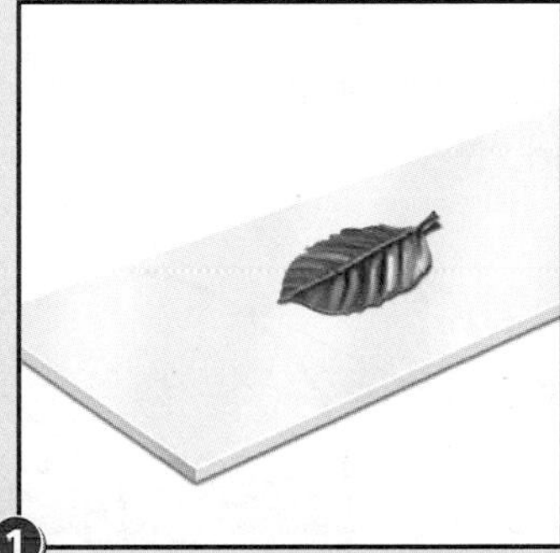

1 Place the specimen in the center of a clean slide.

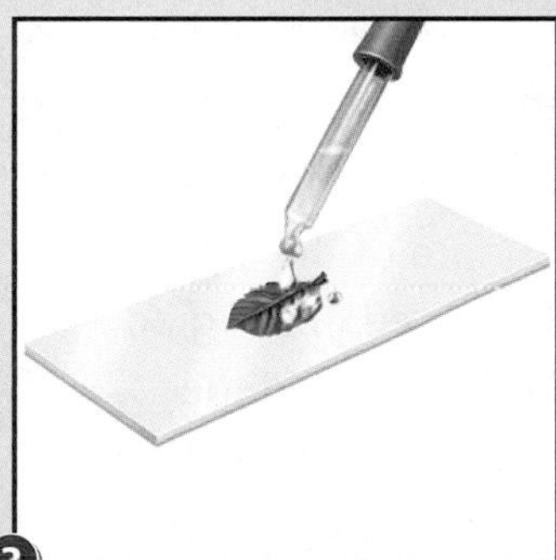

2 Place a drop of water on the specimen.

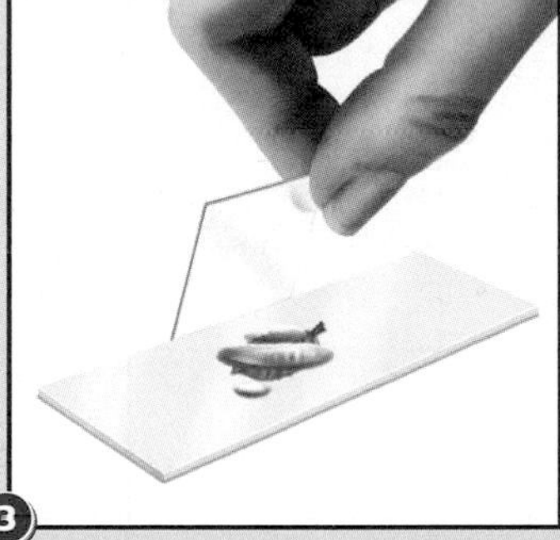

3 Place a cover slip on the slide. Put one edge of the cover slip into the drop of water and slowly lower it over the specimen.

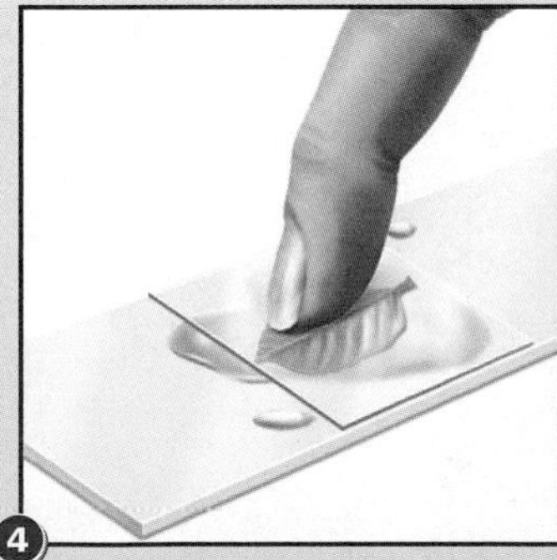

4 Remove any air bubbles from under the cover slip by gently tapping the cover slip.

5 Dry any excess water before placing the slide on the microscope stage for viewing.

Spring Scale (Force Meter)

- Use a spring scale to measure a force pulling on the scale.
- Use a spring scale to measure the force of gravity exerted on an object by Earth.
- To measure a force accurately, a spring scale must be zeroed before it is used. The scale is zeroed when no weight is attached and the indicator is positioned at zero.
- Do not attach a weight that is either too heavy or too light to a spring scale. A weight that is too heavy could break the scale or exert too great a force for the scale to measure. A weight that is too light may not exert enough force to be measured accurately.

Graduated Cylinder

- Use a graduated cylinder to measure the volume of a liquid.
- Be sure that the graduated cylinder is on a flat surface so that your measurement will be accurate.
- When reading the scale on a graduated cylinder, be sure to have your eyes at the level of the surface of the liquid.
- The surface of the liquid will be curved in the graduated cylinder. Read the volume of the liquid at the bottom of the curve, or meniscus (muh-NIHS-kuhs).
- You can use a graduated cylinder to find the volume of a solid object by measuring the increase in a liquid's level after you add the object to the cylinder.

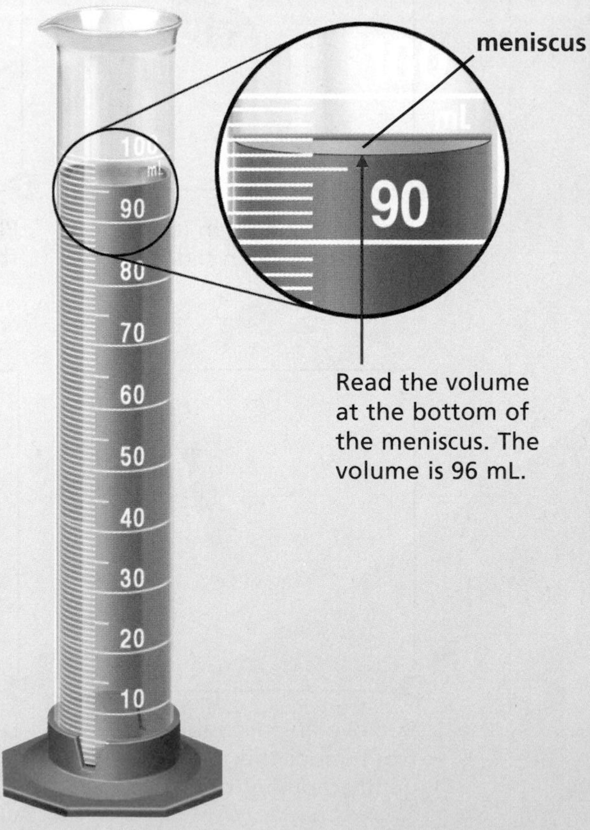

Read the volume at the bottom of the meniscus. The volume is 96 mL.

Metric Rulers

- Use metric rulers or meter sticks to measure objects' lengths.
- Do not measure an object from the end of a metric ruler or meter stick, because the end is often imperfect. Instead, measure from the 1-centimeter mark, but remember to subtract a centimeter from the apparent measurement.
- Estimate any lengths that extend between marked units. For example, if a meter stick shows centimeters but not millimeters, you can estimate the length that an object extends between centimeter marks to measure it to the nearest millimeter.
- **Controlling Variables** If you are taking repeated measurements, always measure from the same point each time. For example, if you're measuring how high two different balls bounce when dropped from the same height, measure both bounces at the same point on the balls—either the top or the bottom. Do not measure at the top of one ball and the bottom of the other.

EXAMPLE

How to Measure a Leaf

1. Lay a ruler flat on top of the leaf so that the 1-centimeter mark lines up with one end. Make sure the ruler and the leaf do not move between the time you line them up and the time you take the measurement.
2. Look straight down on the ruler so that you can see exactly how the marks line up with the other end of the leaf.
3. Estimate the length by which the leaf extends beyond a marking. For example, the leaf below extends about halfway between the 4.2-centimeter and 4.3-centimeter marks, so the apparent measurement is about 4.25 centimeters.
4. Remember to subtract 1 centimeter from your apparent measurement, since you started at the 1-centimeter mark on the ruler and not at the end. The leaf is about 3.25 centimeters long (4.25 cm – 1 cm = 3.25 cm).

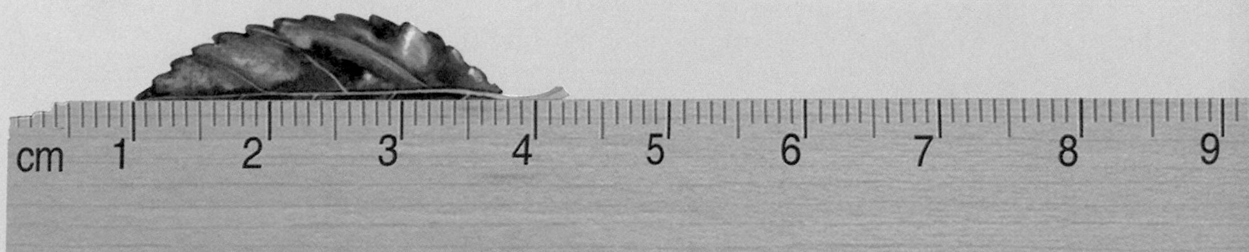

LAB HANDBOOK

Triple-Beam Balance

This balance has a pan and three beams with sliding masses, called riders. At one end of the beams is a pointer that indicates whether the mass on the pan is equal to the masses shown on the beams.

1. Make sure the balance is zeroed before measuring the mass of an object. The balance is zeroed if the pointer is at zero when nothing is on the pan and the riders are at their zero points. Use the adjustment knob at the base of the balance to zero it.
2. Place the object to be measured on the pan.
3. Move the riders one notch at a time away from the pan. Begin with the largest rider. If moving the largest rider one notch brings the pointer below zero, begin measuring the mass of the object with the next smaller rider.
4. Change the positions of the riders until they balance the mass on the pan and the pointer is at zero. Then add the readings from the three beams to determine the mass of the object.

300 g	position of largest rider
90 g	position of middle rider
+ 3 g	position of smallest rider
393 g	mass of beaker

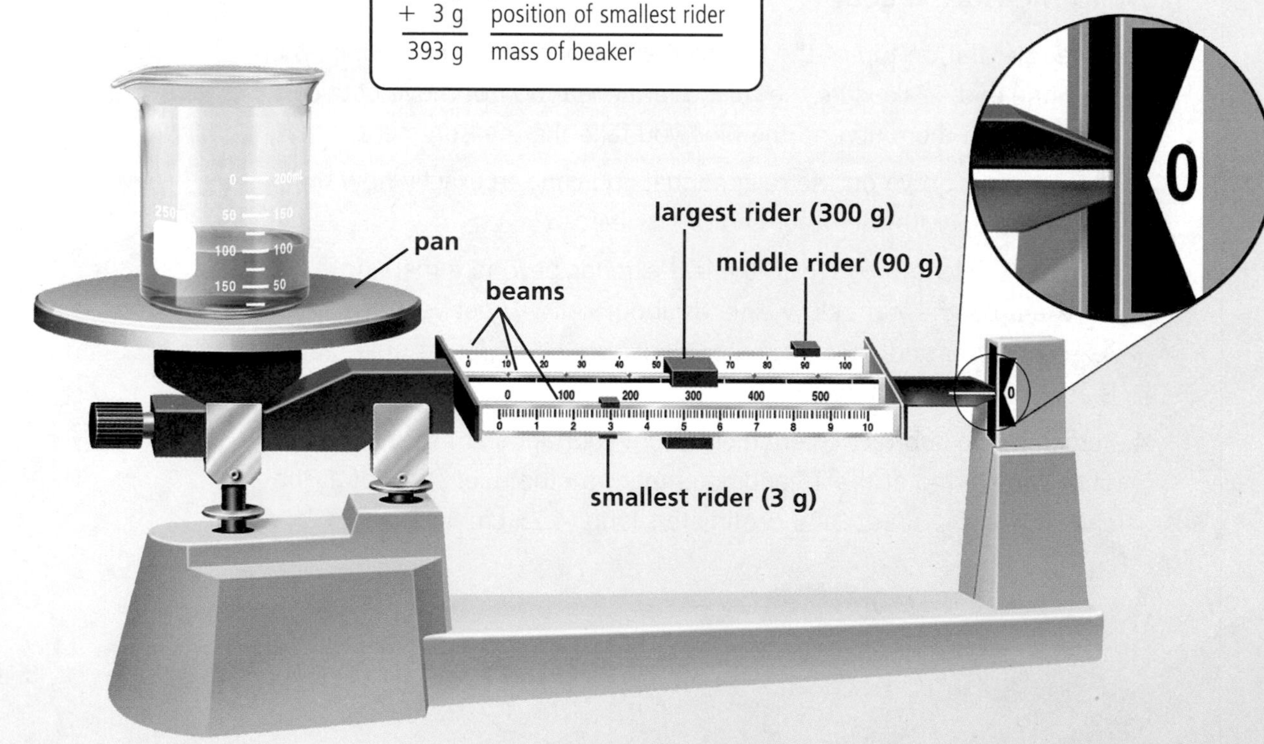

Double-Pan Balance

This type of balance has two pans. Between the pans is a pointer that indicates whether the masses on the pans are equal.

1. Make sure the balance is zeroed before measuring the mass of an object. The balance is zeroed if the pointer is at zero when there is nothing on either of the pans. Many double-pan balances have sliding knobs that can be used to zero them.
2. Place the object to be measured on one of the pans.
3. Begin adding standard masses to the other pan. Begin with the largest standard mass. If this adds too much mass to the balance, begin measuring the mass of the object with the next smaller standard mass.
4. Add standard masses until the masses on both pans are balanced and the pointer is at zero. Then add the standard masses together to determine the mass of the object being measured.

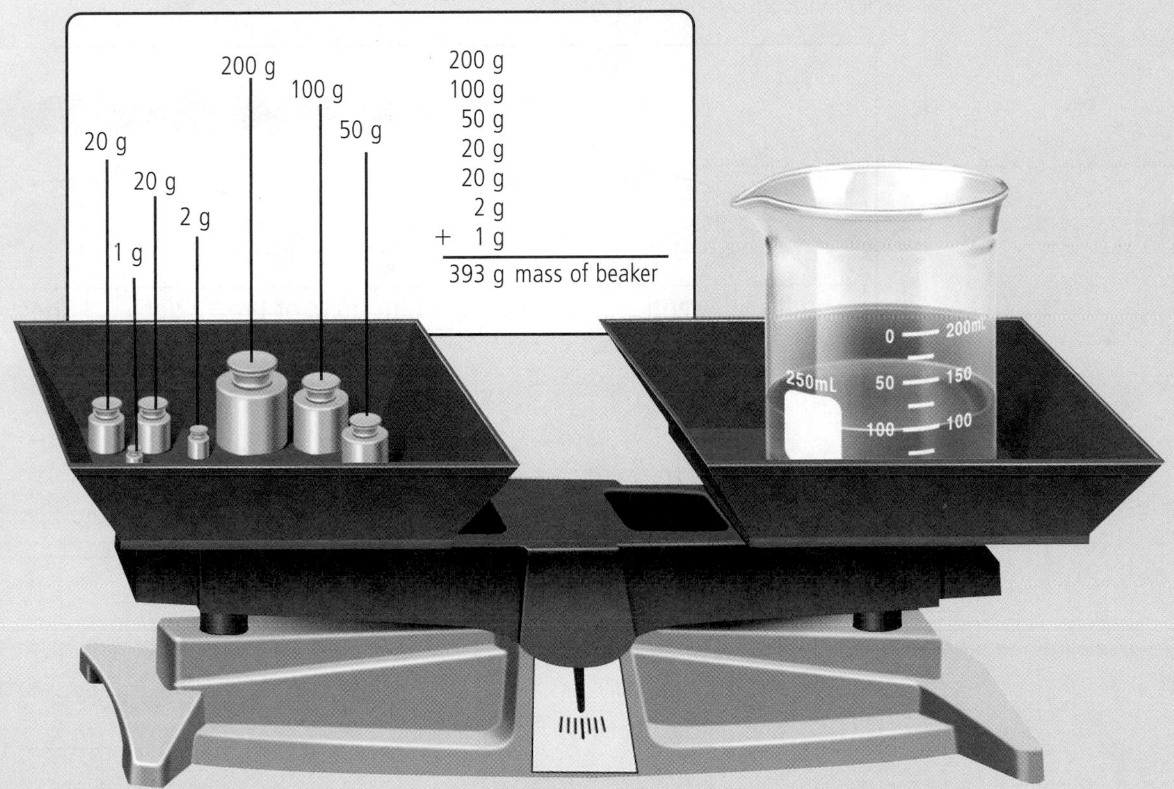

Never place chemicals or liquids directly on a pan. Instead, use the following procedure:

1. Determine the mass of an empty container, such as a beaker.
2. Pour the substance into the container, and measure the total mass of the substance and the container.
3. Subtract the mass of the empty container from the total mass to find the mass of the substance.

R19

The Metric System and SI Units

Scientists use International System (SI) units for measurements of distance, volume, mass, and temperature. The International System is based on multiples of ten and the metric system of measurement.

Basic SI Units		
Property	**Name**	**Symbol**
length	meter	m
volume	liter	L
mass	kilogram	kg
temperature	kelvin	K

SI Prefixes		
Prefix	**Symbol**	**Multiple of 10**
kilo-	k	1000
hecto-	h	100
deca-	da	10
deci-	d	0.1 $(\frac{1}{10})$
centi-	c	0.01 $(\frac{1}{100})$
milli-	m	0.001 $(\frac{1}{1000})$

Changing Metric Units

You can change from one unit to another in the metric system by multiplying or dividing by a power of 10.

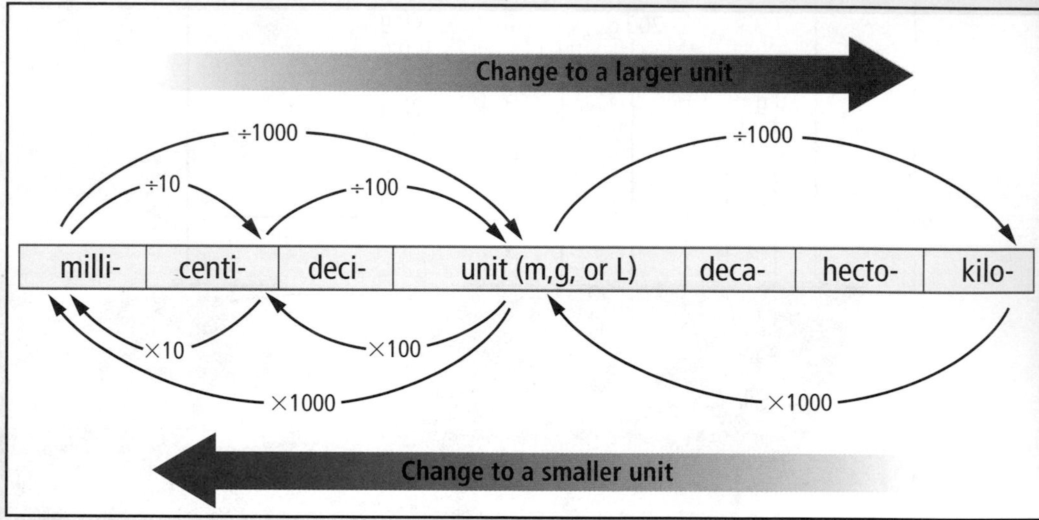

Example

Change 0.64 liters to milliliters.

(1) Decide whether to multiply or divide.

(2) Select the power of 10.

ANSWER 0.64 L = 640 mL

Change to a smaller unit by multiplying.

mL ← × 1000 — **L**

0.64 × 1000 = **640.**

Example

Change 23.6 grams to kilograms.

(1) Decide whether to multiply or divide.

(2) Select the power of 10.

ANSWER 23.6 g = 0.0236 kg

Change to a larger unit by dividing.

g — ÷ 1000 → **kg**

23.6 ÷ 1000 = **0.0236**

Temperature Conversions

Even though the kelvin is the SI base unit of temperature, the degree Celsius will be the unit you use most often in your science studies. The formulas below show the relationships between temperatures in degrees Fahrenheit (°F), degrees Celsius (°C), and kelvins (K).

$$°C = \frac{5}{9}(°F - 32)$$

$$°F = \frac{9}{5}°C + 32$$

$$K = °C + 273$$

See page R42 for help with using formulas.

Examples of Temperature Conversions		
Condition	**Degrees Celsius**	**Degrees Fahrenheit**
Freezing point of water	0	32
Cool day	10	50
Mild day	20	68
Warm day	30	86
Normal body temperature	37	98.6
Very hot day	40	104
Boiling point of water	100	212

LAB HANDBOOK

Converting Between SI and U.S. Customary Units

Use the chart below when you need to convert between SI units and U.S. customary units.

SI Unit	From SI to U.S. Customary			From U.S. Customary to SI		
Length	**When you know**	**multiply by**	**to find**	**When you know**	**multiply by**	**to find**
kilometer (km) = 1000 m	kilometers	0.62	miles	miles	1.61	kilometers
meter (m) = 100 cm	meters	3.28	feet	feet	0.3048	meters
centimeter (cm) = 10 mm	centimeters	0.39	inches	inches	2.54	centimeters
millimeter (mm) = 0.1 cm	millimeters	0.04	inches	inches	25.4	millimeters
Area	**When you know**	**multiply by**	**to find**	**When you know**	**multiply by**	**to find**
square kilometer (km^2)	square kilometers	0.39	square miles	square miles	2.59	square kilometers
square meter (m^2)	square meters	1.2	square yards	square yards	0.84	square meters
square centimeter (cm^2)	square centimeters	0.155	square inches	square inches	6.45	square centimeters
Volume	**When you know**	**multiply by**	**to find**	**When you know**	**multiply by**	**to find**
liter (L) = 1000 mL	liters	1.06	quarts	quarts	0.95	liters
	liters	0.26	gallons	gallons	3.79	liters
	liters	4.23	cups	cups	0.24	liters
	liters	2.12	pints	pints	0.47	liters
milliliter (mL) = 0.001 L	milliliters	0.20	teaspoons	teaspoons	4.93	milliliters
	milliliters	0.07	tablespoons	tablespoons	14.79	milliliters
	milliliters	0.03	fluid ounces	fluid ounces	29.57	milliliters
Mass	**When you know**	**multiply by**	**to find**	**When you know**	**multiply by**	**to find**
kilogram (kg) = 1000 g	kilograms	2.2	pounds	pounds	0.45	kilograms
gram (g) = 1000 mg	grams	0.035	ounces	ounces	28.35	grams

Precision and Accuracy

When you do an experiment, it is important that your methods, observations, and data be both precise and accurate.

low precision

precision, but not accuracy

precision and accuracy

LAB HANDBOOK

Precision

In science, **precision** is the exactness and consistency of measurements. For example, measurements made with a ruler that has both centimeter and millimeter markings would be more precise than measurements made with a ruler that has only centimeter markings. Another indicator of precision is the care taken to make sure that methods and observations are as exact and consistent as possible. Every time a particular experiment is done, the same procedure should be used. Precision is necessary because experiments are repeated several times and if the procedure changes, the results will change.

EXAMPLE

Suppose you are measuring temperatures over a two-week period. Your precision will be greater if you measure each temperature at the same place, at the same time of day, and with the same thermometer than if you change any of these factors from one day to the next.

Accuracy

In science, it is possible to be precise but not accurate. **Accuracy** depends on the difference between a measurement and an actual value. The smaller the difference, the more accurate the measurement.

EXAMPLE

Suppose you look at a stream and estimate that it is about 1 meter wide at a particular place. You decide to check your estimate by measuring the stream with a meter stick, and you determine that the stream is 1.32 meters wide. However, because it is hard to measure the width of a stream with a meter stick, it turns out that you didn't do a very good job. The stream is actually 1.14 meters wide. Therefore, even though your estimate was less precise than your measurement, your estimate was actually more accurate.

Making Data Tables and Graphs

Data tables and graphs are useful tools for both recording and communicating scientific data.

Making Data Tables

You can use a **data table** to organize and record the measurements that you make. Some examples of information that might be recorded in data tables are frequencies, times, and amounts.

EXAMPLE

Suppose you are investigating photosynthesis in two elodea plants. One sits in direct sunlight, and the other sits in a dimly lit room. You measure the rate of photosynthesis by counting the number of bubbles in the jar every ten minutes.

1. Title and number your data table.
2. Decide how you will organize the table into columns and rows.
3. Any units, such as seconds or degrees, should be included in column headings, not in the individual cells.

Table 1. Number of Bubbles from Elodea

Time (min)	Sunlight	Dim Light
0	0	0
10	15	5
20	25	8
30	32	7
40	41	10
50	47	9
60	42	9

Always number and title data tables.

The data in the table above could also be organized in a different way.

Table 1. Number of Bubbles from Elodea

Light Condition	Time (min)						
	0	10	20	30	40	50	60
Sunlight	0	15	25	32	41	47	42
Dim light	0	5	8	7	10	9	9

Put units in column heading.

LAB HANDBOOK

Making Line Graphs

You can use a **line graph** to show a relationship between variables. Line graphs are particularly useful for showing changes in variables over time.

EXAMPLE

Suppose you are interested in graphing temperature data that you collected over the course of a day.

Table 1. Outside Temperature During the Day on March 7

	Time of Day						
	7:00 A.M.	9:00 A.M.	11:00 A.M.	1:00 P.M.	3:00 P.M.	5:00 P.M.	7:00 P.M.
Temp (°C)	8	9	11	14	12	10	6

1. Use the vertical axis of your line graph for the variable that you are measuring—temperature.
2. Choose scales for both the horizontal axis and the vertical axis of the graph. You should have two points more than you need on the vertical axis, and the horizontal axis should be long enough for all of the data points to fit.
3. Draw and label each axis.
4. Graph each value. First find the appropriate point on the scale of the horizontal axis. Imagine a line that rises vertically from that place on the scale. Then find the corresponding value on the vertical axis, and imagine a line that moves horizontally from that value. The point where these two imaginary lines intersect is where the value should be plotted.
5. Connect the points with straight lines.

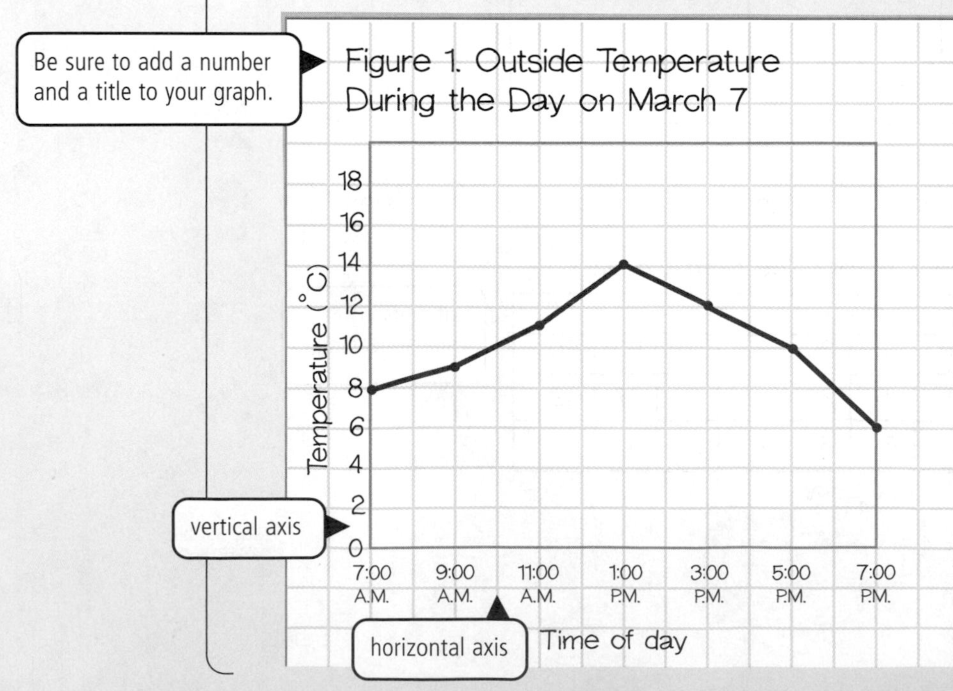

LAB HANDBOOK

Making Circle Graphs

You can use a **circle graph,** sometimes called a pie chart, to represent data as parts of a circle. Circle graphs are used only when the data can be expressed as percentages of a whole. The entire circle shown in a circle graph is equal to 100 percent of the data.

EXAMPLE

Suppose you identified the species of each mature tree growing in a small wooded area. You organized your data in a table, but you also want to show the data in a circle graph.

1. To begin, find the total number of mature trees.

   $56 + 34 + 22 + 10 + 28 = 150$

2. To find the degree measure for each sector of the circle, write a fraction comparing the number of each tree species with the total number of trees. Then multiply the fraction by 360°.

   Oak: $\frac{56}{150} \times 360° = 134.4°$

3. Draw a circle. Use a protractor to draw the angle for each sector of the graph.
4. Color and label each sector of the graph.
5. Give the graph a number and title.

Table 1. Tree Species in Wooded Area

Species	Number of Specimens
Oak	56
Maple	34
Birch	22
Willow	10
Pine	28

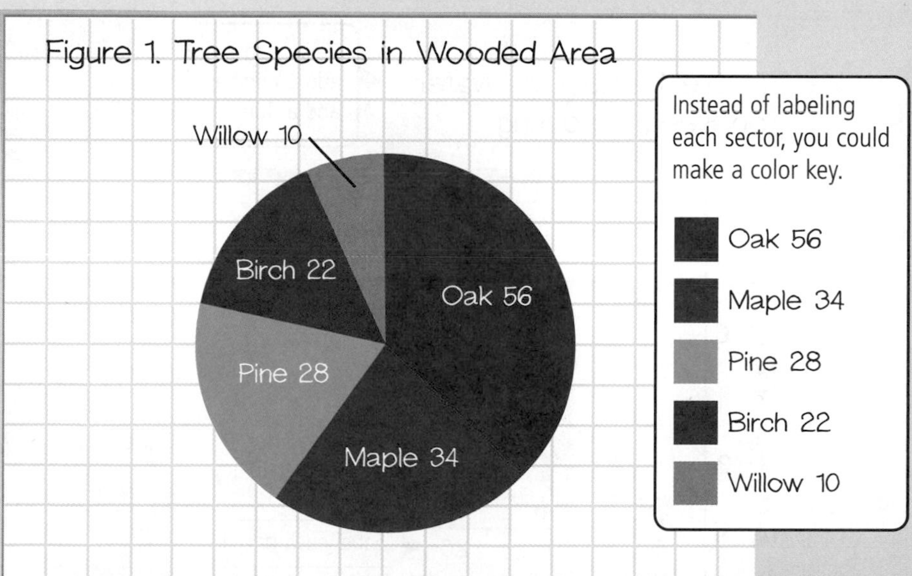

Figure 1. Tree Species in Wooded Area

Instead of labeling each sector, you could make a color key.

Oak 56
Maple 34
Pine 28
Birch 22
Willow 10

LAB HANDBOOK

Bar Graph

A **bar graph** is a type of graph in which the lengths of the bars are used to represent and compare data. A numerical scale is used to determine the lengths of the bars.

EXAMPLE

To determine the effect of water on seed sprouting, three cups were filled with sand, and ten seeds were planted in each. Different amounts of water were added to each cup over a three-day period.

Table 1. Effect of Water on Seed Sprouting

Daily Amount of Water (mL)	Number of Seeds That Sprouted After 3 Days in Sand
0	1
10	4
20	8

1. Choose a numerical scale. The greatest value is 8, so the end of the scale should have a value greater than 8, such as 10. Use equal increments along the scale, such as increments of 2.
2. Draw and label the axes. Mark intervals on the vertical axis according to the scale you chose.
3. Draw a bar for each data value. Use the scale to decide how long to make each bar.

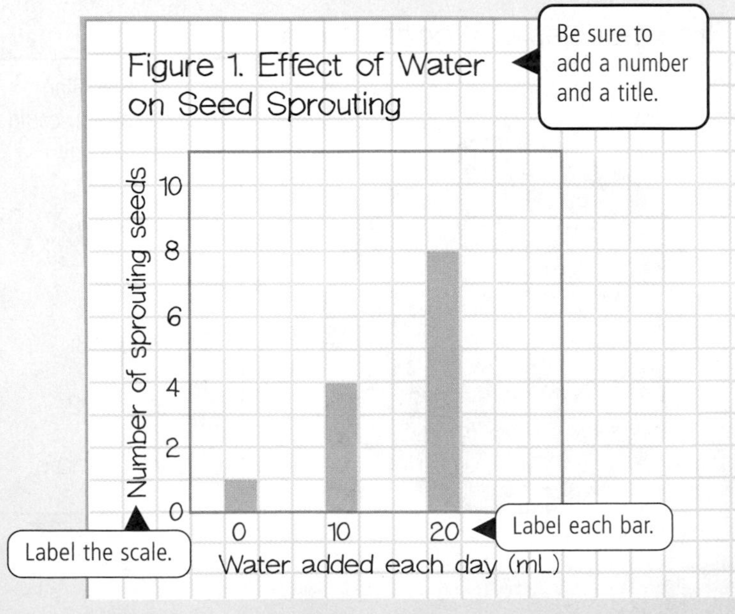

LAB HANDBOOK

Double Bar Graph

A **double bar graph** is a bar graph that shows two sets of data. The two bars for each measurement are drawn next to each other.

EXAMPLE

The same seed-sprouting experiment was repeated with potting soil. The data for sand and potting soil can be plotted on one graph.

1. Draw one set of bars, using the data for sand, as shown below.
2. Draw bars for the potting-soil data next to the bars for the sand data. Shade them a different color. Add a key.

Table 2. Effect of Water and Soil on Seed Sprouting

Daily Amount of Water (mL)	Number of Seeds That Sprouted After 3 Days in Sand	Number of Seeds That Sprouted After 3 Days in Potting Soil
0	1	2
10	4	5
20	8	9

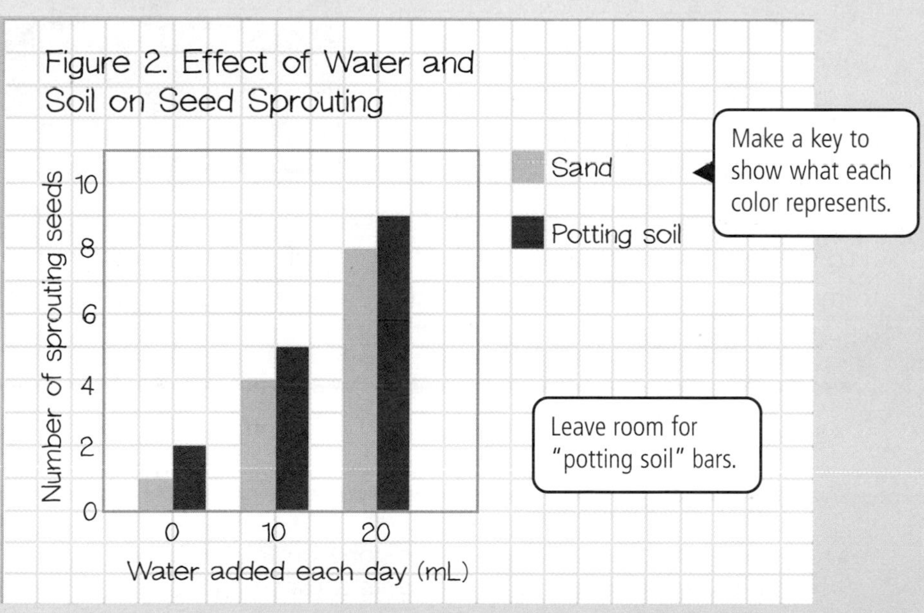

LAB HANDBOOK

Designing an Experiment

Use this section when designing or conducting an experiment.

Determining a Purpose

You can find a purpose for an experiment by doing research, by examining the results of a previous experiment, or by observing the world around you. An **experiment** is an organized procedure to study something under controlled conditions.

1. Write the purpose of your experiment as a question or problem that you want to investigate.
2. Write down research questions and begin searching for information that will help you design an experiment. Consult the library, the Internet, and other people as you conduct your research.

Don't forget to learn as much as possible about your topic before you begin.

EXAMPLE

Middle school students observed an odor near the lake by their school. They also noticed that the water on the side of the lake near the school was greener than the water on the other side of the lake. The students did some research to learn more about their observations. They discovered that the odor and green color in the lake came from algae. They also discovered that a new fertilizer was being used on a field nearby. The students inferred that the use of the fertilizer might be related to the presence of the algae and designed a controlled experiment to find out whether they were right.

Problem

How does fertilizer affect the presence of algae in a lake?

Research Questions

- Have other experiments been done on this problem? If so, what did those experiments show?
- What kind of fertilizer is used on the field? How much?
- How do algae grow?
- How do people measure algae?
- Can fertilizer and algae be used safely in a lab? How?

Research

As you research, you may find a topic that is more interesting to you than your original topic, or learn that a procedure you wanted to use is not practical or safe. It is OK to change your purpose as you research.

LAB HANDBOOK

Writing a Hypothesis

A **hypothesis** is a tentative explanation for an observation or scientific problem that can be tested by further investigation. You can write your hypothesis in the form of an "If . . . , then . . . , because . . ." statement.

Hypothesis

If the amount of fertilizer in lake water is increased, then the amount of algae will also increase, because fertilizers provide nutrients that algae need to grow.

Hypotheses
For help with hypotheses, refer to page R3.

Determining Materials

Make a list of all the materials you will need to do your experiment. Be specific, especially if someone else is helping you obtain the materials. Try to think of everything you will need.

Materials

- 1 large jar or container
- 4 identical smaller containers
- rubber gloves that also cover the arms
- sample of fertilizer-and-water solution
- eyedropper
- clear plastic wrap
- scissors
- masking tape
- marker
- ruler

R29

Determining Variables and Constants

EXPERIMENTAL GROUP AND CONTROL GROUP

An experiment to determine how two factors are related always has two groups—a control group and an experimental group.

1. Design an experimental group. Include as many trials as possible in the experimental group in order to obtain reliable results.
2. Design a control group that is the same as the experimental group in every way possible, except for the factor you wish to test.

Experimental Group: two containers of lake water with one drop of fertilizer solution added to each

Control Group: two containers of lake water with no fertilizer solution added

Go back to your materials list and make sure you have enough items listed to cover both your experimental group and your control group.

VARIABLES AND CONSTANTS

Identify the variables and constants in your experiment. In a controlled experiment, a **variable** is any factor that can change. **Constants** are all of the factors that are the same in both the experimental group and the control group.

Hypothesis
If the amount of fertilizer in lake water is increased, then the amount of algae will also increase, because fertilizers provide nutrients that algae need to grow.

1. Read your hypothesis. The **independent variable** is the factor that you wish to test and that is manipulated or changed so that it can be tested. The independent variable is expressed in your hypothesis after the word *if*. Identify the independent variable in your laboratory report.
2. The **dependent variable** is the factor that you measure to gather results. It is expressed in your hypothesis after the word *then*. Identify the dependent variable in your laboratory report.

Table 1. Variables and Constants in Algae Experiment

Independent Variable	Dependent Variable	Constants
Amount of fertilizer in lake water	Amount of algae that grow	• Where the lake water is obtained • Type of container used • Light and temperature conditions where water will be stored

Set up your experiment so that you will test only one variable.

MEASURING THE DEPENDENT VARIABLE

Before starting your experiment, you need to define how you will measure the dependent variable. An **operational definition** is a description of the one particular way in which you will measure the dependent variable.

Your operational definition is important for several reasons. First, in any experiment there are several ways in which a dependent variable can be measured. Second, the procedure of the experiment depends on how you decide to measure the dependent variable. Third, your operational definition makes it possible for other people to evaluate and build on your experiment.

EXAMPLE 1

An operational definition of a dependent variable can be qualitative. That is, your measurement of the dependent variable can simply be an observation of whether a change occurs as a result of a change in the independent variable. This type of operational definition can be thought of as a "yes or no" measurement.

Table 2. Qualitative Operational Definition of Algae Growth

Independent Variable	Dependent Variable	Operational Definition
Amount of fertilizer in lake water	Amount of algae that grow	Algae grow in lake water

A qualitative measurement of a dependent variable is often easy to make and record. However, this type of information does not provide a great deal of detail in your experimental results.

EXAMPLE 2

An operational definition of a dependent variable can be quantitative. That is, your measurement of the dependent variable can be a number that shows how much change occurs as a result of a change in the independent variable.

Table 3. Quantitative Operational Definition of Algae Growth

Independent Variable	Dependent Variable	Operational Definition
Amount of fertilizer in lake water	Amount of algae that grow	Diameter of largest algal growth (in mm)

A quantitative measurement of a dependent variable can be more difficult to make and analyze than a qualitative measurement. However, this type of data provides much more information about your experiment and is often more useful.

Writing a Procedure

Write each step of your procedure. Start each step with a verb, or action word, and keep the steps short. Your procedure should be clear enough for someone else to use as instructions for repeating your experiment.

If necessary, go back to your materials list and add any materials that you left out.

Procedure

1. Put on your gloves. Use the large container to obtain a sample of lake water.
2. Divide the sample of lake water equally among the four smaller containers.
3. Use the eyedropper to add one drop of fertilizer solution to two of the containers.

Controlling Variables The same amount of fertilizer solution must be added to two of the four containers.

4. Use the masking tape and the marker to label the containers with your initials, the date, and the identifiers "Jar 1 with Fertilizer," "Jar 2 with Fertilizer," "Jar 1 without Fertilizer," and "Jar 2 without Fertilizer."
5. Cover the containers with clear plastic wrap. Use the scissors to punch ten holes in each of the covers.
6. Place all four containers on a window ledge. Make sure that they all receive the same amount of light.

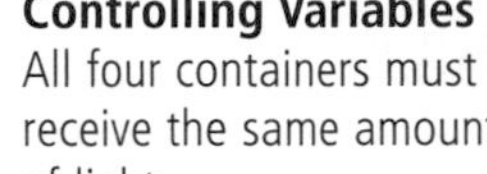

Controlling Variables All four containers must receive the same amount of light.

7. Observe the containers every day for one week.
8. Use the ruler to measure the diameter of the largest clump of algae in each container, and record your measurements daily.

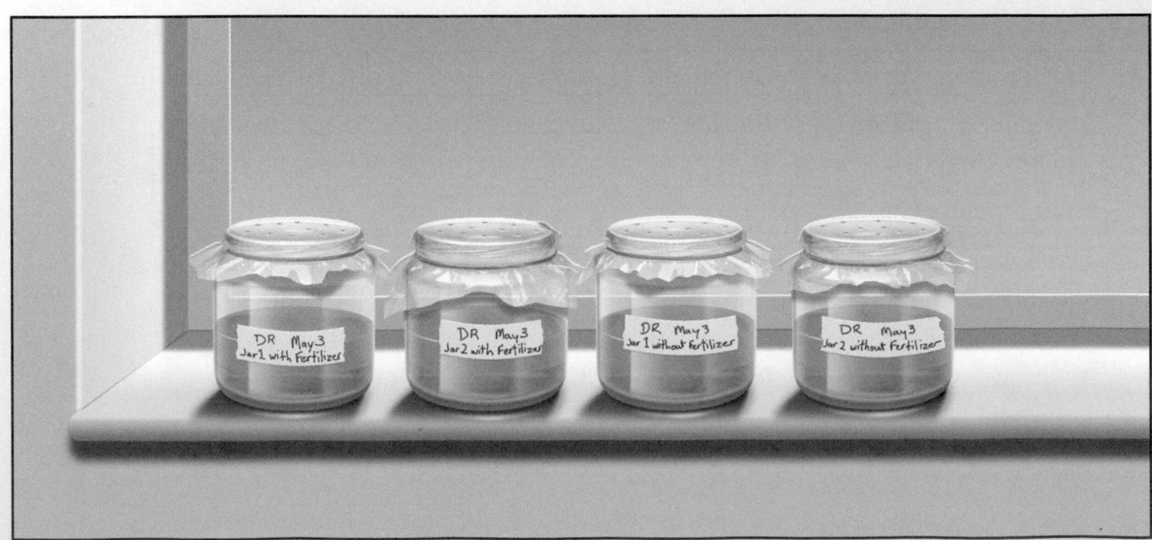

LAB HANDBOOK

Recording Observations

Once you have obtained all of your materials and your procedure has been approved, you can begin making experimental observations. Gather both quantitative and qualitative data. If something goes wrong during your procedure, make sure you record that too.

Observations
For help with making qualitative and quantitative observations, refer to page R2.

For more examples of data tables, see page R23.

Table 4. Fertilizer and Algae Growth

	Experimental Group		Control Group		
Date and Time	Jar 1 with Fertilizer (diameter of algae in mm)	Jar 2 with Fertilizer (diameter of algae in mm)	Jar 1 without Fertilizer (diameter of algae in mm)	Jar 2 without Fertilizer (diameter of algae in mm)	Observations
5/3 4:00 P.M.	0	0	0	0	condensation in all containers
5/4 4:00 P.M.	0	3	0	0	tiny green blobs in jar 2 with fertilizer
5/5 4:15 P.M.	4	5	0	3	green blobs in jars 1 and 2 with fertilizer and jar 2 without fertilizer
5/6 4:00 P.M.	5	6	0	4	water light green in jar 2 with fertilizer
5/7 4:00 P.M.	8	10	0	6	water light green in jars 1 and 2 with fertilizer and in jar 2 without fertilizer
5/8 3:30 P.M.	10	18	0	6	cover off jar 2 with fertilizer
5/9 3:30 P.M.	14	23	0	8	drew sketches of each container

Notice that on the sixth day, the observer found that the cover was off one of the containers. It is important to record observations of unintended factors because they might affect the results of the experiment.

Drawings of Samples Viewed Under Microscope on 5/9 at 100x

Use technology, such as a microscope, to help you make observations when possible.

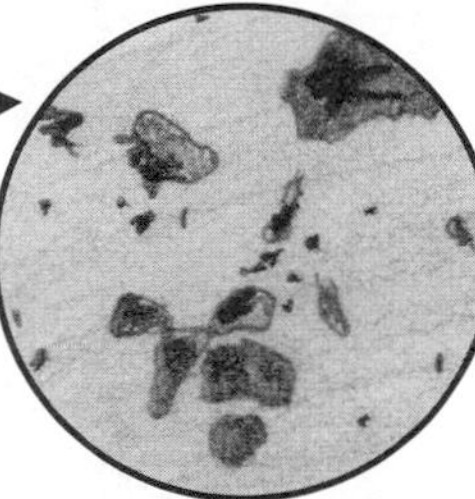
Jar 1 with Fertilizer

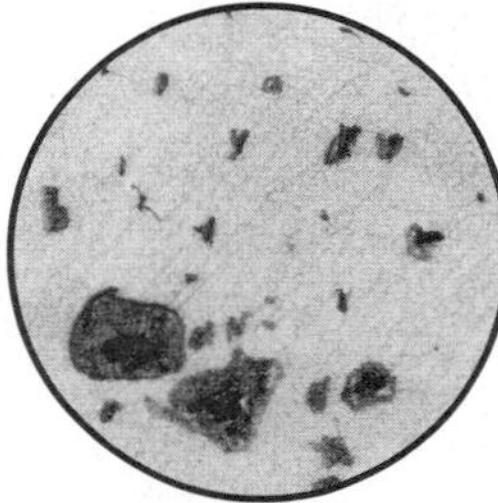
Jar 2 with Fertilizer

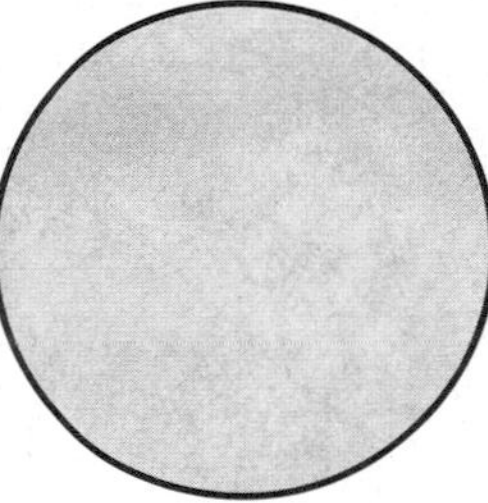
Jar 1 without Fertilizer

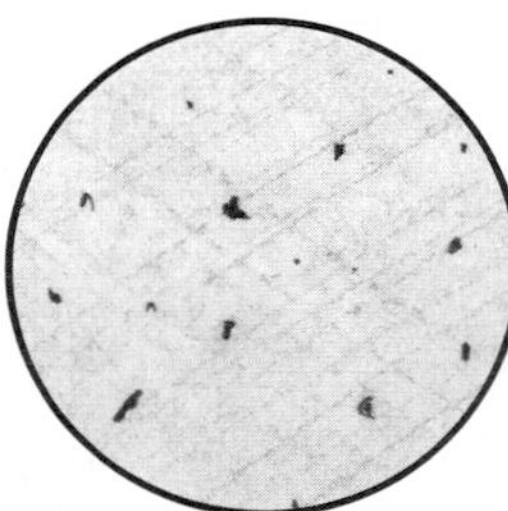
Jar 2 without Fertilizer

LAB HANDBOOK

Summarizing Results

To summarize your data, look at all of your observations together. Look for meaningful ways to present your observations. For example, you might average your data or make a graph to look for patterns. When possible, use spreadsheet software to help you analyze and present your data. The two graphs below show the same data.

EXAMPLE 1

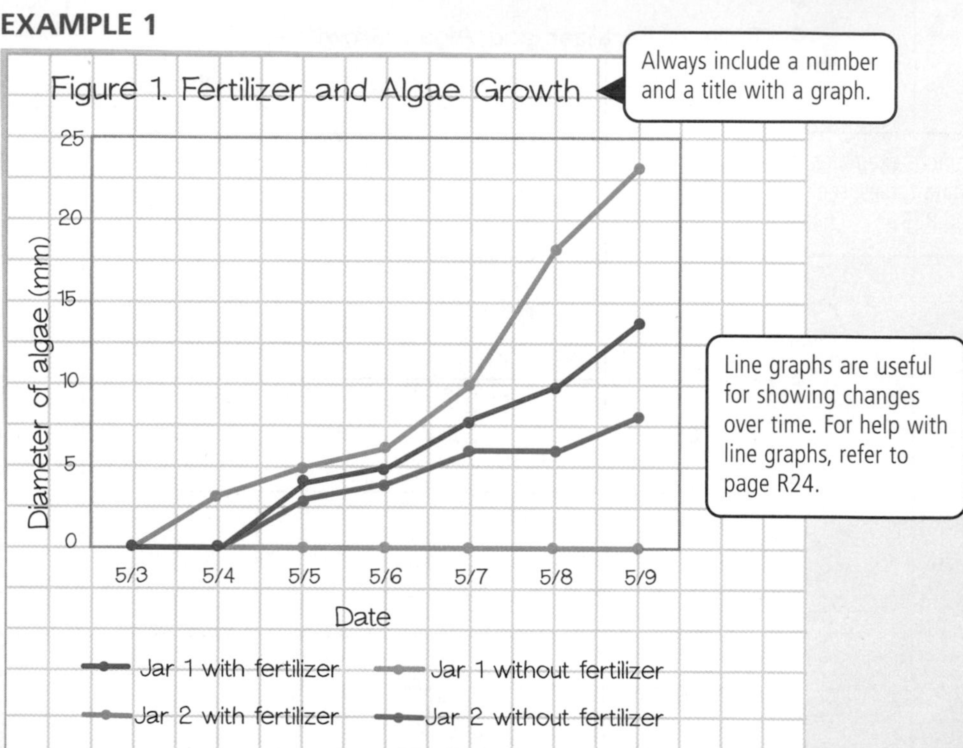

Line graphs are useful for showing changes over time. For help with line graphs, refer to page R24.

EXAMPLE 2

Bar graphs are useful for comparing different data sets. This bar graph has four bars for each day. Another way to present the data would be to calculate averages for the tests and the controls, and to show one test bar and one control bar for each day.

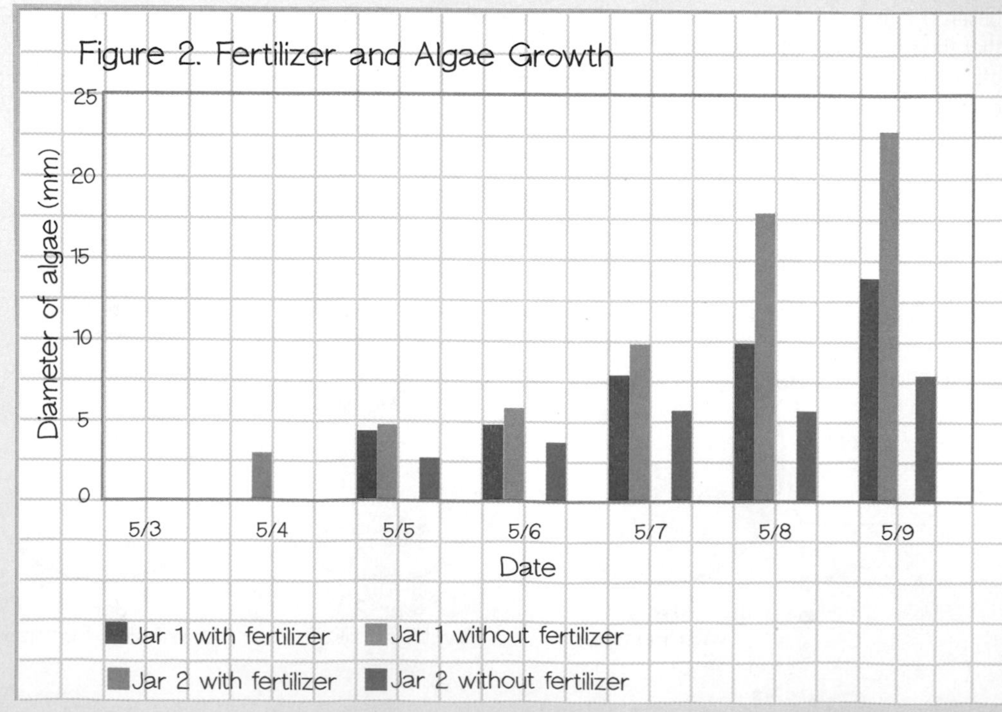

LAB HANDBOOK

Drawing Conclusions

RESULTS AND INFERENCES

To draw conclusions from your experiment, first write your results. Then compare your results with your hypothesis. Do your results support your hypothesis? Be careful not to make inferences about factors that you did not test.

For help with making inferences, see page R4.

Results and Inferences

The results of my experiment show that more algae grew in lake water to which fertilizer had been added than in lake water to which no fertilizer had been added. My hypothesis was supported. I infer that it is possible that the growth of algae in the lake was caused by the fertilizer used on the field.

Notice that you cannot conclude from this experiment that the presence of algae in the lake was due only to the fertilizer.

QUESTIONS FOR FURTHER RESEARCH

Write a list of questions for further research and investigation. Your ideas may lead you to new experiments and discoveries.

Questions for Further Research

- What is the connection between the amount of fertilizer and algae growth?
- How do different brands of fertilizer affect algae growth?
- How would algae growth in the lake be affected if no fertilizer were used on the field?
- How do algae affect the lake and the other life in and around it?
- How does fertilizer affect the lake and the life in and around it?
- If fertilizer is getting into the lake, how is it getting there?

R35

Describing a Set of Data

Means, medians, modes, and ranges are important math tools for describing data sets such as the following widths of fossilized clamshells.

13 mm 25 mm 14 mm 21 mm 16 mm 23 mm 14 mm

Mean

The **mean** of a data set is the sum of the values divided by the number of values.

Example

To find the mean of the clamshell data, add the values and then divide the sum by the number of values.

$$\frac{13\text{ mm} + 25\text{ mm} + 14\text{ mm} + 21\text{ mm} + 16\text{ mm} + 23\text{ mm} + 14\text{ mm}}{7} = \frac{126\text{ mm}}{7} = 18\text{ mm}$$

ANSWER The mean is 18 mm.

Median

The **median** of a data set is the middle value when the values are written in numerical order. If a data set has an even number of values, the median is the mean of the two middle values.

Example

To find the median of the clamshell data, arrange the values in order from least to greatest. The median is the middle value.

13 mm 14 mm 14 mm 16 mm 21 mm 23 mm 25 mm

ANSWER The median is 16 mm.

Mode

The **mode** of a data set is the value that occurs most often.

Example

To find the mode of the clamshell data, arrange the values in order from least to greatest and determine the value that occurs most often.

13 mm   14 mm   14 mm   16 mm   21 mm   23 mm   25 mm

ANSWER The mode is 14 mm.

A data set can have more than one mode or no mode. For example, the following data set has modes of 2 mm and 4 mm:

2 mm   2 mm   3 mm   4 mm   4 mm

The data set below has no mode, because no value occurs more often than any other.

2 mm   3 mm   4 mm   5 mm

Range

The **range** of a data set is the difference between the greatest value and the least value.

Example

To find the range of the clamshell data, arrange the values in order from least to greatest.

13 mm   14 mm   14 mm   16 mm   21 mm   23 mm   25 mm

Subtract the least value from the greatest value.

13 mm is the least value.
25 mm is the greatest value.

25 mm – 13 mm = 12 mm

ANSWER The range is 12 mm.

R37

Using Ratios, Rates, and Proportions

You can use ratios and rates to compare values in data sets. You can use proportions to find unknown values.

Ratios

A **ratio** uses division to compare two values. The ratio of a value a to a nonzero value b can be written as $\frac{a}{b}$.

Example

The height of one plant is 8 centimeters. The height of another plant is 6 centimeters. To find the ratio of the height of the first plant to the height of the second plant, write a fraction and simplify it.

$$\frac{8 \text{ cm}}{6 \text{ cm}} = \frac{4 \times \overset{1}{\cancel{2}}}{3 \times \underset{1}{\cancel{2}}} = \frac{4}{3}$$

ANSWER The ratio of the plant heights is $\frac{4}{3}$.

You can also write the ratio $\frac{a}{b}$ as "a to b" or as $a{:}b$. For example, you can write the ratio of the plant heights as "4 to 3" or as 4:3.

MATH HANDBOOK

Rates

A **rate** is a ratio of two values expressed in different units. A unit rate is a rate with a denominator of 1 unit.

Example

A plant grew 6 centimeters in 2 days. The plant's rate of growth was $\frac{6 \text{ cm}}{2 \text{ days}}$. To describe the plant's growth in centimeters per day, write a unit rate.

Divide numerator and denominator by 2: $\frac{6 \text{ cm}}{2 \text{ days}} = \frac{6 \text{ cm} \div 2}{2 \text{ days} \div 2}$

Simplify: $= \frac{3 \text{ cm}}{1 \text{ day}}$

You divide 2 days by 2 to get 1 day, so divide 6 cm by 2 also.

ANSWER The plant's rate of growth is 3 centimeters per day.

Proportions

A **proportion** is an equation stating that two ratios are equivalent. To solve for an unknown value in a proportion, you can use cross products.

Example

If a plant grew 6 centimeters in 2 days, how many centimeters would it grow in 3 days (if its rate of growth is constant)?

Write a proportion: $\frac{6 \text{ cm}}{2 \text{ days}} = \frac{x \text{ cm}}{3 \text{ days}}$

Set cross products: $6 \cdot 3 = 2x$

Multiply 6 and 3: $18 = 2x$

Divide each side by 2: $\frac{18}{2} = \frac{2x}{2}$

Simplify: $9 = x$

ANSWER The plant would grow 9 centimeters in 3 days.

Using Decimals, Fractions, and Percents

Decimals, fractions, and percentages are all ways of recording and representing data.

Decimals

A **decimal** is a number that is written in the base-ten place value system, in which a decimal point separates the ones and tenths digits. The values of each place is ten times that of the place to its right.

Example

A caterpillar traveled from point *A* to point *C* along the path shown.

ADDING DECIMALS To find the total distance traveled by the caterpillar, add the distance from *A* to *B* and the distance from *B* to *C*. Begin by lining up the decimal points. Then add the figures as you would whole numbers and bring down the decimal point.

$$\begin{array}{r} 36.9 \text{ cm} \\ +\ 52.4 \text{ cm} \\ \hline 89.3 \text{ cm} \end{array}$$

ANSWER The caterpillar traveled a total distance of 89.3 centimeters.

MATH HANDBOOK

Example *continued*

SUBTRACTING DECIMALS To find how much farther the caterpillar traveled on the second leg of the journey, subtract the distance from *A* to *B* from the distance from *B* to *C*.

$$\begin{array}{r} 52.4 \text{ cm} \\ -\ 36.9 \text{ cm} \\ \hline 15.5 \text{ cm} \end{array}$$

ANSWER The caterpillar traveled 15.5 centimeters farther on the second leg of the journey.

Example

A caterpillar is traveling from point *D* to point *F* along the path shown. The caterpillar travels at a speed of 9.6 centimeters per minute.

D E 33.6 cm F

MULTIPLYING DECIMALS You can multiply decimals as you would whole numbers. The number of decimal places in the product is equal to the sum of the number of decimal places in the factors.

For instance, suppose it takes the caterpillar 1.5 minutes to go from *D* to *E*. To find the distance from *D* to *E*, multiply the caterpillar's speed by the time it took.

$$\begin{array}{r} 9.6 \\ \times\ 1.5 \\ \hline 480 \\ 96 \\ \hline 14.40 \end{array} \qquad \begin{array}{rl} 1 & \text{decimal place} \\ +\ 1 & \text{decimal place} \\ \hline & \\ & \\ 2 & \text{decimal places} \end{array}$$

Align as shown.

ANSWER The distance from *D* to *E* is 14.4 centimeters.

DIVIDING DECIMALS When you divide by a decimal, move the decimal points the same number of places in the divisor and the dividend to make the divisor a whole number.

For instance, to find the time it will take the caterpillar to travel from *E* to *F*, divide the distance from *E* to *F* by the caterpillar's speed.

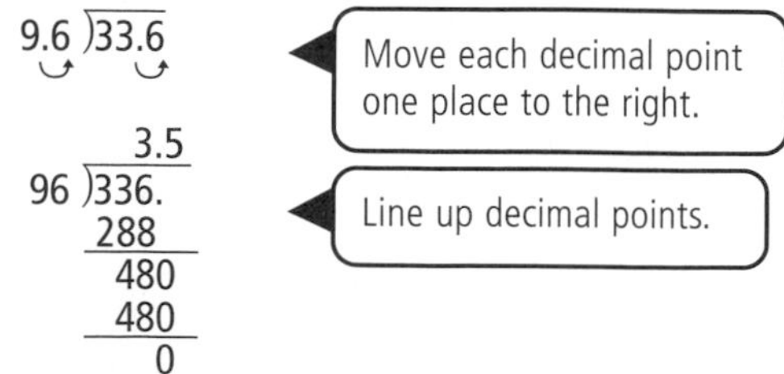

9.6)33.6
Move each decimal point one place to the right.
3.5
96)336.
288
480
480
0
Line up decimal points.

ANSWER The caterpillar will travel from *E* to *F* in 3.5 minutes.

Fractions

A **fraction** is a number in the form $\frac{a}{b}$, where b is not equal to 0. A fraction is in **simplest form** if its numerator and denominator have a greatest common factor (GCF) of 1. To simplify a fraction, divide its numerator and denominator by their GCF.

Example

A caterpillar is 40 millimeters long. The head of the caterpillar is 6 millimeters long. To compare the length of the caterpillar's head with the caterpillar's total length, you can write and simplify a fraction that expresses the ratio of the two lengths.

Write the ratio of the two lengths: $\frac{\text{Length of head}}{\text{Total length}} = \frac{6 \text{ mm}}{40 \text{ mm}}$

Write numerator and denominator as products of numbers and the GCF: $= \frac{3 \times 2}{20 \times 2}$

Divide numerator and denominator by the GCF: $= \frac{3 \times \overset{1}{\cancel{2}}}{20 \times \underset{1}{\cancel{2}}}$

Simplify: $= \frac{3}{20}$

ANSWER In simplest form, the ratio of the lengths is $\frac{3}{20}$.

Percents

A **percent** is a ratio that compares a number to 100. The word *percent* means "per hundred" or "out of 100." The symbol for *percent* is %.

For instance, suppose 43 out of 100 caterpillars are female. You can represent this ratio as a percent, a decimal, or a fraction.

Percent	Decimal	Fraction
43%	0.43	$\frac{43}{100}$

Example

In the preceding example, the ratio of the length of the caterpillar's head to the caterpillar's total length is $\frac{3}{20}$. To write this ratio as a percent, write an equivalent fraction that has a denominator of 100.

Multiply numerator and denominator by 5: $\frac{3}{20} = \frac{3 \times 5}{20 \times 5}$

$= \frac{15}{100}$

Write as a percent: $= 15\%$

ANSWER The caterpillar's head represents 15 percent of its total length.

MATH HANDBOOK

Using Formulas

A mathematical **formula** is a statement of a fact, rule, or principle. It is usually expressed as an equation.

In science, a formula often has a word form and a symbolic form. The formula below expresses Ohm's law.

Word Form	**Symbolic Form**
$\text{Current} = \dfrac{\text{voltage}}{\text{resistance}}$	$I = \dfrac{V}{R}$

The term *variable* is also used in science to refer to a factor that can change during an experiment.

In this formula, *I*, *V*, and *R* are variables. A mathematical **variable** is a symbol or letter that is used to represent one or more numbers.

Example

Suppose that you measure a voltage of 1.5 volts and a resistance of 15 ohms. You can use the formula for Ohm's law to find the current in amperes.

Write the formula for Ohm's law: $I = \dfrac{V}{R}$

***Substitute 1.5 volts for* V *and 15 ohms for* R:** $I = \dfrac{1.5 \text{ volts}}{15 \text{ ohms}}$

Simplify: $I = 0.1 \text{ amp}$

ANSWER The current is 0.1 ampere.

If you know the values of all variables but one in a formula, you can solve for the value of the unknown variable. For instance, Ohm's law can be used to find a voltage if you know the current and the resistance.

Example

Suppose that you know that a current is 0.2 amperes and the resistance is 18 ohms. Use the formula for Ohm's law to find the voltage in volts.

Write the formula for Ohm's law: $I = \dfrac{V}{R}$

***Substitute 0.2 amp for* I *and 18 ohms for* R:** $0.2 \text{ amp} = \dfrac{V}{18 \text{ ohms}}$

Multiply both sides by 18 ohms: $0.2 \text{ amp} \cdot 18 \text{ ohms} = V$

Simplify: $3.6 \text{ volts} = V$

ANSWER The voltage is 3.6 volts.

MATH HANDBOOK

R42

Finding Areas

The area of a figure is the amount of surface the figure covers.

Area is measured in square units, such as square meters (m^2) or square centimeters (cm^2). Formulas for the areas of three common geometric figures are shown below.

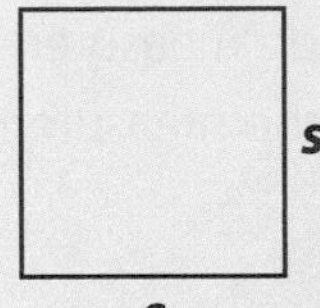

Area = (side length)^2
$A = s^2$

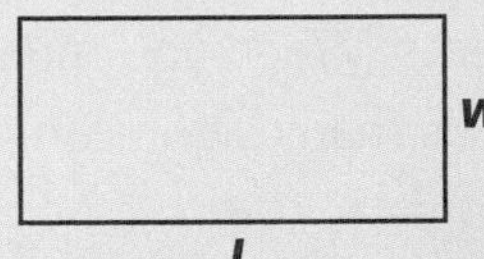

Area = length × width
$A = lw$

h
b

Area = $\frac{1}{2}$ × base × height
$A = \frac{1}{2}bh$

Example

Each face of a halite crystal is a square like the one shown. You can find the area of the square by using the steps below.

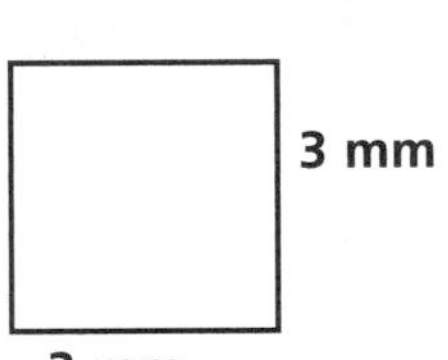

Write the formula for the area of a square: $A = s^2$

Substitute 3 mm for s: $= (3 \text{ mm})^2$

Simplify: $= 9 \text{ mm}^2$

ANSWER The area of the square is 9 square millimeters.

Finding Volumes

The volume of a solid is the amount of space contained by the solid.

Volume is measured in cubic units, such as cubic meters (m^3) or cubic centimeters (cm^3). The volume of a rectangular prism is given by the formula shown below.

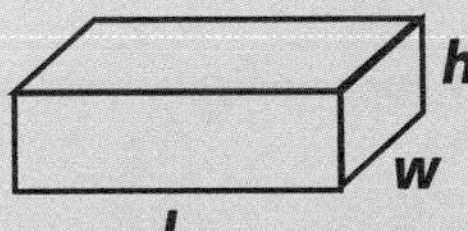

Volume = length × width × height
$V = lwh$

Example

A topaz crystal is a rectangular prism like the one shown. You can find the volume of the prism by using the steps below.

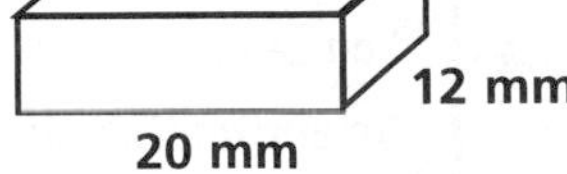

Write the formula for the volume of a rectangular prism: $V = lwh$

Substitute dimensions: $= 20 \text{ mm} \times 12 \text{ mm} \times 10 \text{ mm}$

Simplify: $= 2400 \text{ mm}^3$

ANSWER The volume of the rectangular prism is 2400 cubic millimeters.

MATH HANDBOOK

Using Significant Figures

The **significant figures** in a decimal are the digits that are warranted by the accuracy of a measuring device.

When you perform a calculation with measurements, the number of significant figures to include in the result depends in part on the number of significant figures in the measurements. When you multiply or divide measurements, your answer should have only as many significant figures as the measurement with the fewest significant figures.

Example

Using a balance and a graduated cylinder filled with water, you determined that a marble has a mass of 8.0 grams and a volume of 3.5 cubic centimeters. To calculate the density of the marble, divide the mass by the volume.

Write the formula for density: Density $= \frac{\text{mass}}{\text{Volume}}$

Substitute measurements: $= \frac{8.0 \text{ g}}{3.5 \text{ cm}^3}$

Use a calculator to divide: $\approx 2.285714286 \text{ g/cm}^3$

ANSWER Because the mass and the volume have two significant figures each, give the density to two significant figures. The marble has a density of 2.3 grams per cubic centimeter.

MATH HANDBOOK

Using Scientific Notation

Scientific notation is a shorthand way to write very large or very small numbers. For example, 73,500,000,000,000,000,000,000 kg is the mass of the Moon. In scientific notation, it is 7.35×10^{22} kg.

Example

You can convert from standard form to scientific notation.

Standard Form	**Scientific Notation**
720,000	7.2×10^5
5 decimal places left	Exponent is 5.
0.000291	2.91×10^{-4}
4 decimal places right	Exponent is –4.

You can convert from scientific notation to standard form.

Scientific Notation	**Standard Form**
4.63×10^7	46,300,000
Exponent is 7.	7 decimal places right
1.08×10^{-6}	0.00000108
Exponent is –6.	6 decimal places left

Note-Taking Handbook

Note-Taking Strategies

Taking notes as you read helps you understand the information. The notes you take can also be used as a study guide for later review. This handbook presents several ways to organize your notes.

Content Frame

1. Make a chart in which each column represents a category.
2. Give each column a heading.
3. Write details under the headings.

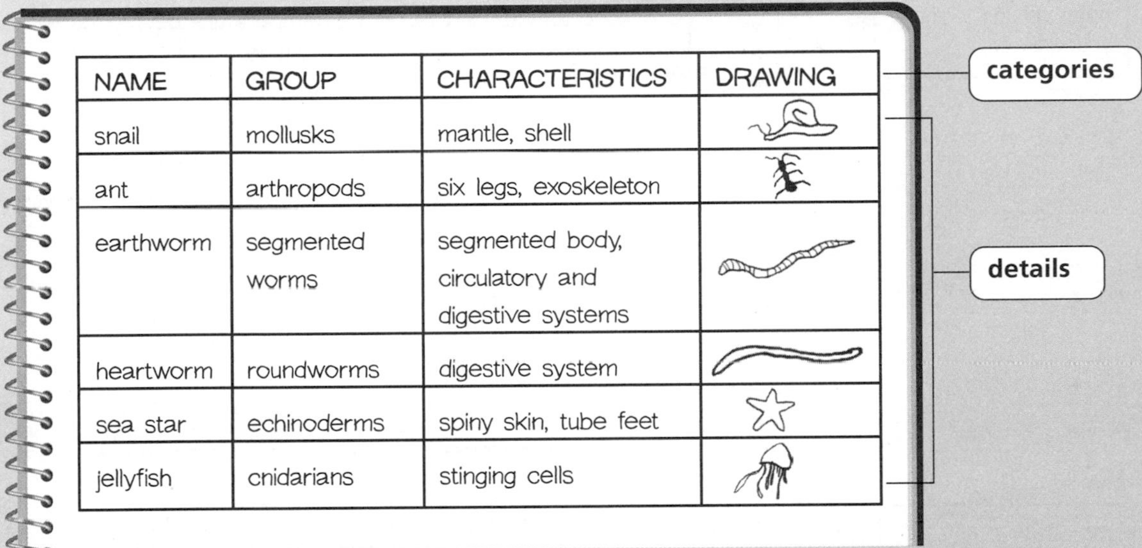

NAME	GROUP	CHARACTERISTICS	DRAWING
snail	mollusks	mantle, shell	
ant	arthropods	six legs, exoskeleton	
earthworm	segmented worms	segmented body, circulatory and digestive systems	
heartworm	roundworms	digestive system	
sea star	echinoderms	spiny skin, tube feet	
jellyfish	cnidarians	stinging cells	

Combination Notes

1. For each new idea or concept, write an informal outline of the information.
2. Make a sketch to illustrate the concept, and label it.

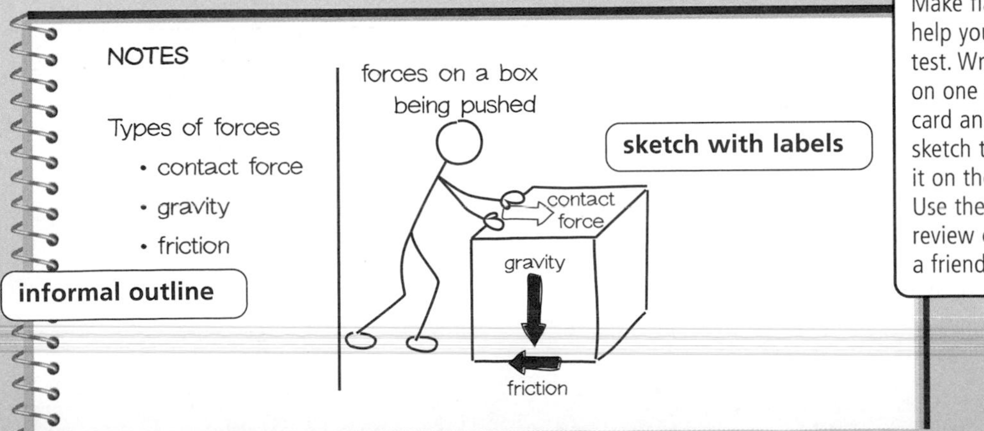

Make flash cards to help you study for a test. Write a concept on one side of each card and draw the sketch that goes with it on the other side. Use the cards to review concepts with a friend.

Main Idea and Detail Notes

1. In the left-hand column of a two-column chart, list main ideas. The blue headings express main ideas throughout this textbook.
2. In the right-hand column, write details that expand on each main idea.

You can shorten the headings in your chart. Be sure to use the most important words.

When studying for tests, cover up the detail notes column with a sheet of paper. Then use each main idea to form a question—such as "How does latitude affect climate?" Answer the question, and then uncover the detail notes column to check your answer.

MAIN IDEAS	DETAIL NOTES
1. Latitude affects climate.	1. Places close to the equator are usually warmer than places close to the poles. 1. Latitude has the same effect in both hemispheres.
2. Altitude affects climate.	2. Temperature decreases with altitude. 2. Altitude can overcome the effect of latitude on temperature.

main idea 1

details abou main idea 1

main idea 2

details abou main idea 2

Main Idea Web

1. Write a main idea in a box.
2. Add boxes around it with related vocabulary terms and important details.

You can find definitions near highlighted terms.

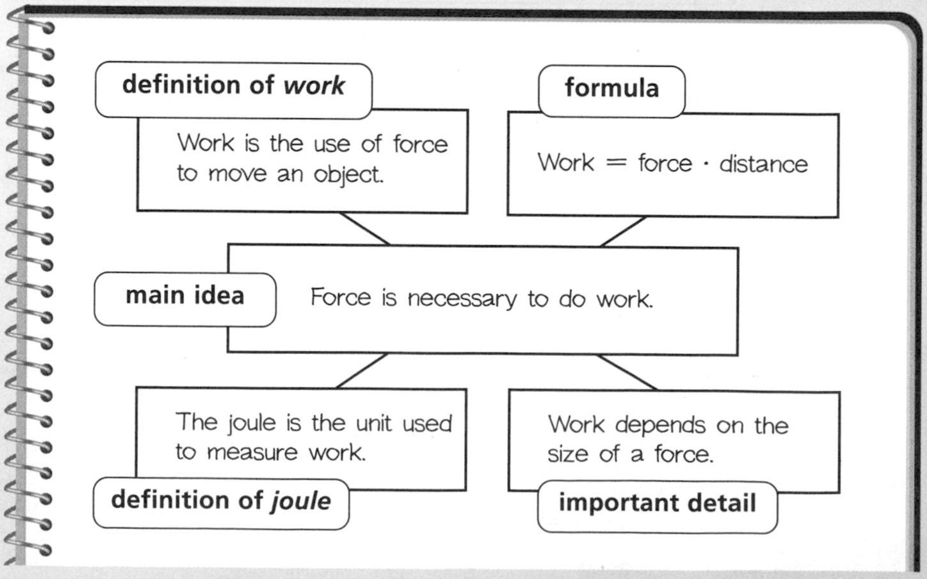

definition of *work*

Work is the use of force to move an object.

formula

Work = force · distance

main idea

Force is necessary to do work.

The joule is the unit used to measure work.

definition of *joule*

Work depends on the size of a force.

important detail

NOTE-TAKING HANDBOOK

Mind Map

1. Write a main idea in the center.
2. Add details that relate to one another and to the main idea.

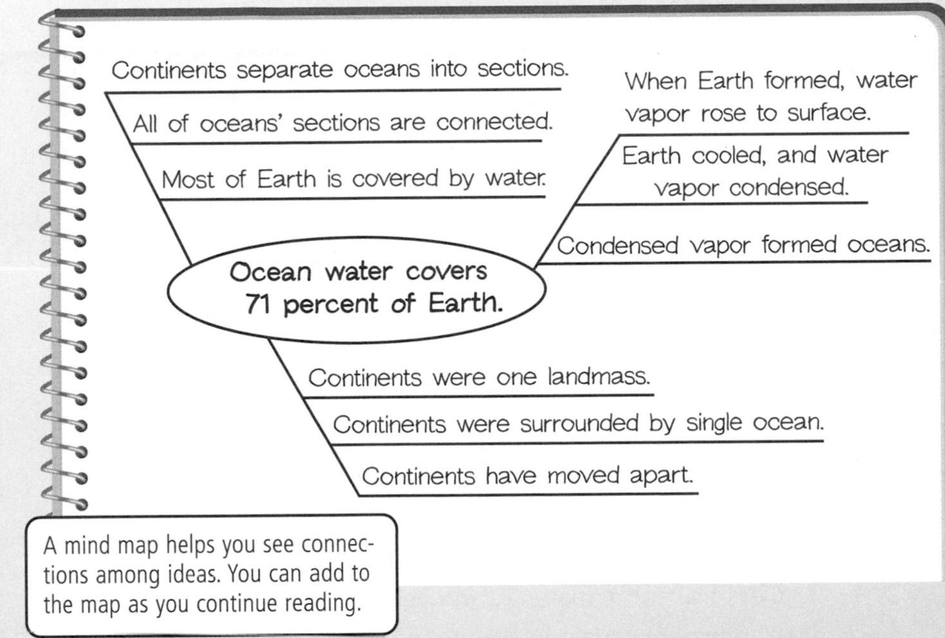

Supporting Main Ideas

1. Write a main idea in a box.
2. Add boxes underneath with information—such as reasons, explanations, and examples—that supports the main idea.

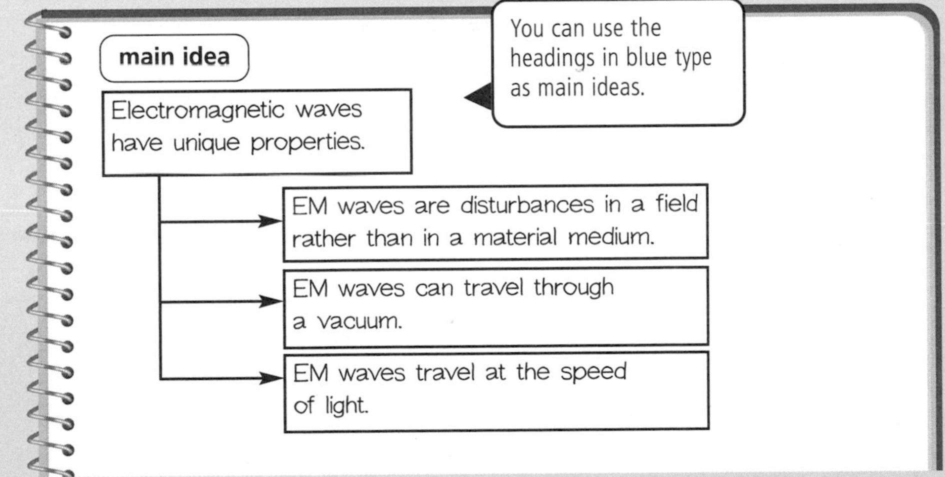

NOTE-TAKING HANDBOOK

Outline

1. Copy the chapter title and headings from the book in the form of an outline.
2. Add notes that summarize in your own words what you read.

Cell Processes

I. Cells capture and release energy. — 1st key idea
   A. All cells need energy. — 1st subpoint of I
   B. Some cells capture light energy. — 2nd subpoint of I
      1. Process of photosynthesis — 1st detail about B
      2. Chloroplasts (site of photosynthesis) — 2nd detail about B
      3. Carbon dioxide and water as raw materials
      4. Glucose and oxygen as products
   C. All cells release energy.
      1. Process of cellular respiration
      2. Fermentation of sugar to carbon dioxide
      3. Bacteria that carry out fermentation
II. Cells transport materials through membranes.
   A. Some materials move by diffusion.
      1. Particle movement from higher to lower concentrations
      2. Movement of water through membrane (osmosis)
   B. Some transport requires energy.
      1. Active transport
      2. Examples of active transport

Correct Outline Form
Include a title.

Arrange key ideas, subpoints, and details as shown.

Indent the divisions of the outline as shown.

Use the same grammatical form for items of the same rank. For example, if A is a sentence, B must also be a sentence.

You must have at least two main ideas or subpoints. That is, every A must be followed by a B, and every 1 must be followed by a 2.

R48

Concept Map

1. Write an important concept in a large oval.
2. Add details related to the concept in smaller ovals.
3. Write linking words on arrows that connect the ovals.

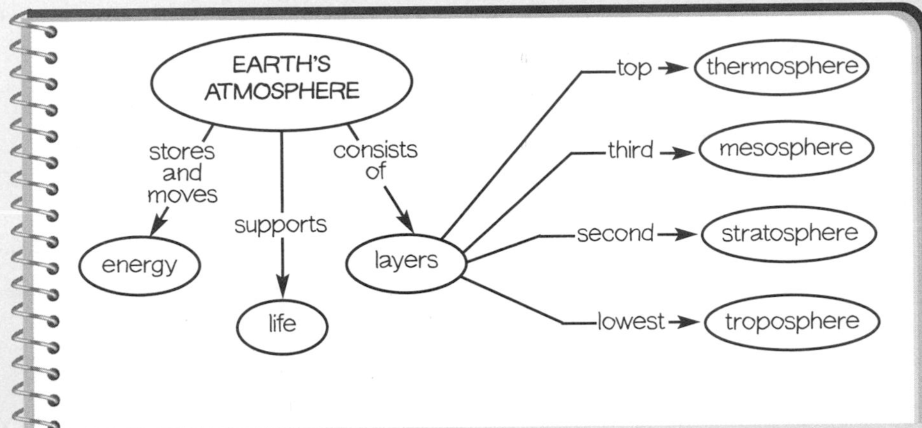

The main ideas or concepts can often be found in the blue headings. An example is "The atmosphere stores and moves energy." Use nouns from these concepts in the ovals, and use the verb or verbs on the lines.

Venn Diagram

1. Draw two overlapping circles, one for each item that you are comparing.
2. In the overlapping section, list the characteristics that are shared by both items.
3. In the outer sections, list the characteristics that are peculiar to each item.
4. Write a summary that describes the information in the Venn diagram.

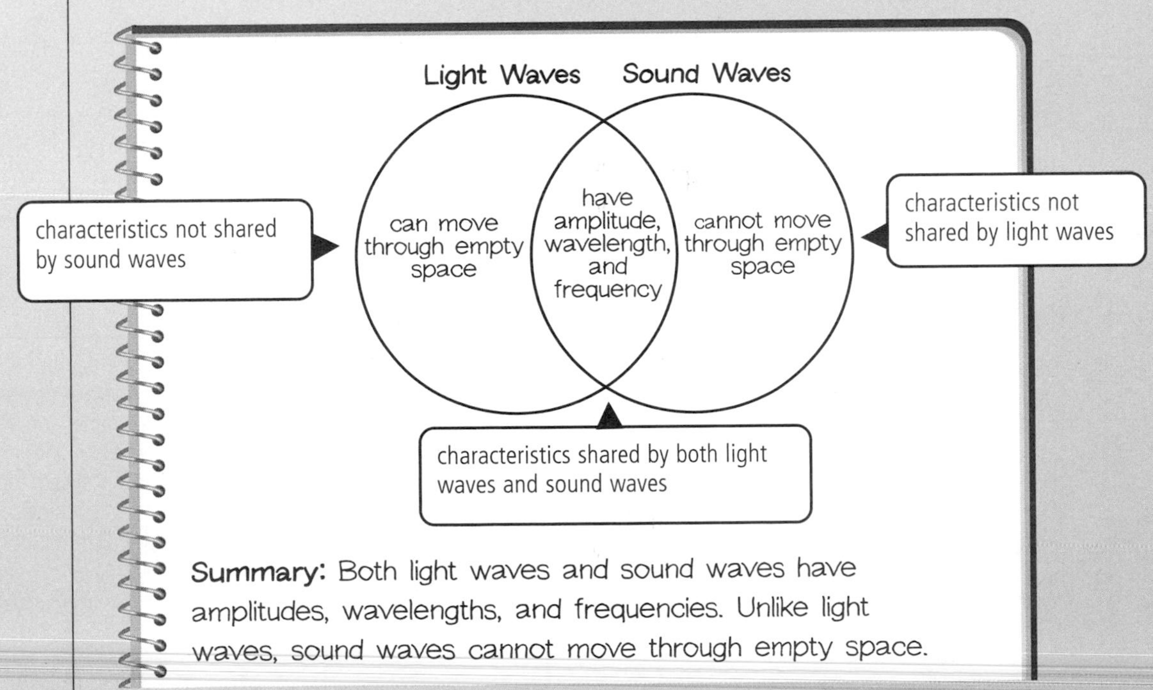

R49

Vocabulary Strategies

Important terms are highlighted in this book. A definition of each term can be found in the sentence or paragraph where the term appears. You can also find definitions in the Glossary. Taking notes about vocabulary terms helps you understand and remember what you read.

Description Wheel

1. Write a term inside a circle.
2. Write words that describe the term on "spokes" attached to the circle.

When studying for a test with a friend, read the phrases on the spokes one at a time until your friend identifies the correct term.

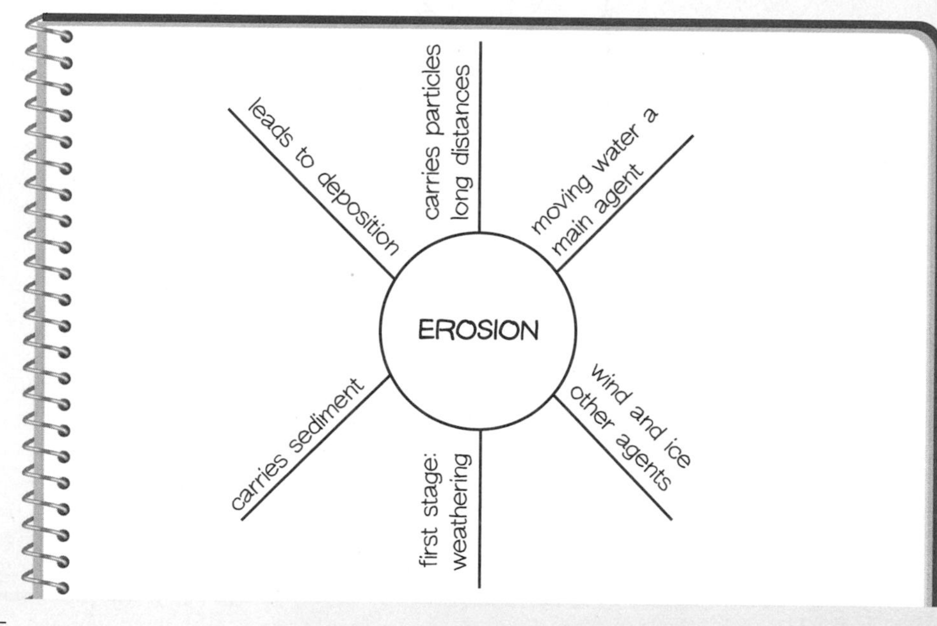

Four Square

1. Write a term in the center.
2. Write details in the four areas around the term.

Definition	Characteristics
any living thing	needs food, water, air; needs energy; grows, develops, reproduces
ORGANISM	
Examples	**Nonexamples**
dogs, cats, birds, insects, flowers, trees	rocks, water, dirt

Include a definition, some characteristics, and examples. You may want to add a formula, a sketch, or examples of things that the term does *not* name.

NOTE-TAKING HANDBOOK

R50

Frame Game

1. Write a term in the center.
2. Frame the term with details.

Include examples, descriptions, sketches, or sentences that use the term in context. Change the frame to fit each new term.

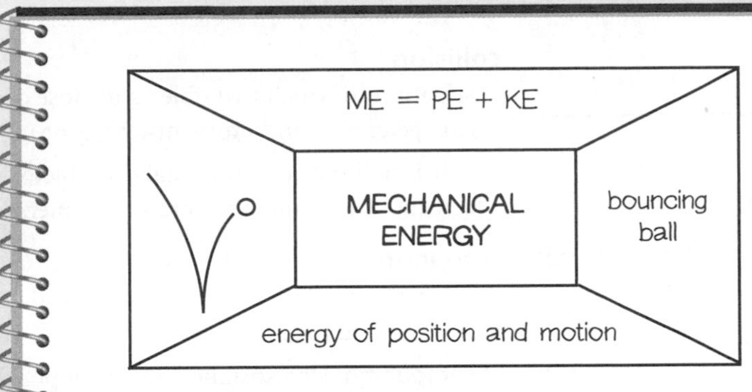

Magnet Word

1. Write a term on the magnet.
2. On the lines, add details related to the term.

You can also use phrases or sentences on the lines.

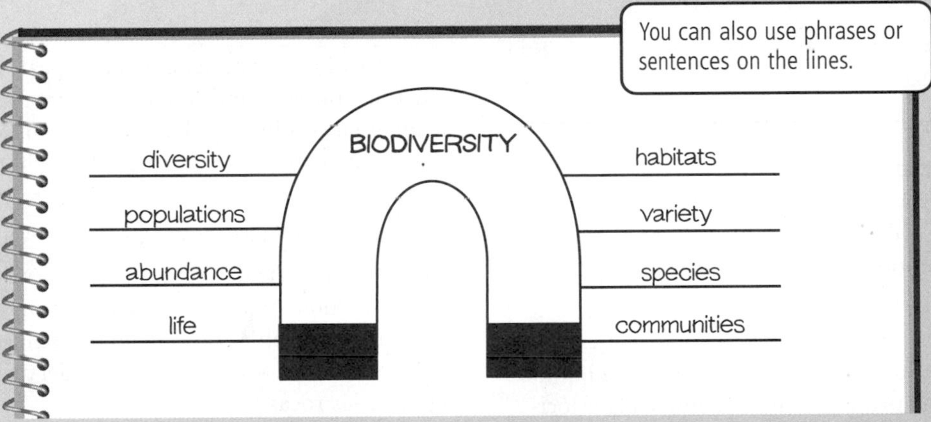

Word Triangle

1. Write a term and its definition in the bottom section.
2. In the middle section, write a sentence in which the term is used correctly.
3. In the top section, draw a small picture to illustrate the term.

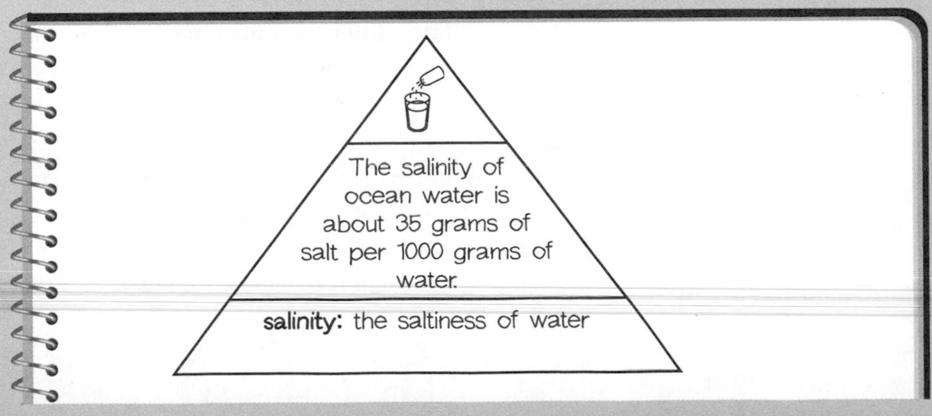

NOTE-TAKING HANDBOOK

Glossary

GLOSSARY

acceleration
The rate at which velocity changes over time. (p. 25)

aceleración La razón a la cual la velocidad cambia con respecto al tiempo.

air resistance
The fluid friction due to air. (p. 89)

resistencia del aire La fricción fluida debida al aire.

atom
The smallest particle of an element that has the chemical properties of that element. (p. xv)

átomo La partícula más pequeña de un elemento que tiene las propiedades químicas del elemento.

Bernoulli's principle
A statement that describes the effects of movement on fluid pressure. According to this principle, an increase in the speed of the motion of a fluid decreases the pressure within the fluid. (p. 100)

principio de Bernoulli Un enunciado que describe los efectos del movimiento sobre la presión de un líquido. De acuerdo a este principio, un aumento en la velocidad del movimiento de un fluido disminuye la presión dentro del líquido

buoyant force
The upward force on objects in a fluid; often called buoyancy. (p. 98)

fuerza flotante La fuerza hacia arriba que ejerce un fluido sobre un objeto inmerso en él, a menudo llamada flotación.

centripetal force (sehn-TRIHP-ih-tuhl)
Any force that keeps an object moving in a circle. (p. 54)

fuerza centrípeta Cualquier fuerza que mantiene a un objeto moviéndose en forma circular.

collision
A situation in which two objects in close contact exchange energy and momentum. (p. 66)

colisión Situación en la cual dos objetos en contacto cercano intercambian energía y momento.

compound
A substance made up of two or more different types of atoms bonded together.

compuesto Una sustancia formada por dos o más diferentes tipos de átomos enlazados.

compound machine
A machine that is made up of two or more simple machines. (p. 164)

máquina compuesta Una máquina que está hecha de dos o más máquinas simples.

cycle
n. A series of events or actions that repeat themselves regularly; a physical and/or chemical process in which one material continually changes locations and/or forms. Examples include the water cycle, the carbon cycle, and the rock cycle.

v. To move through a repeating series of events or actions.

ciclo *s.* Una serie de eventos o acciones que se repiten regularmente; un proceso físico y/o químico en el cual un material cambia continuamente de lugar y/o forma. Ejemplos: el ciclo del agua, el ciclo del carbono y el ciclo de las rocas.

data
Information gathered by observation or experimentation that can be used in calculating or reasoning. *Data* is a plural word; the singular is *datum.*

datos Información reunida mediante observación o experimentación y que se puede usar para calcular o para razonar.

density
A property of matter representing the mass per unit volume. (p. 99)

densidad Una propiedad de la materia que representa la masa por unidad de volumen.

E

efficiency
The percentage of the input work done on a machine that the machine can return in output work. A machine's output work divided by its input work and multiplied by 100. (p. 150)

eficiencia El porcentaje del trabajo de entrada suministrado a una máquina que la máquina puede devolver como trabajo de salida. El trabajo de salida de una máquina dividido por su trabajo de entrada y multiplicado por cien.

element
A substance that cannot be broken down into a simpler substance by ordinary chemical changes. An element consists of atoms of only one type. (p. xv)

elemento Una sustancia que no puede descomponerse en otra sustancia más simple por medio de cambios químicos normales. Un elemento consta de átomos de un solo tipo.

energy
The ability to do work or to cause a change. For example, the energy of a moving bowling ball knocks over pins; energy from food allows animals to move and to grow; and energy from the Sun heats Earth's surface and atmosphere, which causes air to move. (p. xix)

energía La capacidad para trabajar o causar un cambio. Por ejemplo, la energía de una bola de boliche en movimiento tumba los pinos; la energía proveniente de su alimento permite a los animales moverse y crecer; la energía del Sol calienta la superficie y la atmósfera de la Tierra, lo que ocasiona que el aire se mueva.

experiment
An organized procedure to study something under controlled conditions. (p. xxiv)

experimento Un procedimiento organizado para estudiar algo bajo condiciones controladas.

F

fluid
A substance that can flow easily, such as a gas or a liquid. (p. 88)

fluido Una sustancia que fluye fácilmente, como por ejemplo un gas o un líquido.

force
A push or a pull; something that changes the motion of an object. (p. 41)

fuerza Un empuje o un jalón; algo que cambia el movimiento de un objeto.

friction
A force that resists the motion between two surfaces in contact. (p. 85)

fricción Una fuerza que resiste el movimiento entre dos superficies en contacto.

fulcrum
A fixed point around which a lever rotates. (p. 155)

fulcro Un punto fijo alrededor del cual gira una palanca.

G

gravity
The force that objects exert on each other because of their masses. (p. 77)

gravedad La fuerza que los objetos ejercen entre sí debido a sus masas.

H

horizontal
Parallel to the horizon; level.

horizontal Paralelo al horizonte; nivelado.

horsepower hp
The unit of measurement of power for engines and motors. One horsepower equals 745 watts. (p. 132)

caballos de fuerza La unidad de medición de potencia para máquinas y motores. Un caballo de fuerza es igual a 745 vatios.

hypothesis
A tentative explanation for an observation or phenomenon. A hypothesis is used to make testable predictions. (p. xxiv)

hipótesis Una explicación provisional de una observación o de un fenómeno. Una hipótesis se usa para hacer predicciones que se pueden probar.

I

inclined plane
A simple machine that is a sloping surface, such as a ramp. (p. 158)

plano inclinado Una máquina simple que es una superficie en pendiente, como por ejemplo una rampa.

inertia (ih-NUR-shuh)
The resistance of an object to a change in the speed or the direction of its motion. (p. 46)

inercia La resistencia de un objeto al cambio de la velocidad o de la dirección de su movimiento.

joule (jool) J
A unit used to measure energy and work. One calorie is equal to 4.18 joules of energy; one joule of work is done when a force of one newton moves an object one meter. (p. 117)

julio Una unidad que se usa para medir la energía y el trabajo. Una caloría es igual a 4.18 julios de energía; se hace un joule de trabajo cuando una fuerza de un newton mueve un objeto un metro.

kinetic energy (kuh-NEHT-ihk)
The energy of motion. A moving object has the most kinetic energy at the point where it moves the fastest. (p. 122)

energía cinética La energía de movimiento. Un objeto en movimiento tiene la mayor energía cinética en el punto en donde se mueve más rápidamente.

L

law
In science, a rule or principle describing a physical relationship that always works in the same way under the same conditions. The law of conservation of energy is an example.

ley En las ciencias, una regla o un principio que describe una relación física que siempre funciona de la misma manera bajo las mismas condiciones. La ley de la conservación de la energía es un ejemplo.

law of conservation of energy
A law stating that no matter how energy is transferred or transformed, all of the energy is still present in one form or another. (p. 126)

ley de la conservación de la energía Una ley que establece que no importa cómo se transfiera o transmita la energía, toda la energía sigue presente de una forma o de otra.

law of conservation of momentum
A law stating that the amount of momentum a system of objects has does not change as long as there are no outside forces acting on that system. (p. 67)

ley de la conservación del momento Una ley que establece que la cantidad de momento que tiene un sistema de objetos no cambia mientras no haya fuerzas externas actuando sobre el sistema.

lever
A solid bar that rotates, or turns, around a fixed point (fulcrum); one of the six simple machines. (p. 155)

palanca Una barra sólida que da vueltas o gira alrededor de un punto fijo (el fulcro); una de las seis máquinas simples.

machine
Any device that makes doing work easier. (p. 145)

máquina Cualquier aparato que facilita el trabajo.

mass
A measure of how much matter an object is made of. (p. xv)

masa Una medida de la cantidad de materia de la que está compuesto un objeto.

matter
Anything that has mass and volume. Matter exists ordinarily as a solid, a liquid, or a gas. (p. xv)

materia Todo lo que tiene masa y volumen. Generalmente la materia existe como sólido, líquido o gas.

mechanical advantage
The number of times a machine multiplies the input force; output force divided by input force (p. 147)

ventaja mecánica El número de veces que una máquina multiplica la fuerza de entrada; la fuerza de salida dividida por la fuerza de entrada.

mechanical energy
A combination of the kinetic energy and potential energy an object has. (p. 125)

energía mecánica La combinación de la energía cinética y la energía potencial que tiene un objeto.

meter m
The international standard unit of length, about 39.37 inches.

metro La unidad estándar internacional de longitud, aproximadamente 39.37 pulgadas.

R54

molecule
A group of atoms that are held together by covalent bonds so that they move as a single unit. (p. xv)

molécula Un grupo de átomos que están unidos mediante enlaces covalentes de tal manera que se mueven como una sola unidad.

momentum (moh-MEHN-tuhm)
A measure of mass in motion. The momentum of an object is the product of its mass and velocity. (p. 64)

momento Una medida de la masa en movimiento. El momento de un objeto es el producto de su masa y su velocidad.

motion
A change of position over time. (p. 11)

movimiento Un cambio de posición a través del tiempo.

N

nanotechnology
The science and technology of building electronic circuits and devices from single atoms and molecules. (p. 167)

nanotecnología La ciencia y tecnología de fabricar circuitos y aparatos electrónicos a partir de átomos y moléculas individuales.

net force
The overall force acting on an object when all of the forces acting on it are combined. (p. 43)

fuerza neta La fuerza resultante que actúa sobre un objeto cuando todas las fuerzas que actúan sobre él son combinadas .

Newton's first law
A scientific law stating that objects at rest remain at rest, and objects in motion remain in motion with the same velocity, unless acted on by an unbalanced force. (p. 45)

primera ley de Newton Una ley científica que establece que los objetos en reposo permanecen en reposo, y que los objetos en movimiento permanecen en movimiento con la misma velocidad, a menos que actúe sobre ellos una fuerza no balanceada.

Newton's second law
A scientific law stating that the acceleration of an object increases with increased force and decreases with increased mass. (p. 50)

segunda ley de Newton Una ley científica que establece que la aceleración de un objeto aumenta al incrementar la fuerza que actúa sobre él y disminuye al incrementar su masa.

Newton's third law
A scientific law stating that every time one object exerts a force on another object, the second object exerts a force that is equal in size and opposite in direction back on the first object. (p. 57)

tercera ley de Newton Una ley científica que establece que cada vez que un objeto ejerce una fuerza sobre otro objeto, el segundo objeto ejerce una fuerza de la misma magnitud y en dirección opuesta sobre el primer objeto.

orbit
The elliptical path one celestial body follows around another celestial body. An object in orbit has a centripetal force acting on it that keeps the object moving in a circle or other ellipse. (p. 80)

órbita El camino elíptico que un cuerpo celeste sigue alrededor de otro cuerpo celeste. La fuerza centrípeta actúa sobre un objeto en órbita y lo mantiene en un movimiento circular o elíptico.

P, Q

pascal Pa
The unit used to measure pressure. One pascal is the pressure exerted by one newton of force on an area of one square meter, or one N/m^2. (p. 92)

pascal La unidad utilizada para medir presión. Un pascal es la presión ejercida por un newton de fuerza sobre un área de un metro cuadrado, o un N/m^2.

Pascal's principle
A statement that says when an outside pressure is applied at any point to a fluid in a container, that pressure is transmitted throughout the fluid with equal strength. (p. 102)

principio de Pascal Un enunciado que dice que cuando una presión externa es aplicada a cualquier punto de un líquido en un contenedor, esta presión es transmitida a través del fluido con igual fuerza.

position
An object's location. (p. 9)

posición La ubicación de un objeto.

potential energy
Stored energy; the energy an object has due to its position, molecular arrangement, or chemical composition. (p. 122)

energía potencial Energía almacenada; o la energía que tiene un objeto debido a su posición, arreglo molecular o composición química.

GLOSSARY

power
The rate at which work is done. (p. 130)

potencia La razón a la cual se hace el trabajo.

pressure
A measure of how much force is acting on a certain area; how concentrated a force is. Pressure is equal to the force divided by area. (p. 91)

presión Una medida de cuánta fuerza actúa sobre cierta área; el nivel de concentración de la fuerza. La presión es igual a la fuerza dividida entre el área.

pulley
A wheel with a grooved rim that turns on an axle; one of the six simple machines. (p. 156)

polea Una rueda con un canto acanalado que gira sobre un eje; una de las seis máquinas simples.

R

reference point
A location to which another location is compared. (p. 10)

punto de referencia Una ubicación con la cual se compara otra ubicación.

relative motion
The idea that the observation of motion depends on the observer. (p. 13)

movimiento relativo La idea de que la observación del movimiento depende del observador.

robot
A machine that works automatically or by remote control. (p. 169)

robot Una máquina que funciona automáticamente o por control remoto.

S

screw
A simple machine that is an inclined plane wrapped around a cylinder. A screw can be used to raise and lower weights as well as to fasten objects. (p. 159)

tornillo Una máquina simple que es un plano inclinado enrollado alrededor de un cilindro. Un tornillo se puede usar para levantar o bajar pesos y también para sujetar objetos.

second s
A unit of time equal to one-sixtieth of a minute.

segundo Una unidad de tiempo igual a una sesentava parte de un minuto.

simple machine
One of the basic machines on which all other mechanical machines are based. The six simple machines are the lever, inclined plane, wheel and axle, pulley, wedge, and screw. (p. 154)

máquina simple Una de las máquinas básicas sobre las cuales están basadas todas las demás máquinas mecánicas. Las seis máquinas simples son la palanca, el plano inclinado, la rueda y eje, la polea, la cuña y el tornillo.

speed
A measure of how fast something moves through a particular distance over a definite time period. Speed is distance divided by time. (p. 16)

rapidez Una medida del desplazamiento de un objeto a lo largo de una distancia específica en un período de tiempo definido. La rapidez es la distancia dividida entre el tiempo.

system
A group of objects or phenomena that interact. A system can be as simple as a rope, a pulley, and a mass. It also can be as complex as the interaction of energy and matter in the four parts of the Earth system.

sistema Un grupo de objetos o fenómenos que interactúan. Un sistema puede ser algo tan sencillo como una cuerda, una polea y una masa. También puede ser algo tan complejo como la interacción de la energía y la materia en las cuatro partes del sistema de la Tierra.

T, U

technology
The use of scientific knowledge to solve problems or engineer new products, tools, or processes.

tecnología El uso de conocimientos científicos para resolver problemas o para diseñar nuevos productos, herramientas o procesos.

terminal velocity
The final, maximum velocity of a falling object. (p. 89)

velocidad terminal La velocidad máxima final de un objeto en caída libre.

theory
In science, a set of widely accepted explanations of observations and phenomena. A theory is a well-tested explanation that is consistent with all available evidence.

teoría En las ciencias, un conjunto de explicaciones de observaciones y fenómenos que es ampliamente aceptado. Una teoría es una explicación bien probada que es consecuente con la evidencia disponible.

V

variable
Any factor that can change in a controlled experiment, observation, or model. (p. R30)

variable Cualquier factor que puede cambiar en un experimento controlado, en una observación o en un modelo.

vector
A quantity that has both size and direction. (p. 22)

vector Una cantidad que tiene magnitud y dirección.

velocity
A speed in a specific direction. (p. 22)

velocidad Una rapidez en una dirección específica.

vertical
Going straight up or down from a level surface.

vertical Que está dispuesto hacia arriba o hacia abajo de una superficie nivelada.

volume
An amount of three-dimensional space, often used to describe the space that an object takes up. (p. xv)

volumen Una cantidad de espacio tridimensional; a menudo se usa este término para describir el espacio que ocupa un objeto.

W, X, Y, Z

watt W
The unit of measurement for power, which is equal to one joule of work done or energy transferred in one second. For example, a 75 W light bulb converts electrical energy into heat and light at a rate of 75 joules per second. (p. 131)

vatio La unidad de medición de la potencia, el cual es igual a un julio de trabajo realizado o energía transferida en un segundo. Por ejemplo, una bombilla de 75 W convierte energía eléctrica a calor y luz a un ritmo de 75 julios por segundo.

wedge
A simple machine that has a thick end and a thin end. A wedge is used to cut, split, or pierce objects, or to hold objects together. (p. 158)

cuña Una máquina simple que tiene un extremo grueso y otro extremo delgado. Una cuña se usa para cortar, partir o penetrar objetos, o para mantener objetos juntos.

weight
The force of gravity on an object. (p. 79)

peso La fuerza de gravedad sobre un objeto.

wheel and axle
A simple machine that is a wheel attached to a shaft, or axle. (p. 156)

rueda y eje Una máquina simple que es una rueda unida a una flecha, o a un eje.

work
The use of force to move an object over a distance. (p. 115)

trabajo El uso de fuerza para mover un objeto una distancia.

R57

Index

Page numbers for definitions are printed in **boldface** type.
Page numbers for illustrations, maps, and charts are printed in *italics*.

A

acceleration, **25**, 25–31, *26, 34,* 49–55, *53, 60*
  average, 28
  force, mass, and, 50–53, *50, 51, 52, 53,* 56, *56, 67, 70, 86*
  gravitational, 78–79, *79, 80, 81,* 123, 148, *149*
  negative, *26, 29, 30,* 31
  velocity, time, and, 25, *26,* 27–31, *28, 30, 34*
accuracy, **R22**
action and reaction, 57–60, *59, 60, 66, 70,* 85, 158
aeolipile, 109, *109,* 111
air
  density of, 94
  weight of, *94*
air bags, 47
air pressure, 94, 100–101
air resistance, **89,** *89,* 152
altitude, air pressure and, 94
analysis, critical, R8
Archimedes, 109, *109*
area, **R43**
  force, pressure, and, 91–95, *92, 93,* 104
Aristotle, 108, *108*
astronomy, 84
atmosphere (unit of pressure), 94
atoms, **xv,** xvii
  nanotechnology and, 167
automobile. *See* car.
averages, calculating, 120, **R36**
axle. *See* wheel and axle.

B

balanced forces, **43,** 43–47, **45,** 59, **86**
Bernoulli, Daniel, 100
Bernoulli's principle, **100,** 100–101, 104
  in nature, *101*
bias, **R6**
  scientific, R6
  sources of, R6
bicycles
  efficiency of, *152*
  forces acting on, *44*
block-and-tackle system. *See* pulley.
buoyancy, **98,** 98–99, 104, 109
  density and, *99*

C

car
  design of, xxvi–xxvii
  fuel-cell powered, xxvii, *xxvii*
carbon, atomic model, *xxv*
catapult, 108, *108*
cause-and-effect relationship, **R5**
chemical energy, 128
collisions, **66,** *67, 68, 70. See also* momentum.
  of molecules and pressure, 93
computers
  modeling forces, 111
  scientific use of, *xxv*
constants, **R30**
contact force, **xxi,** *42*
control group, 129, R30
critical analysis, R8
  of statements, R8–R9
current (electrical), 135

D

data, analyzing, xx
data tables, making R23
da Vinci, Leonardo, 110
decimals, **R39,** R40
  adding, R39
  dividing, R40
  multiplying, R40
  subtracting, R40
Deep Space 1 (spacecraft), *29*
density, **99**
  air, *94*
  buoyancy and, 99
  mass, volume, and, 99
  water, *95*
design, technological, xxvi–xxvii
direction. *See also* vectors.
  of force, 42, 43, 147, *149*
displacement, 99, 104
distance
  distance-time graph, 20, *21, 30,* 31, *34*
  force, work, and, 115–119
  gravity and, 78, 104
  measuring, 11, 24
  speed and, 16–19, 23, 24, 34
drag, 89. *See also* air resistance.

INDEX

E

Earth
curvature of, 80, *81*
gravity of, 78, 79
mass and weight on, *80*
orbit around Sun, 80
efficiency, **150,** *150,* 150–152, *153*
calculating, 150, *153*
friction and, *150,* 151, 156
ideal mechanical advantage and, 160
Einstein, Albert, 84, 111
electricity, 128, 135, 146, 151
electromagnetic energy, 128
elements, xv
ellipse, *80*
energy, xviii, **xix,** 121–128, 138
chemical, xviii, 128
conservation of, xix, **126,** *126, 127,* 128, *138*
efficiency and, 151, 152
elastic potential, 122
electrical, xviii, 128, 135, 146, 151
electromagnetic, 128
forms of, xviii, 128
kinetic, **122,** *122, 124,* 125–128, *127, 138*
mechanical, **125,** 125–128, *127,* 146, 151
nuclear, 128
potential, **122,** *122, 123,* 125–128, *127, 138,* 148, *149*
power, time, and, 133–136
radiant, 128
sources of, xix
thermal, 128, 151
transfer of, 121–128, *122,* 133–135, *134, 135, 149*
transformation of, 121, 122, 125–128, *127*
engine. *See also* motor.
efficiency of car, 151, 152
power of different types, 132
steam, 109–110, *109, 110,* 132
equator, *10*
evaluating, **R8**
media claims, R8
evidence, collection of, xxiv
experiment, **xxiv.** *See also* lab.
conclusions, drawing, R35
constants, determining, R30
controlled, **R28,** R30
designing, R28–R35
hypothesis, writing, R29
materials, determining, R29
observations, recording, R33
procedure, writing, R32
purpose, determining, R28
results, summarizing, R34
variables, R30–R31, R32
experimental group, R30

F

fact, **R9**
different from opinion, R9
faulty reasoning, **R7**
first law of motion. *See* Newton's laws of motion.
floating. *See* buoyancy.
fluids, **88,** 88–89, *93,* 93–95, 98–103
friction in, 88–89, *89*
pressure in, *93,* 93–95
transmission of force through, 102–103, 104
force, **xxi,** 38–70, **41,** *42*
acceleration, mass, and, *50, 51, 52, 53, 56, 67, 70, 86*
action and reaction, 57–60, *59, 60,* 66, *70,* 85, 158
applied, *116*
area and, 91–95, 104
balanced, *43,* 43–47, *45,* 59, *86*
buoyancy, **98**
centripetal, **54,** *55,* 80, *81*
changing direction of, 42, 43, 147, *149*
contact, **xxi,** *42*
direction of motion changed by, 53–55
distance, work, and, 115–119, *116, 118,* 138
electrical, xxi
friction, xxi, **85,** *85–89*
gravitational, **xxi,** 42, *42,* 77–84
input, 146–147, *146,* 149–153, *155,* 160–162
Internet activity, 39
machines, work, and, 145–152, *146, 147,* 155–162, *155, 172*
magnetic, xxi
mass, distance, and, 77–85, *78,* 104
multiplication of, 146, 147
needed to overcome friction, *86, 87, 104*
net, **43,** 44, *93,* 98, *99, 149*
output, 146–147, *146,* 150–153, *155,* 160–162
physical, xx–xxi
strong, 111
transmission through fluids, 102–103
types of, 42
unbalanced, *43,* 43–47, *45, 47, 70*
formulas, **R42**
fractions, **R41**
frame of reference, *13. See also* motion, observing.
free fall, 83, 90
friction, xxi, *42,* 44, **85,** 85–89, *86, 87,* 110.
air resistance, **89,** *89,* 152
compound machines and, 165
efficiency and, *150,* 151, 152, 156
fluids and, 88–89, *89*
force needed to overcome, *86, 87, 104*
heat and, *87*
reducing, 152
surfaces and, 86–88
weight and, *87*
fuel cell, xxvii, *xxvii*
fulcrum, 109, **155,** *155, 162,* 163, *172*

INDEX

INDEX

G

Galilei, Galileo, 44, 110
  theory of motion, 44–45
gears, *165*
  mechanical advantage of, 165
  used in nanotechnology, *167*
Goddard, Robert H., 111
graphs
  bar, R26, R31
  circle, R25
  distance-time, 20, *21, 30,* 31, *34*
  double bar, R27
  line, *90,* R24, R34
  velocity-time, *30,* 31, *90*
gravity, **xxi,** xxiv, 42, *42, 60,* **77,** 77–84, *78, 79, 104*
  acceleration, mass, and, 78–79, *79,* 148, *149*
  distance, mass, and, 77–85, *78, 104*
  Earth's, 78–81, *80, 81,* 123
  effect on light, *84,* 111
  energy and, 123, 126, *127*
  internet activity, 75
  orbit and, 80–83, *81*
  weight and, 79–80, *80*
  work and, 118–119, *119, 158*
Greek philosophy, theory of motion in, 44, 108

H

hair dryer, energy use of, *135*
heat
  friction and, *87,* 151
horsepower, 132. *See also* power; Watt.
hydraulics, *103, 166*
hydrogen, in fuel cells, xxvii
hypothesis, **xxiv,** xxv, **R3,** R29

I, J, K

ideal mechanical advantage, 160–162
inclined plane, *147,* 148, *149,* 154, **158,** *158,* 159, *160*
  ideal mechanical advantage of, *160,* 160–161
  See also wedge.
inertia, **46,** *47. See also* momentum.
inference, **R4,** R35
infrared waves, 128
International System of Units, R20–R21
jellyfish, movement of, *57,* 58
joule, **117,** *123, 124,* 131
kangaroos, movement of, *60*
kilowatt, 131
kinetic energy, **122,** *122, 124,* 125–128, *127, 138*

L

laboratory equipment
  beakers, R12, *R12*
  double-pan balances, R19, *R19*
  force meter, R16, *R16*
  forceps, R13, *R13*
  graduated cylinder, R16, *R16*
  hot plate, R13, *R13*
  meniscus, R16, *R16*
  microscope, R14–R15, *R14*
  spring scale, R16, *R16*
  ruler, metric, R17, *R17*
  test-tube holder, R12, *R12*
  test-tube racks, R13, *R13*
  test tubes, R12, *R12*
  triple-beam balances, R18, *R18*
labs, R10–R35. *See also* experiments.
  equipment, R12–R19
  safety, R10–R11
laser, chemical reaction and, *xxiv*
latitude, 10, *10*
laws of motion. *See* Newton's laws of motion.
lens, gravitational, *84*
Leonardo da Vinci, 110
lever, 109, *146, 153,* 154, **155,** *162, 172*
  compound machines using, 164, 165, *166, 172*
  first-class, *155*
  force and, *155*
  ideal mechanical advantage of, *162*
  in human body, 163
  real-world examples, *163,* 164
  second-class, *155*
  third-class, *155*
lift, 101
light
  bending by gravity, *84,* 111
  visible, 128
liquids. *See* fluids; hydraulics; water.
location, ways of describing, 9, *10*
longitude, 10, *10*

M

machines, 108–110, **145,** 145–169, *146, 172.*
  compound, **164,** 164–169, *172*
  efficiency of, **150,** *150,* 150–152, **153**
  electronic, 146
  force, work, and, 145–152, *146,* 147, 155–162, *155, 172*
  Internet activity, 143
  mechanical advantage of, **147,** 160–162
  perpetual motion, 109, 110
  robotic, 2–5, *2, 4,* 168, **169**
  simple, 146, **154,** 154–162, 163, *172. See also* inclined plane; lever; pulley; screw; wedge; wheel and axle.
magnetic force, xxi

manufacturing
methods, 167
robots used in, 169
Mars, exploration of, 2–5
Mars Exploration Rover (robot), 3–5, 3
mass, **xv,** 46, 49–53
acceleration, gravitation, and, 78–79, *79,* 148, 149
force, acceleration, and, 50, 51, 52, 53, 56, 67, 70, 86
gravitation, distance, and, 77–85, *78, 104*
inertia and, 46
momentum, velocity, and, 64–65, *65,* 70
volume, density, and, 99
weight and, compared, 79–80, *80*
math skills. *See also* units of measurement.
area, **R43**
averages, 120, **R36, R37**
decimal, **R39,** R40
describing a set of data, R36–R37
eliminating outlying values, 120
formulas, 18, 23, 51, 52, 53, 65, 92, 117, 123, 124, 125, 131, 133, **R42**
fractions, **R41**
line graph, 90
mean, 120, **R36**
median, **R36**
mode, **R37**
percents, 153, **R41**
proportions, **R39**
range, **R37**
rates, **R38**
ratios, 153, **R38**
scientific notation, **R44**
significant figures, 56, **R44**
units, 24
volume, **R43**
matter, xiv–xvii, **xv**
conservation of, xvii
forms of, xvi–xvii
movement of, xvii
particles and, xiv–xv
physical forces and, xx
mean, 120, **R36**
measurement
acceleration, 28, 51
area, 92
density, 99
distance, 11, 24
Earth's gravity, 78
energy, 121
force, 51, 80
International System of Units (SI), R20–R21
mass, 51
power, 131, 132
pressure, 92, 94
speed, 18
temperature, R21
weight, 80
work, 117
mechanical advantage, **147,** 157, 160. *See also* efficiency; machines.
calculating, 147
compound machines, 165
ideal, 160–162
median, **R36**
metric system, R20–R21
changing metric units, R20, *R20*
converting between U.S. customary units, R21, *R21*
temperature conversion, R21, *R21*
microgears, *167*
microgravity, *83*
microscope, R14–R15, *R14*
making a slide or wet mount, R15, *R15*
viewing an object, R15
scanning tunneling (STM), *xxiv*
microtechnology, 166–167
mode, **R37**
molecule, **xv**
air, 94
collision of, 93–95
fluid, *93,* 93–95
nanotechnology and, 166–167
water, *93,* 94, *95*
momentum, **64,** 64–69, *65, 67, 70. See also* inertia.
conservation of, **67,** *68, 69*
transfer of, 66, *67*
velocity, mass, and, 64–65, *65, 70*
Moon
exploration of, 111, *111*
mass and weight on, *80*
orbit around Earth, 80
motion, 6–34, **11,** *34. See also* inertia; Newton's laws of motion.
Aristotle and, 108
Bhaskara and, 109
circular, 54, *55*
direction of, force, and, 53–55, *55, 116*
direction of, work, and, *116*
fluids, and, 88–89, *89,* 100–101, 104
force, work, and, 115–119, *116, 118,* 138, 148–149, *149*
friction and, 42, *42,* 44, 85, 86–87, *86, 87*
Galileo and, 110
Internet activity, 7
Leonardo da Vinci and, 110
observing, 13–14, 34
perpetual, 109, 110, 126
relative, 13–14, *14,* 34
motors, 151. *See also* engine.

N

nanotechnology, 166–167, **167,** *167*
Newton, Sir Isaac, 44, 45, 80, 109, 110
Newton's laws of motion, 44, 45, *60,* 61, *70,* 109, 110, 111
first, 44–47, **45,** *45, 60*
second, 49–55, **50,** *50, 55, 60,* 78
third, **57,** 57–59, *59, 60, 67,* 68–69
newton (unit of force), 51, *92, 117, 123*
note-taking strategies, **R45–R49**
combination notes, 40, 144, R45, *R45*
concept map, R49, *R49*
content frame, R45, *R45*
main idea and detail notes, R46, *R46*
main idea web, 114, 144, R46, *R46*
mind map, R47, *R47*
outline, 8, 144, R48, *R48*
supporting main ideas, 76, 144, R47, *R47*
Venn diagram, R49, *R49*

INDEX

nuclear energy, 128
numbers. *See also* math skills.
  meaningful, 56
  outlying, 120

O

observation, **xxiv, R2,** R5, R33
  qualitative, R2
  quantitative, R2
operational definition, **R31**
opinion, **R9**
  different from fact, R9
orbit, **80,** 80–83, *81. See also* force, centripetal.
  humans in, *83*
  velocity needed to achieve, 82
outliers, 120

P, Q

particle, xiv–xv
Pascal, Blaise, 102
Pascal's principle, **102,** 102–103, *102,* 104
pascal (unit of pressure), **92,** 94
pendulum, *126*
percents, 153, **R41**
physical science, xiii. *See also* science.
  unifying principles of, xiv–xxi
plane, inclined, *147,* 148, *149,* 154, **158,** *158,* 159, *160*
point of view. *See* frame of reference.
pollution, xxvii, 167
position, **9,** 9–14, *34. See also* reference point; motion.
  energy transfer, speed, and, 121
  potential energy and, 122
  ways of describing, *10*
potential energy, **122,** *122, 123,* 125–128, *127, 138,* 148, *149*
power, **130,** 130–135, *132,* 138
  energy, time, and, 133–135, *134, 135,* 138
  everyday usage, *135*
  work, time, and, 130–132, *132,* 138
precision, **R22**
prediction, **xxiv, R3**
pressure, **91,** 91–96
  air, *94*
  area, force, and, 91–95, *92, 93,* 104
  atmospheric, *94*
  in fluids, *93,* 93–95, 100–103, *101, 102, 104*
  water, 93, *95*
prime meridian, *10*
proportions, **R39**
pulley, *147,* 154, **156,** *156, 157, 172*
  block-and-tackle system, using, 157
  fixed, *156,* 157
  mechanical advantage of, 157
  movable, *157*

R

radius, 161
ramp. *See* inclined plane.
range, **R37**
rates, 17, **R38**
ratios, 153, **R38**
reaction, 57–60, *59, 60,* 66, *77,* 85, 158
reasoning, faulty, **R7**
reference point, **10,** *10,* 34
robots, 2–5, *2, 4, 168,* **169,** *169*
  Mars exploration and, 2–5
rocket, 111
rounding numbers, 56

S

safety, R10–R11
  animal, R11
  chemical, R11
  clean up, R11
  directions, R10
  dress code, R10
  electrical, R11
  fire, R10
  glassware, R11
  heating, R10
  icons, R10–R11
  lab, R10–R11
  sharp object, R11
science, nature of, xxii–xxv
Scientific American Frontiers, 2, 4
scientific notation, **R44**
scientific process, xxiii–xxv
  asking questions and, xxiii
  determining what is known and, xxiii
  interpreting results and, xxv
  investigating and, xxiv
  sharing results and, xxv
screw, 154, *158,* **159,** *159*
seat belts, *47*
second law of motion. *See* Newton's laws of motion.
significant figures, 56, **R44**
sinking. *See* buoyancy.
SI units. *See* International System of Units.
skydiving, *89*
slides, making, R15, *R15*
slope (steepness), 148, *149, 160*
spacecraft, *29, 53,* 81–83, *83, 169*
  velocity needed to achieve orbit, 82
speed, 12, **16,** 16–23, *17, 18, 23, 34. See also* velocity.
  average, 19, 23
  instantaneous, 19
  relation to velocity, 22–23
  time, distance, and, 16–23, *17, 18, 20, 21, 24, 34*
  using distance-time graph to show, 20, *21*
Standardized Test Practice, 37, 73, 107, 141, 175
  analyzing data, 73
  analyzing graphics, 175
  interpreting diagrams, 107
  interpreting graphs, 37
  understanding experiments, 141
steam engine, 109–110, *109, 110,* 132

steering wheel, example of wheel and axle, 156
surface area. *See also* pressure.
air resistance and, *89*
systems. *See* energy, conservation of; momentum, conservation of.

T

tables. *See* data tables.
taking notes. *See* note-taking strategies.
technology, nature of, xxvi–xxvii
temperature, unit conversion, R21, *R21*
terminal velocity, 89, 90
thermal energy, 128
third law of motion. *See* Newton's laws of motion.
time
distance, speed, and, 16–23, *17, 18, 20, 21,* 24, 34
distance-time graph, 20, *21, 30,* 31, *34*
energy, power, and, 133–135, *134, 135,* 138
velocity, acceleration, and, 25, *26,* 27–31, *28, 30, 34*
velocity-time graph, *30,* 31, *90*
work, power, and, 130–132, *132,* 138
timeline, understanding forces, 108–111

U

units of measurement, 24. *See also* metric system.
horsepower, 132
joule, 117
kilogram, 51
kilogram-meter per second (kg•m/s), 65
kilometers per hour (km/h), 23
meters per second (m/s), 18, 24
meters per second squared (m/s^2), 28, 51
newton, 51–53
pascal, 92
watt, 131

V

vacuum, acceleration in, 79
variables, **R30,** R31, R32
controlling, R17
dependent, **R30,** R31
independent, **R30**
isolating, 129
vector, **22,** 22
acceleration, 26
of force, 42
momentum, 65
velocity, 22
velocity, **22,** *22,* 25–31, *26, 34. See also* speed.
acceleration, time, and, 25, *26,* 27–31, *28, 30, 34*
average, 23
escape, 82
kinetic energy, mass, and, 124, 126
momentum, mass, and, 64–65, 65, *70*
needed to achieve orbit, 82
relation to speed, 22–23
terminal, *89, 90*
velocity-time graph, *30,* 31, 89, *90*
vocabulary strategies, R50–R51
description wheel, 8, 114, R50, *R50*
four square, 76, 114, R50, *R50*
frame game, R51, *R51*
magnet word, 40, 114, R51, *R51*
word triangle, 144, R51, *R51*
volume, **xv, R43**
buoyancy and, 109
displacement and, 99, 104
mass, density, and, 99

W, X, Y, Z

water, **xvii,** *93,* 94, 95. *See also* buoyancy, hydraulics.
pressure in, *93, 95*
used to power machines, *119*
vapor, xvii
water wheels, *119*
watt (unit of power), **131**
Watt, James, 110, 132
wedge, 154, **158,** *158,* 159
compound machines using, 165, *172*
real-world examples, *158,* 159
used to hold objects together, 159
used to separate objects, 159
weight, **79**
friction and, *87*
gravitation and, 79–80, *80*
mass and, compared, 79–80, *80*
weightlessness, *83*
wet mount, making a, R15, *R15*
whales, *95*
wheel and axle, 154, **156,** *156, 161, 172*
compound machines using, 164–165, *165,* 166, *172*
mechanical advantage of, *161,* 165
real-world examples, 156, *159, 161*
wind, used to power machines, 119
windmills, 119
wings, 101
work, **115,** 115–135, *118, 138*
distance, force, and, 115–119, *116, 118,* 138, 148–149, *149*
energy transfer and, 121, 126, *127*
gravitation and, 118–119, *119*
input, 150–152, *150,* 153
Internet activity, 113
machines, force, and, 145–152, *146, 147,* 155–162, *155,* 172
output, 150–152, *150,* 153
time, power, and, 130–132, *132,* 138

R63

Acknowledgments

Photography

Cover © Brett Froomer/Getty Images; **i** © Brett Froomer/Getty Images; **iii** *left (top to bottom)* Photograph of James Trefil by Evan Cantwell; Photograph of Rita Ann Calvo by Joseph Calvo; Photograph of Linda Carnine by Amilcar Cifuentes; Photograph of Sam Miller by Samuel Miller; *right (top to bottom)* Photograph of Kenneth Cutler by Kenneth A. Cutler; Photograph of Donald Steely by Marni Stamm; Photograph of Vicky Vachon by Redfern Photographics; **vi** © Arthur Tilley/Getty Images; **vii** © Mike Chew/Corbis; **ix** Photographs by Sharon Hoogstraten; **xiv–xv** © Larry Hamill/age fotostock america, inc.; **xvi–xvii** © Fritz Poelking/age fotostock america, inc.; **xviii–xix** © Galen Rowell/Corbis; **xx–xxi** © Jack Affleck/SuperStock; **xxii** AP/Wide World Photos; **xxiii** © David Parker/IMI/University of Birmingham High, TC Consortium/Photo Researchers; **xxiv** *left* AP/Wide World Photos; *right Washington University Record;* **xxv** *top* © Kim Steele/Getty Images; *bottom* Reprinted with permission from S. Zhou et al., *SCIENCE* 291:1944–47. © 2001 AAAS; **xxvi–xxvii** © Mike Fiala/Getty Images; **xxvii** *left* © Derek Trask/Corbis; *right* AP/Wide World Photos; **xxxii** © The Chedd-Angier Production Company; **2–3** Courtesy of NASA/JPL/Caltech; **3** © Stocktrek/Corbis; **4** *top* Courtesy of NASA/JPL/Caltech; *bottom* © The Chedd-Angier Production Company; **6–7** © Lester Lefkowitz/Corbis; **7** Photographs by Sharon Hoogstraten; **9** © Royalty-Free/Corbis; **11** © Globus, Holway & Lobel/Corbis; **12** *top* Photograph by Sharon Hoogstraten; *bottom* © B.P./Getty Images; **14** *top* © Georgina Bowater/Corbis; *bottom* © SuperStock; **15** © Graham Wheatley/ The Military Picture Library/Corbis; **16, 17** Photographs by Sharon Hoogstraten; **18** © Gunter Marx Photography/Corbis; **19** Photograph by Sharon Hoogstraten; **21** © Tom Brakefield/Corbis; **22** © David M. Dennis/Animals Animals; **23** © Kelly-Mooney Photography/Corbis; **24** © Gallo Images/Corbis; **25** © 1986 Richard Megna/Fundamental Photographs, NYC; **27** Photograph by Sharon Hoogstraten; **28** © Royalty-Free/Corbis; **29** Courtesy of NASA/JPL/Caltech; **30** © Robert Essel NYC/Corbis; **32** *top* © Mark Jenkinson/Corbis; *bottom* Photographs by Sharon Hoogstraten; **34** *top* © Globus, Holway & Lobel/Corbis; *center* Photograph by Sharon Hoogstraten; **36** © David M. Dennis/Animals Animals; **38–39** © Arthur Tilley/Getty Images; **39, 41** Photographs by Sharon Hoogstraten; **42** © John Kelly/Getty Images; **43** *left* © AFP/Corbis; *right* © Reuters NewMedia Inc./Corbis; **44** © Michael Kevin Daly/Corbis; **45** *left* © Jim Cummins/Getty Images; *right* © Piecework Productions/Getty Images; **46** Photograph by Sharon Hoogstraten; **48** *left, inset* © Bill Ross/Corbis; *right* Dr. Paula Messina, San Jose State University; **49, 50** Photographs by Sharon Hoogstraten; **52** AP/Wide World Photos; **53** NASA; **54** Photograph by Sharon Hoogstraten; **55** AP/Wide World Photos; **56** *top* Clare Hirn, Jewish Hospital, University of Louisville and ABIOMED; *bottom* John Lair, Jewish Hospital, University of Louisville and ABIOMED; **57** © Danny Lehman/Corbis; **58, 59** Photographs by Sharon Hoogstraten; **60** © Photodisc/Getty Images; *background* © David C. Fritts/Animals Animals; **62** *top* Digital image © 1996 Corbis/Original image courtesy of NASA/Corbis; *bottom* Photographs by Sharon Hoogstraten; **64, 66** Photographs by Sharon Hoogstraten; **68** © TRL Ltd./Photo Researchers; **69** © Charles O'Rear/Corbis; **70** *top* © Photodisc/Getty Images; *bottom* Photographs by Sharon Hoogstraten; **71** © Siede Preis/ Getty Images; **72** Photographs by Sharon Hoogstraten; **74–75** © Mike Chew/Corbis; **75, 77** Photographs by Sharon Hoogstraten; **80, 81** Photographs of models by Sharon Hoogstraten; **80** *left* NASA; *right* © Photodisc/Getty Images; **81** *top, bottom, background* NASA; **82** Photograph by Sharon Hoogstraten; **83** NASA; **84** *left* © Royalty-Free/Corbis; *right* NASA/ESA; **85** © John Beatty/Getty Images; **86, 87** Photographs by Sharon Hoogstraten; **88** *top* © Al Francekevich/Corbis; *bottom* Photograph by Sharon Hoogstraten; **89** © Joe McBride/Getty Images; **90** © NatPhotos/Tony Sweet/Digital Vision; *inset* © Michael S. Yamashita/Corbis; **91** Photograph by Sharon Hoogstraten; **92** © Wilson Goodrich/Index Stock; **93** © Royalty-Free/Corbis; **94** © Philip & Karen Smith/Getty Images; **95** © Ralph A. Clevenger/Corbis; **96** *top* © Stephen Frink/Corbis; *bottom* Photographs by Sharon Hoogstraten; **98, 99, 100** Photographs by Sharon Hoogstraten; **101** Photograph of prairie dogs © W. Perry Conway/Corbis; **103** © Omni Photo Communications Inc./Index Stock; **104** *top, bottom* Photographs by Sharon Hoogstraten; *center* © Royalty-Free/Corbis; **105** Photograph by Sharon Hoogstraten; **106** *left* © Joe McBride/Getty Images; *right* Photograph by Sharon Hoogstraten; **108** *top* © Erich Lessing/Art Resource, New York; *bottom* © Dagli Orti/The Art Archive; **109** *top left* © SPL/Photo Researchers; *top right* Sam Fogg Rare Books & Manuscripts; *bottom* © Dorling Kindersley; **110** *left* © Victoria & Albert Museum, London/Art Resource, New York; *top right* Photo Franca Principe, Institute and Museum of the History of Science; *center right* © Scala/Art Resource, New York; *bottom right* © Dorling Kindersley; **111** *top* © Gerald L. Schad/Photo Researchers; *bottom* NASA; **112–113** © Digital Vision; **113** *top* Image Club Graphics; *center* Photograph by Sharon Hoogstraten; **115, 116** Photographs by Sharon Hoogstraten; **117** © Rob Lewine/Corbis; **118** Photograph by Sharon Hoogstraten; **119** © Reinhard Eisele/Corbis; **120** © Roger Allyn Lee/ SuperStock; **121** Chris Wipperman/KCPDSA; **123** © Patrik Giardino/Corbis; **124** © Tony Anderson/Getty Images; **125** Photograph by Sharon Hoogstraten; **126** © 1988 Paul Silverman/Fundamental Photographs, NYC; **127** © Tony Donaldson/Icon Sports Media; **129** © AFP/Corbis; **130** Photograph by Sharon Hoogstraten; **131** © Pete Saloutos/Corbis; **132** © Digital Vision; **133** Photograph by Sharon Hoogstraten; **134** © Walter Hodges/Corbis; **135** © Grantpix/Index Stock; **136** *top* © David Young-Wolff/PhotoEdit; *bottom* Photographs by Sharon Hoogstraten; **138** © Pete Saloutos/Corbis; **140** Photographs by Sharon Hoogstraten; **142–143** © Balthazar Korab; **145** Photograph by Sharon Hoogstraten; **146** © David Young-Wolff/PhotoEdit; **147** © Joseph Sohm/ ChromoSohm Inc./Corbis; **149** © Brad Wrobleski/Masterfile; **150** © Michael Macor/San Francisco Chronicle/Corbis SABA; **151** Photograph by Sharon Hoogstraten; **152** © Jean-Yves Ruszniewski/Corbis; **153** © Royalty-Free/Corbis; *inset* © Felicia Martinez/ PhotoEdit; **154, 155** Photographs by Sharon Hoogstraten; **156** © Tom Stewart/Corbis; **157** Photograph by Sharon Hoogstraten; **158** *top* © David Butow/Corbis SABA; *bottom* © Peter Beck/Corbis; **159** © Henryk T. Kaiser/Index Stock; **160** © Tony Freeman/ PhotoEdit; **161** © Todd A. Gipstein/Corbis; **163** AP/Wide World Photos; **164** © Tony Freeman/PhotoEdit; **165** © Lester Lefkowitz/ Corbis; **166** Hurst Jaws of Life; **167** © David Parker/Photo Researchers; **168** *top* AP/Wide World Photos; *bottom* © Robert Caputo/Stock Boston; *background* © Royalty-Free/Corbis; **170** *top* © Photodisc/Getty Images; *bottom* Photograph by Sharon Hoogstraten; **172** © ThinkStock/SuperStock; **173** *top left* Photograph by Sharon Hoogstraten; **174** © Tony Freeman/ PhotoEdit; **R28** © Photodisc/Getty Images.

Illustrations and Maps

Accurate Art, Inc. **107, 175**; Steve Cowden **122**; Ampersand Design Group **15**; Houghton Mifflin School Division **47**; MapQuest.com, Inc. **10, 60, 129, 168**; Tony Randazzo/American Artists Rep. Inc. **13**; Dan Stuckenschneider **102, 135, 156, 157, 158, 159, 162, 165, 172, 173**; Dan Stukenschneider based on an illustration by Matt Cioffi **168**.

ACKNOWLEDGMENTS

Content Standards: 5–8

A. Science as Inquiry

As a result of activities in grades 5–8, all students should develop

Abilities Necessary to do Scientific Inquiry

A.1 Identify questions that can be answered through scientific investigations. Students should develop the ability to refine and refocus broad and ill-defined questions. An important aspect of this ability consists of students' ability to clarify questions and inquiries and direct them toward objects and phenomena that can be described, explained, or predicted by scientific investigations. Students should develop the ability to identify their questions with scientific ideas, concepts, and quantitative relationships that guide investigation.

A.2 Design and conduct a scientific investigation. Students should develop general abilities, such as systematic observation, making accurate measurements, and identifying and controlling variables. They should also develop the ability to clarify their ideas that are influencing and guiding the inquiry, and to understand how those ideas compare with current scientific knowledge. Students can learn to formulate questions, design investigations, execute investigations, interpret data, use evidence to generate explanations, propose alternative explanations, and critique explanations and procedures.

A.3 Use appropriate tools and techniques to gather, analyze, and interpret data. The use of tools and techniques, including mathematics, will be guided by the question asked and the investigations students design. The use of computers for the collection, summary, and display of evidence is part of this standard. Students should be able to access, gather, store, retrieve, and organize data, using hardware and software designed for these purposes.

A.4 Develop descriptions, explanations, predictions, and models using evidence. Students should base their explanation on what they observed, and as they develop cognitive skills, they should be able to differentiate explanation from description—providing causes for effects and establishing relationships based on evidence and logical argument. This standard requires a subject matter knowledge base so the students can effectively conduct investigations, because developing explanations establishes connections between the content of science and the contexts within which students develop new knowledge.

A.5 Think critically and logically to make the relationships between evidence and explanations. Thinking critically about evidence includes deciding what evidence should be used and accounting for anomalous data. Specifically, students should be able to review data from a simple experiment, summarize the data, and form a logical argument about the cause-and-effect relationships in the experiment. Students should begin to state some explanations in terms of the relationship between two or more variables.

A.6 Recognize and analyze alternative explanations and predictions. Students should develop the ability to listen to and respect the explanations proposed by other students. They should remain open to and acknowledge different ideas and explanations, be able to accept the skepticism of others, and consider alternative explanations.

A.7 Communicate scientific procedures and explanations. With practice, students should become competent at communicating experimental methods, following instructions, describing observations, summarizing the results of other groups, and telling other students about investigations and explanations.

A.8 Use mathematics in all aspects of scientific inquiry. Mathematics is essential to asking and answering questions about the natural world. Mathematics can be used to ask questions; to gather, organize, and present data; and to structure convincing explanations.

Understandings about Scientific Inquiry

A.9.a Different kinds of questions suggest different kinds of scientific investigations. Some investigations involve observing and describing objects, organisms, or events; some involve collecting specimens; some involve experiments; some involve seeking more information; some involve discovery of new objects and phenomena; and some involve making models.

A.9.b Current scientific knowledge and understanding guide scientific investigations. Different scientific domains employ different methods, core theories, and standards to advance scientific knowledge and understanding.

A.9.c Mathematics is important in all aspects of scientific inquiry.

A.9.d Technology used to gather data enhances accuracy and allows scientists to analyze and quantify results of investigations.

A.9.e Scientific explanations emphasize evidence, have logically consistent arguments, and use scientific principles, models, and theories. The scientific community accepts and uses such explanations until displaced by better scientific ones. When such displacement occurs, science advances.

A.9.f Science advances through legitimate skepticism. Asking questions and querying other scientists' explanations is part of scientific inquiry. Scientists evaluate the explanations proposed by other scientists by examining evidence, comparing evidence, identifying faulty reasoning, pointing out statements that go beyond the evidence, and suggesting alternative explanations for the same observations.

A.9.g Scientific investigations sometimes result in new ideas and phenomena for study, generate new methods or procedures for an investigation, or develop new technologies to improve the collection of data. All of these results can lead to new investigations.

B. Physical Science

As a result of their activities in grades 5–8, all students should develop an understanding of

Properties and Changes of Properties in Matter

B.1.a A substance has characteristic properties, such as density, a boiling point, and solubility, all of which are independent of the amount of the sample. A mixture of substances often can be separated into the original substances using one or more of the characteristic properties.

B.1.b Substances react chemically in characteristic ways with other substances to form new substances (compounds) with different characteristic properties. In chemical reactions, the total mass is conserved. Substances often are placed in categories or groups if they react in similar ways; metals is an example of such a group.

B.1.c Chemical elements do not break down during normal laboratory reactions involving such treatments as heating, exposure to electric current, or reaction with acids. There are more than 100 known elements that combine in a multitude of ways to produce compounds, which account for the living and nonliving substances that we encounter.

Motions and Forces

B.2.a The motion of an object can be described by its position, direction of motion, and speed. That motion can be measured and represented on a graph.

B.2.b An object that is not being subjected to a force will continue to move at a constant speed and in a straight line.

B.2.c If more than one force acts on an object along a straight line, then the forces will reinforce or cancel one another, depending on their direction and magnitude. Unbalanced forces will cause changes in the speed or direction of an object's motion.

Transfer of Energy

B.3.a Energy is a property of many substances and is associated with heat, light, electricity, mechanical motion, sound, nuclei, and the nature of a chemical. Energy is transferred in many ways.

B.3.b Heat moves in predictable ways, flowing from warmer objects to cooler ones, until both reach the same temperature.

B.3.c Light interacts with matter by transmission (including refraction), absorption, or scattering (including reflection). To see an object, light from that object—emitted by or scattered from it—must enter the eye.

B.3.d Electrical circuits provide a means of transferring electrical energy when heat, light, sound, and chemical changes are produced.

B.3.e In most chemical and nuclear reactions, energy is transferred into or out of a system. Heat, light, mechanical motion, or electricity might all be involved in such transfers.

B.3.f The sun is a major source of energy for changes on the earth's surface. The sun loses energy by emitting light. A tiny fraction of that light reaches the earth, transferring energy from the sun to the earth. The sun's energy arrives as light with a range of wavelengths, consisting of visible light, infrared, and ultraviolet radiation.

C. Life Science

As a result of their activities in grades 5–8, all students should develop understanding of

Structure and Function in Living Systems

C.1.a Living systems at all levels of organization demonstrate the complementary nature of structure and function. Important levels of organization for structure and function include cells, organs, tissues, organ systems, whole organisms, and ecosystems.

C.1.b All organisms are composed of cells—the fundamental unit of life. Most organisms are single cells; other organisms, including humans, are multicellular.

C.1.c Cells carry on the many functions needed to sustain life. They grow and divide, thereby producing more cells. This requires that they take in nutrients, which they use to provide energy for the work that cells do and to make the materials that a cell or an organism needs.

C.1.d Specialized cells perform specialized functions in multicellular organisms. Groups of specialized cells cooperate to form a tissue, such as a muscle. Different tissues are in turn grouped together to form larger functional units, called organs. Each type of cell, tissue, and organ has a distinct structure and set of functions that serve the organism as a whole.

C.1.e The human organism has systems for digestion, respiration, reproduction, circulation, excretion, movement, control, and coordination, and for protection from disease. These systems interact with one another.

C.1.f Disease is a breakdown in structures or functions of an organism. Some diseases are the result of intrinsic failures of the system. Others are the result of damage by infection by other organisms.

Reproduction and Heredity

C.2.a Reproduction is a characteristic of all living systems; because no individual organism lives forever, reproduction is essential to the continuation of every species. Some organisms reproduce asexually. Other organisms reproduce sexually.

C.2.b In many species, including humans, females produce eggs and males produce sperm. Plants also reproduce sexually—the egg and sperm are produced in the flowers of flowering plants. An egg and sperm unite to begin development of a new individual. That new individual receives genetic information from its mother (via the egg) and its father (via the sperm). Sexually produced offspring never are identical to either of their parents.

C.2.c Every organism requires a set of instructions for specifying its traits. Heredity is the passage of these instructions from one generation to another.

C.2.d Hereditary information is contained in genes, located in the chromosomes of each cell. Each gene carries a single unit of information. An inherited trait of an individual can be determined by one or by many genes, and a single gene can influence more than one trait. A human cell contains many thousands of different genes.

C.2.e The characteristics of an organism can be described in terms of a combination of traits. Some traits are inherited and others result from interactions with the environment.

National Science Education Standards

Regulation and Behavior

C.3.a All organisms must be able to obtain and use resources, grow, reproduce, and maintain stable internal conditions while living in a constantly changing external environment.

C.3.b Regulation of an organism's internal environment involves sensing the internal environment and changing physiological activities to keep conditions within the range required to survive.

C.3.c Behavior is one kind of response an organism can make to an internal or environmental stimulus. A behavioral response requires coordination and communication at many levels, including cells, organ systems, and whole organisms. Behavioral response is a set of actions determined in part by heredity and in part from experience.

C.3.d An organism's behavior evolves through adaptation to its environment. How a species moves, obtains food, reproduces, and responds to danger are based in the species' evolutionary history.

Populations and Ecosystems

C.4.a A population consists of all individuals of a species that occur together at a given place and time. All populations living together and the physical factors with which they interact compose an ecosystem.

C.4.b Populations of organisms can be categorized by the function they serve in an ecosystem. Plants and some microorganisms are producers—they make their own food. All animals, including humans, are consumers, which obtain food by eating other organisms. Decomposers, primarily bacteria and fungi, are consumers that use waste materials and dead organisms for food. Food webs identify the relationships among producers, consumers, and decomposers in an ecosystem.

C.4.c For ecosystems, the major source of energy is sunlight. Energy entering ecosystems as sunlight is transferred by producers into chemical energy through photosynthesis. That energy then passes from organism to organism in food webs.

C.4.d The number of organisms an ecosystem can support depends on the resources available and abiotic factors, such as quantity of light and water, range of temperatures, and soil composition. Given adequate biotic and abiotic resources and no disease or predators, populations (including humans) increase at rapid rates. Lack of resources and other factors, such as predation and climate, limit the growth of populations in specific niches in the ecosystem.

Diversity and Adaptations of Organisms

C.5.a Millions of species of animals, plants, and microorganisms are alive today. Although different species might look dissimilar, the unity among organisms becomes apparent from an analysis of internal structures, the similarity of their chemical processes, and the evidence of common ancestry.

C.5.b Biological evolution accounts for the diversity of species developed through gradual processes over many generations. Species acquire many of their unique characteristics through biological adaptation, which involves the selection of naturally occurring variations in populations. Biological adaptations include changes in structures, behaviors, or physiology that enhance survival and reproductive success in a particular environment.

C.5.c Extinction of a species occurs when the environment changes and the adaptive characteristics of a species are insufficient to allow its survival. Fossils indicate that many organisms that lived long ago are extinct. Extinction of species is common; most of the species that have lived on the earth no longer exist.

D. Earth and Space Science

As a result of their activities in grades 5–8, all students should develop an understanding of

Structure of the Earth System

D.1.a The solid earth is layered with a lithosphere; hot, convecting mantle; and dense, metallic core.

D.1.b Lithospheric plates on the scales of continents and oceans constantly move at rates of centimeters per year in response to movements in the mantle. Major geological events, such as earthquakes, volcanic eruptions, and mountain building, result from these plate motions.

D.1.c Land forms are the result of a combination of constructive and destructive forces. Constructive forces include crustal deformation, volcanic eruption, and deposition of sediment, while destructive forces include weathering and erosion.

D.1.d Some changes in the solid earth can be described as the "rock cycle." Old rocks at the earth's surface weather, forming sediments that are buried, then compacted, heated, and often recrystallized into new rock. Eventually, those new rocks may be brought to the surface by the forces that drive plate motions, and the rock cycle continues.

D.1.e Soil consists of weathered rocks and decomposed organic material from dead plants, animals, and bacteria. Soils are often found in layers, with each having a different chemical composition and texture.

D.1.f Water, which covers the majority of the earth's surface, circulates through the crust, oceans, and atmosphere in what is known as the "water cycle." Water evaporates from the earth's surface, rises and cools as it moves to higher elevations, condenses as rain or snow, and falls to the surface where it collects in lakes, oceans, soil, and in rocks underground.

D.1.g Water is a solvent. As it passes through the water cycle it dissolves minerals and gases and carries them to the oceans.

D.1.h The atmosphere is a mixture of nitrogen, oxygen, and trace gases that include water vapor. The atmosphere has different properties at different elevations.

D.1.i Clouds, formed by the condensation of water vapor, affect weather and climate.

D.1.j Global patterns of atmospheric movement influence local weather. Oceans have a major effect on climate, because water in the oceans holds a large amount of heat.

D.1.k Living organisms have played many roles in the earth system, including affecting the composition of the atmosphere, producing some types of rocks, and contributing to the weathering of rocks.

Earth's History

D.2.a The earth processes we see today, including erosion, movement of lithospheric plates, and changes in atmospheric composition, are similar to those that occurred in the past. Earth history is also influenced by occasional catastrophes, such as the impact of an asteroid or comet.

D.2.b Fossils provide important evidence of how life and environmental conditions have changed.

Earth in the Solar System

D.3.a The earth is the third planet from the sun in a system that includes the moon, the sun, eight other planets and their moons, and smaller objects, such as asteroids and comets. The sun, an average star, is the central and largest body in the solar system.

D.3.b Most objects in the solar system are in regular and predictable motion. Those motions explain such phenomena as the day, the year, phases of the moon, and eclipses.

D.3.c Gravity is the force that keeps planets in orbit around the sun and governs the rest of the motion in the solar system. Gravity alone holds us to the earth's surface and explains the phenomena of the tides.

D.3.d The sun is the major source of energy for phenomena on the earth's surface, such as growth of plants, winds, ocean currents, and the water cycle. Seasons result from variations in the amount of the sun's energy hitting the surface, due to the tilt of the earth's rotation on its axis and the length of the day.

E. Science and Technology

As a result of activities in grades 5–8, all students should develop

Abilities of Technological Design

E.1 Identify appropriate problems for technological design. Students should develop their abilities by identifying a specified need, considering its various aspects, and talking to different potential users or beneficiaries. They should appreciate that for some needs, the cultural backgrounds and beliefs of different groups can affect the criteria for a suitable product.

E.2 Design a solution or product. Students should make and compare different proposals in the light of the criteria they have selected. They must consider constraints—such as cost, time, trade-offs, and materials needed—and communicate ideas with drawings and simple models.

E.3 Implement a proposed design. Students should organize materials and other resources, plan their work, make good use of group collaboration where appropriate, choose suitable tools and techniques, and work with appropriate measurement methods to ensure adequate accuracy.

E.4 Evaluate completed technological designs or products. Students should use criteria relevant to the original purpose or need, consider a variety of factors that might affect acceptability and suitability for intended users or beneficiaries, and develop measures of quality with respect to such criteria and factors; they should also suggest improvements and, for their own products, try proposed modifications.

E.5 Communicate the process of technological design. Students should review and describe any completed piece of work and identify the stages of problem identification, solution design, implementation, and evaluation.

Understandings about Science and Technology

E.6.a Scientific inquiry and technological design have similarities and differences. Scientists propose explanations for questions about the natural world, and engineers propose solutions relating to human problems, needs, and aspirations. Technological solutions are temporary; technologies exist within nature and so they cannot contravene physical or biological principles; technological solutions have side effects; and technologies cost, carry risks, and provide benefits.

E.6.b Many different people in different cultures have made and continue to make contributions to science and technology.

E.6.c Science and technology are reciprocal. Science helps drive technology, as it addresses questions that demand more sophisticated instruments and provides principles for better instrumentation and technique. Technology is essential to science, because it provides instruments and techniques that enable observations of objects and phenomena that are otherwise unobservable due to factors such as quantity, distance, location, size, and speed. Technology also provides tools for investigations, inquiry, and analysis.

E.6.d Perfectly designed solutions do not exist. All technological solutions have trade-offs, such as safety, cost, efficiency, and appearance. Engineers often build in back-up systems to provide safety. Risk is part of living in a highly technological world. Reducing risk often results in new technology.

E.6.e Technological designs have constraints. Some constraints are unavoidable, for example, properties of materials, or effects of weather and friction; other constraints limit choices in the design, for example, environmental protection, human safety, and aesthetics.

E.6.f Technological solutions have intended benefits and unintended consequences. Some consequences can be predicted, others cannot.

F. Science in Personal and Social Perspectives

As a result of activities in grades 5–8, all students should develop understanding of

Personal Health

F.1.a Regular exercise is important to the maintenance and improvement of health. The benefits of physical fitness include maintaining healthy weight, having energy and strength for routine activities, good muscle tone, bone strength, strong heart/lung systems, and improved mental health. Personal exercise, especially developing cardiovascular endurance, is the foundation of physical fitness.

F.1.b The potential for accidents and the existence of hazards imposes the need for injury prevention. Safe living involves the development and use of safety precautions and the recognition of risk in personal decisions. Injury prevention has personal and social dimensions.

F.1.c The use of tobacco increases the risk of illness. Students should understand the influence of short-term social and psychological factors that lead to tobacco use, and the possible long-term detrimental effects of smoking and chewing tobacco.

F.1.d Alcohol and other drugs are often abused substances. Such drugs change how the body functions and can lead to addiction.

F.1.e Food provides energy and nutrients for growth and development. Nutrition requirements vary with body weight, age, sex, activity, and body functioning.

F.1.f Sex drive is a natural human function that requires understanding. Sex is also a prominent means of transmitting diseases. The diseases can be prevented through a variety of precautions.

F.1.g Natural environments may contain substances (for example, radon and lead) that are harmful to human beings. Maintaining environmental health involves establishing or monitoring quality standards related to use of soil, water, and air.

Populations, Resources, and Environments

F.2.a When an area becomes overpopulated, the environment will become degraded due to the increased use of resources.

F.2.b Causes of environmental degradation and resource depletion vary from region to region and from country to country.

Natural Hazards

F.3.a Internal and external processes of the earth system cause natural hazards, events that change or destroy human and wildlife habitats, damage property, and harm or kill humans. Natural hazards include earthquakes, landslides, wildfires, volcanic eruptions, floods, storms, and even possible impacts of asteroids.

F.3.b Human activities also can induce hazards through resource acquisition, urban growth, land-use decisions, and waste disposal. Such activities can accelerate many natural changes.

F.3.c Natural hazards can present personal and societal challenges because misidentifying the change or incorrectly estimating the rate and scale of change may result in either too little attention and significant human costs or too much cost for unneeded preventive measures.

Risks and Benefits

F.4.a Risk analysis considers the type of hazard and estimates the number of people that might be exposed and the number likely to suffer consequences. The results are used to determine the options for reducing or eliminating risks.

F.4.b Students should understand the risks associated with natural hazards (fires, floods, tornadoes, hurricanes, earthquakes, and volcanic eruptions), with chemical hazards (pollutants in air, water, soil, and food), with biological hazards (pollen, viruses, bacterial, and parasites), social hazards (occupational safety and transportation), and with personal hazards (smoking, dieting, and drinking).

F.4.c Individuals can use a systematic approach to thinking critically about risks and benefits. Examples include applying probability estimates to risks and comparing them to estimated personal and social benefits.

F.4.d Important personal and social decisions are made based on perceptions of benefits and risks.

Science and Technology in Society

F.5.a Science influences society through its knowledge and world view. Scientific knowledge and the procedures used by scientists influence the way many individuals in society think about themselves, others, and the environment. The effect of science on society is neither entirely beneficial nor entirely detrimental.

F.5.b Societal challenges often inspire questions for scientific research, and social priorities often influence research priorities through the availability of funding for research.

F.5.c Technology influences society through its products and processes. Technology influences the quality of life and the ways people act and interact. Technological changes are often accompanied by social, political, and economic changes that can be beneficial or detrimental to individuals and to society. Social needs, attitudes, and values influence the direction of technological development.

F.5.d Science and technology have advanced through contributions of many different people, in different cultures, at different times in history. Science and technology have contributed enormously to economic growth and productivity among societies and groups within societies.

F.5.e Scientists and engineers work in many different settings, including colleges and universities, businesses and industries, specific research institutes, and government agencies.

F.5.f Scientists and engineers have ethical codes requiring that human subjects involved with research be fully informed about risks and benefits associated with the research before the individuals choose to participate. This ethic extends to potential risks to communities and property. In short, prior knowledge and consent are required for research involving human subjects or potential damage to property.

F.5.g Science cannot answer all questions and technology cannot solve all human problems or meet all human needs. Students should understand the difference between scientific and other questions. They should appreciate what science and technology can reasonably contribute to society and what they cannot do. For example, new technologies often will decrease some risks and increase others.

G. History and Nature of Science

As a result of activities in grades 5–8, all students should develop understanding of

Science as a Human Endeavor

G.1.a Women and men of various social and ethnic backgrounds—and with diverse interests, talents, qualities, and motivations—engage in the activities of science, engineering, and related fields such as the health professions. Some scientists work in teams, and some work alone, but all communicate extensively with others.

G.1.b Science requires different abilities, depending on such factors as the field of study and type of inquiry. Science is very much a human endeavor, and the work of science relies on basic human qualities, such as reasoning, insight, energy, skill, and creativity—as well as on scientific habits of mind, such as intellectual honesty, tolerance of ambiguity, skepticism, and openness to new ideas.

Nature of Science

G.2.a Scientists formulate and test their explanations of nature using observation, experiments, and theoretical and mathematical models. Although all scientific ideas are tentative and subject to change and improvement in principle, for most major ideas in science, there is much experimental and observational confirmation. Those ideas are not likely to change greatly in the future. Scientists do and have changed their ideas about nature when they encounter new experimental evidence that does not match their existing explanations.

G.2.b In areas where active research is being pursued and in which there is not a great deal of experimental or observational evidence and understanding, it is normal for scientists to differ with one another about the interpretation of the evidence or theory being considered. Different scientists might publish conflicting experimental results or might draw different conclusions from the same data. Ideally, scientists acknowledge such conflict and work towards finding evidence that will resolve their disagreement.

G.2.c It is part of scientific inquiry to evaluate the results of scientific investigations, experiments, observations, theoretical models, and the explanations proposed by other scientists. Evaluation includes reviewing the experimental procedures, examining the evidence, identifying faulty reasoning, pointing out statements that go beyond the evidence, and suggesting alternative explanations for the same observations. Although scientists may disagree about explanations of phenomena, about interpretations of data, or about the value of rival theories, they do agree that questioning, response to criticism, and open communication are integral to the process of science. As scientific knowledge evolves, major disagreements are eventually resolved through such interactions between scientists.

History of Science

G.3.a Many individuals have contributed to the traditions of science. Studying some of these individuals provides further understanding of scientific inquiry, science as a human endeavor, the nature of science, and the relationships between science and society.

G.3.b In historical perspective, science has been practiced by different individuals in different cultures. In looking at the history of many peoples, one finds that scientists and engineers of high achievement are considered to be among the most valued contributors to their culture.

G.3.c Tracing the history of science can show how difficult it was for scientific innovators to break through the accepted ideas of their time to reach the conclusions that we currently take for granted.

Project 2061
Benchmarks for Science Literacy

1. The Nature of Science

By the end of the 8th grade, students should know that

1.A The Scientific World View

1.A.1 When similar investigations give different results, the scientific challenge is to judge whether the differences are trivial or significant, and it often takes further studies to decide. Even with similar results, scientists may wait until an investigation has been repeated many times before accepting the results as correct.

1.A.2 Scientific knowledge is subject to modification as new information challenges prevailing theories and as a new theory leads to looking at old observations in a new way.

1.A.3 Some scientific knowledge is very old and yet is still applicable today.

1.A.4 Some matters cannot be examined usefully in a scientific way. Among them are matters that by their nature cannot be tested objectively and those that are essentially matters of morality. Science can sometimes be used to inform ethical decisions by identifying the likely consequences of particular actions but cannot be used to establish that some action is either moral or immoral.

1.B Scientific Inquiry

1.B.1 Scientists differ greatly in what phenomena they study and how they go about their work. Although there is no fixed set of steps that all scientists follow, scientific investigations usually involve the collection of relevant evidence, the use of logical reasoning, and the application of imagination in devising hypotheses and explanations to make sense of the collected evidence.

1.B.2 If more than one variable changes at the same time in an experiment, the outcome of the experiment may not be clearly attributable to any one of the variables. It may not always be possible to prevent outside variables from influencing the outcome of an investigation (or even to identify all of the variables), but collaboration among investigators can often lead to research designs that are able to deal with such situations.

1.B.3 What people expect to observe often affects what they actually do observe. Strong beliefs about what should happen in particular circumstances can prevent them from detecting other results. Scientists know about this danger to objectivity and take steps to try and avoid it when designing investigations and examining data. One safeguard is to have different investigators conduct independent studies of the same questions.

1.C The Scientific Enterprise

1.C.1 Important contributions to the advancement of science, mathematics, and technology have been made by different kinds of people, in different cultures, at different times.

1.C.2 Until recently, women and racial minorities, because of restrictions on their education and employment opportunities, were essentially left out of much of the formal work of the science establishment; the remarkable few who overcame those obstacles were even then likely to have their work disregarded by the science establishment.

1.C.3 No matter who does science and mathematics or invents things, or when or where they do it, the knowledge and technology that result can eventually become available to everyone in the world.

1.C.4 Scientists are employed by colleges and universities, business and industry, hospitals, and many government agencies. Their places of work include offices, classrooms, laboratories, farms, factories, and natural field settings ranging from space to the ocean floor.

1.C.5 In research involving human subjects, the ethics of science require that potential subjects be fully informed about the risks and benefits associated with the research and of their right to refuse to participate. Science ethics also demand that scientists must not knowingly subject coworkers, students, the neighborhood, or the community to health or property risks without their prior knowledge and consent. Because animals cannot make informed choices, special care must be taken in using them in scientific research.

1.C.6 Computers have become invaluable in science because they speed up and extend people's ability to collect, store, compile, and analyze data, prepare research reports, and share data and ideas with investigators all over the world.

1.C.7 Accurate record-keeping, openness, and replication are essential for maintaining an investigator's credibility with other scientists and society.

3. The Nature of Technology

By the end of the 8th grade, students should know that

3.A Technology and Science

3.A.1 In earlier times, the accumulated information and techniques of each generation of workers were taught on the job directly to the next generation of workers. Today, the knowledge base for technology can be found as well in libraries of print and electronic resources and is often taught in the classroom.

3.A.2 Technology is essential to science for such purposes as access to outer space and other remote locations, sample collection and treatment, measurement, data collection and storage, computation, and communication of information.

3.A.3 Engineers, architects, and others who engage in design and technology use scientific knowledge to solve practical problems. But they usually have to take human values and limitations into account as well.

3.B Design and Systems

3.B.1 Design usually requires taking constraints into account. Some constraints, such as gravity or the properties of the materials to be used, are unavoidable. Other constraints, including economic, political, social, ethical, and aesthetic ones, limit choices.

3.B.2 All technologies have effects other than those intended by the design, some of which may have been predictable and some not. In either case, these side effects may turn out to be unacceptable to some of the population and therefore lead to conflict between groups.

3.B.3 Almost all control systems have inputs, outputs, and feedback. The essence of control is comparing information about what is happening to what people want to happen and then making appropriate adjustments. This procedure requires sensing information, processing it, and making changes. In almost all modern machines, microprocessors serve as centers of performance control.

3.B.4 Systems fail because they have faulty or poorly matched parts, are used in ways that exceed what was intended by the design, or were poorly designed to begin with. The most common ways to prevent failure are pretesting parts and procedures, overdesign, and redundancy.

3.C Issues in Technology

3.C.1 The human ability to shape the future comes from a capacity for generating knowledge and developing new technologies—and for communicating ideas to others.

3.C.2 Technology cannot always provide successful solutions for problems or fulfill every human need.

3.C.3 Throughout history, people have carried out impressive technological feats, some of which would be hard to duplicate today even with modern tools. The purposes served by these achievements have sometimes been practical, sometimes ceremonial.

3.C.4 Technology has strongly influenced the course of history and continues to do so. It is largely responsible for the great revolutions in agriculture, manufacturing, sanitation and medicine, warfare, transportation, information processing, and communications that have radically changed how people live.

3.C.5 New technologies increase some risks and decrease others. Some of the same technologies that have improved the length and quality of life for many people have also brought new risks.

3.C.6 Rarely are technology issues simple and one-sided. Relevant facts alone, even when known and available, usually do not settle matters entirely in favor of one side or another. That is because the contending groups may have different values and priorities. They may stand to gain or lose in different degrees, or may make very different predictions about what the future consequences of the proposed action will be.

3.C.7 Societies influence what aspects of technology are developed and how these are used. People control technology (as well as science) and are responsible for its effects.

4. The Physical Setting

By the end of the 8th grade, students should know that

4.A The Universe

4.A.1 The sun is a medium-sized star located near the edge of a disk-shaped galaxy of stars, part of which can be seen as a glowing band of light that spans the sky on a very clear night. The universe contains many billions of galaxies, and each galaxy contains many billions of stars. To the naked eye, even the closest of these galaxies is no more than a dim, fuzzy spot.

4.A.2 The sun is many thousands of times closer to the earth than any other star. Light from the sun takes a few minutes to reach the earth, but light from the next nearest star takes a few years to arrive. The trip to that star would take the fastest rocket thousands of years. Some distant galaxies are so far away that their light takes several billion years to reach the earth. People on earth, therefore, see them as they were that long ago in the past.

4.A.3 Nine planets of very different size, composition, and surface features move around the sun in nearly circular orbits. Some planets have a great variety of moons and even flat rings of rock and ice particles orbiting around them. Some of these planets and moons show evidence of geologic activity. The earth is orbited by one moon, many artificial satellites, and debris.

4.A.4 Large numbers of chunks of rock orbit the sun. Some of those that the earth meets in its yearly orbit around the sun glow and disintegrate from friction as they plunge through the atmosphere—and sometimes impact the ground. Other chunks of rocks mixed with ice have long, off-center orbits that carry them close to the sun, where the sun's radiation (of light and particles) boils off frozen material from their surfaces and pushes it into a long, illuminated tail.

4.B The Earth

4.B.1 We live on a relatively small planet, the third from the sun in the only system of planets definitely known to exist (although other, similar systems may be discovered in the universe).

4.B.2 The earth is mostly rock. Three-fourths of its surface is covered by a relatively thin layer of water (some of it frozen), and the entire planet is surrounded by a relatively thin blanket of air. It is the only body in the solar system that appears able to support life. The other planets have compositions and conditions very different from the earth's.

4.B.3 Everything on or anywhere near the earth is pulled toward the earth's center by gravitational force.

4.B.4 Because the earth turns daily on an axis that is tilted relative to the plane of the earth's yearly orbit around the sun, sunlight falls more intensely on different parts of the earth during the year. The difference in heating of the earth's surface produces the planet's seasons and weather patterns.

4.B.5 The moon's orbit around the earth once in about 28 days changes what part of the moon is lighted by the sun and how much of that part can be seen from the earth—the phases of the moon.

4.B.6 Climates have sometimes changed abruptly in the past as a result of changes in the earth's crust, such as volcanic eruptions or impacts of huge rocks from space. Even relatively small changes in atmospheric or ocean content can have widespread effects on climate if the change lasts long enough.

4.B.7 The cycling of water in and out of the atmosphere plays an important role in determining climatic patterns. Water evaporates from the surface of the earth, rises and cools, condenses into rain or snow, and falls again to the surface. The water falling on land collects in rivers and lakes, soil, and porous layers of rock, and much of it flows back into the ocean.

4.B.8 Fresh water, limited in supply, is essential for life and also for most industrial processes. Rivers, lakes, and groundwater can be depleted or polluted, becoming unavailable or unsuitable for life.

4.B.9 Heat energy carried by ocean currents has a strong influence on climate around the world.

4.B.10 Some minerals are very rare and some exist in great quantities, but—for practical purposes—the ability to recover them is just as important as their abundance. As minerals are depleted, obtaining them becomes more difficult. Recycling and the development of substitutes can reduce the rate of depletion but may also be costly.

4.B.11 The benefits of the earth's resources—such as fresh water, air, soil, and trees—can be reduced by using them wastefully or by deliberately or inadvertently destroying them. The atmosphere and the oceans have a limited capacity to absorb wastes and recycle materials naturally. Cleaning up polluted air, water, or soil or restoring depleted soil, forests, or fishing grounds can be very difficult and costly.

4.C Processes that Shape the Earth

4.C.1 The interior of the earth is hot. Heat flow and movement of material within the earth cause earthquakes and volcanic eruptions and create mountains and ocean basins. Gas and dust from large volcanoes can change the atmosphere.

4.C.2 Some changes in the earth's surface are abrupt (such as earthquakes and volcanic eruptions) while other changes happen very slowly (such as uplift and wearing down of mountains). The earth's surface is shaped in part by the motion of water and wind over very long times, which act to level mountain ranges.

4.C.3 Sediments of sand and smaller particles (sometimes containing the remains of organisms) are gradually buried and are cemented together by dissolved minerals to form solid rock again.

4.C.4 Sedimentary rock buried deep enough may be reformed by pressure and heat, perhaps melting and recrystallizing into different kinds of rock. These re-formed rock layers may be forced up again to become land surface and even mountains. Subsequently, this new rock too will erode. Rock bears evidence of the minerals, temperatures, and forces that created it.

4.C.5 Thousands of layers of sedimentary rock confirm the long history of the changing surface of the earth and the changing life forms whose remains are found in successive layers. The youngest layers are not always found on top, because of folding, breaking, and uplift of layers.

4.C.6 Although weathered rock is the basic component of soil, the composition and texture of soil and its fertility and resistance to erosion are greatly influenced by plant roots and debris, bacteria, fungi, worms, insects, rodents, and other organisms.

4.C.7 Human activities, such as reducing the amount of forest cover, increasing the amount and variety of chemicals released into the atmosphere, and intensive farming, have changed the earth's land, oceans, and atmosphere. Some of these changes have decreased the capacity of the environment to support some life forms.

4.D Structure of Matter

4.D.1 All matter is made up of atoms, which are far too small to see directly through a microscope. The atoms of any element are alike but are different from atoms of other elements. Atoms may stick together in well-defined molecules or may be packed together in large arrays. Different arrangements of atoms into groups compose all substances.

4.D.2 Equal volumes of different substances usually have different weights.

4.D.3 Atoms and molecules are perpetually in motion. Increased temperature means greater average energy, so most substances expand when heated. In solids, the atoms are closely locked in position and can only vibrate. In liquids, the atoms or molecules have higher energy, are more loosely connected, and can slide past one another; some molecules may get enough energy to escape into a gas. In gases, the atoms or molecules have still more energy and are free of one another except during occasional collisions.

4.D.4 The temperature and acidity of a solution influence reaction rates. Many substances dissolve in water, which may greatly facilitate reactions between them.

4.D.5 Scientific ideas about elements were borrowed from some Greek philosophers of 2,000 years earlier, who believed that everything was made from four basic substances: air, earth, fire, and water. It was the combinations of these "elements" in different proportions that gave other substances their observable properties. The Greeks were wrong about those four, but now over 100 different elements have been identified, some rare and some plentiful, out of which everything is made. Because most elements tend to combine with others, few elements are found in their pure form.

4.D.6 There are groups of elements that have similar properties, including highly reactive metals, less-reactive metals, highly reactive nonmetals (such as chlorine, fluorine, and oxygen), and some almost completely nonreactive gases (such as helium and neon). An especially important kind of reaction between substances involves combination of oxygen with something else—as in burning or rusting. Some elements don't fit into any of the categories; among them are carbon and hydrogen, essential elements of living matter.

4.D.7 No matter how substances within a closed system interact with one another, or how they combine or break apart, the total weight of the system remains the same. The idea of atoms explains the conservation of matter: If the number of atoms stays the same no matter how they are rearranged, then their total mass stays the same.

4.E Energy Transformations

4.E.1 Energy cannot be created or destroyed, but only changed from one form into another.

4.E.2 Most of what goes on in the universe—from exploding stars and biological growth to the operation of machines and the motion of people—involves some form of energy being transformed into another. Energy in the form of heat is almost always one of the products of an energy transformation.

4.E.3 Heat can be transferred through materials by the collisions of atoms or across space by radiation. If the material is fluid, currents will be set up in it that aid the transfer of heat.

4.E.4 Energy appears in different forms. Heat energy is in the disorderly motion of molecules; chemical energy is in the arrangement of atoms; mechanical energy is in moving bodies or in elastically distorted shapes; gravitational energy is in the separation of mutually attracting masses.

4.F Motion

4.F.1 Light from the sun is made up of a mixture of many different colors of light, even though to the eye the light looks almost white. Other things that give off or reflect light have a different mix of colors.

4.F.2 Something can be "seen" when light waves emitted or reflected by it enter the eye—just as something can be "heard" when sound waves from it enter the ear.

4.F.3 An unbalanced force acting on an object changes its speed or direction of motion, or both. If the force acts toward a single center, the object's path may curve into an orbit around the center.

4.F.4 Vibrations in materials set up wavelike disturbances that spread away from the source. Sound and earthquake waves are examples. These and other waves move at different speeds in different materials.

4.F.5 Human eyes respond to only a narrow range of wavelengths of electromagnetic radiation—visible light. Differences of wavelength within that range are perceived as differences in color.

4.G Forces of Nature

4.G.1 Every object exerts gravitational force on every other object. The force depends on how much mass the objects have and on how far apart they are. The force is hard to detect unless at least one of the objects has a lot of mass.

4.G.2 The sun's gravitational pull holds the earth and other planets in their orbits, just as the planets' gravitational pull keeps their moons in orbit around them.

4.G.3 Electric currents and magnets can exert a force on each other.

5. The Living Environment

By the end of the 8th grade, students should know that

5.A Diversity of Life

5.A.1 One of the most general distinctions among organisms is between plants, which use sunlight to make their own food, and animals, which consume energy-rich foods. Some kinds of organisms, many of them microscopic, cannot be neatly classified as either plants or animals.

5.A.2 Animals and plants have a great variety of body plans and internal structures that contribute to their being able to make or find food and reproduce.

5.A.3 Similarities among organisms are found in internal anatomical features, which can be used to infer the degree of relatedness among organisms. In classifying organisms, biologists consider details of internal and external structures to be more important than behavior or general appearance.

5.A.4 For sexually reproducing organisms, a species comprises all organisms that can mate with one another to produce fertile offspring.

5.A.5 All organisms, including the human species, are part of and depend on two main interconnected global food webs. One includes microscopic ocean plants, the animals that feed on them, and finally the animals that feed on those animals. The other web includes land plants, the animals that feed on them, and so forth. The cycles continue indefinitely because organisms decompose after death to return food material to the environment.

5.B Heredity

5.B.1 In some kinds of organisms, all the genes come from a single parent, whereas in organisms that have sexes, typically half of the genes come from each parent.

5.B.2 In sexual reproduction, a single specialized cell from a female merges with a specialized cell from a male. As the fertilized egg, carrying genetic information from each parent, multiplies to form the complete organism with about a trillion cells, the same genetic information is copied in each cell.

5.B.3 New varieties of cultivated plants and domestic animals have resulted from selective breeding for particular traits.

5.C Cells

5.C.1 All living things are composed of cells, from just one to many millions, whose details usually are visible only through a microscope. Different body tissues and organs are made up of different kinds of cells. The cells in similar tissues and organs in other animals are similar to those in human beings but differ somewhat from cells found in plants.

5.C.2 Cells repeatedly divide to make more cells for growth and repair. Various organs and tissues function to serve the needs of cells for food, air, and waste removal.

5.C.3 Within cells, many of the basic functions of organisms—such as extracting energy from food and getting rid of waste—are carried out. The way in which cells function is similar in all living organisms.

5.C.4 About two-thirds of the weight of cells is accounted for by water, which gives cells many of their properties.

5.D Interdependence of Life

5.D.1 In all environments—freshwater, marine, forest, desert, grassland, mountain, and others—organisms with similar needs may compete with one another for resources, including food, space, water, air, and shelter. In any particular environment, the growth and survival of organisms depend on the physical conditions.

5.D.2 Two types of organisms may interact with one another in several ways: They may be in a producer/consumer, predator/prey, or parasite/host relationship. Or one organism may scavenge or decompose another. Relationships may be competitive or mutually beneficial. Some species have become so adapted to each other that neither could survive without the other.

5.E Flow of Matter and Energy

5.E.1 Food provides molecules that serve as fuel and building material for all organisms. Plants use the energy in light to make sugars out of carbon dioxide and water. This food can be used immediately for fuel or materials or it may be stored for later use. Organisms that eat plants break down the plant structures to produce the materials and energy they need to survive. Then they are consumed by other organisms.

5.E.2 Over a long time, matter is transferred from one organism to another repeatedly and between organisms and their physical environment. As in all material systems, the total amount of matter remains constant, even though its form and location change.

5.E.3 Energy can change from one form to another in living things. Animals get energy from oxidizing their food, releasing some of its energy as heat. Almost all food energy comes originally from sunlight.

5.F Evolution of Life

5.F.1 Small differences between parents and offspring can accumulate (through selective breeding) in successive generations so that descendants are very different from their ancestors.

5.F.2 Individual organisms with certain traits are more likely than others to survive and have offspring. Changes in environmental conditions can affect the survival of individual organisms and entire species.

5.F.3 Many thousands of layers of sedimentary rock provide evidence for the long history of the earth and for the long history of changing life forms whose remains are found in the rocks. More recently deposited rock layers are more likely to contain fossils resembling existing species.

6. The Human Organism

By the end of the 8th grade, students should know that

6.A Human Identity

6.A.1 Like other animals, human beings have body systems for obtaining and providing energy, defense, reproduction, and the coordination of body functions.

6.A.2 Human beings have many similarities and differences. The similarities make it possible for human beings to reproduce and to donate blood and organs to one another throughout the world. Their differences enable them to create diverse social and cultural arrangements and to solve problems in a variety of ways.

6.A.3 Fossil evidence is consistent with the idea that human beings evolved from earlier species.

6.A.4 Specialized roles of individuals within other species are genetically programmed, whereas human beings are able to invent and modify a wider range of social behavior.

6.A.5 Human beings use technology to match or excel many of the abilities of other species. Technology has helped people with disabilities survive and live more conventional lives.

6.A.6 Technologies having to do with food production, sanitation, and disease prevention have dramatically changed how people live and work and have resulted in rapid increases in the human population.

6.B Human Development

6.B.1 Fertilization occurs when sperm cells from a male's testes are deposited near an egg cell from the female ovary, and one of the sperm cells enters the egg cell. Most of the time, by chance or design, a sperm never arrives or an egg isn't available.

6.B.2 Contraception measures may incapacitate sperm, block their way to the egg, prevent the release of eggs, or prevent the fertilized egg from implanting successfully.

6.B.3 Following fertilization, cell division produces a small cluster of cells that then differentiate by appearance and function to form the basic tissues of an embryo. During the first three months of pregnancy, organs begin to form. During the second three months, all organs and body features develop. During the last three months, the organs and features mature enough to function well after birth. Patterns of human development are similar to those of other vertebrates.

6.B.4 The developing embryo—and later the newborn infant—encounters many risks from faults in its genes, its mother's inadequate diet, her cigarette smoking or use of alcohol or other drugs, or from infection. Inadequate child care may lead to lower physical and mental ability.

6.B.5 Various body changes occur as adults age. Muscles and joints become less flexible, bones and muscles lose mass, energy levels diminish, and the senses become less acute. Women stop releasing eggs and hence can no longer reproduce. The length and quality of human life are influenced by many factors, including sanitation, diet, medical care, sex, genes, environmental conditions, and personal health behaviors.

6.C Basic Functions

6.C.1 Organs and organ systems are composed of cells and help to provide all cells with basic needs.

6.C.2 For the body to use food for energy and building materials, the food must first be digested into molecules that are absorbed and transported to cells.

6.C.3 To burn food for the release of energy stored in it, oxygen must be supplied to cells, and carbon dioxide removed. Lungs take in oxygen for the combustion of food and they eliminate the carbon dioxide produced. The urinary system disposes of dissolved waste molecules, the intestinal tract removes solid wastes, and the skin and lungs rid the body of heat energy. The circulatory system moves all these substances to or from cells where they are needed or produced, responding to changing demands.

6.C.4 Specialized cells and the molecules they produce identify and destroy microbes that get inside the body.

6.C.5 Hormones are chemicals from glands that affect other body parts. They are involved in helping the body respond to danger and in regulating human growth, development, and reproduction.

6.C.6 Interactions among the senses, nerves, and brain make possible the learning that enables human beings to cope with changes in their environment.

6.D Learning

6.D.1 Some animal species are limited to a repertoire of genetically determined behaviors; others have more complex brains and can learn a wide variety of behaviors. All behavior is affected by both inheritance and experience.

6.D.2 The level of skill a person can reach in any particular activity depends on innate abilities, the amount of practice, and the use of appropriate learning technologies.

6.D.3 Human beings can detect a tremendous range of visual and olfactory stimuli. The strongest stimulus they can tolerate may be more than a trillion times as intense as the weakest they can detect. Still, there are many kinds of signals in the world that people cannot detect directly.

6.D.4 Attending closely to any one input of information usually reduces the ability to attend to others at the same time.

6.D.5 Learning often results from two perceptions or actions occurring at about the same time. The more often the same combination occurs, the stronger the mental connection between them is likely to be. Occasionally a single vivid experience will connect two things permanently in people's minds.

6.D.6 Language and tools enable human beings to learn complicated and varied things from others.

6.E Physical Health

6.E.1 The amount of food energy (calories) a person requires varies with body weight, age, sex, activity level, and natural body efficiency. Regular exercise is important to maintain a healthy heart/lung system, good muscle tone, and bone strength.

6.E.2 Toxic substances, some dietary habits, and personal behavior may be bad for one's health. Some effects show up right away, others may not show up for many years. Avoiding toxic substances, such as tobacco, and changing dietary habits to reduce the intake of such things as animal fat increases the chances of living longer.

6.E.3 Viruses, bacteria, fungi, and parasites may infect the human body and interfere with normal body functions. A person can catch a cold many times because there are many varieties of cold viruses that cause similar symptoms.

6.E.4 White blood cells engulf invaders or produce antibodies that attack them or mark them for killing by other white cells. The antibodies produced will remain and can fight off subsequent invaders of the same kind.

6.E.5 The environment may contain dangerous levels of substances that are harmful to human beings. Therefore, the good health of individuals requires monitoring the soil, air, and water and taking steps to keep them safe.

6.F Mental Health

6.F.1 Individuals differ greatly in their ability to cope with stressful situations. Both external and internal conditions (chemistry, personal history, values) influence how people behave.

6.F.2 Often people react to mental distress by denying that they have any problem. Sometimes they don't know why they feel the way they do, but with help they can sometimes uncover the reasons.

8. The Designed World

By the end of the 8th grade, students should know that

8.A Agriculture

8.A.1 Early in human history, there was an agricultural revolution in which people changed from hunting and gathering to farming. This allowed changes in the division of labor between men and women and between children and adults, and the development of new patterns of government.

8.A.2 People control the characteristics of plants and animals they raise by selective breeding and by preserving varieties of seeds (old and new) to use if growing conditions change.

8.A.3 In agriculture, as in all technologies, there are always trade-offs to be made. Getting food from many different places makes people less dependent on weather in any one place, yet more dependent on transportation and communication among far-flung markets. Specializing in one crop may risk disaster if changes in weather or increases in pest populations wipe out that crop. Also, the soil may be exhausted of some nutrients, which can be replenished by rotating the right crops.

8.A.4 Many people work to bring food, fiber, and fuel to U.S. markets. With improved technology, only a small fraction of workers in the United States actually plant and harvest the products that people use. Most workers are engaged in processing, packaging, transporting, and selling what is produced.

8.B Materials and Manufacturing

8.B.1 The choice of materials for a job depends on their properties and on how they interact with other materials. Similarly, the usefulness of some manufactured parts of an object depends on how well they fit together with the other parts.

8.B.2 Manufacturing usually involves a series of steps, such as designing a product, obtaining and preparing raw materials, processing the materials mechanically or chemically, and assembling, testing, inspecting, and packaging. The sequence of these steps is also often important.

8.B.3 Modern technology reduces manufacturing costs, produces more uniform products, and creates new synthetic materials that can help reduce the depletion of some natural resources.

8.B.4 Automation, including the use of robots, has changed the nature of work in most fields, including manufacturing. As a result, high-skill, high-knowledge jobs in engineering, computer programming, quality control, supervision, and maintenance are replacing many routine, manual-labor jobs. Workers therefore need better learning skills and flexibility to take on new and rapidly changing jobs.

8.C Energy Sources and Use

8.C.1 Energy can change from one form to another, although in the process some energy is always converted to heat. Some systems transform energy with less loss of heat than others.

8.C.2 Different ways of obtaining, transforming, and distributing energy have different environmental consequences.

8.C.3 In many instances, manufacturing and other technological activities are performed at a site close to an energy source. Some forms of energy are transported easily, others are not.

8.C.4 Electrical energy can be produced from a variety of energy sources and can be transformed into almost any other form of energy. Moreover, electricity is used to distribute energy quickly and conveniently to distant locations.

8.C.5 Energy from the sun (and the wind and water energy derived from it) is available indefinitely. Because the flow of energy is weak and variable, very large collection systems are needed. Other sources don't renew or renew only slowly.

8.C.6 Different parts of the world have different amounts and kinds of energy resources to use and use them for different purposes.

8.D Communication

8.D.1 Errors can occur in coding, transmitting, or decoding information, and some means of checking for accuracy is needed. Repeating the message is a frequently used method.

8.D.2 Information can be carried by many media, including sound, light, and objects. In this century, the ability to code information as electric currents in wires, electromagnetic waves in space, and light in glass fibers has made communication millions of times faster than is possible by mail or sound.

8.E Information Processing

8.E.1 Most computers use digital codes containing only two symbols, 0 and 1, to perform all operations. Continuous signals (analog) must be transformed into digital codes before they can be processed by a computer.

8.E.2 What use can be made of a large collection of information depends upon how it is organized. One of the values of computers is that they are able, on command, to reorganize information in a variety of ways, thereby enabling people to make more and better uses of the collection.

8.E.3 Computer control of mechanical systems can be much quicker than human control. In situations where events happen faster than people can react, there is little choice but to rely on computers. Most complex systems still require human oversight, however, to make certain kinds of judgments about the readiness of the parts of the system (including the computers) and the system as a whole to operate properly, to react to unexpected failures, and to evaluate how well the system is serving its intended purposes.

8.E.4 An increasing number of people work at jobs that involve processing or distributing information. Because computers can do these tasks faster and more reliably, they have become standard tools both in the workplace and at home.

8.F Health Technology

8.F.1 Sanitation measures such as the use of sewers, landfills, quarantines, and safe food handling are important in controlling the spread of organisms that cause disease. Improving sanitation to prevent disease has contributed more to saving human life than any advance in medical treatment.

8.F.2 The ability to measure the level of substances in body fluids has made it possible for physicians to make comparisons with normal levels, make very sophisticated diagnoses, and monitor the effects of the treatments they prescribe.

8.F.3 It is becoming increasingly possible to manufacture chemical substances such as insulin and hormones that are normally found in the body. They can be used by individuals whose own bodies cannot produce the amounts required for good health.

9. The Mathematical World

By the end of the 8th grade, students should know that

9.A Numbers

9.A.1 There have been systems for writing numbers other than the Arabic system of place values based on tens. The very old Roman numerals are now used only for dates, clock faces, or ordering chapters in a book. Numbers based on 60 are still used for describing time and angles.

9.A.2 A number line can be extended on the other side of zero to represent negative numbers. Negative numbers allow subtraction of a bigger number from a smaller number to make sense, and are often used when something can be measured on either side of some reference point (time, ground level, temperature, budget).

9.A.3 Numbers can be written in different forms, depending on how they are being used. How fractions or decimals based on measured quantities should be written depends on how precise the measurements are and how precise an answer is needed.

9.A.4 The operations + and – are inverses of each other—one undoes what the other does; likewise x and ÷ .

9.A.5 The expression *a/b* can mean different things: *a* parts of size *1/b* each, *a* divided by *b,* or *a* compared to *b.*

9.A.6 Numbers can be represented by using sequences of only two symbols (such as 1 and 0, on and off); computers work this way.

9.A.7 Computations (as on calculators) can give more digits than make sense or are useful.

9.B Symbolic Relationships

9.B.1 An equation containing a variable may be true for just one value of the variable.

9.B.2 Mathematical statements can be used to describe how one quantity changes when another changes. Rates of change can be computed from differences in magnitudes and vice versa.

9.B.3 Graphs can show a variety of possible relationships between two variables. As one variable increases uniformly, the other may do one of the following: increase or decrease steadily, increase or decrease faster and faster, get closer and closer to some limiting value, reach some intermediate maximum or minimum, alternately increase and decrease indefinitely, increase or decrease in steps, or do something different from any of these.

9.C Shapes

9.C.1 Some shapes have special properties: triangular shapes tend to make structures rigid, and round shapes give the least possible boundary for a given amount of interior area. Shapes can match exactly or have the same shape in different sizes.

9.C.2 Lines can be parallel, perpendicular, or oblique.

9.C.3 Shapes on a sphere like the earth cannot be depicted on a flat surface without some distortion.

9.C.4 The graphic display of numbers may help to show patterns such as trends, varying rates of change, gaps, or clusters. Such patterns sometimes can be used to make predictions about the phenomena being graphed.

9.C.5 It takes two numbers to locate a point on a map or any other flat surface. The numbers may be two perpendicular distances from a point, or an angle and a distance from a point.

9.C.6 The scale chosen for a graph or drawing makes a big difference in how useful it is.

9.D Uncertainty

9.D.1 How probability is estimated depends on what is known about the situation. Estimates can be based on data from similar conditions in the past or on the assumption that all the possibilities are known.

9.D.2 Probabilities are ratios and can be expressed as fractions, percentages, or odds.

9.D.3 The mean, median, and mode tell different things about the middle of a data set.

9.D.4 Comparison of data from two groups should involve comparing both their middles and the spreads around them.

9.D.5 The larger a well-chosen sample is, the more accurately it is likely to represent the whole. But there are many ways of choosing a sample that can make it unrepresentative of the whole.

9.D.6 Events can be described in terms of being more or less likely, impossible, or certain.

9.E Reasoning

9.E.1 Some aspects of reasoning have fairly rigid rules for what makes sense; other aspects don't. If people have rules that always hold, and good information about a particular situation, then logic can help them to figure out what is true about it. This kind of reasoning requires care in the use of key words such as if, and, not, or, all, and some. Reasoning by similarities can suggest ideas but can't prove them one way or the other.

9.E.2 Practical reasoning, such as diagnosing or troubleshooting almost anything, may require many-step, branching logic. Because computers can keep track of complicated logic, as well as a lot of information, they are useful in a lot of problem-solving situations.

9.E.3 Sometimes people invent a general rule to explain how something works by summarizing observations. But people tend to overgeneralize, imagining general rules on the basis of only a few observations.

9.E.4 People are using incorrect logic when they make a statement such as "If *A* is true, then *B* is true; but *A* isn't true, therefore *B* isn't true either."

9.E.5 A single example can never prove that something is always true, but sometimes a single example can prove that something is not always true.

9.E.6 An analogy has some likenesses to but also some differences from the real thing.

10. Historical Perspectives

By the end of the 8th grade, students should know that

10.A Displacing the Earth from the Center of the Universe

10.A.1 The motion of an object is always judged with respect to some other object or point and so the idea of absolute motion or rest is misleading.

10.A.2 Telescopes reveal that there are many more stars in the night sky than are evident to the unaided eye, the surface of the moon has many craters and mountains, the sun has dark spots, and Jupiter and some other planets have their own moons.

10.F Understanding Fire

10.F.1 From the earliest times until now, people have believed that even though millions of different kinds of material seem to exist in the world, most things must be made up of combinations of just a few basic kinds of things. There has not always been agreement, however, on what those basic kinds of things are. One theory long ago was that the basic substances were earth, water, air, and fire. Scientists now know that these are not the basic substances. But the old theory seemed to explain many observations about the world.

10.F.2 Today, scientists are still working out the details of what the basic kinds of matter are and of how they combine, or can be made to combine, to make other substances.

10.F.3 Experimental and theoretical work done by French scientist Antoine Lavoisier in the decade between the American and French revolutions led to the modern science of chemistry.

10.F.4 Lavoisier's work was based on the idea that when materials react with each other many changes can take place but that in every case the total amount of matter afterward is the same as before. He successfully tested the concept of conservation of matter by conducting a series of experiments in which he carefully measured all the substances involved in burning, including the gases used and those given off.

10.F.5 Alchemy was chiefly an effort to change base metals like lead into gold and to produce an elixir that would enable people to live forever. It failed to do that or to create much knowledge of how substances react with each other. The more scientific study of chemistry that began in Lavoisier's time has gone far beyond alchemy in understanding reactions and producing new materials.

10.G Splitting the Atom

10.G.1 The accidental discovery that minerals containing uranium darken photographic film, as light does, led to the idea of radioactivity.

10.G.2 In their laboratory in France, Marie Curie and her husband, Pierre Curie, isolated two new elements that caused most of the radioactivity of the uranium mineral. They named one radium because it gave off powerful, invisible rays, and the other polonium in honor of Madame Curie's country of birth. Marie Curie was the first scientist ever to win the Nobel prize in two different fields—in physics, shared with her husband, and later in chemistry.

10.I Discovering Germs

10.I.1 Throughout history, people have created explanations for disease. Some have held that disease has spiritual causes, but the most persistent biological theory over the centuries was that illness resulted from an imbalance in the body fluids. The introduction of germ theory by Louis Pasteur and others in the 19th century led to the modern belief that many diseases are caused by microorganisms—bacteria, viruses, yeasts, and parasites.

10.I.2 Pasteur wanted to find out what causes milk and wine to spoil. He demonstrated that spoilage and fermentation occur when microorganisms enter from the air, multiply rapidly, and produce waste products. After showing that spoilage could be avoided by keeping germs out or by destroying them with heat, he investigated animal diseases and showed that microorganisms were involved. Other investigators later showed that specific kinds of germs caused specific diseases.

10.I.3 Pasteur found that infection by disease organisms—germs—caused the body to build up an immunity against subsequent infection by the same organisms. He then demonstrated that it was possible to produce vaccines that would induce the body to build immunity to a disease without actually causing the disease itself.

10.I.4 Changes in health practices have resulted from the acceptance of the germ theory of disease. Before germ theory, illness was treated by appeals to supernatural powers or by trying to adjust body fluids through induced vomiting, bleeding, or purging. The modern approach emphasizes sanitation, the safe handling of food and water, the pasteurization of milk, quarantine, and aseptic surgical techniques to keep germs out of the body; vaccinations to strengthen the body's immune system against subsequent infection by the same kind of microorganisms; and antibiotics and other chemicals and processes to destroy microorganisms.

10.I.5 In medicine, as in other fields of science, discoveries are sometimes made unexpectedly, even by accident. But knowledge and creative insight are usually required to recognize the meaning of the unexpected.

10.J Harnessing Power

10.J.1 Until the 1800s, most manufacturing was done in homes, using small, handmade machines that were powered by muscle, wind, or running water. New machinery and steam engines to drive them made it possible to replace craftsmanship with factories, using fuels as a source of energy. In the factory system, workers, materials, and energy could be brought together efficiently.

10.J.2 The invention of the steam engine was at the center of the Industrial Revolution. It converted the chemical energy stored in wood and coal, which were plentiful, into mechanical work. The steam engine was invented to solve the urgent problem of pumping water out of coal mines. As improved by James Watt, it was soon used to move coal, drive manufacturing machinery, and power locomotives, ships, and even the first automobiles.

11. Common Themes

By the end of the 8th grade, students should know that

11.A Systems

11.A.1 A system can include processes as well as things.

11.A.2 Thinking about things as systems means looking for how every part relates to others. The output from one part of a system (which can include material, energy, or information) can become the input to other parts. Such feedback can serve to control what goes on in the system as a whole.

11.A.3 Any system is usually connected to other systems, both internally and externally. Thus a system may be thought of as containing subsystems and as being a subsystem of a larger system.

11.B Models

11.B.1 Models are often used to think about processes that happen too slowly, too quickly, or on too small a scale to observe directly, or that are too vast to be changed deliberately, or that are potentially dangerous.

11.B.2 Mathematical models can be displayed on a computer and then modified to see what happens.

11.B.3 Different models can be used to represent the same thing. What kind of a model to use and how complex it should be depends on its purpose. The usefulness of a model may be limited if it is too simple or if it is needlessly complicated. Choosing a useful model is one of the instances in which intuition and creativity come into play in science, mathematics, and engineering.

11.C Constancy and Change

11.C.1 Physical and biological systems tend to change until they become stable and then remain that way unless their surroundings change.

11.C.2 A system may stay the same because nothing is happening or because things are happening but exactly counterbalance one another.

11.C.3 Many systems contain feedback mechanisms that serve to keep changes within specified limits.

11.C.4 Symbolic equations can be used to summarize how the quantity of something changes over time or in response to other changes.

11.C.5 Symmetry (or the lack of it) may determine properties of many objects, from molecules and crystals to organisms and designed structures.

11.C.6 Cycles, such as the seasons or body temperature, can be described by their cycle length or frequency, what their highest and lowest values are, and when these values occur. Different cycles range from many thousands of years down to less than a billionth of a second.

11.D Scale

11.D.1 Properties of systems that depend on volume, such as capacity and weight, change out of proportion to properties that depend on area, such as strength or surface processes.

11.D.2 As the complexity of any system increases, gaining an understanding of it depends increasingly on summaries, such as averages and ranges, and on descriptions of typical examples of that system.

12. Habits of Mind

By the end of the 8th grade, students should know that

12.A Values and Attitudes

12.A.1 Know why it is important in science to keep honest, clear, and accurate records.

12.A.2 Know that hypotheses are valuable, even if they turn out not to be true, if they lead to fruitful investigations.

12.A.3 Know that often different explanations can be given for the same evidence, and it is not always possible to tell which one is correct.

12.B Computation and Estimation

12.B.1 Find what percentage one number is of another and figure any percentage of any number.

12.B.2 Use, interpret, and compare numbers in several equivalent forms such as integers, fractions, decimals, and percents.

12.B.3 Calculate the circumferences and areas of rectangles, triangles, and circles, and the volumes of rectangular solids.

12.B.4 Find the mean and median of a set of data.

12.B.5 Estimate distances and travel times from maps and the actual size of objects from scale drawings.

12.B.6 Insert instructions into computer spreadsheet cells to program arithmetic calculations.

12.B.7 Determine what unit (such as seconds, square inches, or dollars per tankful) an answer should be expressed in from the units of the inputs to the calculation, and be able to convert compound units (such as yen per dollar into dollar per yen, or miles per hour into feet per second).

12.B.8 Decide what degree of precision is adequate and round off the result of calculator operations to enough significant figures to reasonably reflect those of the inputs.

12.B.9 Express numbers like 100, 1,000, and 1,000,000 as powers of 10.

12.B.10 Estimate probabilities of outcomes in familiar situations, on the basis of history or the number of possible outcomes.

12.C Manipulation and Observation

12.C.1 Use calculators to compare amounts proportionally.

12.C.2 Use computers to store and retrieve information in topical, alphabetical, numerical, and key-word files, and create simple files of their own devising.

12.C.3 Read analog and digital meters on instruments used to make direct measurements of length, volume, weight, elapsed time, rates, and temperature, and choose appropriate units for reporting various magnitudes.

12.C.4 Use cameras and tape recorders for capturing information.

12.C.5 Inspect, disassemble, and reassemble simple mechanical devices and describe what the various parts are for; estimate what the effect that making a change in one part of a system is likely to have on the system as a whole.

12.D Communication Skills

12.D.1 Organize information in simple tables and graphs and identify relationships they reveal.

12.D.2 Read simple tables and graphs produced by others and describe in words what they show.

12.D.3 Locate information in reference books, back issues of newspapers and magazines, compact disks, and computer databases.

12.D.4 Understand writing that incorporates circle charts, bar and line graphs, two-way data tables, diagrams, and symbols.

12.D.5 Find and describe locations on maps with rectangular and polar coordinates.

12.E Critical-Response Skills

12.E.1 Question claims based on vague attributions (such as "Leading doctors say...") or on statements made by celebrities or others outside the area of their particular expertise.

12.E.2 Compare consumer products and consider reasonable personal trade-offs among them on the basis of features, performance, durability, and cost.

12.E.3 Be skeptical of arguments based on very small samples of data, biased samples, or samples for which there was no control sample.

12.E.4 Be aware that there may be more than one good way to interpret a given set of findings.

12.E.5 Notice and criticize the reasoning in arguments in which (1) fact and opinion are intermingled or the conclusions do not follow logically from the evidence given, (2) an analogy is not apt, (3) no mention is made of whether the control groups are very much like the experimental group, or (4) all members of a group (such as teenagers or chemists) are implied to have nearly identical characteristics that differ from those of other groups.